AGN SURVEYS
IAU Colloquium 184

COVER ILLUSTRATION:

The cover shows an optical and near-infrared color-color diagram for stars (top) and galaxies (bottom) observed by both the Sloan Digital Sky Survey (SDSS) and the Two Micron All-Sky Survey (2MASS)

Ž. Ivezić, these proceedings, pg. 141

A SERIES OF BOOKS ON RECENT DEVELOPMENTS IN
ASTRONOMY AND ASTROPHYSICS

Publisher

THE ASTRONOMICAL SOCIETY OF THE PACIFIC
390 Ashton Avenue, San Francisco, California, USA 94112-1722
Phone: (415) 337-1100   E-Mail: catalog@astrosociety.org
Fax: (415) 337-5205   Web Site: www.astrosociety.org

ASP CONFERENCE SERIES - EDITORIAL STAFF
Managing Editor: D. H. McNamara   LaTeX-Computer Consultant: T. J. Mahoney
Associate Managing Editor: J. W. Moody   Production Manager: Enid L. Livingston
Production Assistant: Andrea Weaver

PO Box 24453, Room 211 - KMB, Brigham Young University, Provo, Utah, 84602-4463
Phone: (801) 422-2111   Fax. (801) 378-4049   E Mail: pasp@byu.edu

ASP CONFERENCE SERIES PUBLICATION COMMITTEE:
Alexei V. Filippenko   Geoffrey Marcy
Ray Norris   Donald Terndrup
Frank X. Timmes   C. Megan Urry

A listing of all of the ASP Conference Series Volumes and IAU Volumes
published by The ASP may be found at the back of this volume

ASTRONOMICAL SOCIETY OF THE PACIFIC
CONFERENCE SERIES

Volume 284

AGN SURVEYS
IAU Colloquium 184

Proceedings of a Colloquium held at
Byurakan, the Republic of Armenia
18-22 June, 2001

Edited by

Richard F. Green
National Optical Astronomy Observatories, Tucson, Arizona, USA

Edward Ye. Khachikian
*Byurakan Astrophysical Observatory, Armenian National Academy of
Sciences, Byurakan, The Republic of Armenia*

and

David B. Sanders
Institute for Astronomy, University of Hawaii, Honolulu, Hawaii, USA

© 2002 by Astronomical Society of the Pacific. All Rights Reserved

No part of the material protected by this copyright notice may be reproduced or utilized in any form or by any means – graphic, electronic, or mechanical including photocopying, taping, recording or by any information storage and retrieval system, without written permission from the publisher.

Library of Congress Cataloging in Publication Data
Main entry under title

Card Number: 2002115616
ISBN: 1-58381-127-3

ASP Conference Series - First Edition

Printed in United States of America by Sheridan Books, Chelsea, Michigan

Contents

Dedication	xi
Preface	xiii
SOC/LOC	xv
List of Participants	xvii
Conference Photograph	xxii

Part 1. Optical Surveys for AGN

Markarian's First Survey (FBS) – Some Most Interesting Objects 3
    E.Ye. Khachikian
The Population of AGNs in Nearby Galaxies 13
    L. Ho
The Second Byurakan Survey - The Final Results. The Multiwavelength AGN Survey in the SBS Area 23
    J.A. Stepanian and V.H. Chavushyan
The Hamburg All-Sky Bright QSO Surveys 33
    D. Reimers and L. Wisotzki
Objective Prism Emission-Line Searches for Active Galactic Nuclei ... 43
    C. Gronwell, V.L. Sarajedini and J.J. Salzer
The US and Other Optically Selected Bright QSO Samples 53
    K.J. Mitchell and P.D. Usher
Luminosity Function and Evolution of Optically Selected QSOs 59
    L. Wisotzki
QSOs from a Variability-and-Proper Motion Survey 69
    H. Meusinger, J. Brunzendorf, R.-D. Scholz and M. Irwin
Spectroscopic and Variability Surveys for AGN in the Groth Survey Strip   75
    V. Sarajedini
Spectral Variability of Quasasrs in the Optical Band 81
    D. Trèvese and F. Vagnetti
AGNs in Shakhbazian Compact Groups 87
    A.S. Amirkhanian, A.G. Egikian, H. Tiersch and D. Stoll
A Morphological Optical Survey of Nearby AGN 89
    C.S. Boschetti, S. Ciroi, J. Funes, A. Omizzolo, P. Rafanelli, G.M. Richter, A. Rifatto and J. Vennik

Byurakan Surveys: Density of Bright AGN 92
L.K. Erastova

Comparison of ELG and UV Galaxies from the SBS Survey 94
M.V. Gyulzadian

Seven Samples of SBS Galaxies in Selected Fields 97
S.A. Hakopian and S.K. Balayan

Properties of the Low-z NELGs from the VPM Survey 99
H. Meusinger and J. Brunzendorf

The Blue stellar Objects of the First Byurakan Survey 101
A.M. Mickaelian

The New BL Lac Candidates from the FBS 103
O.Kh. Torosyan

Probable Associations of BL Lac objects with Zwicky and Abell Clusters 106
O.Kh. Torosyan

Part 2. Infrared and Submillimeter Surveys for AGN

Spectroscopic Diagnostics for AGNs 111
S. Veilleux

The 2MASS Red AGN Survey 127
R.M. Cutri, B.O. Nelson, P.J. Francis and P.S. Smith

The Optical, Infrared and Radio Properties of Extragalactic Sources Observed by SDSS, 2MASS and FIRST Surveys 137
Ž. Ivezić, R.H. Becker, M. Blanton, X. Fan, K. Finlator, J.E. Gunn, P. Hall, R.S.J. Kim, G.R. Knapp, J. Loveday, R.H. Lupton, K. Menou, V. Narayanan, G.R. Richards, C.M. Rockosi, D. Schlegel, D.P. Schneider, I. Strateva, M.A. Strauss, D. Vanden Berk, W. Voges, B. Yanny, for the SDSS Collaboration

Dubious Deductions fom AGN Survey Data 147
R. Antonucci

Infrared Surveys for AGN 157
H.E. Smith

Unveiling the Evolution of Type I AGNs in the IR (15μm) – as Seen by ISO in the ELIAS-S1 Region 167
I. Matute, F. La France, C. Gruppioni, F. Pozzi and C. Lari

Testing the Unified Model with an Infrared Selected Sample of Seyferts 173
H.R. Schmitt, J.S. Ulvestad, R.R.J. Antonucci, C.J. Clarke, J.E. Pringle and A.L. Kinney

Results from ISOCAM Deep Surveys: An Answer on the AGN Contribution to the Cosmic Infrared Background 179
H. Aussel

Blazars from the CLASS Survey 189
   M.J.M. Marchã and A. Caccianiga
Discovery of Active Galactic Nuclei in Mid- and Far-Infrared Deep Surveys
   with ISO 195
   Y. Taniguchi
A New MIR/submm Diagnostic for Dust-Enshrouded AGN 205
   M. Haas
The Nature of the Faint Far-Infrared Extragalactic Source Population:
   Optical/NIR and Radio Follow-up Observatioons of ISOPHOT Deep-
   Field Sources using the Keck, Subaru, and VLA Telescopes 213
   Y. Kakazu, D.B. Sanders, R.D. Joseph, L.L. Cowie, T. Murayama,
   Y. Taniguchi, S. Veilleux, M.S. Yun, K. Kawara, Y. Sofue, Y. Sato,
   H. Okuda, K. Wakamatsu, T. Matsumoto and H. Matsuhara
IRAS03158+4227 – a ULIRG in a Widely Separated Pair of Galaxies . 215
   H. Meusinger, B. Stecklum and J. Brunzendorf
The Byurakan-IRAS galaxy (*BIG*) Sample: The Redshift Survey 217
   A.M. Mickaelian, S.K. Balayan and S.A. Hakopian
Search for Obscured IRAS Galaxies 220
   A.M. Mickaelian, S.A. Hakopian and S.K. Balayan

Part 3. X-Ray Surveys for AGN

The Obscured AGN Population Probed by X-ray Observations 225
   K. Iwasawa
The AGN Content of Hard X-ray Surveys 235
   A. Comastri, C. Vignali and M. Brusa, on behalf of the HELLAS and
   HELLAS2XMM consortia
AGN Populations from Optical Identification of ASCA Surveys 245
   M. Akiyama, Y. Ueda and K. Ohta
The X-ray Variability of High-Redshift QSOs 251
   J. Manners, O. Almaini and A. Lawrence
New Results from the REX Survey 257
   A. Caccianiga, M.J.M. Marchã, T. Maccacaro, A. Wolter, R. Della Ceca
   and I.M. Gioia
Optical Identification of X-ray Sources in a High X-ray Flux Sensitivity
   Area from the RASS 259
   R. Mújica, F.-J. Zickgraf, V. Chavushyan, Y. Juárez, A. Serrano,
   I. Appenzeller and J. Krautter
AGN Search from Multicolor Photometric Observations of Faint ROSAT
   X-ray sources in a One Square Degree Field 261
   S. Xue, X. Zhou and H. Zhang

Part 4. Radio Surveys for AGN

Surveys of Parsec-scale Radio Structures in AGN 265
    L.I. Gurvits

Radio AGN Surveys 275
    C. de Breuck, W. van Breugel, H. Röttgering and C. Carilli

AGN Selection by Size of Dominant Emission 281
    P. Augusto and J.I. Gonzalez-Serrano

The Orientation of the Seyfert Nucleus in Mrk 348 289
    S. Antón, A. Thean, I. Browne and A. Pedlar

Optical Identification of Weak and Compact Radio Sources 291
    V.S. Artyukh, M.A. Hovhannisyan, A.P. Mahtesyan and V.H. Movsesyan

The FIRST-APM QSO Survey (FAQS) in the SBS Region. Current Status 293
    V. Chavushyan, R. Mújica, J.R. Valdés, L. Carrasco, J. Stepanian and O. Verekhodanov

Energy Density and Radiation Losses in Giant Radio Galaxies 295
    M. Jamrozy and J. Machalski

Compact jets in 100 AGNs with the Strongest Broad-band Variability of 1-22 GHz spectra in 1997-2001 297
    Yu.A. Kovalev, Y.Y. Kovalev, N.A. Nizhelsky and A.V. Bogdantsov

Survey and Analysis of 1-22 GHz Spectra for the Full Sample of 660 AGNs North of Declination $-30°$ 299
    Y.Y. Kovalev, N.A. Nizhelsky, Yu.A. Kovalev, G.V. Zhekanis and A.V. Bogdantsov

Different Types of Radio Sources and Possible Evolution of Radio Galaxies 301
    G.A. Ohanian

5-GHz VLBI Imaging Observations of 7 Equatorial AGNs 304
    Z.-Q. Shen, D.R. Jiang, Y.J. Chen and T.-S. Wan

Decametric AGNs: FIRST and NVSS Maps and Radio Spectra 306
    O.V. Verkhodanov, N.V. Verkhodanova and H. Andernach

IRAS F02044+0957: A Radio Source in an Interacting System of Galaxies 308
    O.V. Verkhodanov, V.H. Chavushyan R. Mújica, J.R. Valdés and S.A. Trushkin

Photometric Study of Radio Galaxies in the RATAN-600 "Cold" Survey 310
    O.V. Verkhodanov, Yu.N. Parijskij, N.S. Soboleva, A.I. Kopylov, A.V. Temirova, O.P. Zhelenkova and W.M. Goss

System to Estimate Ages and Redshifts for Radio Galaxies 312
    O.V. Verkhodanov, A.I. Kopylov, N.V. Verkhodanova, O.P. Zhelenkova, V.N. Chernenkov, Yu.N. Parijskij, N.S. Soboleva and A.V. Temirova

Part 5. AGN Phenomena

Surveys of High-Redshift QSO Hosts 317
  J.B. Hutchings

Are There AGNS in the Nearby Dwarf Galaxies ? 325
  I.D. Karachentsev and V.E. Karachentseva

AGN and the Demographics of Supermassive Black Holes 335
  R.F. Green

The Formation and Feeding of Massive Black Holes in the Early Universe 343
  W.J. Duschl and P.A. Strittmatter

Quasar Variability: New Surveys and New Models 351
  M.R.S. Hawkins

Rapid Variations in the Broad Hβ Profile of the Radio Galaxy 3C 390.3:
    Possible Evidence for Turbulence in the Accretion Disk 357
  N.S. Asatrian, E.Ye. Khachikian and P. Notni

Rapid Variations in the Broad Hα Profile of the Seyfert Galaxy Markarian
    6: Possible Evidence for Turbulence in the Accretion Disk 359
  N.S. Asatrian, E.Ye. Khachikian and P. Notni

Spectral Variability of Some Seyfert Galaxies 361
  E.K. Denissyuk, V.N. Gaisina and R.R. Valiullin

Analysis of Color Variability of BL Lac during the 1997 and 1999 Outbursts 363
  V.A. Hagen-Thorn, V.M. Larionov, A.V. Hagen-Thorn, S.G. Jorstad and
  G.O. Temnov

AGN from the Perspective of New Approaches in Astrophysics 365
  S.G. Iskudarian

Helical Structures in Seyfert Galaxies 367
  A.V. Moiseev, V.L. Afanasiev, S.N. Dodonov, V.V. Mustsevoi and
  S.S. Khrapov

Correlation of Optical and X-ray Radiation of NGC 4151 and 3C 390.3.
    Preliminary Results 369
  O.A. Novikova and N.G. Bochkarev

Energy Releases from Accreting Superdense Compact Objects in AGNs 371
  A.A. Sadoyan

Intermediate Resolution Hβ Spectroscopy and Photometric Monitoring of
    3C 390.3. I. Further Evidence of a Nuclear Accretion Disk 373
  A.I. Shapovalova, A.N. Burenkov, O.I. Spiridonova, V.V. Vlasuyk,
  V.P. Mikhailov, L. Carrasco, V.H. Chavushyan, J.R. Valdes, F. Legrand,
  V.T. Doroshenko, V.M. Lyuty, N.G. Bochkarev, A-M. Dumont, S. Collin,
  O. Kurtanidze and M.G. Nikolashvili

Rapid Variations in the Seyfert 1 Galaxy Mrk 474 375
  R.R. Valiullin

Part 6. Future Projects

Using the NASA/IPAC Extragalactic Database (NED) and Federated Virtual Observatory Archives for Multiwavelength Studies of AGNs   379
   J.M. Mazzarella and the NED Team

New Statistical Methods for Analysis of Large Surveys: Distributions and Correlations   389
   V. Petrosian

Digitization of the FBS: Its Future Use and Expected Results   399
   A.M. Mickaelian

Part 7. Summary

Conference Summary   411
   D.B. Sanders

Author Index   423

We dedicate this volume to the memory of
BENIAMIN EGISHEVICH MARKARIAN

(1913 - 1985)

Preface

IAU Colloquium 184, *AGN Surveys*, was hosted by the Byurakan Astrophysical Observatory on 18-22 June, 2001. It was dedicated to the memory of Beniamin Markarian. He was a world-recognized pioneer in systematic surveying for UV-excess galaxies. His vision, physical insight, consummate skill as an observer, and exacting standards served as an example to his colleagues and a model for subsequent major survey enterprises.

The Byurakan Astrophysical Observatory was founded in 1946 through the initiative and vision of its first director, Academician of the National Academy of Sciences of Armenia, Victor Ambartsumian. Ambartsumian was Markarian's thesis supervisor and invited him at that time to join the scientific staff as a Senior Research Associate. In 1950, Ambartsumian and Markarian were awarded the USSR State Prize for their study of stellar associations, and related work on star formation and distribution of interstellar dust. Markarian served as the project scientist for the 1-m Schmidt telescope, put into operation in 1960.

The Byurakan Schmidt telescope was built because of its potential for systematic pursuit of two related phenomena in galaxies: nuclear activity and powerful star formation. Ambartsumian charged Markarian with the responsibility for carrying out a slitless spectroscopic survey for galaxies with UV excess and/or strong emission lines. The First Byurakan Survey demonstrated beyond doubt the effectiveness of the thin objective prism over a wide field of view for isolating such galaxies. In 1968, Khachikian and Weedman made the first spectroscopic observations from Markarian's lists, generating tremendous excitement with the rich return of active objects and effectively doubling the number of known Seyfert galaxies. They were subsequently able to use their spectral studies to develop the initial classification system for Seyfert galaxies.

The scientific realization of Markarian's efforts came in the First and Second Byurakan Surveys. It is widely known that these surveys covered large, contiguous areas, and pushed the limits of photographic sensitivity to generate large, statistically significant samples of active galaxies in the local universe. The First Byurakan Survey was carried out from 1965-1980 by Markarian, together with the late Valentin Lipovetsky and Jivan Stepanian. The Second Byurakan Survey used Kodak fine-grained emulsions for greater depth, and was completed over the period 1974-1991 at first by Markarian and J. Stepanian and then with V. Lipovetsky,L. Erastova, V. Chavushian and S. Balayan. A major contribution of the SAO 6-m telescope was the spectroscopic investigation of Markarian galaxies.

Markarian clearly commanded the respect of his observatory colleagues and professional peers. Excerpts from contributed reminiscences reflect that view:

"We can truly credit Markarian and his galaxies with providing the link between AGN and quasars, now such a crucial part of our 'unified schemes' for

understanding objects throughout the Universe...His credentials as a brilliant astronomical observer are well known, but I also want to attest that he was a kind and gentle person, with a very broad range of interests outside of astronomy."
– Dan Weedman

"He developed a new very effective way of searching of active extragalactic objects, but this is not just a realization of a successful idea, it is a titanic manual job, a tremendous volume of which is even hard to imagine, and also an inexplicable paradoxical intuition of a researcher behind it." – Anatoly Zasov.

This IAU Colloquium was a fitting tribute to the heritage of AGN surveying at the Byurakan Observatory. It brought together those working on major surveys for AGN, now expanded to scales only dreamed of in the days of photographic plates and visual inspections. Full confrontation of the AGN phenomenon and related starburst activity requires samples collected from radio, IR, optical, and X-ray surveys. The excitement of the meeting is reflected in the reviews and contributions, with early looks at the implications of 2MASS, SDSS, and hard X-ray surveys.

Large-scale surveys of AGN broadly address the accretion history of the Universe. These data are uniquely valuable to a range of key current questions in astrophysics: To what degree have AGNs produced the diffuse radiation backgrounds from X-ray through far-IR? To what extent are heavily obscured AGNs major constituents of the total population, and how does that fraction change with cosmic time? How much of the range of observed AGN phenomena arises from changes in viewing angle vs. genuine diversity in circum-nuclear structure? How are the formation and evolution of supermassive black holes and their activity related to the formation and evolution of their host galaxies? What is the nature and formation of double and multiple nuclei in active galaxies?

The volume is organized, for convenience, by the primary energy band of the investigations reported. The deep inter-relationship of the sections testifies to the inherently multi-wavelength nature of the exploration of the AGN phenomenon. This topic continues to attract vigorous intellectual attention, a tribute both to its challenge and to its heritage.

We are grateful to the Scientific and Local Organizing Committees for their generous efforts in making the Colloquium a success and a pleasure.

Richard F. Green, Tucson, AZ, USA
Edward Ye. Khachikian, Byurakan, Republic of Armenia
David B. Sanders, Honolulu, HI, USA & Garching, Germany

June, 2002

Scientific Organizing Committee

Francesco Bertola (Italy)
Brian J. Boyle (Australia)
Richard F. Green (USA)
Edward Ye. Khachikian (Co-Chair, Armenia)
Gunther Hasinger (Germany)
Areg M. Mickaelian (Armenia)
Dieter Reimers (Germany)
Brigitte Rocca-Volmerange (France)
David B. Sanders (Co-Chair, USA, Germany)
Govind Swarup (India)
Yervant Terzian (USA)
Philippe Veron (France)
Daniel W. Weedman (USA)

Local Organizing Committee

Areg Mickaelian (Chair)
Susanna Hakopian (Secretary)
A.S. Amirkhanian
S.K. Balayan
K.S. Gigoyan
A.L. Gyulbudaghian
E.R. Hovhanissian
V.Kh. Khachatrian
A.P. Mahtessian
V.H. Movsissian
E.H. Nikoghossian
G.A. Ohanian

Participants

Masayuki Akiyama, Subaru Telescope, NAOJ, Hilo, HI, USA
⟨akiyama@subaru.naoj.org⟩

Omar Almaini, University of Edinburgh, Royal Observatory, Edinburgh, UK
⟨omar@roe.ac.uk⟩

Arthur Amirkhanyan, BAO, Armenian National Academy of Sciences, Byurakan, Armenia ⟨amir@bao.sci.am⟩

Sonia Anton, Observatorio Astronomico de Lisboa da U.L., Lisboa, Portugal
⟨Sonia.Anton@oal.ul.pt⟩

Robert Antonucci, University of California (UCSB), Santa barbara, CA, USA
⟨antonucci@physics.ucsb.edu⟩

Tigran Arshakian, BAO, Armenian National Academy of Sciences, Byurakan, Armenia ⟨tigar@bao.sci.am⟩

Norair Asatrian, BAO, Armenian National Academy of Sciences, Byurakan, Armenia ⟨asat@bao.sci.am⟩

Pedro Augusto, Universidade da Madeira, Funchal, Portugal
⟨augusto@uma.pt⟩

Hervé Aussel, IfA, University of Hawaii, Honolulu, HI, USA
⟨aussel@ifa.hawaii.edu⟩

Smbat Balayan, BAO, Armenian National Academy of Sciences, Byurakan, Armenia ⟨sbalayan@bao.sci.am⟩

Francesco Bertola, University of Padova, Padova, Italy ⟨bertola@pd.astro.it⟩

Nikolai Bochkarev, Sternberg Astronomical Institute, Moscow, Russia
⟨boch@sai.msu.ru⟩

Carla Boschetti, Padova University, Padova, Italy ⟨boschetti@pd.astro.it⟩

Brian Boyle, Anglo-Australian Observatory, Epping, NSW, Australia
⟨hmw@aaoepp.gov.au⟩

Marcella Brusa, Observatorio Astronomico di Bologna, Bologna, Italy
⟨marcella@anastasia.bo.astro.it⟩

Alessandro Caccianiga, Observatorio Astronomico di Lisboa, Lisbon, Portugal
⟨caccia@oal.ul.pt⟩

Vahram Chavushian, Instituto Nacional de Astrofisica, Optida y Electrocica, Puebla, Mexico ⟨vahram@inaoep.mx⟩

Andrea Comastri, Observatorio Astronomico di Bologna, Bologna, Italy
⟨comastri@bo.astro.it⟩

Roc Cutri, IPAC, Caltech, Pasadena, CA, USA ⟨roc@ipac.caltech.edu⟩

Carlos De Breuck, Institut d'Astrophysique de Paris, Paris, France
⟨debreuck@iap.fr⟩

Serguei Dodonov, Special Astrophysical Observatory, Karachai-Cherkessian Rep., Russia ⟨dodo@sao.ru⟩

Participants

Wolfgang Duschl, Institut für Theoretische Astrophysik, Heidelberg, Germany
⟨wjd@ita.uni-heidelberg.de⟩

Anahit Egikian, BAO, Armenian National Academy of Sciences, Byurakan, Armenia ⟨aegikian@bao.sci.am⟩

Lidia Erastova, BAO, Armenian National Academy of Sciences, Byurakan, Armenia ⟨lke@bao.sci.am⟩

Richard Green, National Obtical Astronomy Observatory, Tucson, AZ, USA
⟨green@noao.edu⟩

Caryl Gronwall, Hohns Hopkins University, Baltimore, MD, USA
⟨caryl@pha.jhu.edu⟩

Leonid Gurvits, Joint Institute for VLBI in Europe, Dwingeloo, The Netherlands ⟨lgurvits@jive.nl⟩

Armen Gyulbudaghian, BAO, Armenian National Academy of Sciences, Byurakan, Armenia ⟨agyulb@bao.sci.am⟩

Marietta Gyulzadian, BAO, Armenian National Academy of Sciences, Byurakan, Armenia ⟨mgyulz@bao.sci.am⟩

Martin Haas, Max-Planck-Institut für Astronomie, Heidelberg, Germany
⟨haas@mpia-hd.mpg.de⟩

Susanna Hakopian, BAO, Armenian National Academy of Sciences, Byurakan, Armenia ⟨susanaha@bao.sci.am⟩

Günther Hasinger, Astrophysikalische Institut Potsdam, Potsdam, Germany
⟨ghasinger@aip.de⟩

Michael Hawkins, Royal Observatory, Edinburgh, Scotland, UK
⟨mrsh@roe.ac.uk⟩

Luis Ho, Carnegie Observatories, Pasadena, CA, USA ⟨lho@ociw.edu⟩

Martik Hovhannissian, BAO, Armenian National Academy of Sciences, Byurakan, Armenia ⟨martin@bao.sci.am⟩

John Hutchings, Dominion Astrophysical Observatory, Victoria, Canada
⟨john.hutchings@nrc.ca⟩

Kate Isaak, Cavendish Laboratory, University of Cambridge, Cambridge, UK
⟨Isaak@mrao.cam.ac.uk⟩

Sofik Iskudarian, BAO, Armenian National Academy of Sciences, Byurakan, Armenia ⟨sofik@bao.sci.am⟩

Garik Israelian, Instituto de Astrofisica de Canarias, Tenerife, Canary Islands, Spain ⟨gil@iac.es⟩

Nina Ivanova, BAO, Armenian National Academy of Sciences, Byurakan, Armenia

Željko Ivezić, Princeton University, Princeton, NJ, USA
⟨ivezic@astro.princeton.edu⟩

Kazushi Iwasawa, Institute of Astronomy, Cambridge, UK ⟨ki@ast.cam.ac.uk⟩

Marek Jamrozy, Jagiellonian University, Krakow, Poland
⟨jamrozy@oa.uj.edu.pl⟩

Yuko Kakazu, Tohoku University, Sendai, Japan ⟨kakazu@ifa.hawaii.edu⟩

Arsen Kalloghlian, BAO, Armenian National Academy of Sciences, Byurakan, Armenia ⟨astro@bao.sci.am⟩

Seiji Kameno, National Observatory of Japan, Tokyo, Japan ⟨kameno@hotaka.mtk.nao.ac.jp⟩

Igor Karachentsev, Special Astrophysical Observatory, Karachai-Cherkessian Rep., Russia ⟨ikar@luna.sao.ru⟩

Edward Khachikian, BAO, Armenian National Academy of Sciences, Byurakan, Armenia ⟨ekhach@bao.sci.am⟩

Yuri Kovalev, P.N. Lebedev Physical Institute, Moscow, Russia ⟨ykovalev@avunda.asc.rssi.ru⟩

Valery Larionov, St. Petersburg State University, St.-Petersburg, Russia ⟨VML@VL1104.spb.edu⟩

Abraham Mahtessian, BAO, Armenian National Academy of Sciences, Byurakan, Armenia ⟨amahtes@bao.sci.am⟩

Vigen Malumian, BAO, Armenian National Academy of Sciences, Byurakan, Armenia ⟨malumian@bao.sci.am⟩

James Manners, Edinburgh University, Edinburg, Scotland, UK ⟨jcm@roe.ac.uk⟩

Maria Marcha, CAAUL, Universidade de Lisboa, Lisbon, Portugal ⟨mmarcha@oal.ul.pt⟩

Israel Matute, Univiversita' degli studi Roma Tre', Rome, Italy ⟨matute@fis.uniroma3.it⟩

Joseph Mazzarella, IPAC, Caltech, Pasadena, CA, USA ⟨mazz@ipac.caltech.edu⟩

Helmut Meusinger, Thueringer Landessternwarte Tautenburg, Tautenburg, Germany ⟨meus@tls-tautenburg.de⟩

Areg Mickaelian, BAO, Armenian National Academy of Sciences, Byurakan, Armenia ⟨aregmick@bao.sci.am⟩

Kenneth Mitchell, NASA, Goddard Space Flight Center, Davidsonville, MD, USA ⟨mitchell@gscmail.gsfc.nasa.gov⟩

Alexi Moiseev, Special Astrophysical Observatory, Karachai-Cherkessian Rep., Russia ⟨moisav@sao.ru⟩

Tigran Movsessian, BAO, Armenian National Academy of Sciences, Byurakan, Armenia ⟨tigmov@bao.sci.am⟩

Vardin Movsissian, BAO, Armenian National Academy of Sciences, Byurakan, Armenia ⟨vmovses@bao.sci.am⟩

Raul Mujica, INAOE, Puebla, Mexico ⟨rmujica@inaoeop.mx⟩

Neil Nagar, Osservatorio AStrofisico di Arcetri, Florence, Italy ⟨neil@arcetri.astro.it⟩

Elena Nikoghossian, BAO, Armenian National Academy of Sciences, Byurakan, Armenia ⟨elena@bao.sci.am⟩

Olga Novikova, Lomonosov Moscow State University, Moscow, Russia
⟨novikova@lnfm1.sai.msu.ru⟩

Gabriel Ohanian, BAO, Armenian National Academy of Sciences, Byurakan, Armenia ⟨gohanian@bao.sci.am⟩

Philip Outram, University of Durham, Durham, UK
⟨phil.outram@durham.ac.uk⟩

Tomokai Oyama, University of Tokyo, Tokyo, Japan ⟨oyamatm@cc.nao.ac.jp⟩

Vazgen Panajyan, BAO, Armenian National Academy of Sciences, Byurakan, Armenia ⟨vpanajia@bao.sci.am⟩

Vahé Petrosian, Stanford University, Palo Alto, CA, USA
⟨vahe@astronomy.stanford.edu⟩

Dieter Reimers, Universität Hamburg, Hamburg, Germany
⟨dreimers@hs.uni-hamburg.de⟩

Agatino Rifatto, Astronomical Observatory of Capodimonte-Naples, Naples, Italy ⟨rifatto@na.astro.it⟩

Brigitte Rocca-Volmerange, Institut d'Astrophysique de Paris, Paris, France
⟨rocca@iap.fr⟩

Avetis Sadoyan, Yerevan State University, Yerevan, Armenia
⟨asadoyan@www.physdep.r.am⟩

David Sanders, IfA, Univeristy of Hawaii, Honolulu, HI, USA
⟨sanders@ifa.hawaii.edu⟩

Vicki Sarajedini, University of Florida, Gainesville, FL, USA
⟨vicki@astro.ufl.edu⟩

Henrique Schmitt, National Radio Astronomy Observatoy, Socorro, NM, USA
⟨hschmitt@aoc.nrao.edu⟩

Alla Shapovalova, Special Astrophysical Observatory, Karachai-Cherkessian Rep., Russia ⟨ashap@sao.ru⟩

Zhi-Qiang Shen, Institute of Space and Astronautical Science, Kanagawa, Japan ⟨zshen@vsop.isas.ac.jp⟩

Harding Smith, University of California, San Diego, CA, USA
⟨hsmith@ucsd.edu⟩

Jivan Stepanian, UNAM, Instituto de Astronomia, Mexico
⟨jstep@astroscu.unam.mx⟩

Govind Swarup, NCRA, Tata Institute of Fundamental Research, Pune, India
⟨gswarup@ncra.tifr.res.in⟩

Yoshiaki Taniguchi, Tohoku University, Sendai, Japan
⟨tani@astroa.astr.tohoku.ac.jp⟩

Yervant Terzian, Cornell University, Ithaca, NY, USA
⟨terzian@astrosun.tn.cornell.edu⟩

Ofelia Torossian, BAO, Armenian National Academy of Sciences, Byurakan, Armenia ⟨ofelia@bao.sci.am⟩

Dario Trevese, Universita' di Roma "La Sapienza", Rome, Italy
⟨ Dario.Trevese@roma1.infn.it ⟩

Yoshihiro Ueda, Institute of Space and AStronautical Science, Kanagawa, Japan ⟨ ueda@astro.isas.ac.jp ⟩

Fausto Vagnetti, Universita' di Roma Tor Vergata, Rome, Italy
⟨ vagnetti@roma2.infn.it ⟩

Rashit Valiullin, Fesenkov Astrophysical Institute, Almaty, Kazakhstan
⟨ rashit@afi.academ.alma-ata.su ⟩

Sylvain Veilleux, University of Maryland, College Park, MD, USA
⟨ veilleux@astro.umd.edu ⟩

Oleg Verkhodanov, Special Astrophysical Observatory, Karachai-Cherkessian Rep., Russia ⟨ vo@sao.ru ⟩

Natalia Verkhodanova, Special Astrophysical Observatory, Karachai-Cherkessian Rep., Russia ⟨ vo@sao.ru ⟩

Philippe Véron, Observatoire de Haute-Provence, Saint-Michel l'Observatoire, France ⟨ veron@obs-hp.fr ⟩

Wolfgang Voges, Max-Planck-Institut für Extraterrestrische Physik, Garching, Germany ⟨ wvoges@mpe.mpg.de ⟩

Daniel Weedman, National Science Foundation, Arlington, VA, USA
⟨ dweedman@nsf.gov ⟩

Lutz Wisotzki, Universität Potsdam, Potsdam, Germany
⟨ lutz@astro.physik.uni-potsdam.de ⟩

Suijian Xue, BAO, Chinese Academy of Sciences, Beijing, China
⟨ xue@bac.pku.edu.cn ⟩

Anahit Yeghiazaryan, BAO, Armenian National Academy of Sciences, Byurakan, Armenia ⟨ anahit@bao.sci.am ⟩

Zhenlong Zou, BAO, Chinese Academy of Sciences, Beijing, China
⟨ zzl@bao.bao.ac.cn ⟩

Conference Photo

Part 1
Optical Surveys for AGN

Ed Khachikian

Jivan Stepanian describing the 1m Schmidt Telescope

Markarian's First Survey (FBS) – Some Most Interesting Objects

E. Ye. Khachikian
Ambartsumian Byurakan Astrophysical Observatory, Byurakan 378433, Armenia, e-mail: ekhach@bao.sci.am

Abstract.

In this review I try to show the important role of the Byurakan Observatory in the discovery and detailed investigations of ultlaviolet excess (UV) galaxies, which are now the center of attention of many observatories. Most new Seyfert (Sy) galaxies by the end of the 1960s, when no more than ten of this type of galaxy were known, were discovered among UV galaxies. In addition, many unusual and interesting galaxies were also discovered among UV galaxies: double and multiple nucleus galaxies, double nucleus Sy galaxies, galaxies with jets, galaxies consisting only of Superassociations (giant HII regions), Sargent-Searle objects, BL Lacs and QSOs possessing a different type of activity according to Ambartsumian. The material presented here overlaps considerably that of some other talks delivered by the author in IAU Symposia Nos. 121 and 194.

1. Brief History

Markarian was born on November 29, 1913. In 1938 he graduated from the Yerevan State University in mathematics. Until 1941 he was a post-graduate student of Prof. V. Ambartsumian. In 1944 he defended his thesis; he continued his scientific work in the Byurakan Observatory until his death on September 29, 1985. During 1976-1979 he was the President of IAU Commission 29 "Galaxies".

It is well known that the idea of Activity of Galaxies and their nuclei was suggested by Ambartsumian in the middle 1950's and reported at the Solvay Conference in 1958. This idea was based on the following astronomical observations and discoveries:

Radio-sources (1949), which were later (1954) identified mainly with double-nuclei galaxies (Radiogalaxies),

Seyfert galaxies (1943),

Haro blue galaxies (1956).

Ambartsumian identified different forms of activity, which are well-known among astronomers, but many of them, especially young ones, do not imagine that Ambartsumian was the first to describe these forms. At the same time in Byurakan Observatory Ambartsumian started with co-workers the quest for blue objects connected by filaments or jets with nearby elliptical galaxies. As a result, a number of blue low luminosity galaxies but brighter than dwarf galaxies were detected. In 1968 Allan Stockton obtained the spectra of some of these objects, which turned out to be emission-line galaxies just like Haro galaxies. I should like to note here Markarian's article (1963), published in the Contributions of the BAO and very rarely cited, where he described a number of bright galaxies having earlier type spectral characteristics, which do not correspond to their integrated colors or morphological types. So, it was natural to ask: is it possible to find blue galaxies with UV excess among fainter ones? For the solution of this problem Ambartsumian ordered for the 1-m Schmidt telescope three objective prisms of the same size and with refraction angles: 1.5, 3.0 and 4.0 degrees. Then he made Markarian an offer to search for blue galaxies. In 1965 Markarian began this job and in 1967 the first list of so-called UV-excess galaxies had already been published in the journal "Astrofizika".

Markarian used the 1.5-degree prism, which gives on the plate about 2500 Angstrom/mm near $H\beta$ and about 1800 Angstrom/mm near $H\gamma$. The first list contains 70 UV-excess objects. On the basis of a general view of the spectra, Markarian divided the objects into two main groups: "s" and "d". "s" objects have narrow and sharp spectra like those of stars. "d" objects have diffuse spectra with weak continuous spectra like compact associations of blue stars and gaseous nebulae. To describe the intensity of the continuum, he used the number 1 for the strongest ones and the number 3 for the weakest. Differently mixed symbols have also been used. If Markarian guessed the presence of emission lines, he noted it by the letter "e". The FBS is the most famous work done with the 1-m Schmidt telescope. More than 2000 photographic plates covering about 17,000 square degrees of sky were obtained. Each plate (approximately $4^\circ\!.1 \times 4^\circ\!.1$) contains low dispersion spectra of more than 15,000 objects. The FBS covers completely the northern part of the sky and also part of the southern ($\delta > -15^o$). The limiting visual magnitude of UV galaxies in the FBS is about $17^m\!.0 - 17^m\!.5$).

The selection of UV objects was carried out by the gap in the continuum objective prism spectrum near 5300 Å/ thanks to the observing technique. This gap divides the spectrum into two parts: blue and red. By means of the brightness of blue part of the spectrum Markarian could choose the UV excess galaxies. In all, during 15 years about 1500 UV-excess objects (including blue stars as it became clear later) were discovered. Because of the low dispersion, it was very difficult for Markarian to distinguish important details in the spectra, estimate the intensity of lines, their widths, the redshifts of objects and so on. Without these data it was very difficult to judge (understand) the physical nature of UV-excess galaxies. It was necessary to observe these interesting objects with a slit spectrograph and higher dispersion.

2. Spectroscopy

I was lucky to be the first to observe almost all the galaxies from the first list of UV galaxies with the largest optical telescopes in the USA (1967-68). I was double lucky when it became clear that numbers 1, 3, 6, 9, 10, 42 and 67 turned out to be Seyfert galaxies! I would like to emphasize once more that the detailed spectral investigation of Markarian objects indicated (Khachikian 1968, Weedman & Khachikian 1968, 1969) that *over 85% of them have emission lines with their intensity being directly dependent on the value of the UV excess. One can conclude that the presence of strong UV continuum is closely associated with the formation of the emission spectrum and the more intense the continuous*

spectrum in the visible ultraviolet is, the more intense are the emission lines. It also became evident that the spectra of those objects differ, nevertheless, essentially from each other as to the excitation degree of the emission lines and their widths. On the basis of slit spectra, Markarian objects have been classified into five groups (Khachikian 1968):
–Narrow line spectra both in emission and absorption.
–Narrow, strong emission lines only.
–Strong and diffuse emission lines, [OIII] lines much stronger than the hydrogen lines (now - Sy2).
–Very broad H lines, narrow forbidden lines (Sy1),
–No strong emission lines (probably BL Lac type).

The most important result from the study of UV objects probably was *the discovery of Seyfert type galaxies among them. About 10% of UV objects turned out to be Seyfert galaxies.* **In the first list of UV-excess galaxies alone (70 objects) there are 7 Sy type galaxies (Khachikian & Weedman, 1971a).**
Thanks to that the number of Seyfert galaxies increased very quickly, which gave Dr. Weedman and me the possibility to divide them into two types (Khachikian & Weedman 1971a):
1. *Galaxies with broad hydrogen lines and narrow forbidden lines (Seyfert 1).*
2. *Galaxies with hydrogen and forbidden lines both broad (Seyfert 2).*

Besides that, the intensity ratio of the [OIII] ($\lambda(5007, 4959)$) lines to $H\beta$ for Sy1 is < 1, and for Sy 2 > 10. So we suggested a very simple method to distinguish Sy1 from Sy2: it is necessary to observe only the [OIII] lines ($\lambda(5007, 4959)$) and $H\beta$. As they are situated practically in the same spectral region, you don't need to determine the spectral sensitivity of the system. The systematic observations of UV objects from the Markarian lists with the purpose of measuring their redshifts and clarifying the activity type were done by Arakelian et al. (1970). They confirmed most of the results obtained by Weedman and Khachikian.

3. Morphology

Among UV galaxies there are all Hubble types, but still, in many cases, unusual forms: Zwicky galaxies, Sargent-Searle objects, radio, X-ray and gamma-ray sources. But as it is now clear, most important is to study the strucure of the central parts of UV galaxies. The morphology of the central parts can be divided into the following groups:
Starlike galaxies, Galaxies with star-like nuclei (in general, spirals), Double nuclei galaxies, Multi nuclei galaxies, Galaxies with bulges, A few galaxies show jets starting from the nucleus. (see Figs. 1–6 in Khachikian, 1987)

4. Some Most Interesting Objects

Markarian 1,3,6: the first Sy2 galaxies discovered in the FBS. It is interesting to note that Mark 6 was the classical Sy2 galaxy before 1968. The appearance of additional new emisson components of the hydrogen lines in the spectra of

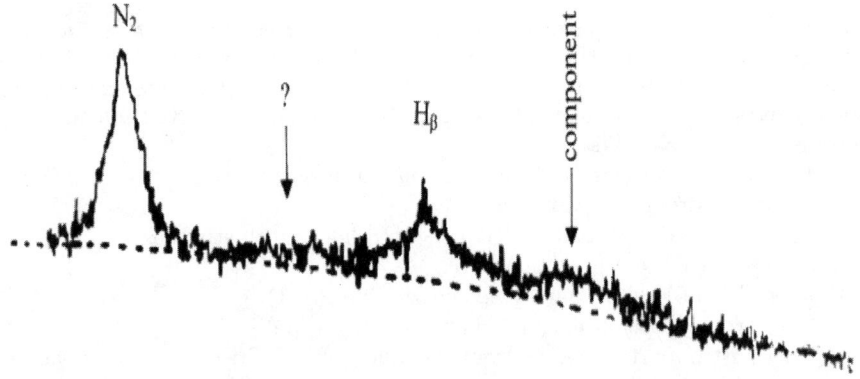

Figure 1. Spectrum of Mark 6, with blue-shifted velocity components apparent.

AGN was first discovered in Markarian 6 in 1969 (Khachikian & Weedman, 1970, 1971). During one year (between February, 1968, and January, 1969) in the spectrum of this galaxy new broad emission components of $H\alpha$, $H\beta$ and $H\gamma$ were detected. Their blue-shifted velocity is 3000 km/sec. What do they look like now? You can know from the posters of Asatrian et al. in this volume (and in Asatrian et al. 1999a,b). The high dispersion (28 Å/mm) spectrum of Mark 6 is shown for the first time in Fig. 1, where the blue-shifted component is well seen.

After the explosion in the nucleus of this galaxy, it was classifed as Sy 1.5, but actually it is probably just a short event in the history of a Sy2 galaxy. Therefore, there are very few of this type of Sy galaxy. So this "new" classification is nonsense. The nature of this type of explosion is still a puzzle. One of the possible new models is suggested in Ambartsumian et al. 1998.

Markarian 7 and 8. Probably a physical pair of very unusual galaxies that very much look like the same shape and consist of five bright superassociations (SA) (Khachikian 1972, Khachikian & Burenkov 1990) forming two almost rectilinear segments in a figure resembling the letter "V" (Mark 8) and an upside-down letter "V" (Mark 7). Actually they are nests of SA, namely, star formation regions, each included in the diffuse envelope. The absolute magnitude of the brightest SA in Mark 8 is M= -18 and in Mark 7 M= - 16.4. Similar morphological structures are in **Markarian 35** (Burenkov & Khachikian, 1986), Mark 297 (Burenkov 1987), Mark 325 (Burenkov, Khachikian, & Nazarova, 1990). All these galaxies show at the same time very interesting dynamical properties.

Markarian 9 and 10 are the first Sy1 galaxies discovered among Markarian galaxies (Khachikian 1968). The most important property of these galaxies is their very high brightness M_V= -21.6 and M_V = -21.3, which filled in the gap in brightnss between giant normal galaxies and QSOs.

Markarian 94. In the second list of UV galaxies, this object was classified as d1e. The spectral and photometric investigation of this object (Arp & Khachikian, 1974) showed that **Mark.94 is not the nucleus of a galaxy, but just the SA in III Zw 0834 +51**. The redshift of this galaxy (z = 0.0022) is about equal to that of Mark 94, z = 0.0025 (Sargent 1972) and z = 0.0030 (Arp & Khachikian, 1974). The prime-focus photograph taken with the 200-inch Hale telescope shows (Arp & Khachikian, 1974) that Mark 94 is located in an area of disordered spiral structure on the edge of the bar of the galaxy. The integrated absolute magnitude of the galaxy is M_{pg} = -16.6, and for Mark 94 M_B = -13.5. The diameter of Mark 94 is about 600 pc. The most interesting is, as became clear later on (Sahakian & Khachikian, 1975), that **Markarian 94 is the prototype of a new class of objects (more than 50 among the UV galaxies from the first six lists of the FBS)**. Representatives of this type of object are Mark 40, 49, 59, 71, 86, 108, 201, 256, 295, 325, 330, and so on. The brightness of these objects falls in the range from -11.4 to -18.7. The majority of them are connected with irregular galaxies (more than 65%). Another important characteristic of this type of object is their quite different spectra: i) both strong continuous spectrum and emission lines; ii) strong continuum and weak emission lines; iii) weak continuum and strong emission lines; iv) both weak continuum and emission lines. The location of these objects in the galaxy also varies (Sahakian & Khachikian, 1975). They can be the nucleus of the UV galaxy; a part of the galaxy (like Mark 94, 59, 298); an isolated UV galaxy (Mark 116= I Zw 18), which is almost the same category as a Sargent-Searle object. All these facts are not understandable completely.

Markarian 231. A unique object, and according to D. Weedman, Mark 231 is the prototype that includes all physical processes associated with active galaxies: it contains both a starburst and an AGN which are highly obscured by dust; its luminosity at various wavelengths shows how such a galaxy would appear at high redshift. Mark 231 provides an excellent example for comparing low redshift and high redshift phenomena. (Weedman, 1999).

Markarian 273. The unique ULIRG double Sy type nucleus galaxy with a long jet to the south (about 40 kpc), which originates from the southern nucleus (Khachikian et al., 1985). The separation of the main bright nuclei is about 800 pc. But the most interesting peculiarity of this galaxy is the double structure of the northern nucleus: the separation between them is only 70 pc! (Knapen et al. 1998). More detailed optical observations suggest that there are multiple nuclei in this galaxy (Koroviakovskii et al., 1981; Mazzarella & Boroson, 1993). The galaxies Mark 423 with a starlike nucleus and multinucleus Mark 773 have jets. Fig. 2 presents a picture of this galaxy obtained with the Keck Telescope in the K-band and kindly given to me by Dr. B. T. Soifer.

Markarian 266. The unique double Seyfert type nucleus galaxy, with unusual physical properties. The detailed spectrophotometric investigation of this galaxy first was presented in Khachikian et. al. (1980). Now there are a lot of new observations of Mark 266. I should like to note the important work of Mazzarella et al. in 1988 and 1993 on Markarian galaxies. It was shown that Mark 266 is a radio source: both optical nuclei are radio sources, at the same time, just in the center of the galaxy between the two optical nuclei there are two radio sources (Fig. 3). The orientation of these two radio nuclei is very interesting: they are elongated in the direction perpendicular to the direction of optical nuclei!

The distance between them is less than one arcsecond. **But the most interesting is that the spectrophotometric investigation of Mark 266**

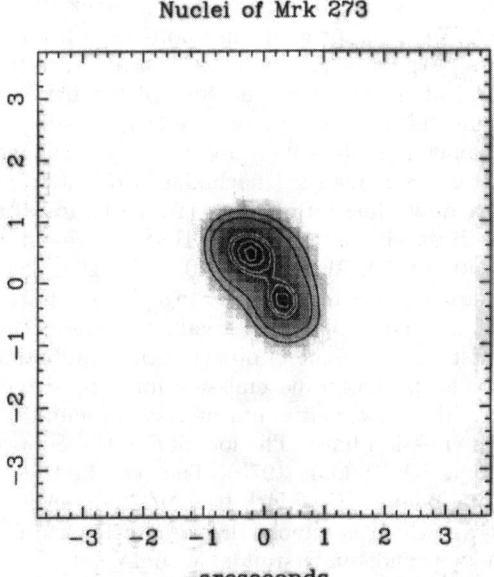

Figure 2. Keck K-band image of Mark 273.

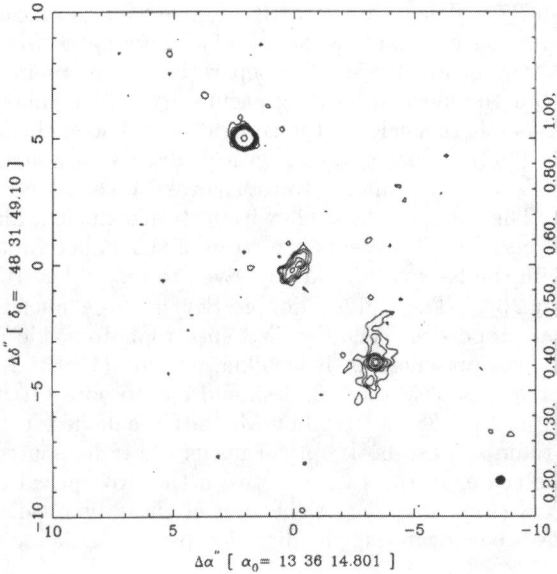

Figure 3. VLA image of the central region of Mark 266 showing the double radio nuclei.

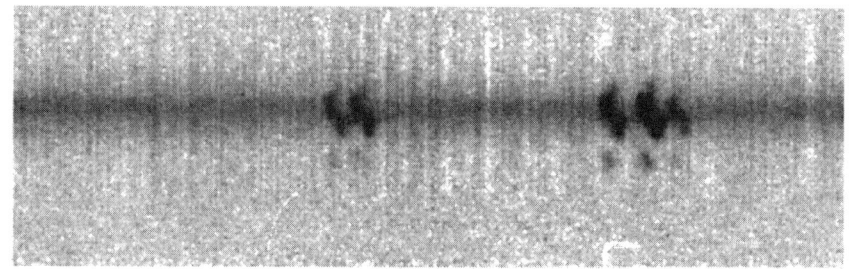

Figure 4. Spectrum of Mark 266. The P.A. of the slit is 120 and crosses the center of the galaxy just between the optical nuclei.

carried out with the Palomar Observatory 5-m by Terzian, Nortgren and me (not yet published) shows that just in the same direction symmetrical to both sides of these radio sources there are two condensations discovered by means of emission lines! (Fig.4).

The distance between these objects is about 15 arcsec. Figs. 5–7 show the results of recent observations with the 2.6m telescope of BAO of Mark 266 in an $H\alpha$ filter and the field of radial velocities obtained with the "Tiger" system in the $H\alpha$ and [N II] lines, respectively. One can see the variations in the profiles of these lines over the central part of Mark 266. The difference of radial velocities of these two new objects is about 500 km/s, which shows that they rotate around the center of Mark 266. More detailed discription of the results of this investigation of Mark 266 will be published elsewhere.

5. Concluding Remarks

There are many other interesting objects among UV-galaxies (Mark 116 = I Zw18 - probably one of the most metal-deficient objects; Mark 64 - first quiet QSO; Mark 421 and 501- BL Lac objects; Mark 205 and 474 - Sy 2 galaxies, very much looks like Mark 9; Mark 5,13,19 and others - galaxies with bright and narrow emission lines; Mark 11, 26, 41, 292 - without any strong emission lines; Mark 133- without N1, N2 lines; double systems - Mark 220 and 221, Mark 261 and 262, Mark 305 (BL Lac object) and Mark 306 (with very strong emission lines) and so on. But because of the limits of space I have no possibility to continue the list of interesting objects from the FBS. I would just like to stress that about 10% of UV galaxies are double and multi-nuclei galaxies. The number of this type of active galaxy is now so high that the majority of them, no doubt, are real double or multi-nucleus galaxies, but not double or multi galaxy systems. Therefore **they are not a result of colliding or interaction of two or more independent galaxies**. The fact that the majority of double nuclei show either form of activity also speaks in favor of this opinion. I would like to finish my review with the following words of Margaret Burbidge: "It is

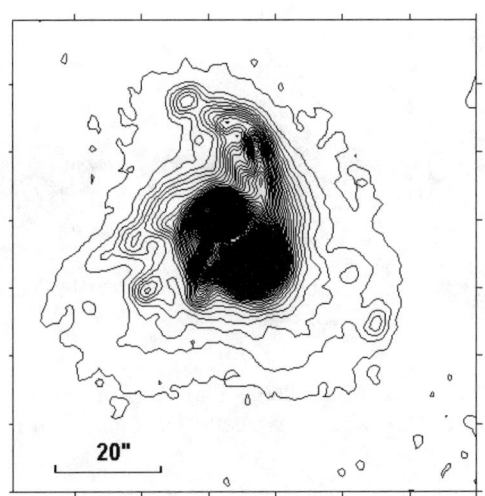

Figure 5. Hα contour image of Mark 6.

very fashionable at the moment to talk about "mergers" of galaxies to explain unusual morphology, but it seems to me that there are too many of these double nuclei for them ever to be accounted for as "mergers" of previously separate galaxies." (Khachikian 1988).

References

Ambartsumian, V. A., Khachikian, E. Ye., & Yengibarian, N., 1998, Astrofizika, 41, 321.
Arakelian, M. A.. Dibay, E. A., & Yesipov, V. F., 1970, Afz, 6, 39.
Arp, H. & Khachikian, E. Ye., 1974, Astrofizika, 10, 173.
Asatrian, N. S., Khachikian, E. Ye., & Notny, P., 1999a, in IAU Symp. No.121, Eds. E. Khachikian, K. Fricke, & J. Melnick, p.406.
Asatrian, N. S. et al., 1999b, in IAU Symp. No.121, Eds.
E. Khachikian, K. Fricke, & J. Melnick, p.409.
Burenkov, A., 1987, in IAU Symp. No.121, Eds. E. Khachikian, K. Fricke, & J. Melnick, Byurakan, Armenia, p.587.
Burenkov, A. & Khachikian, E. Ye., 1986 , Astrofizika, 24, 349.
Burenkov, A., Khachikian, E. Ye., Nazarova, E., 1990, Proc. Special Astrophy. Obs., Russian AS, 32, 16.
Khachikian, E. Ye., 1968, AJ, 73, 891.
Khachikian, E. Ye., 1972, Afz., 8, 529.
Khachikian, E. Ye., 1987, in IAU Symp. No.121 Eds. E. Khachikian et al., Byurakan, p.65.

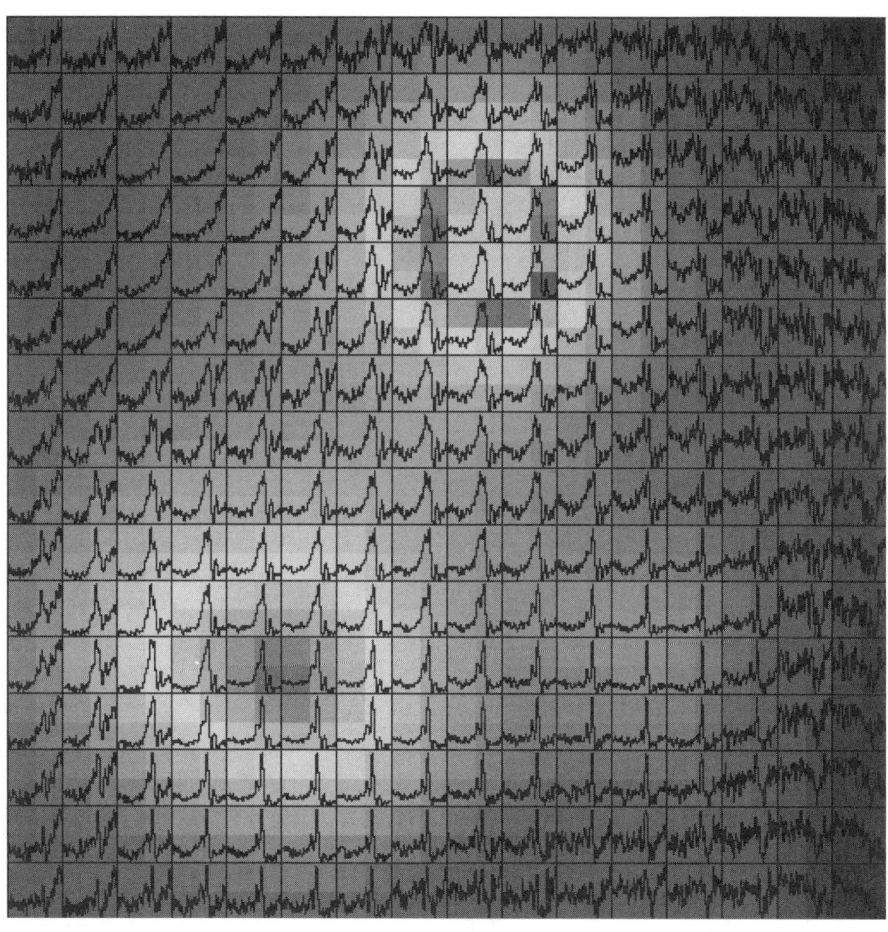

Figure 6. Multi-pupil Hα + [N II] observation of Mark 6.

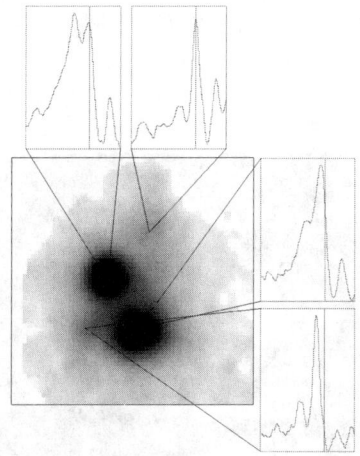

Figure 7. Averaged profiles of 4 regions of Mark 266. The vertical line corresponds to λ6750Å.

Khachikian, E. Ye., 1988, New Ideas in Astronomy, Eds. F. Bertola, J. Sulentic, B. Madore, p.115, (in "Discussions").
Khachikian, E. Ye. & Burenkov, A., 1983, Afz, 19, 826.
Khachikian, E. Ye. & Burenkov, A., 1990, Astrofizika, 32, 245.
Khachikian, E. Ye., Petrosian, A R., & Sahakian, K. A., 1980, Astrofiz., 16, 621.
Khachikian, E. Ye., Petrosian, A., & Sahakian, K., 1985, Afz, 22, 229.
Khachikian, E Ye. & Weedman, D. W., 1968, Astrofizika, 4, 587.
Khachikian, E. Ye. & Weedman, D. W., 1969, Astrofizika, 5, 113.
Khachikian, E. Ye. & Weedman, D. W., 1971a, Astrofizika, 7, 389.
Khachikian, E. Ye. & Weedman, D. W., 1971b, ApJ, 164, L109.
Knapen J., Laine, S., Yates, J., Robinson, A., Richards, A., Doyon, R., & Nadeau, D., 1997, Ap.J.Lett.,**490**,L29.
Koroviakovskii, Y. P., Khachikian, E Ye., Petrosian, A., & Sahakian, K., 1981, Afz, 17, 231.
Markarian, B. E, 1967, Astrofizika, 3, 55.
Markarian, B. E., 1963, Contributions of BAO.
Mazzarella, J. M. & Boroson, T. A. ApJS, 85, 27.
Mazzarella J.M., Gaume, R., Aller, H. & Hughes, P., 1988, ApJ, **333**, 166.
Sahakian, K. A. & Khachikian, E. Ye., 1975, Astrofizika, 11, 207.
Sargent, W., 1972, ApJ, 173, 7.
Weedman, 1999, in IAU Symp. No.194, Eds. Terzian, Weedman, & Khachikian, p.191.

The Population of AGNs in Nearby Galaxies

Luis C. Ho

The Observatories of the Carnegie Institution of Washington, 813 Santa Barbara St., Pasadena, CA 91101, U.S.A.

Abstract. This contribution reviews the properties of nuclear activity in nearby galaxies, with emphasis on results obtained from current optical surveys and multiwavelength follow-up observations thereof.

1. Introduction

In a meeting which celebrates different techniques of surveying AGNs, it is appropriate to remind ourselves of the properties of the AGN population in nearby galaxies. Nearby AGNs are important for at least two reasons. First, they inform us of the faint end of the local ($z \approx 0$) AGN luminosity function, a fundamental constraint on a variety of statistical considerations of the AGN population. Second, as the evolutionary endpoints of quasars, they present an opportunity to study black hole accretion in a unique regime of parameter space. This paper limits itself to three topics. Section 2 summarizes some general statistics resulting from optical searches for nearby AGNs. The interpretation of these results largely hinges on our understanding of the physical origin of low-ionization nuclear emission-line regions (LINERs; Heckman 1980), which is the subject of Section 3. Finally, Section 4 draws some inferences on the demographics of massive black holes.

2. Statistics of Nearby AGNs from the Palomar Survey

Most AGN surveys rely on selection criteria that isolate some previously known characteristics of these objects. Common strategies to find quasars, for example, employ color cuts to highlight the UV excess typically present in AGN spectra, or objective prism plates to identify objects with strong emission lines. Other wavelength-specific techniques to find candidate AGNs include combing areas of the sky in the radio or X-rays. Infrared-based methods use temperature selection to cull sources "warmer" than might be expected for star-forming galaxies. While all of these techniques have been successful, each introduces biases specific to the wavelength. And all require follow-up optical spectroscopy to confirm the AGN identification, to classify its type, and to determine its redshift.

A more direct, less biased approach is to spectroscopically survey every object within a particular region of the sky to a given optical magnitude limit. This, of course, is an expensive route, and in practice one is confined to go deep over only a small solid angle or stay relatively shallow over a wider area. In addition, when covering a wide area, a morphological cut has to be made so that one does not waste a lot of time observing foreground stars.

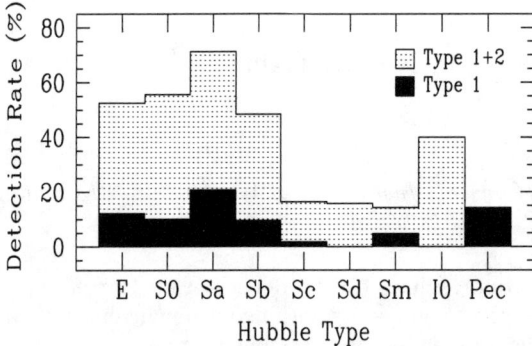

Fig. 1. Detection rate of AGNs as a function of Hubble type in the Palomar survey. "Type 1" AGNs (those with broad Hα) are shown separately from the total population (types 1 and 2). Adapted from Ho et al. (1997b).

The results presented in the rest of this paper stem from the the Palomar optical spectroscopic survey of nearby galaxies (Filippenko & Sargent 1985; Ho, Filippenko & Sargent 1995). In brief, the Palomar 200-inch telescope was employed to take moderate-dispersion, high-quality spectra of 486 bright ($B_T \leq$ 12.5 mag), northern ($\delta > 0°$) galaxies, with the primary aim of conducting an accurate census of the AGN population in the nearby ($z \approx 0$) universe. The Palomar survey, the most sensitive of its kind (see Ho 1996 for a comparison with previous optical studies), produced a comprehensive, homogeneous catalog of spectral classifications of nearby galaxies (Ho et al. 1997a, 1997b, 1997c). The main results of the survey are the following. (1) AGNs are very common in nearby galaxies (Fig. 1). At least 40% of all galaxies brighter than $B_T = 12.5$ mag emit AGN-like spectra. The emission-line nuclei are classified as Seyferts, LINERs, or transition objects (LINER/H II), and most have very low luminosities compared to traditionally studied AGNs. The luminosities of the Hα emission line range from 10^{37} to 10^{41} erg s^{-1}, with a median value of $\sim 10^{39}$ erg s^{-1}. (2) The detectability of AGNs depends strongly on the morphological type of the galaxy, being most common in early-type systems (E–Sbc). The detection rate of AGNs reaches 50%–75% in ellipticals, lenticulars, and bulge-dominated spirals but drops to $\lesssim 20\%$ in galaxies classified as Sc or later. (3) LINERs make up the bulk (2/3) of the AGN population and a sizable fraction (1/3) of all galaxies. (4) A significant number of objects show a faint, broad (FWHM $\approx$ 1000–4000 km s^{-1}) Hα emission line that qualitatively resembles emission arising from the conventional broad-line region of "classical" Seyfert 1 nuclei and quasars.

3. The Nature of LINERs

If LINERs[1] are powered by a nonstellar central source, then they clearly would be the most abundant type of AGNs in nearby galaxies. However, ever since their discovery, the physical origin of LINERs has been controversial. The AGN

---

[1] Note that this paper is concerned only with compact, *nuclear* LINERs ($r \lesssim 200$ pc), which are most relevant to the AGN issue. LINER-like spectra are often also observed in extended nebulae such as those associated with cooling flows, nuclear outflows, and circumnuclear disks.

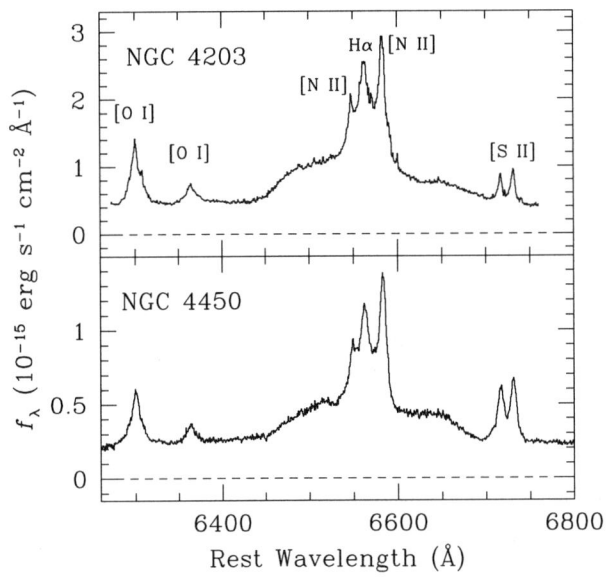

Fig. 2. Double-peaked broad Hα lines detected in STIS spectra of NGC 4203 (Shields et al. 2000) and NGC 4450 (Ho et al. 2000).

interpretation for LINERs is only one of several possible explanations, as recently reviewed by Barth (2001). The most uncertain subset are the type 2 LINERs and transition objects, whose energetics may arise, at least in part, from stellar processes.

Ho (1999a) used several lines of evidence to argue that a significant fraction of LINERs are genuine AGNs. Some of these are repeated below, along with updated information when available.

(1) The host galaxies of LINERs are similar to those of Seyfert nuclei. Both classes reside in bulge-dominated hosts (E–Sbc), quite distinctive from starforming H II nuclei, which are found mainly in Hubble types Sc and later. To the extent that bulges invariably contain massive black holes (e.g., Magorrian et al. 1998; Kormendy & Gebhardt 2001; R. Green, these proceedings), this is consistent with the idea that LINERs are associated with accretion processes.

(2) Some of the best candidates for massive black holes are in LINER galaxies. Well-known examples include M81, M84, M87, and the Sombrero galaxy. The rapidly growing list of objects with kinematically determined black hole masses (see, e.g., Kormendy & Gebhardt 2001; Ho 2001b) continues to support this.

(3) LINERs contain broad-line regions. The Palomar survey discovered that 15%–25% of the LINER population are "type 1" LINERs — LINERs with a directly visible broad component of Hα emission (Ho et al. 1997c). The broad-line component, however, is generally rather weak; the measurement is extremely challenging, requiring careful subtraction of the underlying starlight and deblending of complex line profiles. The robustness of the broad Hα detections, therefore, remained to be evaluated by independent observations. Rix et al. (2002) recently used STIS on *HST* to obtain nuclear spectra of a statistically complete subsample of emission-line nuclei selected from the original Palomar survey. The small aperture (0″.2) used in these observations excludes

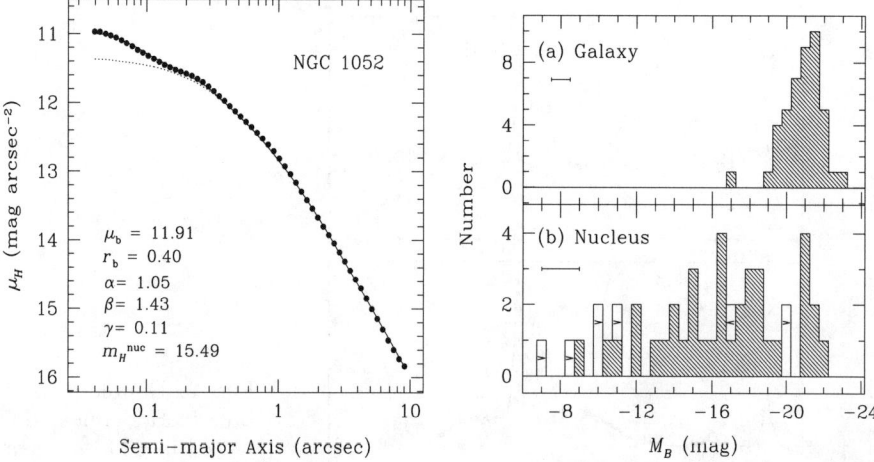

Fig. 3. (*Left*) NICMOS/*HST* H-band central surface brightness profile of NGC 1052. The central point source is extracted after modeling the underlying bulge light. Adapted from Ravindranath et al. (2001). (*Right*) Optical (*B*) absolute magnitudes for Seyfert 1 galaxies. The top panel plots the integrated emission from the galaxy, and the bottom panel shows the light from the nucleus alone, as determined from *HST* photometry. Adapted from Ho & Peng (2001).

most of the contaminating bulge light and enables a much more straightforward assessment of the presence of broad lines. These data, and other related observations using *HST*, have shown that the statistics of broad Hα emission derived from the Palomar survey are essentially robust (Rix et al. 2002).

However, the *HST* spectra revealed that in some objects (Fig. 2) the broad line does show a qualitative difference in the form of a double-peaked, asymmetric profile, highly reminiscent of similar structures found in some broad-line radio galaxies that are commonly interpreted as signatures of a relativistic accretion disk (e.g., Chen, Halpern, & Filippenko 1989). Double-peaked broad lines are supposed to be quite rare; according to Eracleous & Halpern (1994), they are seen in only 10% of radio-loud AGNs (∼1% of all bright AGNs). Given the relatively high detection frequency of these objects in serendipitous *HST* observations (M81: Bower et al. 1996; NGC 4203: Shields et al. 2000; NGC 4450: Ho et al. 2000; NGC 4579: Barth et al. 2001), we suspect that this phenomenon is much more common than previously thought. Indeed, Ho et al. (2000) discuss how double-peaked broad lines might arise naturally in nearby nuclei with low-accreting black holes.

(4) Compact nuclei are common in LINERs. But they are not easy to measure. Emission from the host galaxy invariably swamps the nuclear component. The contrast problem is not as severe at some wavelengths compared to others (e.g., UV), but generally it never goes away. In order to reliably quantify the nuclear fluxes from these sources, one needs observations not only with good sensitivity but also high angular resolution. The second point is crucial: generally angular resolutions of $\lesssim 1''$ are necessary to unambiguously disentable a tiny central core

embedded in a bright bulge. The requisite data, although in principle not that challenging to obtain with modern facilities, have been subject to systematic analysis only quite recently.

Figure 3 illustrates the practical challenge. The left panel shows the central surface brightness profile of the well-studied LINER NGC 1052 measured in the H band with NICMOS on *HST* (Ravindranath et al. 2001). The central point source is faint, but quite well defined at $r \lesssim 0\rlap{.}''1$. It almost certainly would have escaped notice in typical ground-based images. Even with *HST* resolution, however, note that it is not trivial to extract the point-source magnitude unambiguously in the presence of a cuspy bulge profile. The right panel in the figure generalizes this problem to a well-defined sample of nearby Seyfert 1 galaxies recently studied by Ho & Peng (2001). It shows that although the contrast problem is most acute in LINERs, it is nonetheless quite important as well in Seyfert galaxies. The top panel, which plots the integrated optical (B) absolute magnitudes of the entire galaxy, gives the familiar result that Seyfert galaxies typically have $M_B \approx -21$ mag, roughly corresponding to L^* (e.g., Ho et al. 1997b). On the other hand, the true magnitude of the Seyfert *nucleus* — the quantity most pertinent to the AGN phenomenon and most analogous to observations of quasars — ranges from $M_B \approx -21$ to -9, a factor up to 10^5 fainter than the integrated luminosity. Clearly, any quantitative discussion of the physical properties of low-luminosity AGNs, be they LINERs or Seyferts, would be meaningless if this were not taken into account.

The incidence of compact nuclei in LINER galaxies depends on wavelength, but in general the detection rate is quite high, on the order of 50% or greater. The most robust statistics come from radio and X-ray observations, since these bands are least affected by dust obscuration.

Several recent studies have exploited the high angular resolution of the VLA to search for compact radio cores in the Palomar galaxies (Van Dyk & Ho 1998; Nagar et al. 2000; Filho, Barthel, & Ho 2000; Ho & Ulvestad 2001). The VLA can deliver fairly sensitive (rms $\sim$ 50–100 μJy) radio continuum images efficiently in "snapshot" mode. In general the observations have been done at 6, 3.6, or 2 cm, with resolutions of $\sim 1''$. The main results are: (a) compact cores are quite common, being present in $\sim$50%–80% of the objects; (b) they are weak, typically $P_{6cm} \approx 10^{18} - 10^{21}$ W Hz^{-1}; (c) a sizable fraction of them have flat or even inverted spectra; and (d) they have relatively simple structures, usually well described by a single unresolved core. Many sources remain unresolved when examined at milli-arcsecond resolution using VLBI techniques (e.g., Wrobel, Fassnacht, & Ho 2001; Filho, Barthel, & Ho 2001; Ulvestad & Ho 2002, in preparation), although some show jetlike linear extensions (Falcke et al. 2000).

The Palomar galaxies have also been the subject of intense scrutiny in the X-rays, both in the soft-energy band using *ROSAT* (Komossa, Böhringer, & Huchra 1999; Roberts & Warwick 2000; Halderson et al. 2001) and in the hard-energy band using *ASCA* (Terashima, Ho, & Ptak 2000; Terashima et al. 2001). The coarse angular resolution of these satellites, however, severely limits one's ability to reliably measure the weak signal from the nucleus. This is the domain of *Chandra*. A large, well-defined subsample of the Palomar galaxies is currently being imaged with ACIS. The preliminary findings, reported by Ho et al. (2001), suggest that compact X-ray sources astrometrically coincident with the radio

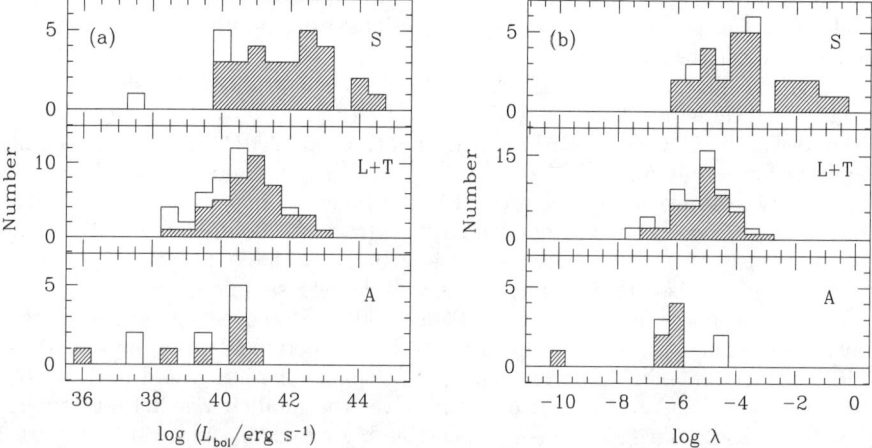

Fig. 4. Distribution of (a) nuclear bolometric luminosities and (b) Eddington ratios $\lambda \equiv L_{\rm bol}/L_{\rm Edd}$ for Seyferts (S), LINERs and transition objects (L+T), and absorption-line nuclei (A). Open histograms denote upper limits. Adapted from Ho (2002, in preparation).

cores are again commonplace. They are detected universally in type 1 LINERs and Seyferts, relatively often in type 2 LINERs and Seyferts, but very rarely in transition nuclei. These statistics need to be verified with the full sample, which is forthcoming (Ho et al. 2002, in preparation).

(5) LINERs are highly sub-Eddington systems. Recent advances in black hole mass determinations in nearby galaxies gives us an opportunity to compare the nuclear luminosities described above with the corresponding Eddington luminosities. Ho (2002, in preparation) estimated nuclear bolometric luminosities for ~100 Palomar galaxies based on X-ray measurements. As shown in Figure 4a, the majority of the objects have $L_{\rm bol} \lesssim 10^{42}$ erg s^{-1}, with the interesting trend that Seyferts are systematically more luminous than LINERs (including transition objects), which in turn are brighter than absorption-line nuclei (objects with no nuclear optical emission lines). More revealing still is the trend with $\lambda \equiv L_{\rm bol}/L_{\rm Edd}$ (Fig. 4b). The distributions of λ systematically shift to lower values following the sequence S → L+T → A. Although LINERs and Seyferts broadly overlap, note that *all* LINERs are characterized by $\lambda \lesssim 10^{-3}$.

(6) The spectral energy distributions (SEDs) are peculiar. Specifically, the SEDs generically lack the optical-UV "big blue bump" (Ho 1999b; Ho et al. 2000), a near-universal feature of unobscured high-luminosity AGNs usually attributed to thermal emission from an optically thick, geometrically thin accretion disk. Another attribute of the SEDs of low-luminosity AGNs, especially of LINERs, is that they are typically "radio loud," defined here by the convention that the radio-to-optical luminosity ratio exceed some fiducial value, say $R > 10$, as normally adopted in quasar studies. In fact, Ho (2001b) finds that among the ~40 nearby galaxies with kinematically determined black hole masses, essentially

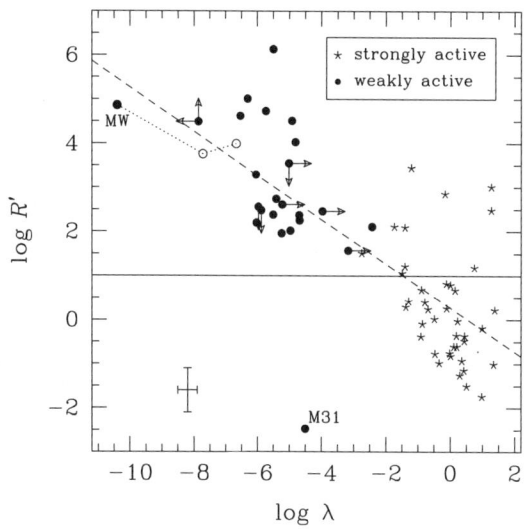

Fig. 5. Distribution of the nuclear radio-to-optical luminosity ratio R' vs. $\lambda \equiv L_{\rm bol}/L_{\rm Edd}$. The *solid line* marks the formal division between radio-loud and radio-quiet objects, $R' = 10$. The *dashed line* is the best-fitting linear regression line. Adapted from Ho (2001b).

all are radio loud (Fig. 5). Since many of these objects are spiral galaxies, this result challenges the conventional wisdom that radio loudness is an attribute unique to giant ellipticals (see Ho & Peng 2001 for a related discussion in the context of Seyfert galaxies). Moreover, Figure 5 shows that the degree of radio loudness evidently increases systematically with decreasing accretion rate, here parameterized by λ.

(7) LINERs have nonstandard central engines. As further elaborated in Ho et al. (2000) and Ho (2001a), LINERs share several charactcristics expected of massive black holes fed by an advection-dominated accretion flow (see reviews by Narayan, Mahadevan, & Quataert 1998; Quataert 2001) instead of a canonical thin accretion disk. The most pertinent of these are (a) the low Eddington ratios and the inferred low mass accretion rates, (b) the absence of the big blue bump in the SEDs, (c) the characteristic radio loudness and its dependence on $L_{\rm bol}/L_{\rm Edd}$, (d) the prevalance of double-peaked broad emission lines and the disk structure they imply, and (e) the low-ionization state of the line-emitting gas, a possible consequence of the particular form of the SEDs.

(8) The nature of type 2 LINERs and transition objects is uncertain. An important unsolved problem is what fraction of the narrow-lined sources should be considered genuine AGNs. To be sure, AGN-like LINER 2s do exist. Barth, Filippenko, & Moran (1999a) discovered that the LINER 2 nucleus of NGC 1052 shows prominent broad-line emission when viewed in scattered light. The unification scheme, popular for Seyfert galaxies, evidently applies to at least some LINERs. If one assumes that all LINERs strictly adhere to the simplest form of the unified model (that all type 2 sources are simply obscured type 1 sources), and that the ratio of LINER 2s to LINER 1s is the same as that of Seyfert 2s to Seyfert 1s in the Palomar survey, then the true AGN fraction among all LINERs is estimated to be between 45% and 65%, depending on whether transitions objects are excluded or included, respectively (Ho 1996, 1999a). This estimate may be overly optimistic for at least two reasons. First, the spectropolarimetric observations of these very faint sources are demanding and still quite limited

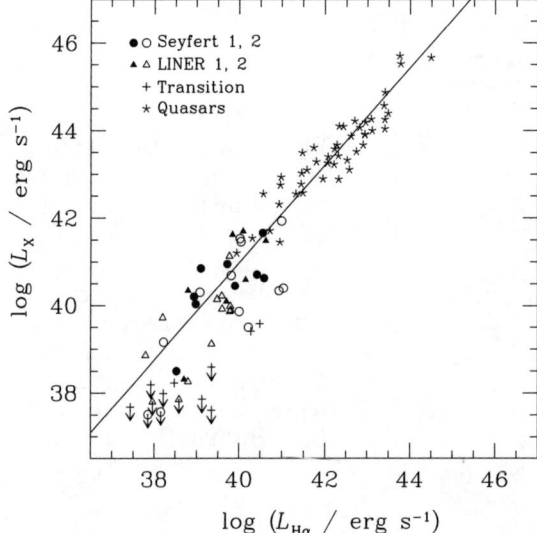

Fig. 6. Correlation between 2–10 keV X-ray luminosity and Hα luminosity for AGNs of various types. Adapted from Ho et al. (2001).

(Barth et al. 1999a, 1999b), and so it may be unwise to generalize these results prematurely. And second, young, massive stars have been found to contribute significantly to the energy budget of some narrow-lined low-ionization objects (e.g., Maoz et al. 1998; Barth & Shields 2000). The first caveat may be partly offset by the fact that some genuine low-luminosity AGNs (e.g., M84, M87, Sombrero) may *intrinsically* lack a broad-line region (Barth 2001).

The resolution of this problem requires access to data which are sensitive to a spectral region that is universally present in AGNs and is relatively unaffected by dust obscuration or modest photoelectric absorption. To be effective, the data must have high angular resolution; to survey large samples, necessary for statistical confidence, the data need to be gathered in an efficient mode. As described above (item 4), the current generation of radio and hard X-ray data both meet these requirements; we give greater weight to the latter because it carries a larger fraction of the AGN bolometric luminosity. Figure 6, taken from the preliminary results of the *Chandra*/ACIS survey of Ho et al. (2001), illustrates the potential of the X-ray data. Type 1 AGNs (LINERs, low-luminosity Seyferts, classical Seyferts, and quasars) obey a well-defined, linear correlation between hard X-ray (2 – 10 keV) and Hα luminosity, down to $L_X(2\text{--}10 \text{ keV}) \approx 3 \times 10^{38}$ erg s^{-1}. The $L_X - L_{H\alpha}$ correlation strongly supports the hypothesis that all of these sources, which span $\sim$8 orders of magnitude in luminosity, share the same physical origin.

Note that the majority of the type 2 sources (LINERs and Seyferts) *do* have compact X-ray cores. They follow the type 1 objects in Figure 6, albeit with somewhat greater scatter and offset toward lower $L_X/L_{H\alpha}$ by about a factor of 10. Interestingly, the X-ray hardness ratios of these sources suggest that absorption is not the culprit for their X-ray weakness. The transition objects, on the other hand, appear distinctly different as a class: most are undetected in X-rays, with upper limits as low as $L_X(2\text{--}10 \text{ keV}) \approx 3 \times 10^{37}$ erg s^{-1}. The incidence of compact radio cores in these objects (Filho et al. 2000) also appears to be lower than in LINERs or Seyferts.

4. Implications for Black Hole Demography

To the extent that an AGN signature signifies accretion onto a massive black hole, a local AGN census gives us a lower limit on the fraction of nearby galaxies hosting massive black holes. If we accept that all LINERs, transition objects, and Seyferts are genuine AGNs, then the AGN statistics from the Palomar survey (Fig. 1) imply that black holes exist in >40% of all galaxies with $B_T \leq 12.5$ mag. For bulge-dominated systems (E–Sbc), this fraction climbs to >50%–75% — not inconsistent with the 100% claimed by direct kinematic studies. But how confident are we about the AGN-ness of nearby emission-line nuclei? The status of objects classified as Seyferts does not seem to be in dispute (for reasons that reflect historical bias more than concrete evidence). In any case, LINERs make up the majority of the population, and so deservedly receive closer scrutiny. The evidence presented in § 3 shows that type 1 LINERs almost surely must be AGNs. The verdict for type 2 LINERs is much less clear-cut at the moment. While some sources have been shown to be powered by hot stars, others clearly have all the hallmark attributes of bona fide AGNs. Perhaps the most telling statistic is the incidence of compact radio and hard X-ray cores; although not yet well quantified, the preliminary results indicate that the detection rate is quite high in both bands ($\gtrsim 50\%$). On the other hand, the "transition objects," postulated to be composite LINER/H II systems, may very well turn out to be unrelated to AGNs. Or if present, the AGN component must be significantly weaker than in "regular" LINERs. If we conservatively exclude all transition objects from the AGN pool, then the overall AGN fraction should be reduced by 13% (Ho et al. 1997b).

Acknowledgments. L.C.H. acknowledges financial support through NASA grants from the Space Telescope Science Institute (operated by AURA, Inc., under NASA contract NAS5-26555).

References

Barth, A. J. 2001, in Issues in Unification of AGNs, ed. R. Maiolino, A. Marconi, & N. Nagar (San Francisco: ASP), in press
Barth, A. J., Filippenko, A. V., & Moran, E. C. 1999a, ApJ, 515, L61
——. 1999b, ApJ, 525, 673
Barth, A. J., Ho, L. C., Filippenko, A. V., Rix, H.-W., & Sargent, W. L. W. 2001, ApJ, 546, 205
Barth, A. J., & Shields, J. C. 2000, PASP, 112, 753
Bower, G. A., Wilson, A. S., Heckman, T. M., & Richstone, D. O. 1996, AJ, 111, 1901
Chen, K., Halpern, J. P., & Filippenko, A. V. 1989, ApJ, 339, 742
Eracleous, M., & Halpern, J. P. 1994, ApJS, 90, 1
Falcke, H., Nagar, N. M., Wilson, A. S., & Ulvestad, J. S. 2000, ApJ, 542, 197
Filho, M. E., Barthel, P. D., & Ho, L. C. 2000, ApJS, 129, 93
——. 2001, MNRAS, submitted
Filippenko, A. V., & Sargent, W. L. W. 1985, ApJS, 57, 503

Halderson, E. L., Moran, E. C., Filippenko, A. V., & Ho, L. C. 2001, AJ, 122, 637
Heckman, T. M. 1980, A&A, 87, 152
Ho, L. C. 1996, in The Physics of LINERs in View of Recent Observations, ed. M. Eracleous et al. (San Francisco: ASP), 103
———. 1999a, Adv. Space Res., 23 (5-6), 813
———. 1999b, ApJ, 516, 672
———. 2001a, in Issues in Unification of AGNs, ed. R. Maiolino, A. Marconi, & N. Nagar (San Francisco: ASP), in press
———. 2001b, ApJ, in press
Ho, L. C., et al. 2001, ApJ, 549, L51
Ho, L. C., Filippenko, A. V., & Sargent, W. L. W. 1995, ApJS, 98, 477
———. 1997a, ApJS, 112, 315
———. 1997b, ApJ, 487, 568
Ho, L. C., Filippenko, A. V., Sargent, W. L. W., & Peng, C. Y. 1997c, ApJS, 112, 391
Ho, L. C., & Peng, C. Y. 2001, ApJ, 555, 650
Ho, L. C., Rudnick, G., Rix, H.-W., Shields, J. C., McIntosh, D. H., Filippenko, A. V., Sargent, W. L. W., & Eracleous, M. 2000, ApJ, 541, 120
Ho, L. C., & Ulvestad, J. S. 2001, ApJS, 133, 77
Komossa, S., Böhringer, H., & Huchra, J. P. 1999, A&A, 349, 88
Kormendy, J., & Gebhardt, K. 2001, in The 20th Texas Symposium on Relativistic Astrophysics, ed. H. Martel & J. C. Wheeler (AIP), in press
Magorrian, J., et al. 1998, AJ, 115, 2285
Maoz, D., Koratkar, A. P., Shields, J. C., Ho, L. C., Filippenko, A. V., & Sternberg, A. 1998, AJ, 116, 55
Nagar, N. M., Falcke, H., Wilson, A. S., & Ho, L. C. 2000, ApJ, 542, 186
Narayan, R., Mahadevan, R., & Quataert, E. 1998, in The Theory of Black Hole Accretion Discs, ed. M. A. Abramowicz, G. Björnsson, & J. E. Pringle (Cambridge: Cambridge Univ. Press), 148
Quataert, E. 2001, in Probing the Physics of AGNs by Multiwavelength Monitoring, ed. B. M. Peterson et al. (San Francisco: ASP), 71
Ravindranath, S., Ho, L. C., Peng, C. Y., Filippenko, A. V., & Sargent, W. L. W. 2001, AJ, 122, 653
Rix, H.-W., et al. 2002, in preparation
Roberts, T. P., & Warwick, R. S. 2000, MNRAS, 315, 98
Shields, J. C., Rix, H.-W., McIntosh, D. H., Ho, L. C., Rudnick, G., Filippenko, A. V., Sargent, W. L. W., & Sarzi, M. 2000, ApJ, 534, L27
Terashima, Y., Ho, L. C., & Ptak, A. F. 2000, ApJ, 539, 161
Terashima, Y., Iyomoto, N., Ho, L. C., & Ptak, A. F. 2001, ApJS, in press
Van Dyk, S. D., & Ho, L. C. 1998, in IAU Symp. 184, The Central Regions of the Galaxy and Galaxies, ed. Y. Sofue (Dordrecht: Kluwer), 489
Wrobel, J. M., Fassnacht, C. D., & Ho, L. C. 2001, ApJ, 553, L23

The Second Byurakan Survey – The Final Results. The Multiwavelength AGN Survey in the SBS Area

J.A. Stepanian

Instituto de Astronomia, Universidad Nacional Autonoma de México, A.P. 70-264, México D.F. 04510, México.

V.H. Chavushyan

Instituto Nacional de Astrofísica Optica y Electrónica, A.P. 51 y 216, C.P. 72000, Puebla, Pue., México.

Abstract.
On the basis of the Second Byurakan Survey (SBS) we have produced the largest and most homogeneous new complete samples of faint Markarian galaxies, Sy galaxies and bright QSOs. The General Catalogue of the SBS Survey, which is completed and prepared for publication, contains the data for more than 3600 objects, half of which are star-like objects while the other half are galaxies. Nearly 2500 slit spectra were obtained for about 2100 SBS objects. 561 new QSOs and AGNs in the magnitude range $15.5 < B < 19.5$ with redshifts $0.0 < z < 3.2$ were spectroscopically confirmed. A sample of 17 BLL has been isolated as well.

The volume for reliable investigation of faint Markarian UVX galaxies has been extended out to a distance of about $\sim 500~Mpc$, more than 50 times deeper than in the FBS. Faint Markarian UVX galaxies from the SBS survey comprise $\sim 12\%$ of field galaxies, slightly greater than in the First Markarian survey; $\sim 10\%$ of them (~ 120) turned out to be Sy galaxies.

A complete sample of faint ($B \leq 17.0$, $z \leq 0.1$) Sy galaxies has been constructed. In the spatial volume with radius of about 500 Mpc, the relative number of AGNs is 1.2% of field galaxies. One of the most complete and representative samples of bright QSOs (~ 130) in the magnitude range $15.7 < B \leq 17.5$ with redshifts $0.3 \leq z \leq 2.2$ has been constructed as well.

The objects in the combined complete sample of SBS Sy galaxies and QSOs might be roughly divided into two groups: classical Sy galaxies and QSOs in agreement with the previous classification scheme and non–classical SyG and QSOs for which the existing classification scheme is not sufficient for their classification. The presence of narrow line QSOs (NLQSOs) or so–called QSO II type objects is established. There is a continuous transition of all properties between all types of Sy galaxies and QSOs, from Liners through Sy2, NLSy1s, BLSy1 to NLQSOs and BLQSOs.

Preliminary results from the Multiwavelength AGN Survey (MWAGN) in X-ray, Radio and IR wavebands in the SBS area are presented. The MWAGN survey in the SBS area shows that the ROSAT or FIRST sur-

veys by themselves cannot represent the whole AGN/QSO sample. The
ROSAT Bright sources along with the FIRST sources comprise only 25%
of the whole SBS QSO sample. The IRAS survey is effective for a redshift
less than $z < 0.05$. All classical Sy2s but only $\sim 50\%$ of classical Sy1s are
IRAS sources. Luminous and Ultraluminous IRAS SBS AGNs comprise
$\sim 25\%$ of the whole AGN sample. The greater part of SBS AGN/QSOs
are neither ROSAT nor FIRST nor IRAS LIG or ULIG sources. They
are optically emitting AGN/QSOs.

The multiwavelength "complete" sample of bright QSOs in the SBS
area allows us to conclude that the fast evolution of bright QSOs is more
likely the result of a selection effect than the result of fast cosmological
evolution. The LogN-B relation for these QSOs reduces to the value
$LogN - B = 0.67 \pm 0.005$.

A direct empirical estimate of the effect of overlapping images is
obtained from the MWAGN survey. The effect of overlap comprises $\sim 5\%$
for QSOs brighter than $B \leq 17.0$ and $\sim 10\%$ for QSOs brighter than
$B \leq 17.5$.

1. Introduction

The tremendous success of the Markarian survey initiated a number of other
extragalactic thin–prism surveys and initiated a new direction in extragalactic astronomy–the systematic search for peculiar objects using low–dispersion
spectroscopy.

J.Stepanian started his work on the First Byurakan Survey in 1974 in Byurakan Observatory. At the end of 1974 in parallel with the FBS, we undertook
a new survey –the Second Byurakan Survey. V. Chavushyan joined the SBS
survey in 1986. 27 years after the beginning of the SBS survey we may report:
Beniamin Egishevich Markarian, the complete sample of faint SBS Sy galaxies
and QSOs is compiled. Your rule of life, the exactitude and the scrupulousness
regarding any question, especially in the field of your main work of the First
Byurakan Survey, were continued in the Second Byurakan Survey. A short summary of today's state of the SBS survey, and a brief description of the Byurakan
surveys are presented below.

2. Byurakan Surveys (FBS and SBS)

The First Byurakan Survey (FBS). Begining in the mid–1960s and continuing through 1980, the first large-scale objective-prism survey for galaxies with
blue and UVX in their continuum radiation was conducted by Markarian. The
observatons of the First Byurakan survey, also commonly known as the Markarian survey, were carried out with the famous Byurakan 1-m Schmidt telescope
with the use of a low-dispersion (1800 Å/mm at Hγ), 1.5deg objective prism.
More than 2500 photographic plates which covered nearly 17000 deg^2 were obtained. The FBS was completed in 1980 and published in a series of 15 papers
(Markarian 1967, Markarian et al. 1981).

The 1500 Markarian galaxies contained in the FBS have provided the principal base from which the major types of AGNs have been discovered, classified, and studied in detail by numerous workers. The FBS resulted in a complete sample of AGNs down to a limiting magnitude $15.^m2$. Markarian galaxies comprise 10% of field galaxies, and about 10% of Markarian galaxies turned out to be Sy galaxies, so 1% of field galaxies were found to be Sy galaxies. The FBS remains perhaps the best known source of AGN in the Local Universe.

The Second Byurakan Survey (SBS). The SBS survey, which is the continuation of the Markarian survey, was aimed to reach fainter limiting magnitudes. The primary goal of the SBS survey was to extend the Markarian survey as deep as possible to obtain a large, well–defined sample of AGNs and QSOs that were selected in a reasonably uniform fashion.

The SBS started in 1974, and plate searching was completed in 1991. A total area of 1000 deg^2 within the contiguous strip defined by $7^h40^m < \alpha < 17^h15^m$, $+49° < \delta < +61°$ has been observed. The SBS was conducted with the same 40-52 inch Schmidt telescope, but with the combination of a set of three objective prisms, 1.5, 3 and 4 degree, respectively, in combination with more modern (in 1974) IIIaJ and IIIaF emulsions sensitized in heated nitrogen. The use of both emulsions extends the wavelength range of sensitivity, increases the uniformity of discoveries, and permits the acquisition of spectra for AGNs down to $B \sim 19.5$, which is about 2-3 magnitudes fainter than was achieved in the FBS.

The SBS survey area is covered by 65 contiguous fields with a size of 4×4 degrees (plate size 16×16 cm), which were obtained in three strips along R.A. with centers of +59, +55 and +51 degrees. To increase the effectiveness and homogeneity of the survey, all the fields were covered by sets of two to five objective prism plates with IIIaF+1.5 degree prism (3500-7500 Å) and IIIaJ+1.5 degree prism (3500-5400 Å). For the main part of them, sets of two to five objective prism plates with IIIaJ+3 degree prism+GG495 filter (4950-5400 Å), and IIIaF+4 degree prism+RG610 filter (6300-7000 Å) were obtained. Some fields were covered with IV-N+4 degree prism+RG8 filter (6900-8500 Å). As a result, the red, blue–green and UV parts of the spectra with practically identical spectral resolution (~ 1000 Å) were available for reliable investigation. The SBS plates were scanned by eye and candidates were selected on the basis of evidence of a strong UV continuum, emission features, or unusual spectral energy distribution. A selection of ~ 3600 objects of which 1800 galaxies (~ 1000 UVX and ~ 800 ELG without significant UV excess) and 1800 stellar objects with excess ultraviolet emission is the main result of the SBS survey.

Medium-resolution spectral observations ($R \sim 5-10$ Å and $S/N \sim 15-25$) were obtained for about 2100 SBS objects from 1977 to 2001. The spectroscopic and photometric observations were made on the 6-m and 1-m telescopes of the SAO (Russia) on the 4.5-m MMT (USA), the 2.6-m Byurakan (Armenia) and 2.1-m GHO (Mexico) telescopes.

The first list of SBS objects was published in 1983 (Markarian & Stepanian 1983). In total there are seven lists published (Stepanian 1994, and references therein). The main parameters for all studied objects were published in a series of papers (Markarian & Stepanian 1984a,b; Markarian et.al. 1985, 1986; Stepanian et.al. 1988-2001, Stepanian 1994- 1998).

So far, the identifications of 578 new AGN/QSOs, $\sim$ 1000 galaxies with narrow emission lines and $\sim$ 1000 galactic stars, the vast majority of which are hot WD and subdwarfs, have been spectroscopically confirmed in a area of $\sim$ 1000 deg^2 of the SBS survey. The present status of the SBS survey is illustrated in Table 1.

Table 1. **The Second Byurakan Survey. Present (2001) Status**

Spectroscopy	Total $\sim$ 2500 spectra		Photometry $\sim$ 500	
SBS Stellar $\sim$ 1250 sp.		SBS QSOs N=441		SBS stars N=810
SBS QSOs	Number	Other source	Number	SBS stars	Number
BALQSO	15	Radio	200	WD	$\sim$ 300
DampQSO	5	X-ray	200	subdwarf	$\sim$ 200
Ly forest	15	Gamma-ray	2	HBB+NHB	$\sim$ 100
BLLac	17	BLAZAR	3	F/G	$\sim$ 100
Abs. QSOs	35	IRAS	20	Continual	27
Grav. lenses	3			C2+dMe	45
Other	360			CV	25
				Composite	17
SBS Galaxies $\sim$ 1250 sp.		SBS SyG N=120		Classified $\sim$ 800	
SBS SyG	Number	Classified	Number	Other sour.	Number
Classical SyG	62	Liners	97	IRAS	$\sim$ 600
Sy1	26	SBN	350	Radio	$\sim$ 200
Sy1.5	8	BCDG	130	X-ray	$\sim$ 50
Sy2	28	HII	100	Close-bin.	$\sim$ 250
Non-class.SyG	58	ELG	600		
NLSy1	40	Abs.Gal	20		
Sy/Lnr	18				

The subsample of relatively bright $B \leq 17.5$ stellar objects (N=820) was investigated completely. B and V band CCD photometric measurements were made for all $B \leq 17.5$ spectroscopically–confirmed new SBS QSOs. A complete sample of 182 optically selected SBS QSOs was isolated.

A similar investigation was done for the SBS galaxy sample. Slit spectra were obtained for $\sim$ 800 galaxies with $B \leq 17.0$ which comprise nearly 70% of the completeness level. A complete sample of faint SBS Sy galaxies was isolated. B and V band CCD photometric measurments were made for nearly 80% of them.

We have produced the largest and most homogeneous, new combined complete optical sample of SBS QSOs and Sy galaxies, which comprises $\sim$ 280 objects in an area of 1000 deg^2 of the Northern sky . This is perhaps the first deep ($z \sim 0.15$) representative sample of faint AGNs compiled in a uniform and homogeneous fashion.

3. The Complete Optical Samples of SBS AGN/QSOs

The total SBS AGN/QSO optical sample contains 578 objects (120 SyG, 441 QSO and 17 BLL) in the magnitude range $15.5 < B < 19.5$ and redshift range

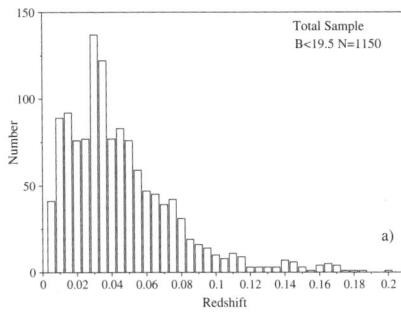 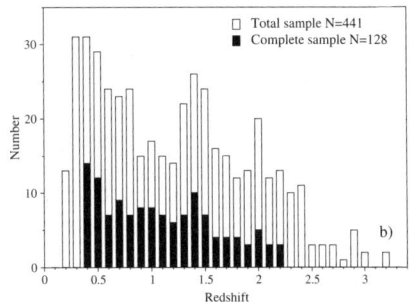

Figure 1. Redshift distribution: a) SBS Galaxies, b) SBS QSOs.

$0.0 < z < 3.2$ (75 with $0.0 < z \leq 0.1$, 75 with $0.1 < z < 0.3$, 373 with $0.3 \leq z \leq 2.2$ and 38 with $2.2 < z < 3.2$). We divide the sample of SBS AGN/QSOs into two groups by redshift and magnitude where the subsample of isolated objects is methodologically complete.

3.1. The Complete Sample of SBS Sy Galaxies. Magnitude Range 13.0<B ≤ 17.0, Redshift Range 0.0<z ≤0.1

The surface density of SBS galaxies corresponds to a value of ∼ 1.84 galaxy per deg^2 to a limit of $B \sim 19$. The more precise value of the surface density of SBS UVX, ELG without UV continua and SBS Sy galaxies as a function of limiting magnitude is presented in Table 2.

Table 2. **The surface density of SBS UVX, ELG and SyG**

	Total number	Surface density					
		$B \leq 15.0$	$B \leq 15.5$	$B \leq 16.0$	$B \leq 16.5$	$B \leq 17.0$	$B \leq 19.0$
SBS total	1840	0.12	0.27	0.53	0.88	1.25	1.84
SBS UVX	1010	0.07	0.13	0.27	0.49	0.72	1.01
SBS ELG	830	0.05	0.14	0.26	0.39	0.53	0.83
SBS SyG	120	0.008	0.014	0.028	0.053	0.085	0.012

The SBS UVX galaxies comprise ∼ 12% of field galaxies, and around 10% of them are Sy galaxies. The proportion of the SBS ELG without UV excess is a little less than the proportion of SBS UVX galaxies. Together they bring the number of ELG with strong emission to no less than 20% of field galaxies.

120 SBS Sy galaxies were found; 62 classical and 58 non–classical Sy galaxies. The proportion of SBS SyG among the field galaxies in a volume of about 500 Mpc radius is around 1.2%, slightly greater than in a volume of ∼ 100 Mpc radius, where the value of ∼ 1% was first obtained on the basis of the Markarian survey. In addition, with the nearly 100 LINERs discovered in the SBS, AGN comprise no less than 2% of field galaxies. The redshift distribution of the total SBS galaxy sample is shown in Fig.1a.

A complete sample of 71 new faint SBS Sy galaxies in the magnitude range $13.0 < B \leq 17.0$ with redshift $z \leq 0.1$ was isolated. It included: 40 classical Sy galaxies (17 Sy1, 5 Sy1.5, 18 Sy2) and 31 non–classical Sy galaxies (15 NLSy1, 16 transition (Sy2/LINER) objects). We were not able to classify the remainder of the objects with the use of the standard classification scheme, where the line width and the line ratio are used, therefore we include them in the class of non-classical Sy galaxies.

Numerous non-classical Sy galaxies were found in the last decade among the ROSAT and IRAS optically identified objects. The subsample of so-called NLSy1 is dominant among the ROSAT detected AGNs. The subsample of non-classical Sy2 and Sy2/LINER composites is detected among the IRAS AGNs. These objects show emission line widths and emission line ratios not typical for classical Sy galaxies.

The subsample of SBS NLSy1 galaxies shows line ratios of $[OIII]\lambda 5007/H\beta$ from 0 to 10, and FWHM line widths from 300–3000 km s^{-1}, which fills the gap between Sy2s and BLSy1s. Some of the SBS Sy1 galaxies show very broad (or multicomponent) $H\alpha$ emission lines along with an absence of $H\beta$ or other emission lines. Some SBS Sy2s show narrow emission lines $H\alpha$ and $[NII]\lambda 6584$, with $H\alpha/[NII]\lambda 6584 \sim 1$, but $H\beta$ or $[OIII]\lambda 5007$ are very weak or not seen at all. Some objects show the signature of composite Sy2/LINER spectra or LINER/Starburst spectra. **There is continuity from classical BLSy1s through NLSy1s, Sy2s, through LINERs and Starbursts.**

3.2. The Complete Sample of SBS QSOs. Magnitude Range 15.7<B ≤17.5, Redshift Range 0.3 ≤ z≤2.2

It is well known that in an optical UVX survey, the waveband 3500-5400 Å is used for AGN/QSO detection, where selection effects are minimized in the redshift range $0.3 < z < 2.2$. The complete sample of SBS QSOs in the redshift range $0.3 \le z \le 2.2$ and magnitude range $15.7 < B \le 17.5$ contains ~ 130 objects. The redshift distribution of the total and complete samples of SBS AGN/QSOs is shown in Fig. 1b.

As in the case of SBS Sy galaxies, the complete sample of SBS QSOs also might be roughly divided into two groups: QSOs with broad emission lines $FWHM > 3000$ km s^{-1} (BLQSOs) and QSOs with relatively narrow emission lines $FWHM < 3000$ km s^{-1} (NLQSOs or QSO II).

It seems to us that the NLQSOs are the continuation of NLSy1 to the class of more luminous objects, the so–called QSO IIs. Examples of similar objects were recently found among the IRAS selected ROSAT galaxies (Morgan et al. 1996). The spectra of some SBS NLQSOs are shown in Fig. 2.

The differential surface density of bright SBS QSOs in the magnitude ranges $16.0 < B < 16.5$, $16.5 < B < 17.0$ and $17.0 < B < 17.5$ is 0.016, 0.34, and 0.71, respectively. The LogN-B relation of SBS QSOs is $LogN - B = 0.65 \pm 0.004$ in the magnitude range $15.7 < B \le 17.5$.

4. Multiwavelength AGN (QSOs+SyG) Survey in the SBS Area in X-ray, Optical, Infrared and Radio Wavebands

In past decades we were able to work only in the optical window. As a result, the majority of observational data and cosmological models were based on the results of optically complete samples of AGN/QSOs. The first sample of AGNs (Markarian survey) and the first sample of bright QSOs (BQS, Schmidt & Green 1983) have been used as a basis for cosmological calculation until now. During

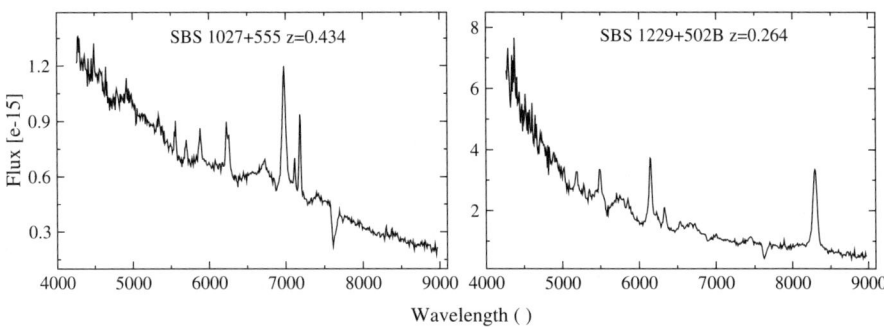

Figure 2. Examples of Narrow-Line SBS QSOs (QSO IIs).

the last two decades, these surveys were re-examined many times in regard to their completeness.

In the last decade, the IRAS, ROSAT, NVSS, FIRST, 2dF, SDSS and 2MASS data completely changed the astronomical database. A huge number of new X-ray, IR, radio, etc, AGN/QSOs were discovered, many of which were not present in optical complete samples. Of course, the combination of different selection techniques will eventually produce a more complete flux limited sample of AGN/QSOs than a single parameter survey.

The multiwavelength AGN survey (MWAGN), aimed for the construction of a combined deep sample of AGNs based on the SBS survey in X-ray, optical, IR and radio wavebands, was undertaken in the last few years. We have used the complete sample of SBS optically selected AGNs as the basis for the compilation of a combined multiwavelength complete sample of bright AGN/QSOs. It is the first attempt to create a sample of AGN/QSOs with a well defined flux limit, where a wide range of the electromagnetic spectrum is used. Nearly 15000 optical counterparts of ROSAT, FIRST and IRAS sources were identified to the magnitude limit of DSS1. A few thousand relatively bright objects with optical counterparts with $B \leq 17.5$ were isolated for spectoscopic observation.

The subsample of optically bright $B \leq 17.5$ and ROSAT Bright ($f_x > 2.4 \times 10^{12}$ erg cm^{-2} s^{-1}) sources in the SBS area were spectroscopically completely investigated. A dozen ROSAT and optically bright new AGN/QSOs were found.

The Radio-Optical survey in the SBS area was undertaken by V.Chavushyan. FIRST sources were indentified with APM stellar optical counterparts, the bright part of which ($B \leq 17.5$) was spectroscopically completely investigated. A sample of about 30 optically bright ($B \leq 17.5$) new AGN/QSOs was discovered. The preliminary result of the multiwavelength FIRST/APM QSO survey in the SBS area was published by Chavushyan et al. (1999, 2001).

The investigation of Infrared sources in the SBS area is underway on the SPM (Mexico) 2.1-m telescope (I.Cruz-Gonzalez, D.Dultzin-Hacyan).

Preliminary results of the MWAGN survey in the SBS area are presented in Table 3.

Twelve new bright (B≤17.5) QSOs with redshifts $0.3 \leq z \leq 2.2$ which are ROSAT and/or FIRST sources were discovered in the course of the MWAGN survey in the SBS area. The resulting LogN-B relation becomes $LogN - B =$

Table 3. The complete sample of SBS QSOs and QSOs of the MWAGN survey in the SBS area. Magnitude range 16.0<B ≤ 17.5, redshift range 0.3<z≤2.2

Mag. range	SBS Optical survey			ROSAT		FIRST		MWAGN
	N Tot.	Missed in SBS ROSAT	Missed in SBS FIRST	N Tot.	Detected in SBS	N Tot.	Detected in SBS	Total number
16.0-16.5	16	-	1	2	2	3	2	17
16.5-17.0	34	1	2	11	10	5	3	37
17.0-17.5	71	3	5	15	12	15	10	79
	LogN=0.65±0.004+C					LogN=0.67±0.005+C		

0.67 ± 0.005. Of course, this magnitude range is too narrow to make the critical cosmological conclusion about the behavior of the LogN-B curve, but the MWAGN data may allow us to conclude that the fast evolution of bright QSOs is more likely the result of selection effects than the result of cosmological evolution.

The effect of overlap is empirically obtained. The MWAGN data allow us to make a direct estimate. Part of the AGN/QSOs missed in the SBS which are ROSAT or FIRST sources are in fact AGNs contaminated by nearby objects. For objects brighter than $B \leq 17.0$ the effect of overlap is $\sim 5\%$; for $B \leq 17.5$ magnitude the effect of overlap is $\sim 10\%$.

The MWAGN survey in the SBS area may allow us to compile a complete sample of AGN/QSOs with a well defined flux limit: in the optical range $B \leq 17.5$, in the soft (0.5-2 keV) X-ray band $f_x > 2.4 \times 10^{-12}$ erg cm^{-2} s^{-1}, in the IRAS band $f(60\mu) > 0.6$ Jy and Radio fluxes at 1.4 GHz greater than 1 mJy.

The MWAGN survey in the SBS area is not finished yet, and the combined complete sample of AGN/QSOs in the SBS area is not compiled yet, but a few important conclusions might be drawn from the current data:

1. **On the basis of the Second Byurakan Survey (SBS) we have produced the largest and most homogeneous new complete samples of faint Markarian Galaxies, Sy galaxies and bright QSOs.**

2. **The volume of reliable investigation of UVX galaxies, extended out to a distance of ~ 500 Mpc, is more than 50 times deeper than in the FBS. The proportion of UVX galaxies among field galaxies in the volume of about 500 Mpc is about 12%, which is slightly greater than in the FBS survey ($\sim$100 Mpc, 10%) .**

3. **The first complete sample of faint (B$\leq 17\overset{m}{.}0$, $z <0.1$) Sy galaxies is compiled. The relative number of SBS SyG consists of about 10% of SBS faint UVX galaxies or 1.2% of field galaxies. A complete sample of relatively bright SBS QSOs in the magnitude range 15.7¡B$\leq$17.5 and redshift range $0.3 \leq z \leq 2.2$ was constructed as well .**

4. **The existing classification of AGN/QSOs is not sufficient for the classification of the entire class of newly detected AGN/QSOs. The sample of SBS AGN/QSOs might be divided into two groups: classical AGN/QSOs, consistent with the previous classification**

scheme and non-classical AGN/QSOs, which needed new classification criteria. The presence of narrow line QSOs (NLQSOs) or so–called QSO II type objects is established. Narrow Line QSOs are the continuation of NLSy1 to the class of more luminous objects (QSO II class) .

5. There is continuity from classical BLSy1s through NLSy1s, Sy2s, to LINERs and Starbursts. The subsample of SBS NLS1 galaxies shows line ratios of [OIII]/Hβ from 0 to 10, and FWHM line widths from 300-3000 km s^{-1}, which fills the gap between Sy2s and BLSy1s.

6. The preliminary result of the Multiwavelength AGN survey in the SBS area shows that the ROSAT, FIRST or IRAS surveys alone cannot represent the whole AGN sample. Only 25% of SBS QSOs are ROSAT or FIRST sources. The IRAS survey is effective for redshifts less than z <0.05. All classical Sy2s, but only ~50% of classical Sy1s are IRAS sources. Luminous and Ultraluminous IRAS SBS AGNs comprise ~25% of the whole AGN sample. Most SBS AGN/QSOs are neither ROSAT nor FIRST nor IRAS LIG or ULIG sources. They are optically emitting AGN/QSOs.

7. The multiwavelength complete sample of bright SBS QSOs allows us to conclude that the fast evolution of bright QSOs is more likely the result of selection effects than the result of cosmological evolution.

Acknowledgments. J.S. is grateful to the Organizing Committee especially to the chairman of the Local Organizing Committee, A. Mikaelian, and to UNAM colleagues, D. Dultzin-Hacyan, I.Cruz Gonzalez and Rafael Costero through the help of whom I had the possibility to participate in the Colloquium. We extend our thanks to the Byurakan co-workers and high quality technical specialists without whom the SBS couldn't be arranged and continued: Aram Astvazaturian (Varpet Aram), S.Karapetian, H. Sarkisian and others.

J.S. thanks G. Kojoian, D. Weedman and R. West who provided the IIIaJ plates at the beginning of the SBS survey. We thank Maarten Schmidt who in 1986 arranged the investigation of the SBS objects with R. Green and C. Foltz. Thank to R. Green and C. Foltz, many of the SBS objects were investigated in more detail. Thanks to Donald Osterbrock and his collaborators, the investigation of the SBS Sy galaxies was continued.

Thanks to V. Afanasév, we had a chance to continue our work in the Special Astrophysical observatory (Russia). Due to the help of Irene Cruz Gonzalez and Deborah Dultzin-Hacyan, the investigation of SBS objects are continuing now in UNAM (Mexico).

J.Stepanian acknowledges the CONACYT Catedras Patrimoniales project EXP.EX-000287. V.Chavushyan acknowledges CONACYT research grant 28499-E.

References

Chavushyan V.H., et al., 2001, ASP. Conf.Ser. 102, 232, Ed. by Clowes et al.
Markarian, B.E., 1967, Astrofizika, 3, 55
Markarian, B.E., et al. 1981, Astrofizika, 17, 619
Markarian, B.E., & Stepanian, J.A., 1983, Astrofizika, 19, 639
Markarian, B.E., & Stepanian, J.A., 1984a;b, Astrofizika, 20, 21, 513
Markarian, B.E., Stepanian, J.A., & Erastova, L.K. 1985, Astrofizika, 23, 439
Markarian, B.E., Stepanian, J.A., & Erastova, L.K. 1986, Astrofizika, 25, 345
Morgan et al. 1996, ApJS, 106, 341
Schmidt, M. & Green, R.F., 1983, Ap.J., 269, 352.
Stepanian, J.A., Lipovetsky, V.A., & Erastova, L.K., 1988, Astrofizika, 29, 247
Stepanian, J.A., Lipovetsky, V.A., & Erastova, L.K., 1990a, Astrofizika, 32, 441
Stepanian, J.A., Lipovetsky, V.A., Shapovalova, A.I., & Erastova, L.K., 1990b, Astrofizika, 33, 89
Stepanian, J.A., Lipovetsky, V.A., Shapovalova, A.I., Erastova, L.K., & Chavushyan, V.H., 1990c;d; Astrofizika, 33, 199, 351
Stepanian, J.A., Lipovetsky, V.A., Chavushyan, V.H., Erastova, L.K., & Shapovalova, A.I., 1991a;b; Astrofizika, 34, 205, 315
Stepanian, J.A., Lipovetsky, V.A., Chavushyan, V.H., Erastova, L.K., & Balayan, S.K., 1993, Bull. Spec. Astrophys Obs. (Izv. SAO), 36, 5
Stepanian, J.A., Chavushyan, V.H., Carrasco, L., Tovmassian, H.M., & Erastova, L.K., 1999, PASP, 111, 1099
Stepanian, J.A., 1994, Doctoral thesis, Nizhnij Arkhys
Stepanian, J.A, IAU Symp 194 "Activity in Galaxies and related phenomena" 1998 Kluwer Acad. Publ. P.O.Box 17, 3300 AA Dordrecht, The Netherlands

The Hamburg All-Sky Bright QSO Surveys

Dieter Reimers

Hamburger Sternwarte, Gojenbergsweg 112, D-21029 Hamburg, Germany

Lutz Wisotzki

Universität Potsdam, Am Neuen Palais 10, D-14469 Potsdam, Germany

Abstract. Over the past decade, our team has pursued a quest to search for the optically brightest quasars in the entire extragalactic sky, by means of automated quasar selection using digitised objective prism spectra. We give a brief overview over the observational database and selection strategies and present some of the major survey highlights.

1. Introduction

Bright quasars are important targets for a variety of astrophysical studies, interesting in their own but also valuable tools to probe the intervening matter by absorption-line spectroscopy and gravitational lensing. However, bright quasars are also very rare, and to sample the brightest part of the quasar population, substantial fractions of the extragalactic sky have to be covered with efficient surveying techniques. Both of our wide-angle surveys were conceived in this spirit: The Northern hemisphere 'Hamburg Quasar Survey' (HQS) and its Southern counterpart, the 'Hamburg/ESO Survey' (HES); both projects are now practically completed. The applications of such surveys can generally be numbered under two categories: (1) The discovery aspect, where statistical completeness is not important; and (2) the completeness aspect, where well-defined flux-limited samples are to be constructed. In the following we pay attention to both aspects.

2. Basic Survey Properties and Project Status

2.1. Common Properties

Both surveys are based on the same basic technical layout, using Schmidt telescopes equipped with objective prisms to produce deep unwidened and unfiltered spectral plates. The blue-sensitive Kodak IIIa-J photographic emulsion determines the spectral range to the red, while to the blue it is limited by the atmosphere; $3200 \lesssim \lambda \lesssim 5400$ Å. The magnitude range for processable spectra is roughly $13 \lesssim B \lesssim 18$, although strongly dependent on observing conditions, in particular on the seeing. The plates were digitised with the Hamburg PDS 1010G microdensitometer. An example of the result of the digitisation procedure is shown in Fig. 1. The total survey area is about $\sim 22\,000$ deg^2. By

Direct plate scan from UKST Schmidt telescope:
Slit 50 μm × 50 μm, step 25 μm × 25 μm (*The Digitized Sky Survey*)

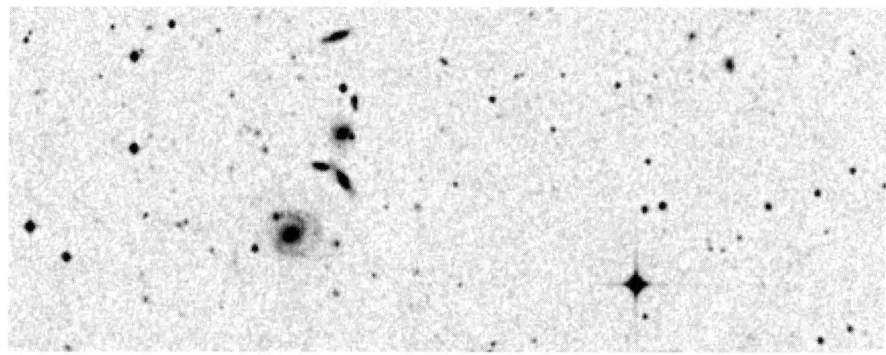

PDS scan of ESO Schmidt objective prism plate:
Slit 30 μm × 30 μm, step 20 μm × 20 μm

Figure 1. The top panel shows the direct data for a small survey area, taken from the DSS. Below is shown the corresponding area in the spectral data.

the end of 2001, the pixel data of fully digitised objective prism plates will be available over the web. Check the survey homepage[1].

2.2. Hamburg Quasar Survey (HQS):

The HQS is based on the 80 cm former Hamburg Schmidt telescope on Calar Alto, Spain. Prism dispersion is 1390 Å/mm and yields a seeing-limited spectral resolution of typically 50 Å FWHM at Hγ. The HQS has reached full coverage of the Northern extragalactic sky, defined as $\delta > 0°$, $|b| > 20°$. The formally covered sky area is 13 600 deg^2, or 567 Schmidt fields. Quasar candidates are part of a search for objects with 'blue continuum shape' (Hagen et al., 1995). Follow-up spectroscopy was mainly conducted at Calar Alto observatory and

---

[1] www.hs.uni-hamburg.de/english/arbgeb/extgalqso/surveys.html

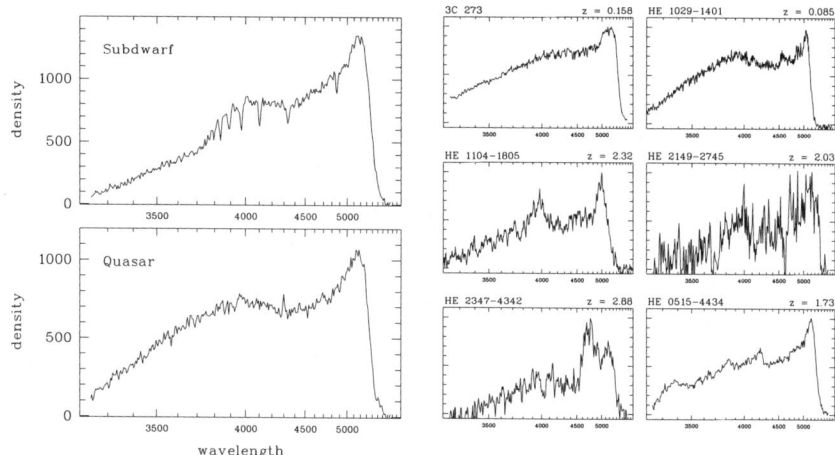

Figure 2. Example digitised objective prism spectra. Left: comparison of two UV excess objects. The top spectrum is a blue star with conspicuous Balmer absorption, the bottom is a low-redshift QSO ($z = 0.15$). Right: 'Mugshots' of several confirmed QSOs.

has yielded so far 720 confirmed new QSOs with $0 \lesssim z \lesssim 3.3$ (Engels et al. (1998); Hagen et al. (1999)). The HQS is a 'discovery survey', as no effort for completeness has been made. Besides quasars, a large variety of odd objects has been identified. Many additional AGN were found through the Hamburg/RASS identification project (Bade et al., 1998).

2.3. Hamburg / ESO survey (HES):

The HES was installed in 1990 as an ESO key programme. The ESO 1 m Schmidt telescope on La Silla, Chile, was used to take prism plates with a dispersion of 450 Å/mm, corresponding to a seeing-limited spectral resolution of ~ 15 Å FWHM at Hγ. The Southern extragalactic sky is fully covered, although more constrained to high Galactic latitudes than the HQS. The survey area is mainly defined by $\delta < 2\overset{\circ}{.}5$ and $E(B-V) < 0.15$, covering 380 Schmidt fields or ~ 9000 deg^2. QSO search was conducted with a multitude of selection criteria (Wisotzki et al., 2000). To date, 1220 QSOs were spectroscopically confirmed at ESO, with $0 \lesssim z \lesssim 3.3$ (Reimers et al. (1996b); Wisotzki et al. (2000), Reimers et al., in prep.). Flux-limited subsamples for statistical analyses have been constructed: 48 QSOs in 611 deg^2 (Köhler et al., 1997); 415 QSOs in 3700 deg^2 (Wisotzki et al., 2000); 862 QSOs in ~ 7500 deg^2 (Wisotzki et al., in prep.). The remaining ~ 4000 QSO candidates will be observed in 2002–2003 with the 6dF facility at the UK Schmidt telescope.

2.4. Quasar Selection

Quasars are known to have spectra that differ strongly from those of stars, most prominently by their ultraviolet excess. On the other hand, there has always been a suspicion that UV excess surveys might miss a substantial fraction of the QSO population, especially when the UV criterion is applied in a very restrictive way. In the HES, we have established a highly efficient two-step selection procedure allowing us to formulate the UV excess condition in an extremely relaxed way, effectively regarding everything bluer than $(U - B) = -0.17$ as a QSO candidate. The high quality of the spectra permits the estimation of colours with an accuracy of better than 0.1 mag rms, another important property to assure that QSOs with only a moderate UV excess will be included. Further selection criteria exploit additional colour information, in particular the presence of a superimposed host galaxy as in low-redshift Seyferts. We emphasise that contrary to popular belief, emission line detection plays a minor role – in fact, *all* our $z < 2.6$ QSOs are colour-selected. Only at the highest accessible redshifts it is that colour selection becomes inefficient because of Lyα moving out of the spectral range, plus additional Lyman forest blanketing. At $z > 2.6$, the selection has therefore to rely on feature detection algorithms.

The crucial second step in the selection procedure is the elimination of the unavoidable stellar contamination. Here the high spectral resolution of the HES prism spectra is an asset: As illustrated in Fig. 2, faint blue stars show conspicuous Balmer absorption lines readily visible in the prism spectra (and detectable by automated techniques); such objects do not enter the final QSO candidate lists. Recall that at $B \simeq 16$, more than 90 % of classical UV excess objects are stars, and this fraction is even higher for the relaxed UV excess criterion used in the HES. Nevertheless, the second stage of eliminating the stellar content is so efficient that the QSO success rate in follow-up spectroscopy is of the order of 70 %, without sacrifices in completeness.

3. Highlights

3.1. Observations of UV-bright QSOs

Bright high-redshift quasars are perfect background sources for absorption line studies of the Intergalactic Medium. They are also very rare. Véron-Cetty & Véron (2001) list 93 (17) QSOs with $z > 2$ ($z > 3$) and $B < 17$. Of these, $\sim 40\%$ were discovered by HQS and HES. In order to facilitate far-UV spectroscopy with HST and FUSE, an additional requirement is an *unabsorbed line of sight*, which in turn eliminates the major fraction of available high-z quasars (Moller & Jakobsen, 1990). Our true all-sky approach opened the possibility to systematically search for new targets.

The first known high-redshift QSO with a clean UV line of sight was HS 1700+6416 ($z = 2.72$; Reimers et al. (1989)). Its rich absorption line spectrum was measured by HST and showed over 50 new EUV absorption lines (Reimers et al. (1992); see Fig. 3), yielding valuable information about the chemical evolution of the intergalactic medium.

The first QSO bright enough for detailed spectroscopy at $\lambda_{\rm rest} < 304$ (the wavelength of the He II Lyα emission line) was HE 2347−4342 at $z = 2.885$. The HST spectrum revealed a complex pattern of He II-Lyα absorption 'voids' and 'troughs' (see Fig. 4), in part much too strong to be explained by the normal hydrogen Lyman forest. Instead, we interpreted this as evidence for intergalactic Helium being not completely ionised, possibly as consequence of

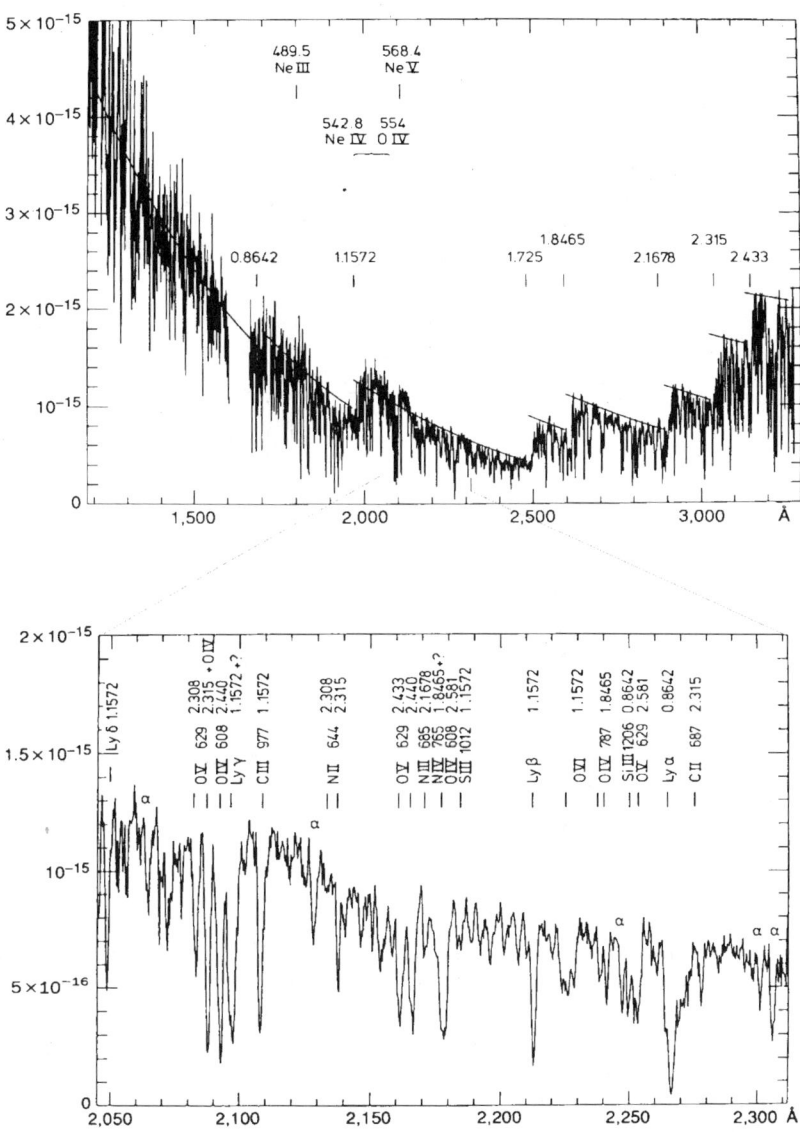

Figure 3. Spectrum of HS 1700+6416 ($z = 2.72$) taken with the Faint Object Spectrograph of the Hubble Space Telescope. Notice the 'Lyman Valley', 7 optically thin Lyman limit systems, and the wealth of intrinsic EUV high ionization metal absorption lines seen for the first time in this QSO (from Reimers et al. 1992).

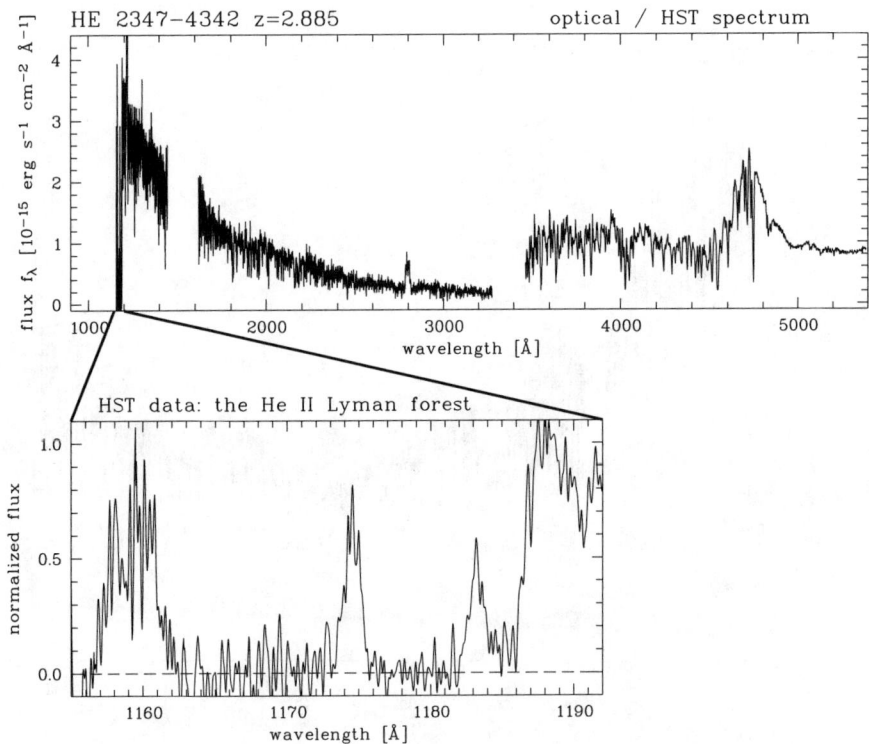

Figure 4. Optical and HST spectroscopy of the UV-bright QSO HE 2347−4342. The bottom panel shows the small section below the λ304 He II Lyα emission line.

only recently started cosmic reionisation of He II to He III (Reimers et al., 1997). The observation of the reionisation phase of the universe has been confirmed by FUSE observations of the He II 304 Å forest of this quasar (Kriss et al. (2001))

High-resolution spectroscopy of the extremely bright QSO HE 0515−4414 ($V = 14.9$, $z = 1.73$) has enabled us to detect the Warm-Hot-Intergalactic medium via O VI absorption lines (Reimers et al. (2001)). Two additional $z > 1.5$ QSOs from the HQS have been observed in 2001 with HST at high resolution for the same purpose.

Further UV-bright QSOs include HS 0747+4259 ($z = 1.90$), HS 1103+6416 ($z = 2.19$) and HS 1307+4716 ($z = 2.13$). A review of UV observations of high-redshift QSOs has been given by Reimers et al. (1998b). Together with the above, these objects constitute 100 % of known high-redshift QSOs bright enough for high-resolution UV spectroscopy. A useful byproduct of these observations are reliable (i.e. Lyα forest-corrected) EUV spectral energy distributions; it seems that most are highly incompatible with canonically assumed EUV SEDs (Reimers et al. (1998b)).

3.2. Gravitational Lenses in the Hamburg Surveys

The optical depth to QSO image splitting induced by gravitational lensing is of the order of $\sim 10^{-3}$ for high-redshift quasars. Because of the so-called magnification bias, the lens fraction is substantially enhanced in bright QSO samples, by up to a factor ~ 10 – this means that at least $\sim 1\%$ of the $z \gtrsim 1$ HQS and HES QSOs are probably lensed. Several interesting cases were already found.

Most prominent is HE 1104−1805, also known as the 'Double Hamburger'. This bright wide separation lens is an excellent target for monitoring, time delay determination, and microlensing studies (Wisotzki et al. (1993, 1998)); it furthermore offers two nearby lines of sight to study the spatial extent of intervening absorbers (Lopez et al. (1999); see also Fig. 4). Another four lensed QSOs (all doubles) were discovered serendipitously, and two recent small imaging follow-up studies found two further lensed QSOs out of only ~ 80 objects. This is a dramatic confirmation of the effect of magnification bias: With 6 newly discovered and several rediscovered lensed QSOs, HQS and HES form together already the second-largest existing single lens survey. Efforts to systematically exploit the lens content of HQS and HES are under way.

3.3. The Bright End of the QSO Luminosity Function

A flux-limited subsample, constructed from about 60% of the HES, comprises 415 QSOs and Seyfert 1 nuclei with broad emission lines in an effective area of $3700\,\mathrm{deg}^2$ (Wisotzki et al. (2000)). Average limiting magnitude is $B_J \lesssim 17.2$, i.e. approx. 1 magnitude brighter than the survey detection limit. This constitutes by far the largest existing sample of very bright QSOs. It shows also a high degree of completeness: (1) The redshift distribution peaks at low z, and there is no evidence for z-dependent incompleteness. (2) A cross-comparison with other surveys showed that all known QSOs within the survey area and (B, z) limits were selected by the HES criteria. (3) The inferred surface density of QSOs is perfectly consistent with many other samples. It is, however, 1.5× higher than in the Palomar-Green survey, but the extreme degree of incompleteness of the PG that others suggested in the past could be ruled out. The statistical analysis of the sample yielded a luminosity function where the bright tail flattens considerably towards lower z, incompatible with Pure Luminosity Evolution (Wisotzki (2000)). Fig. 5 shows the nonparametrically constructed cumulative luminosity function in five redshift shells and compares these data to the prediction from 'standard' PLE. Especially at low redshifts there is a significant excess of high-luminosity QSOs over the predictions.

3.4. The Local AGN Population

Despite their proximity, the statistical properties of AGN with $z \approx 0$ are not well constrained. This has the main reason that most optical QSO surveys are limited to *point sources*, but another important effect is that host galaxies contribute to the photometry and make interpretations of number counts difficult. To overcome these problems, special strategies were developed in the HES surveying procedure: (1) No morphological discrimination against extended sources is exerted; all objects go through the same filters. (2) A specific 'colour' criterion for Seyfert-like spectra has been installed. (3) Magnitudes are measured

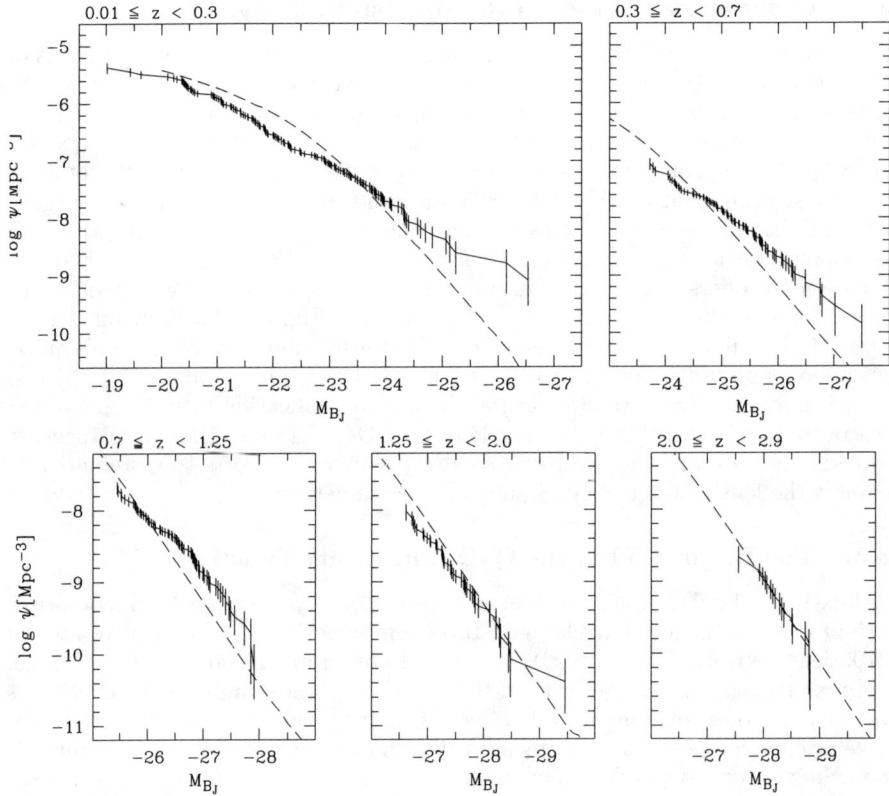

Figure 5. Segments of the cumulative QSO luminosity function, as constructed from the flux-limited HES subsample. Overplotted are the predictions from Pure Luminosity Evolution.

in an aperture of the size of the seeing disk – the sample is limited by *nuclear magnitude*.

These properties make the HES unique among optical AGN surveys, as it is unbiased to sample the local AGN population. Applications so far include the discovery that the local luminosity function of QSOs and Seyfert 1 nuclei is nearly a featureless power law (Köhler et al. (1997); cf. also the lowest redshift panel of Fig. 5) and the definition of complete samples for QSO host galaxy studies. Combining the statistical analyses performed on complete QSO samples with the measurement of QSO host galaxy parameters, we have recently been able to construct the luminosity function of QSO host galaxies (Wisotzki et al. (2001); Fig. 6). It is sufficiently similar to that of 'inactive' field galaxies to suggest that the parent population of QSOs are completely average galaxies.

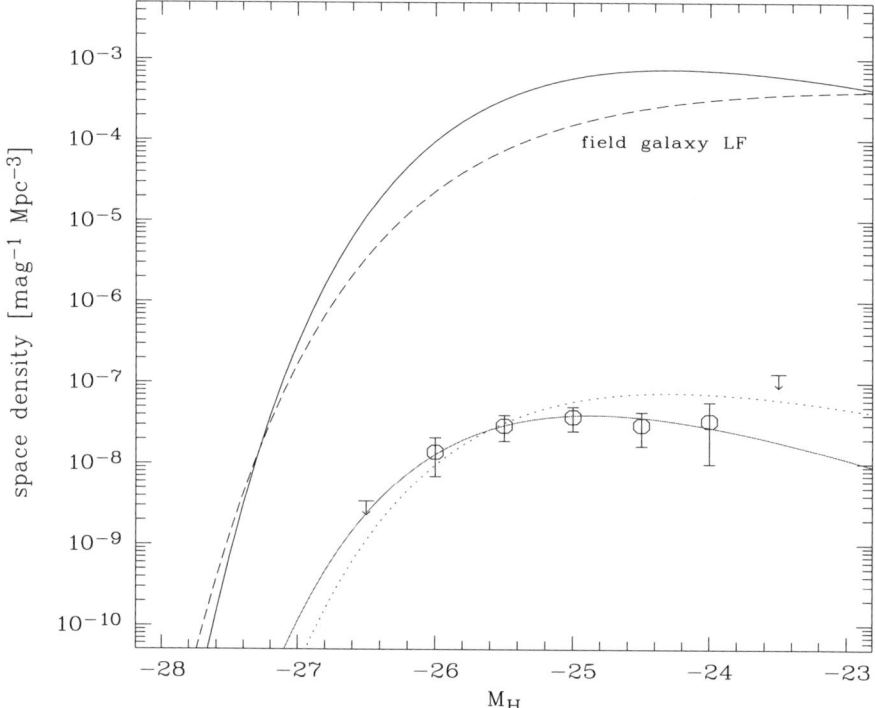

Figure 6. H-band luminosity function of QSO host galaxies (HGLF), in comparison to recent determinations of the NIR LF of field galaxies. The HGLF at $z = 0$ is nearly compatible with the field galaxy LF, downscaled by a factor $10^3 \ldots 10^4$.

3.5. Non-AGN Applications

Hamburg / SAO Survey for Emission Line Galaxies: A semi-automated search for BCD and starburst galaxies in an 1700 deg² HQS subarea, has so far generated 559 published objects (e.g., Ugryumov et al. (2001)).

Serendipitous search for magnetic white dwarfs: This is a byproduct of our QSO candidate search: White Dwarfs with strong magnetic fields are not recognised as stars in the prism spectra and therefore included as QSO candidates in the follow-up spectroscopy (Reimers et al. (1996a, 1998a)).

The stellar component of the Hamburg/ESO survey: Recently a systematic exploitation of the entire digital database of $\sim 10^7$ spectra was started, including a fully automated search for white dwarfs (Christlieb et al. (2001b)), carbon stars (Christlieb et al. (2001a)), cataclysmic variables etc.

A search for extremely metal-poor halo stars: This project makes full use of the spectral resolution provided by HES: Automated selection of stars with no detectable metal absorption lines (Christlieb et al. (2000)).

References

Bade N., Engels D., Voges W., Beckmann V., Boller T., Cordis L., Dahlem M., Englhauser J., Molthagen K., Nass P., Studt J., Reimers D., 1998, A&AS 127, 145

Christlieb N., Reimers D., Wisotzki L., Reetz J., Gehren T., Beers T., 2000, The First Stars, eds. A. Weiss, T. G. Abel,, V. Hill, ESO Astrophysics Symposia, ESO Astrophysics Symposia, 49

Christlieb N., Green P. J., Wisotzki L., Reimers D., 2001a, A&A 375, 366

Christlieb N., Wisotzki L., Homeier D., Koester D., Reimers D., Heber U., 2001b, A&A 366, 898

Engels D., Hagen H.-J., Cordis L., Köhler S., Wisotzki L., Reimers D., 1998, A&AS 128, 507

Hagen H. J., Groote D., Engels D., Reimers D., 1995, A&AS 111, 195

Hagen H.-J., Engels D., Reimers D., 1999, A&AS 134, 483

Köhler T., Groote D., Reimers D., Wisotzki L., 1997, A&A 325, 502

Kriss G. A., Shull J. M., Oegerle W., et al., 2001, Science 293, 1112

Lopez S., Reimers D., Rauch M., Sargent W. L. W., Smette A., 1999, ApJ 513, 598

Moller P., Jakobsen P., 1990, A&A 228, 299

Reimers D., Clavel J., Groote D., Engels D., Hagen H. J., Naylor T., Wamsteker W., Hopp U., 1989, A&A 218, 71

Reimers D., Vogel S., Hagen H.-J., Engels D., Groote D., Wamsteker W., Clavel J., Rosa M. R., 1992, Nature 360, 561

Reimers D., Jordan S., Koester D., Bade N., Köhler T., Wisotzki L., 1996a, A&A 311, 572

Reimers D., Köhler T., Wisotzki L., 1996b, A&AS 115, 235

Reimers D., Köhler S., Wisotzki L., Groote D., Rodriguez-Pascual P., Wamsteker W., 1997, A&A 327, 890

Reimers D., Jordan S., Beckmann V., Christlieb N., Wisotzki L., 1998a, A&A 337, L13

Reimers D., Köhler S., Hagen H.-J., Wisotzki L., 1998b, Ultraviolet astrophysics. Beyond the IUE Final Archive, ESA SP 413, 579

Reimers D., Baade R., Hagen H.-J., Lopez S., 2001, A&A 374, 871

Ugryumov A. V., Engels D., Kniazev A. Y., et al., 2001, A&A 374, 907

Véron-Cetty M.-P., Véron P., 2001, A&A 374, 92

Wisotzki L., 2000, A&A 353, 853

Wisotzki L., Köhler T., Kayser R., Reimers D., 1993, A&A 278, L15

Wisotzki L., Wucknitz O., Lopez S., Sørensen A., 1998, A&A 339, L73

Wisotzki L., Christlieb N., Bade N., Beckmann V., Köhler T., Vanelle C., Reimers D., 2000, A&A 358, 77

Wisotzki L., Kuhlbrodt B., Jahnke K., 2001, QSO hosts and their environments, ed. I. Marquez, in press, astro-ph/0103112

Objective-Prism Emission-Line Searches for Active Galactic Nuclei

Caryl Gronwall

Dept. of Physics & Astronomy, Johns Hopkins University, Baltimore MD 21218

Vicki L. Sarajedini

Dept. of Astronomy, University of Florida, Gainesville, FL 32611

John J. Salzer

Astronomy Dept., Wesleyan University, Middletown, CT 06459

Abstract. Objective-prism surveys for UV-excess and emission-line objects, especially the First and Second Byurakan Surveys, have been central to the study of active galactic nuclei (AGNs). We review previous line-selected surveys for AGNs and discuss their contribution to our understanding of the AGN phenomena. In addition, we present results from the KPNO International Spectroscopic Survey, a modern digital objective-prism survey for emission line objects. This survey is discovering substantial numbers of new AGNs, in particular low-luminosity AGNs and LINERs.

1. Introduction

Objective-prism surveys revolutionized our knowledge of active galactic nuclei (AGNs) in the local universe. In particular, many of the known nearby Seyfert galaxies were discovered by Markarian in the First Byurakan Survey (FBS; see Khachikian this volume.) Galaxies were selected for inclusion in the FBS based on the presence of a UV excess. This selection criterion is biased against the discovery of more heavily reddened AGN such as Seyfert 2's and LINER galaxies. A number of line-selected surveys have been carried out over the past 20 years in order to select well-defined samples of local AGNs. In this paper, we review existing line-selected surveys for local AGNs carried out using objective-prism techniques (see Stepanian, this volume, for a discussion of the UV+line selected Second Byurakan Survey). We then present a discussion of a new, modern digital objective-prism survey for emission-line galaxies (ELGs), the KPNO International Spectroscopic Survey (KISS). KISS probes several magnitudes deeper than existing surveys, and due to its selection via $H\alpha$ emission, is less biased against Seyfert 2 and LINER galaxies than previous surveys. KISS provides one of the best, well-defined samples of local AGNs for comparison to higher redshift studies.

2. Previous Line-Selected Surveys

The use of line-selected objective-prism surveys to detect active and star-forming galaxies in the local universe has a long and fruitful history (e.g., Smith 1975; Smith et al. 1976). The University of Michigan (UM) survey (MacAlpine et al. 1977; MacAlpine & Lewis 1981) selected objects via their [OIII]λ5007 emission line. 349 ELGs were detected in 667 sq. degrees. This is a number density of 0.52 per square degree. In Lists 4 & 5 of the UM survey (for which complete follow-up spectroscopy is available from Salzer et al. 1989) 166 ELGs are found in 325 sq. degrees. Approximately 10% of the ELGs detected in the UM survey are AGNs: 9 Seyfert 1's and 7 Seyfert 2's. Wasilewski (1983) also utilized the [OIII]-line as his selection criterion and found 96 ELGs in 825 sq. degrees, or 0.18 per square degree. Wasilewski found that 8% are AGNs with 1 Seyfert 1 and 7 Seyfert 2 galaxies being detected. The Case survey (Pesch & Sanduleak 1983; Stephenson et al. 1992) used two selection criteria, both UV-excess and [OIII]-line emission and cataloged 1440 ELGs in 1551 sq. degrees or 0.94 per sq. degree. Salzer et al. (1995) obtained follow-up spectroscopy for a complete subsample of 176 of the Case galaxies and found that only 6% are AGNs: 2 Seyfert 1's, 7 Seyfert 2's, and 2 LINERs. Most recently the Universidad Complutense de Madrid (UCM; Zamorano et al. 1994, 1996) survey utilized selection via Hα-emission to discover 263 ELGs in 471 sq. degrees or 0.56 per square degree. Follow-up spectroscopy by Gallego et al. (1997) found that 14 (or 5%) of these are AGNs, including 5 Seyfert 1's and 9 Seyfert 2's. The completeness limits of all of the above surveys range from $B = 15$ to 17.

Most of the previous line-selected surveys discussed above have been done in the blue (selecting via [OIII]λ5007 emission) and all used photographic plates. There are a number of biases introduced by these methods. First, selection in the blue limits the sensitivity of the surveys to more heavily reddened Seyfert 2's and low-ionization LINER galaxies. Second, because the standard IIIa-J emulsions used in most photographic surveys cut off at $\sim$ 5350 Å, the redshift depth of these surveys is limited to $z \sim 0.065$. A modern, digital survey using CCDs to substantially improve survey depth (to fainter magnitudes and higher redshifts) plus selecting via a redder line (Hα) would greatly reduce these biases. Such a survey is now available, the KPNO International Spectroscopic Survey (KISS).

3. The KPNO International Spectroscopic Survey

The KPNO International Spectroscopic Survey (KISS) was initiated in 1994 as a collaborative effort between astronomers from Russia, Ukraine, and the US who shared a common interest in the study of galaxian activity. The survey was envisioned as being the "next generation Markarian survey" for active and star-forming galaxies. By combining the traditional objective-prism survey method for emission line detection with modern detector technology and computer-based analysis, KISS is able to discover Seyfert Galaxies (Sy 1/Sy 2), LINER's, Starburst Nucleus Galaxies (SBNs), HII Galaxies, and Blue Compact Dwarfs (BCDs) at least 2 magnitudes fainter than previous work. The survey is thus a unique resource for probing the nature of star-forming and active galaxies in the nearby universe. A full description of the survey can be found in Salzer et al. (2000, 2001).

The KISS survey is conducted with CCD detectors on the 24-inch Burrell Schmidt on Kitt Peak. For the first survey strip, a 2048^2 CCD was used, which afforded a field-of-view of 70 arcmin (1.18°) square, with a scale of 2.03 arcsec/pixel. This one-degree-wide strip was chosen to coincide with the Century Redshift survey (Geller et al. 1997) covering a strip at a constant declination of 29° from RA = 12^h 15^m to 17^h which encompasses an area of 62 deg^2. The second survey strip goes through the Boötes void at a constant declination of 43° from RA = 12^h to 16^h 15^m covering an area of 66 deg ^{2}. For the second strip a 2048×4096 CCD with 1.45 arcsec/pixel resolution was used. The KISS survey data consists of broadband B and V images (for astrometric and photometric calibration), and spectral (objective prism) data. The spectra were taken in the red (using a blocking filter to restrict the wavelength range to 6400-7200 Å) to detect Hα emission. One of the important features of the KISS project is its objective criteria: the search and measurement process is carried out entirely by computer. A complete package of IRAF scripts and executables analyzes every source in the field (typically 5000 – 7000 objects) automatically; the output of these routines is a catalog of photometric and astrometric data, and an extracted spectrum for every object in the field. In addition, a list of ELG candidates is produced.

Our survey technique has proven to be quite successful. In our first red spectral strip we have cataloged 1128 ELG candidates. This is about *18.1 per square degree.* In the second survey strip, we have detected 1030 ELG candidates or 15.6 per square degree. For comparison, the entire Markarian survey cataloged 1500 UV-excess galaxies (0.1 galaxy per square degree), and had a completeness limit of m$_B$ = 15.2 (Mazzarella & Balzano 1986), while the deeper UM (median m$_B$ = 16.9; MacAlpine et al. 1977) and UCM (median m$_B$=16.5; Zamarano et al. 1994) surveys both detected only 0.5 ELGs per square degree.

3.1. Spectroscopic Follow-up

We have obtained follow-up spectroscopy for 725 of 1128 ELG candidates selected via Hα emission from the first survey strip using the WIYN 3.5-m, the KPNO 2.1-m, the MDM 2.4-m, the APO 3.5-m, and the Lick 3-m telescopes as well as the Hobby Eberly Telescope. These spectra provide redshifts, Hα equivalent widths, and line fluxes for various important emission lines including Hβ, [O III]$\lambda\lambda$4959,5007, HeIλ5876, [O I]λ6300, Hα, [N II]$\lambda\lambda$6548,6583, and [S II]$\lambda\lambda$6717,6731. Measurements of the Balmer decrement allow us to directly measure the extinction (A_V) for each of these galaxies. The follow-up spectra also allow us to identify AGNs in our sample: Seyfert 1s are identified via their broad permitted emission lines, while Seyfert 2's and LINERs are distinguished primarily via the [OIII]/Hβ line ratio with LINERs having [OIII]/Hβ < 3 (e.g., Veilleux & Osterbrock 1987). Representative spectra of AGN in the KISS sample are shown in Figures 1 & 2.

We find that 91% (663) of the KISS candidate ELGs are confirmed emission-lines galaxies. About 11% of the galaxies are active galactic nuclei with 67 (10 Seyfert 1's, 19 Seyfert 2's, and 38 LINER galaxies) detected via Hα with redshifts less than 0.095, and an additional 10 (1 Seyfert 1, 5 Seyfert 2's and 4 QSO's)

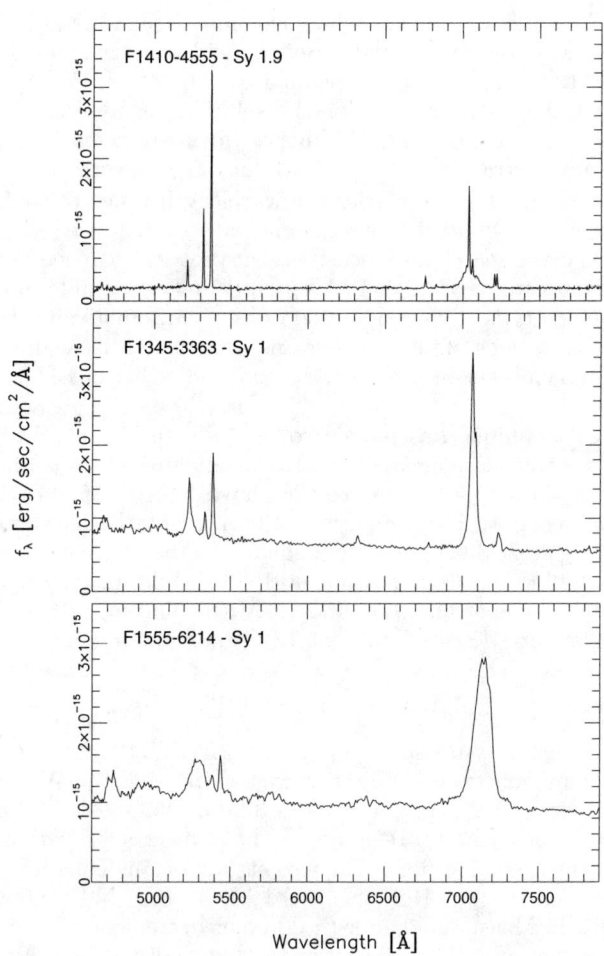

Figure 1. Representative spectra of KISS Seyfert 1 galaxies.

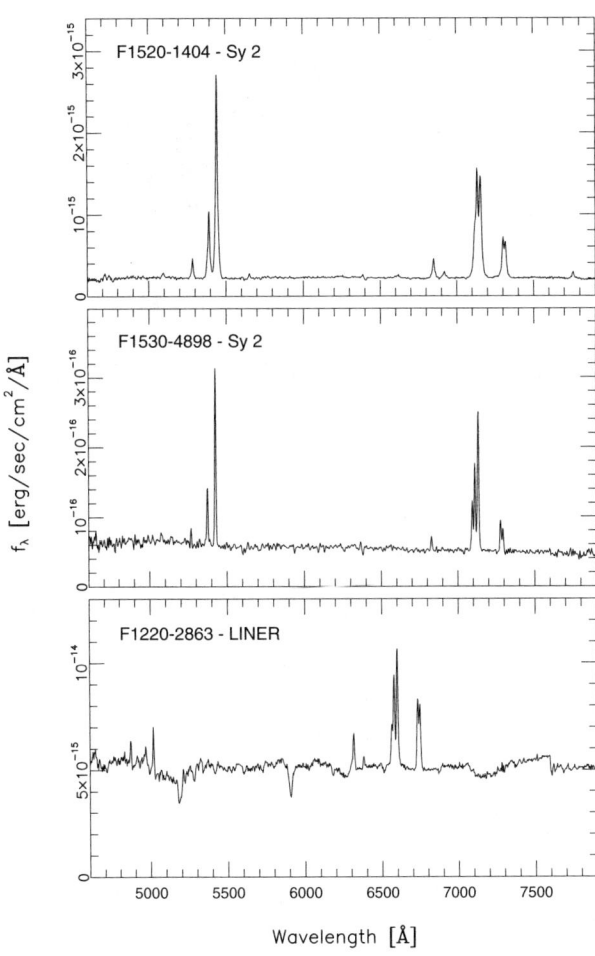

Figure 2. Representative spectra of KISS Seyfert 2 and LINER galaxies.

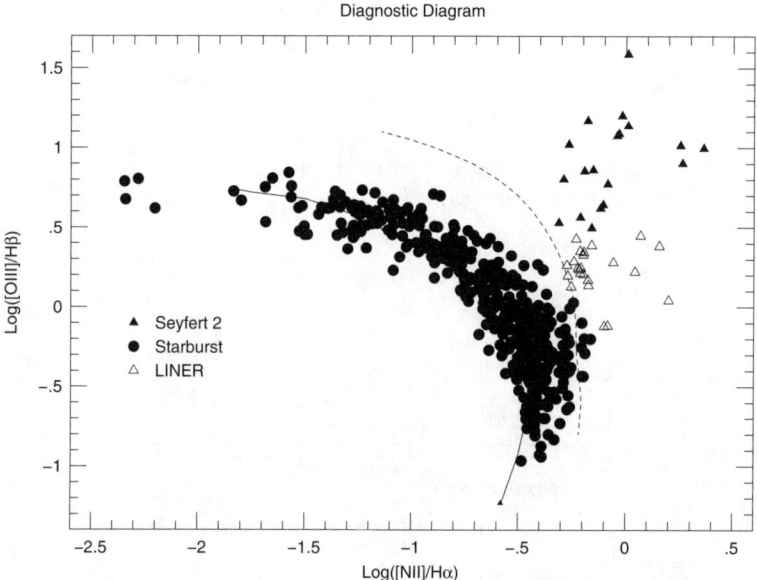

Figure 3. Line diagnostic diagram plotting the logarithm of [O III]λ5007/Hβ against the logarithm of [N II]λ6583/Hα. Solid circles represent star-forming ELGs, open triangles are LINERS, and the solid triangles are Seyfert 2 galaxies. The solid line represents an HII model sequence at various metallicities from 0.1 $Z_\odot$ at the upper left to 2 $Z_\odot$ at the lower right (Dopita & Evans 1986).

detected at higher redshifts where another line has redshifted into our filter. A line diagnostic diagram plotting the logarithm of [O III]λ5007/Hβ against the logarithm of [N II]λ6583/Hα is shown in Figure 3. The solid line represents the HII sequence with low metallicity, high-ionization ELGs in the upper left and high metallicity, low-ionization objects in the lower right part of the sequence. The star-forming ELGs follow the HII sequence allowing us to classify them from BCDs in the upper left to starburst nuclei in the lower right. AGNs are clearly separated from star-forming galaxies. We find many more high metallicity, low-ionization objects than found with traditional [O III]λ5007-selected surveys. The data make it clear that Hα-selected surveys are much more effective at detecting the full range of ionization and metallicity present in star-forming galaxies than [O III] or [O II] observations.

3.2. Properties of KISS AGN

We will concentrate here on the 67 Hα-selected AGN which constitute a well-defined local sample of AGN. Note that because our survey has a sharp upper-

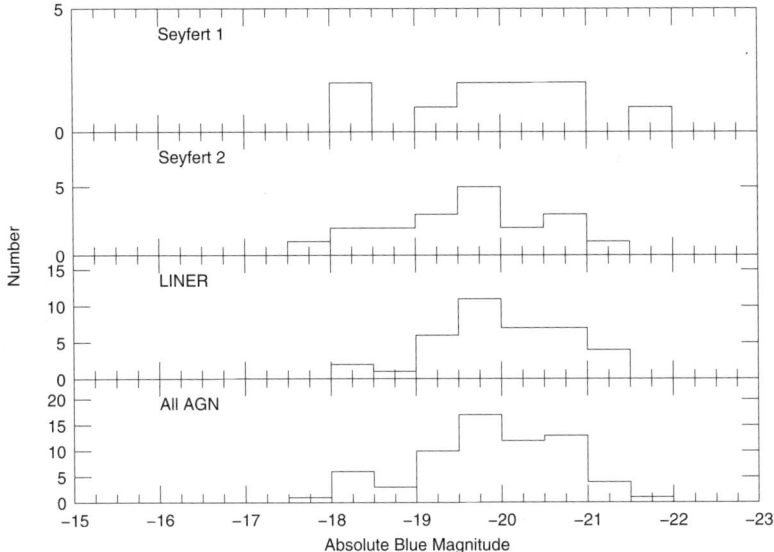

Figure 4. $B - V$ colors of Seyfert 1's, Seyfert 2's, LINER's and all KISS AGN.

wavelength cutoff due to the blocking filter used, the sample is *volume-limited* for the more luminous objects and all but 8 of the AGNs fall in this category. Figure 4 shows the integrated absolute magnitude distributions of the AGNs in the local KISS sample. KISS detects AGNs with M_B ranging from -21.8 to -18 (assuming $H_0 = 75$ km s^{-1} Mpc^{-1}). The Seyfert 1, Seyfert 2, and LINER luminosity distributions look similar. Figure 5 shows the $B - V$ color distributions of the KISS AGNs. As expected, Seyfert 2 and LINER distributions are significantly redder (median $B - V = 0.92$ and 0.93 respectively) than the Seyfert 1 color distribution (median $B - V = 0.70$). Because of their redder colors, an Hα-selected survey is much more sensitive to Seyfert 2's and LINERs than a blue (e.g., UV excess or [OIII]-selected) survey. This is borne out by comparing the 2:1 LINER/Sy 2 ratio in the KISS survey to the complete lack of LINERs discovered in the [OIII]-selected surveys discussed in Section 2.

We have also calculated the B-band luminosity function for the Seyfert galaxies in our sample. Because our direct images lack the angular resolution required to resolve the nucleus, this is an *integrated* luminosity function. However, it is directly comparable to the integrated local Seyfert luminosity function of Huchra & Burg (1992). Figure 6 shows the blue luminosity function for both the KISS sample and that of Huchra & Burg. The two LFs are consistent, with the KISS LF extending to fainter luminosities. Note that for this calculation we

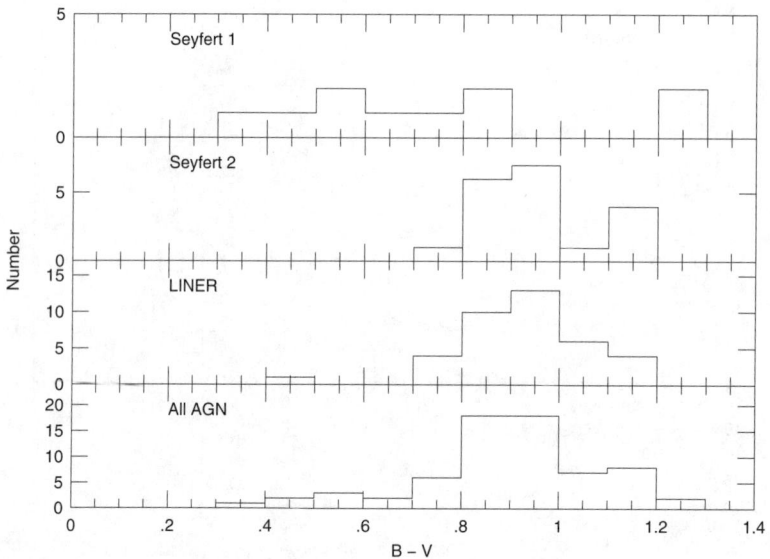

Figure 5. $B - V$ colors of Seyfert 1's, Seyfert 2's, LINER's and all KISS AGN.

have not accounted for the possibility that there are AGNs in the fraction of our sample for which we have no follow-up spectra. Because we have preferentially obtained spectra for the more luminous galaxies in our sample, we expect that such a correction would be small, but it could increase our volume densities slightly. Further discussion of the LF and its implications for AGN evolution when compared to other local and higher redshift AGN LFs will be discussed in a future paper (Sarajedini, Gronwall & Salzer 2002).

3.3. Multiwavelength Properties

We have also cross-correlated our $H\alpha$-selected sample of ELGs from both our first and second survey strips with surveys done in the x-ray, radio, and the far-infrared. A comparison of the KISS sample with the ROSAT All-sky Survey (Voges et al. 1999) finds that 17 of the 2158 KISS ELGs are detected by ROSAT. These are all AGNs: 13 Sy 1's, 2 Sy 2's, and 2 LINERs. The x-ray detected AGNs are primarily the most luminous AGNs detected by KISS. A cross-correlation of the KISS sample with the 1.4 GHz FIRST survey (White et al. 1997) done with the VLA reveals that 178 of the 2158 KISS ELGs are detected by FIRST. We have follow-up spectra for all of these sources and find 5 Sy 1's, 28 Sy 2's, and 30 LINERs. Similarly, 154 of the 2158 KISS ELGs are found in the IRAS Faint Source Catalog (Moshir et al. 1992) and follow-spectra for these sources

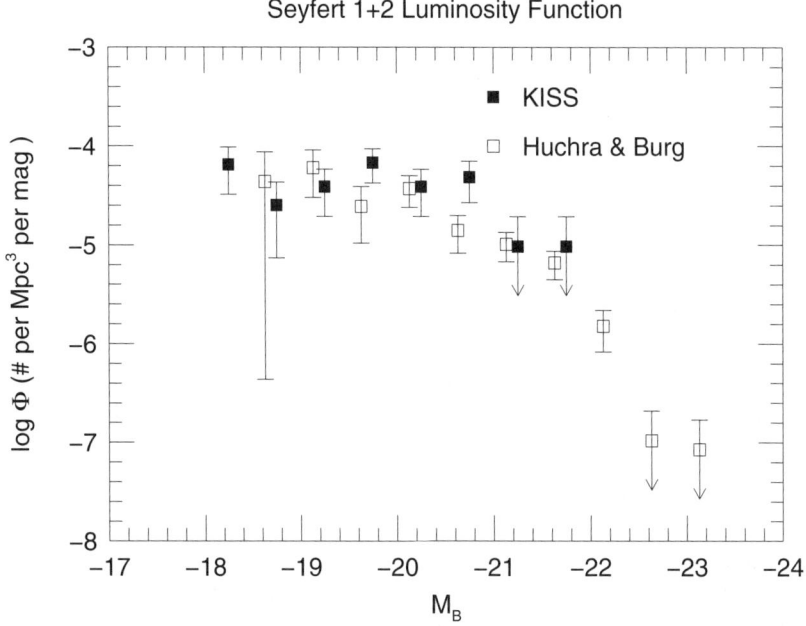

Figure 6. Integrated blue luminosity function for Seyfert galaxies in the KISS AGN sample. The KISS luminosity function is shown in solid squares while the luminosity function from Huchra & Burg (1992) is shown with open squares.

have detected 1 Sy 1, 9 Sy 2's, and 12 LINERs. The KISS ELGs follow the well-known radio-IR correlation discovered by Condon et al. (1991). This correlation is thought to be due to the fact that both the radio and far-IR emission are dominated by star formation. Somewhat surprisingly, we find that the majority of the AGNs in our sample follow the same radio-IR correlation as the star-forming galaxies, implying that the dominant ionizing source in these objects might be young stars, not the active nucleus. We are currently investigating this possiblity. We also plan to use these data to test the validity of IR color diagnostics for AGN activity.

4. Summary

The KISS sample represents a substantial step forward in our understanding of the AGN population in the local universe. It is essentially volume-limited and less biased against redder Seyfert 2 and LINER galaxies, and provides a well-defined sample of nearby ($z < 0.1$) AGN with which to study the statistical properties of AGN. In the first KISS survey strip covering 62 square degrees, 10

Seyfert 1's, 19 Seyfert 2's, and 38 LINER's were discovered. KISS provides an essential database of local AGNs for comparison to higher redshift samples.

Acknowledgments. CG is grateful to the American Astronomical Society and the International Astronomical Union for travel grants which allowed her to participate in this meeting. She would also like to thank the Local Organizing Committee for their wonderful hospitality.

References

Condon, J.J., Anderson, M.L., & Helou, G. 1991, ApJ, 376, 95
Dopita, M.A., & Evans, I.N. 1986, ApJ, 307, 431
Gallego, J., Zamorano, J., Rego, M., & Vitores, A. G. 1997, ApJ, 475, 502
Geller, M. J. et al. 1997, AJ, 114, 2205
Huchra, J. & Burg, R. 1992, ApJ, 393, 90
Khachikian, E. Ye. 2001, this volume
MacAlpine, G. M., Smith, S. B., & Lewis, D. W. 1977, ApJS, 34, 95
MacAlpine, G. M., & Williams, G. A. 1981, ApJS, 34, 95
Mazzarella, J. M., & Balzano, V. A. 1986, ApJS, 62, 751
Moshir, M. et al. 1992, *Explanatory Supplement to the IRAS Faint Source Survey, Version 2*, JPL D-10015, (Pasadena: JPL)
Pesch, P. & Sanduleak, N. 1983, ApJS, 51, 171
Salzer, J. J., MacAlpine, G. M., & Boroson, T. A. 1989, ApJS, 70, 479
Salzer, J. J., Moody, J. W., Rosenberg, J. L., Gregory, S. A., & Newberry, M. V. 1995, AJ, 109, 2376
Salzer, J. J., Gronwall, C., Lipovetsky, V. A., Kniazev, A., Moody, J. W., Boroson, T. A., Thuan, T. X., Izotov, Y. I., Herrero, J. L., & Frattare, L. M. 2000, AJ, 120, 80
Salzer, J. J., Gronwall, C., Lipovetsky, V. A., Kniazev, A., Moody, J. W., Boroson, T. A., Thuan, T. X., Izotov, Y. I., Herrero, J. L., & Frattare, L. M. 2001, AJ, 121, 66
Sanduleak, N. & Pesch, P. 1984, ApJS, 55, 517
Sarajedini, V.L., Gronwall, C., & Salzer, J.J. 2002, in preparation
Smith, M. G. 1975, ApJ, 202, 591
Smith, M. G., Aguirre, C., & Zemelman, M. 1976, ApJS, 32, 217
Stepanian, J.A. 2001, this volume
Stephenson, C. B., Pesch, P., & MacConnell, D. J. 1992, ApJS, 32, 217
Veilleux, S. & Osterbrock, D. E. 1987, ApJS, 63, 295
Voges, W. et al. 1999, A&A, 349, 389
Wasilewski, A.J. 1983, ApJ, 272, 68
White, R. L., Becker, R. H., Helfand, D. J., & Gregg, M. D. 1997, ApJ, 475, 479
Zamorano, J., et al. 1996, ApJS, 104, 99
Zamorano, J., et al. 1994, ApJS, 95, 387

The US and Other Optically Selected Bright QSO Samples

Kenneth J. Mitchell
Science Applications International Corporation, 4600 Powder Mill Rd., Suite 400, Beltsville, MD, 20705-2675

Peter D. Usher
Dept. of Astronomy and Astrophysics, The Pennsylvania State Univ., University Park, PA 16802

Abstract. The US Bright Quasar Sample (UBQS) is a color-selected sample of quasars with $B < 17$ that has been constructed with an eye towards completeness within well-defined selection limits. The redshift distribution of the UBQS shows an interesting spike at $z \sim 0.55$. The significance of this enhancement increases when the UBQS is combined with other bright quasar samples which also show evidence of high levels of completeness. Reconstitution of this combined bright quasar sample after removal of the emission-line flux from the B magnitudes indicates that the $z \sim 0.55$ spike is not caused by an emission-line selection bias. Rather, the largest effect of these sample corrections would be to depress the high end of the optical luminosity function that is derivable from this combined sample.

1. Introduction

Bright quasar samples make unique contributions to the study of the cosmological evolution of the quasar population by establishing the optical luminosity function at low redshifts and by determining the bright end of the optical luminosity function at all redshifts. However, historically the numbers of bright quasars in well-defined, complete samples have been relatively small because of both their intrinsically low surface density on the sky and the problems with selection effects in the necessarily large-scale bright-quasar surveys. Thus, the accuracy of the derived redshift-dependent quasar optical luminosity function at low redshifts and high luminosities has been uncertain.

The 19 member UBQS (Usher & Mitchell 2000; herein UM2000) was established to help bolster the numbers of bright ($B < 17$) quasars in complete samples and to help further explore the properties of bright quasar surveys in general. The UBQS is derived from the US color-excess survey (Usher 1981; Usher & Mitchell 1990) which features manual selection of candidates from 3-color photographic plates. The US selection techniques have proven to be highly reliable and complete for the selection of unresolved objects with colors that are bluer (in $B-V$) and/or more ultraviolet (in $U-B$) than the metal-deficient halo subdwarfs (e.g. Mitchell, Warnock, & Usher 1984; Mitchell 1998). Although the

UBQS covers a modest ~ 200 square degrees, it provides a useful comparison with past bright-quasar survey work, and it can also be used to help test the properties of the modern, large-scale, digital optical/near-IR surveys reported on at this conference: the Sloan Digital Sky Survey (Ivezic 2002), the 2 Micron All Sky Survey (Cutri 2002), and the Hamburg/ESO Survey (HES: Wisotzki 2002) all of which show great promise for establishing large, complete samples of color-selected bright quasars.

This paper reviews the status of the UBQS and other pre-existing bright quasar samples. It also extends the results of UM2000 by exploring the effects on bright quasar samples when emission line flux is explicitly accounted for and removed from the quasar B magnitudes.

2. A Combined Bright Quasar Sample

Since the numbers of quasars in existing individual bright samples are relatively small, it is necessary to combine samples to enhance statistical accuracy. The approach taken by UM2000 is to combine only those bright quasar samples (or subsamples thereof with $B < 16.95$ mag) which show signs of absolute completeness. Of the six bright quasar samples extant at the time, it was found that only three share the properties that: (a) their surface densities are close to the maximum detected values; and (b) their redshift distributions have shapes that are mutually consistent and that generally agree with the predictions of recent empirical models of the evolving quasar luminosity function (La Franca & Cristiani 1997); these three are the UBQS, an early version of the HES (HES97: Köhler et al. 1997), and the Edinburgh Quasar Survey (EQS: Goldschmidt et al. 1992). The other three extant bright quasar samples were not chosen for combination by UM2000 for the following reasons: (1) the BQS (Schmidt and Green 1983) has low counts, and the $(U-B)$ color limit biases the redshift distribution; (2) the LBQS (Hewett, Foltz, & Chaffee 1995) has low counts, a redshift distribution underpopulated at low redshift, and a claimed bright-magnitude limit that vacillates; (3) the HBQS (Cristiani et al. 1995) has very low counts. This relatively straightforward approach for the combination of quasar samples is complementary to those that strive to include more samples by attempting to correct for differing sample selection effects and/or errors; such effects can cause some quasar samples to deviate markedly in their overall properties from others. For example, the LBQS and HBQS were used in the La Franca & Cristiani (1997) study of multiple quasar samples; however, the BQS was not – a curiosity, since La Franca & Cristiani's methodology seems particularly well-suited to account for the well-defined selection effects of the BQS and since that sample potentially carries such a large weight at the brightest magnitudes.

The combined bright quasar sample constructed from the UBQS, EQS, and HES97 covers ~ 1115 square degrees and contains 53 quasars with $B < 16.95$ mag and $0.2 < z < 2.2$ ($0.3 < z < 2.2$ for the EQS). These limits ensure selection completeness for quasars with starlike morphology on 1.2m Schmidt imagery according to the morphology criterion established by Fabian & Usher (1996). The surface densities of the combined sample closely match those predicted by the Luminosity Dependent Luminosity Evolution (LDLE) model of La Franca and Cristiani (1997). The shape of the observed redshift distribution (shown in

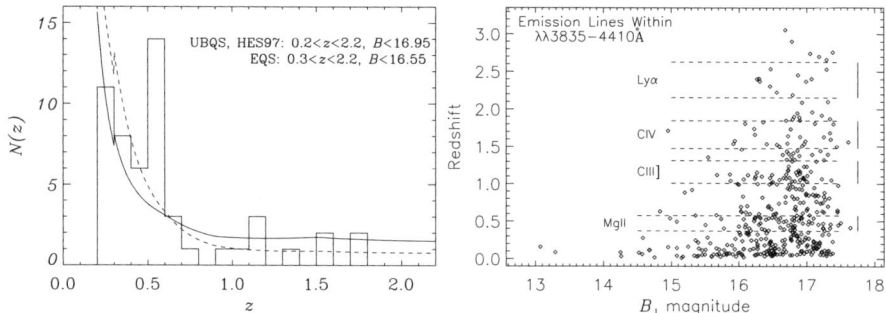

Figure 1. *Left:* Redshift distribution of the combined bright quasar sample compared to LDLE (dashed curve) and Pure Luminosity Evolution (PLE: continuous curve) model predictions. *Right:* Hubble daigram of the HES00 sample. Redshift ranges are noted where the emission lines enter the > 85% response level of the B_J bandpass.

Figure 1, left side) also matches the general behavior of the LDLE model, except for the noticeable excess numbers in the $0.5 < z \leq 0.6$ bin. The subject of this "spike" in the redshift distribution is taken up in the next section.

Since the UM2000 study, a new version of the HES sample has been published (HES00: Wisotzki et al. 2000). The size of this bright sample is impressive: 415 quasars with $B_J < 17.5$. However, as indicated in Figure 1 (right side), the HES00 Hubble diagram contains apparent horizontal striation. The unusual distribution of objects can be traced to a preference for the higher luminosity quasars to be located at redshifts where the major quasar emission lines enter the > 85% response level of the B_J bandpass. Unlike the B magnitudes of the HES97 sample, the HES00 sample uses B_J magnitudes derived directly from the survey objective prism spectra. The suspect structure in the Hubble diagram might be an indication that the emission lines, via uncalibrated emulsion non-linearities or other effects, have biased the derived B_J magnitudes. For this reason the HES00 sample is not considered here; instead, the even larger HES quasar sample (Wisotzki 2002) is awaited. These considerations serve as a reminder of the potential effects that emission lines can have on quasar magnitudes and selection.

3. The $z \sim 0.55$ "Spike": Real or Statistical Artifact?

An intriguing feature of the combined-sample redshift distribution of Figure 1 (left side) is the obvious excess of quasars in the $0.5 < z \leq 0.6$ bin. This excess consists of ~ 10 of the 14 quasars in that bin, as estimated both from the trends in the observed data and by comparisons with the model distributions. UM2000 assess the significance of the spike using the conservative Kolmogorov-Smirnov (KS) and Cramer-von Mises (CvM) statistical tests to compare the observed and modeled distributions. While the PLE model was rejected at the 99.9% level, LDLE could only be rejected at the $\sim 80\%$ level, with the conclusion that the spike could just be a statistical artifact. The less conservative Chi-square test rejected both PLE and LDLE at the > 99% level.

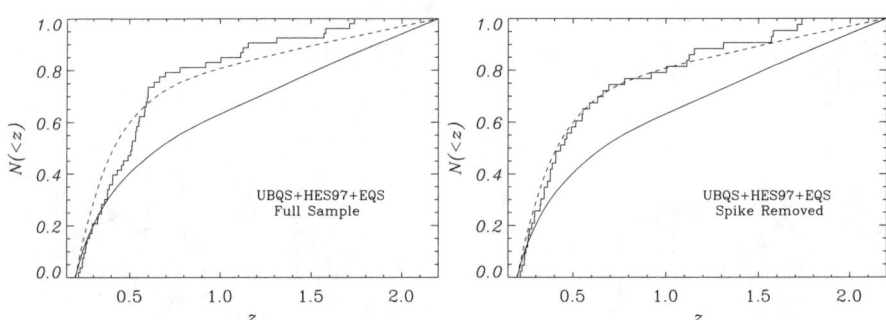

Figure 2. *Left:* Normalized, cumulative redshift distribution of the combined bright quasar sample compared to LDLE (dashed curve) and PLE (continuous curve) model predictions. *Right:* Same, except the excess within the $z \sim 0.55$ "spike" has been removed.

Another view of the data is provided in Figure 2 which shows the observed and modeled cumulative redshift distributions (normalized to unity) that are the basis for the KS and CvM tests. In the full-sample plot (on the left) the data are seen to be a better match, overall, to the LDLE model prediction than to the PLE model, in agreement with the statistical results. However the match is not entirely satisfactory due to the effects of the $z \sim 0.55$ spike. The plot on the right shows the results when the spike is removed: i.e. 10 quasars with $0.5 < z \leq 0.6$ have been randomly removed. The agreement between the shape of the remaining redshift distribution and the LDLE prediction is seen to be excellent. But note that the removal of the 10 quasars reduces the combined-sample surface densities by $\sim 20\%$ on average, so that the original agreement between the surface densities of the full combined sample and those of the LDLE model would no longer hold.

The evidence is not conclusive either for or against the reality of the $z \sim 0.55$ spike. The conservative statistical tests indicate that the agreement between the data (spike included) and the LDLE model is acceptable, if not entirely satisfying. However, if the addition of further complete bright quasar samples prove the spike to be real, then the quasar luminosity function at $z \sim 0.55$ is uniquely distorted, and it could be an indication of very large scale structure. Not only would the current LDLE and other evolutionary models need modifications to account for the spike, but they would also need to be modified to correctly account for both the quasar surface density and redshift distribution outside of the spike.

4. Effects of Removing the Emission-line Flux from the Combined-Sample B Magnitudes

Emission-lines contribute flux to quasar magnitudes at certain redshifts and, unless accounted for, can lead to sample bias. Such a bias, caused by the MgII emission line entering the B bandpass, could be one reason for the enhanced numbers of $z \sim 0.55$ quasars in the combined UBQS, HES97, and EQS sample. UM2000 made an initial estimate that most of the spike is not likely caused

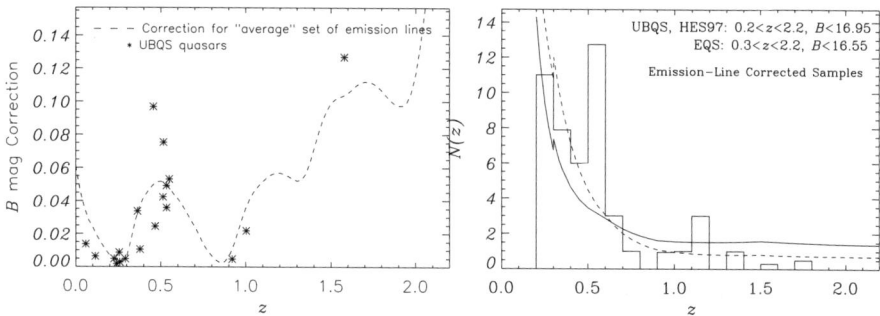

Figure 3. *Left:* B magnitude corrections for the individual UBQS quasars due to removal of emission-line flux. The dotted line represents the correction for an assumed set of average emission-line EWs. *Right:* Redshift distribution of the "corrected" combined bright quasar sample after objects with $B_{corr} > B_{lim}$ have been removed. Lines are as in Figure 1.

by the effects of the MgII line. This selection effect is more fully investigated here, given the potential consequences of detecting a real spike in the quasar redshift distribution. The method used corrects the B magnitudes by removing any emission-line flux, and then removes from the combined sample any quasars with corrected magnitudes, B_{corr}, that are fainter than the faint magnitude limit B_{lim} of their parent sample. This is a different approach than attempting to account for emission-line flux via the K-correction.

The magnitude correction due to the removal of emission line flux in the B bandpass, S_B, is:

$$\Delta B = 2.5\log[1 + \sum_l \{(EW_l/\lambda_l^\alpha)(\int G_l(\lambda)S_B(\lambda)\mathrm{d}\lambda / \int \lambda^{-\alpha}S_B(\lambda)\mathrm{d}\lambda)\}]$$

where EW_l is the equivalent width of emission line l, λ_l is the central wavelength of line l, G_l is the Gaussian line profile of line l centered on λ_l, and α is the slope of the quasar continuum. The corrected magnitude is then $B_{corr} = B + \Delta B$.

All of these parameters have been measured or can be estimated for the UBQS, HES97, and EQS quasars in order to calculate ΔB. For example, the ΔB calculated for the individual UBQS quasars based on measured EWs are shown in Figure 3 (left-hand side). Also shown in that same plot is the ΔB relation for a set of emission lines (Balmer, MgII, CIII], CIV, SiIV+OIV], Lyα) with estimated "average" EWs. Measured EWs have not been published for the HES97 and EQS quasars and so a crude, Gaussian-like probability distribution centered on these average EWs is assumed for their EW strengths. Based on this EW probability distribution and the quasar's redshift, Monte-Carlo calculations then provide estimated probabilities of excluding each quasar from the sample based on the probability, P, that $B_{corr} > B_{lim}$. Most quasars have P = 0; quasars with non-zero P have been kept in the combined sample, but with a reduced weight (= 1 − P).

The results indicate that a weighted total of ∼ 4.6 quasars of the original 53 should be excluded from the corrected combined sample, an overall reduction of $\sim 10\%$. 99% of this loss is attributed to the effects of 2 lines: MgII excludes 1.4 quasars, and CIV excludes 3.2 quasars. The resulting redshift distribution for the corrected, combined sample is shown in Figure 3 (right-hand side), and it can be compared with the original redshift distribution of Figure 1 (left-hand side). The $z \sim 0.55$ spike has lost only $\sim 14\%$ of its members, and it remains prominent in the corrected sample. On the other hand 3.2 of the original 4 high-redshift ($z > 1.5$) quasars have been excluded from the corrected combined sample, a reduction of 80%! KS and CvM statistical tests indicate that the match with LDLE remains statistically acceptable, albeit dubious.

It is concluded that the $z \sim 0.55$ spike in the bright quasar redshift distribution is very likely *not* caused by emission-line bias in the B magnitudes, and thus it remains a candidate as an unusual feature in the quasar luminosity function. The largest effects of correcting the B magnitudes has been at high redshifts, where the resulting sample corrections depress the bright end of the optical luminosity function that is derivable from the combined sample. This is in essential agreement with one of the cautionary notes struck by Wampler & Ponz (1985).

References

Cristiani, S., et al. 1995, A&AS, 112, 347

Cutri, R. M. 2002, these proceedings

Fabian, D., & Usher, P.D. 1996, AJ, 111, 645

Goldschmidt, P., Miller, L., La Franca, F., & Cristiani, S. 1992, MNRAS, 256, 65P

Hewett, P.C., Foltz, C.B., & Chaffee, F.C. 1995, AJ, 109, 1498

Ivezic, Z. 2002, these proceedings

Köhler, T., Groote, D., Reimers, D., & Wisotzki, L. 1997, A&A, 325, 502

La Franca, F., & Cristiani, S. 1997, AJ, 113, 1517 Err.: 1998, AJ, 115, 1688

Mitchell, K.J. 1998, ApJ, 494, 256

Mitchell, K.J., Warnock, A. III, & Usher, P.D. 1984, ApJ, 287, L3

Schmidt, M. & Green, R.F. 1983, ApJ, 269, 352

Usher, P.D. 1981, ApJS, 46, 117

Usher, P.D., & Mitchell, K.J. 1990, ApJS, 74, 885

Usher, P.D., & Mitchell, K.J. 2000, AJ, 120, 1683

Wampler, E.J. & Ponz, D. 1985, ApJ, 298, 448

Wisotzki, L. et al. 2000, A&A, 358, 77

Wisotzki, L. 2002, these proceedings

Luminosity Function and Evolution of Optically Selected QSOs

Lutz Wisotzki

Universität Potsdam, Am Neuen Palais 10, D-14469 Potsdam, Germany

Abstract. I summarise a few recent results on the evolution of optically selected QSOs, with special emphasis on the notoriously difficult but physically important extremes at low and high luminosities and redshifts. It seems that quasar evolution is a more complex phenomenon than has often been assumed, and the new generation of surveys is just about to make this visible.

1. Introduction

Luminosity functions are an important tool to study the population properties of Active Galactic Nulcei. The strong evolution of nuclear activity in galaxies over cosmic epochs is most evidently visible in the changing luminosity function, implying that the space density of present-day luminous AGN is only $\sim 10^{-3}$ of that of high-redshift quasars. Precise quantitative measurements of the AGN luminosity function, however, have proven to be surprisingly difficult. Comparing the quasar LF at different redshifts inevitably requires the usage of different surveys, each with its own selection criteria and ideosyncrasies. Especially at extremely high and low luminosities and redshifts, statistical coverage is still poor, even though current surveys are now churning out new quasars by the thousands.

A good knowledge of the AGN luminosity function is also a prerequisite to understand the thermal history of the early universe. QSOs provide a significant fraction of the metagalactic UV radiation field that keep the intergalactic medium ionised, yet current estimates are highly uncertain due to the poorly constrained contribution of low-luminosity AGN.

2. The Optical QSO Luminosity Function

2.1. General Remarks

Quasars are still most easily found at optical wavelengths, due to the existence of large-format detectors and the fact that optical spectra are required to get redshifts. Optical QSO surveys have therefore been leading in shaping the overall picture of the QSO luminosity function (QLF) and its evolution. Particularly influential was the work of Boyle and collaborators (Boyle et al. (1988)) who argued that the shape of the QLF remains invariant with redshift, and evolution manifests itself as a mere shift in characteristic luminosity (Pure Luminosity

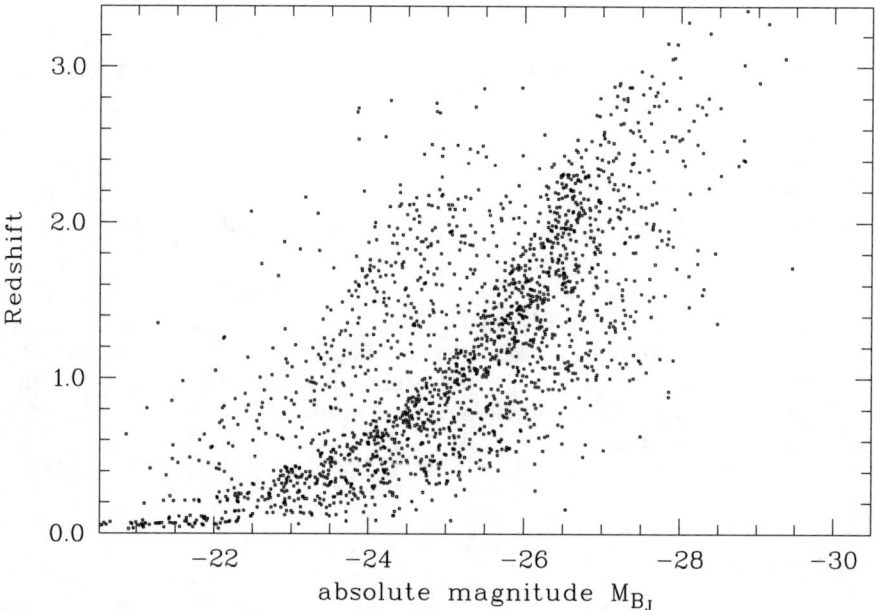

Figure 1. Distribution of redshifts and absolute blue magnitudes for the combined sample.

Evolution, PLE). PLE characteristics have also been found for AGN evolution at X-ray (Jones et al. (1997)) and radio wavelengths (Dunlop & Peacock (1990)), albeit with much poorer statistics than in the optical.

More recently, the applicability of simple PLE models have been questioned based on new survey data. Hewett et al. (1993) showed that the PLE model predicts too few bright and too many faint low-redshift ($z \lesssim 1$) quasars. This fact has been independently confirmed by our Hamburg/ESO survey results (Köhler et al. (1997); Wisotzki (2000a); cf. also Reimers & Wisotzki, these proceedings), where the mismatch between PLE predictions and actual data is highly significant. It can be safely said PLE as a global description of quasar evolution is ruled out because the QLF changes its shape with cosmic time.

2.2. Semi-Parametric Analysis

We have recently conducted a new attempt to model the evolving QLF, working on a merged sample of six optical QSO surveys. Altogether, the sample contains 1946 objects, and coverage of the Hubble diagram is more or less complete for $M_{B_J} < -23$ and $z < 3$, although the scarcity of low-luminosity AGN at high redshifts and of high-luminosity AGN at low redshifts is still prominent (Fig. 1). We do not consider for the moment the very high redshift range, for which the data are even scarcer, although much progress is currently made (see also below).

When modelling the QLF, one is confronted with the choice between non-parametric and parametric estimators. The former has the virtue of being free of premeditated assumptions, but is unfortunately prone to a number of numerical biases such as evolution within redshift bins (see also Petrosian, these proceedings). Parametric model fitting, on the other hand, avoids these biases but may be too inflexible to describe the actual data (especially when small samples demand a small number of parameters). To overcome these limitations, we have devised a hybrid 'semi-parametric' scheme (Wisotzki (1998)), featuring the following properties: (1) The QLF is described as polynomial in $\mu \propto \log L$. (2) Evolution is parameterised as a polynomial in the variable $\zeta = \zeta(z)$. (3) The general expression for an evolving QLF is a bivariate polynomial of μ and ζ,

$$\log \phi = \sum_{i=0}^{n} \sum_{j=0}^{m} C_{ij} \mu^i \zeta^j .$$

This 'free form' LF representation was introduced by Peacock & Gull (1981) for the analysis of radio surveys, but has rarely been used otherwise. Its main advantage is that given samples of sufficient size, almost any QLF shape can be recovered by fitting the above form to the data. The result is a QLF estimate that is an unbinned continuous function, yet free of the above mentioned biases. Notice that the fitting parameters in this method are not interesting in themselves and basically meaningless in physical terms, especially for a high-order QLF (this is why the method could be called 'semi-parametric').

Applying the method to our combined sample, we derive the following constraints (see also Fig. 2): (1) No global model with an invariant QLF is acceptable – both PLE as well as its counterpart PDE (Pure Density Evolution) are ruled out with high significance. (2) An acceptable fit is achieved only with a high-order luminosity-dependent density evolution scheme. (3) The evolution rates depend both on luminosity and on redshift. (4) Evolution at $z < 1$ is much faster than at $z > 1$. (5) There is no evidence for significant slow-down of evolution around $z \approx 2$ as suggested by earlier analyses and also required in modelling the evolution of X-ray selected AGN.

2.3. Evolution in Cosmological Epochs

The fact that a global fit to our $0 < z < 3$ sample is only achieved by employing a complicated QLF model and many free parameters indicates that the underlying physical processes have not been isolated. It is therefore interesting to find that at least for our current sample the description becomes very simple if the sample is split into low- and high-redshift subsets, each containing roughly 50 % of the sample.

'High' Redshift ($1 \lesssim z \lesssim 3$): The QLF shows strong curvature with a steep high-luminosity and a shallow low-luminosity tail. The shape is well-approximated by a 3rd-order polynomial in $log\phi(M)$ (in fact better than the with the canonic 'double power law') and is nearly invariant within the covered redshift range. Surprisingly, both PLE and PDE are equally acceptable. While 20 free-form coefficients were not enough for the *global* description, a simple 5-parameter model is perfectly adequate for half the sample. This explains why the PLE model performed so well in the past and was still favoured in the recent analysis by Boyle et al. (2000): These samples are dominated by intermediate-redshift QSOs (e.g., > 70 % of the AAT sample are at $1 < z < 2.2$), and for

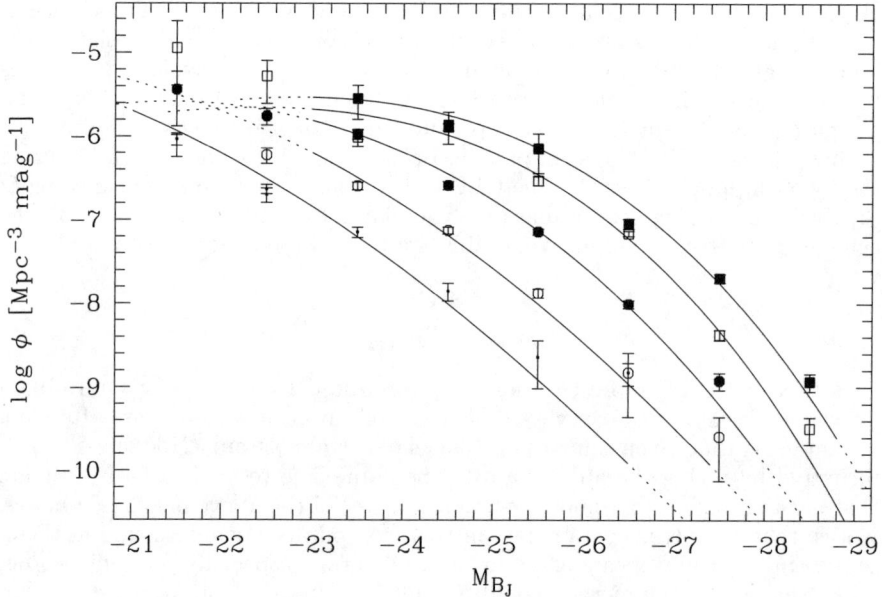

Figure 2. Semiparametric estimate of the evolving QLF in five redshift shells ($z < 0.3$, $0.3 < z < 0.7$, $0.7 < z < 1.25$, $1.25 < z < 2$, $2 < z < 2.9$).

these redshifts PLE indeed is an acceptable description (although PDE does just as well).

Low Redshift ($z \lesssim 1$): The low-redshift QLF is almost a single power law, with very little curvature at least within the classical QSO domain. If the sample is restricted in absolute magnitude to $M_{B_J} \lesssim -23$, an excellent description is achieved with pure density evolution, while PLE performs much poorer. This shows an interesting parallel to the results of Miyaji et al. (2000) who found that PDE was *almost* a good fit to their X-ray sample and failed only because of a slight overproduction of $z > 1$ AGN.

3. High-Luminosity AGN at High Redshifts

General wisdom says that the peak of the QSO space density lies between $z \simeq 2$ and $z \simeq 3$. However, if quasar luminosities depend in some way on the masses of their underlying host galaxies, the location of the actual maximum may well depend on the luminosity range sampled. There is some evidence that this is indeed the case, in the sense that the peak is shifted towards higher redshifts for the highest luminosity QSOs. If confirmed, this would rule out at least the naive version of a hierarchical model with QSO luminosities being proportional to the corresponding dark matter haloes.

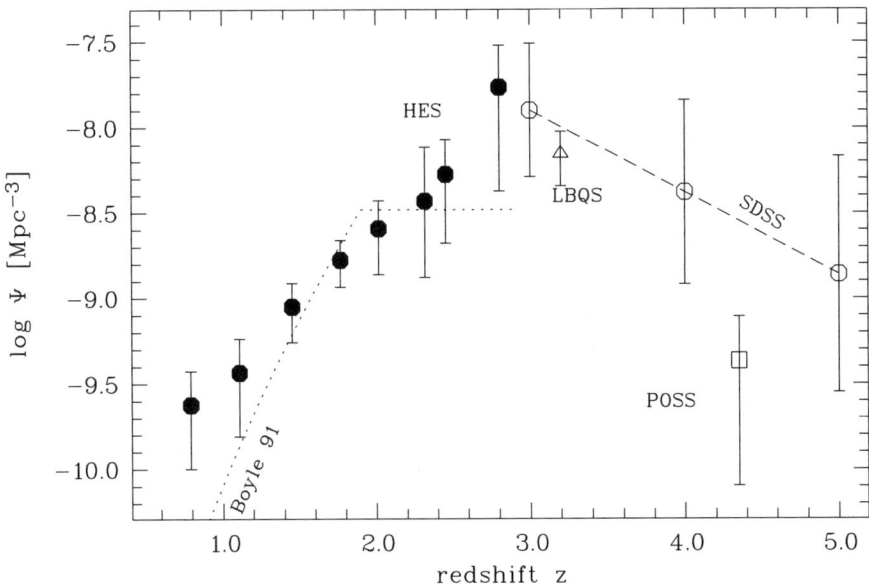

Figure 3. Evolution of the space densities of the most luminous QSOs. The SDSS points have been constructed from the parametric model by Fan et al. (2001); the error bars include the extrapolation of that model to $M_B = -28$.

Unfortunately, the maximum itself has so far been escaped detection. The main constraint for extremely luminous QSOs with $M_B < -28$ at $z < 3$ comes from the Hamburg/ESO survey (Wisotzki (2000a)) from where it appears that their evolution proceeds at almost constant rate until $z \simeq 3$ (Fig. 3), with no indication of a slow-down. At even higher redshifts, the results were quite contentious until recently the first results from SDSS (Fan et al. (2001)) demonstrated that the decline towards $z = 5$ is definitely real. However, Fan et al. also found that the high-redshift high-luminosity tail of the QLF is much flatter than at $z \simeq 2$ – this is equivalent to saying that the evolution rate depends on luminosity. Accordingly, very high-luminosity QSOs have a slower decline, or alternatively they may have a maximum at higher z. Figure 3 shows that for $M_B < -28$, the SDSS and HES results are not in conflict, as they nicely bracket the probable maximum. It is nevertheless puzzling to realise that this peak corresponds to a cosmic time span of less than 1 Gyr FWHM.

Finally, one additional source of uncertainty should be mentioned: in order to combine low- and high-redshift space densities, the spectral energy distributions need to be known; usually they are approximated as power laws $f_\nu \propto \nu^\alpha$ with $\alpha = -0.5$ or by composite QSO spectra. Such approximations are a non-negligible source of uncertainty, and past specifications have even lead to a systematic overestimate of high-redshift QSO luminosities by up to 0.5 mag at $z \simeq 2$ (Wisotzki (2000b)). Consequently, the inferred evolution rates are significantly reduced and now in fact compatible with what is found in X-ray surveys. Much more work needs to be done on this aspect in the future.

4. High-Luminosity AGN in the Local Universe

Before discussing the results, one important operational caveat should be mentioned that has mostly been neglected: Source photometry of low-redshift AGN using standard techniques will inevitably lead to significant host galaxy contributions, *even* in the B band, *even* for high-luminosity AGN. In the HES, all flux measurements are essentially *nuclear* rather than total magnitudes, thereby largely eliminating host galaxy bias, but other low-redshift samples are less well defined in terms of their AGN luminosities. Comparing the resulting surface and space densities is therefore difficult and should be regarded with suspicion unless host galaxy contributions are explicitly accounted for.

Qualitatively it is now clear that the bright end of the QLF flattens towards low redshifts; this is confirmed by several surveys (Goldschmidt & Miller (1998); Wisotzki (2000a); Grazian et al. (2000); see Fig. 2). In the same manner as above for the SDSS results, this implies a reduced evolution rate for most luminous QSOs. Physical origins for this effect are not yet known. It has been speculated that radio-quiet and radio-loud populations might evolve with different rates, which would explain the very high fraction of RLQs in the Palomar/Green Bright Quasar Survey (Kellerman et al. (1989)). However, our ongoing radio follow-up of the Hamburg/ESO survey shows that the RLQ fraction in the HES is $\sim$ 10–15 % even for high-luminosity AGN at low redshifts, in stark contrast to the BQS. Furthermore, RQQ and RLQ have indistinguishable redshift distributions. The conclusion so far is that the radio-loud population cannot be responsible for the flattening of the QLF.

5. Low-Luminosity AGN in the Local Universe

Although an abundant species, an accurate assessment of the contribution of low-luminosity AGN to the QLF is fraught with technical difficulties, mainly arising from the fact that $L_{host} \gtrsim L_{nuc}$: How to disentangle nuclear and host luminosity contributions? How to define limiting survey magnitudes? How to combine different surveys? Several heroic attempts in the past have tried to account for these problems at least in a statistical manner (e.g., Cheng et al. (1985); Marshall (1987)), but mostly these were lacking adequate data. Recall also that magnitudes from galaxy surveys are meaningless when used to construct a *nuclear* LF (e.g., the Sy 1 galaxy LF derived in the CfA survey, Huchra & Burg (1992) is mostly a host galaxy luminosity function).

In the last years, a few authors have estimates the local LF of 'Type 1' AGN taking care of these problems. First, the local QLF from the Hamburg/ESO survey (Köhler et al. (1997)) included individual host subtraction. The LF was found to be close to a single power law down to $M_B \simeq -18$, but the statistics at low luminosities were very poor. More recently, the optical LF of low-redshift X-ray-selected AGN was constructed by Londish et al. (2000) using HST imaging and proper subtraction of host galaxy contributions. The

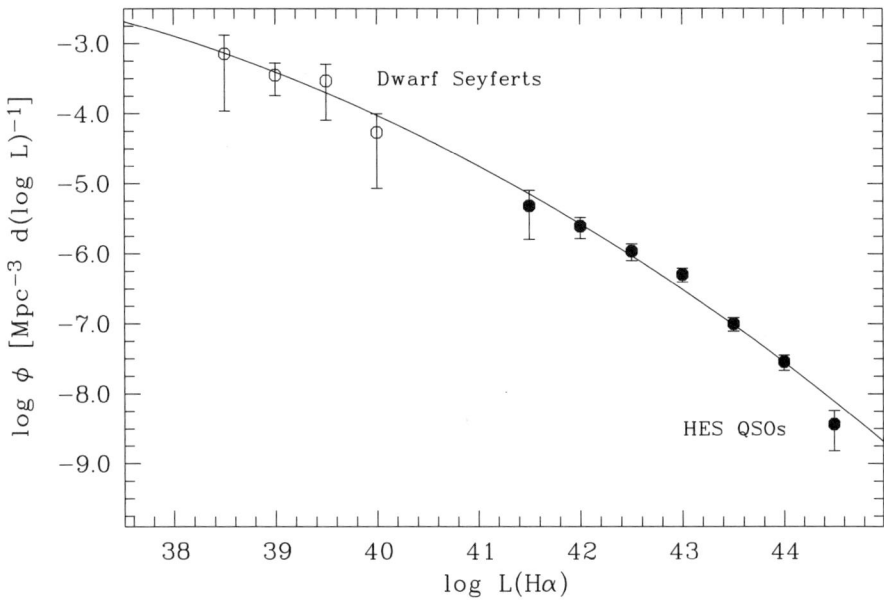

Figure 4. Hα luminosity function of low-redshift AGN. Note the smooth connection between high-luminosity quasars and low-luminosity Seyferts. The solid line is a free-form polynomial fit to the data.

power law slope for $M_B \lesssim -20$ was confirmed, but there is a significant turnover towards low luminosities.

The Hα Luminosity Function of Local AGN It is known that the Hα line flux is well correlated with blue continuum magnitude. Furthermore, it is physically expected to reflect the number of UV photons intercepted by the BLR. We adopt therefore the integrated luminosity of the broad component of Hα as an alternative measure of nuclear power in type 1 AGN. Its importance lies in the fact that this quantity can be separated spectroscopically from the host galaxy, which is relatively easy to accomplish in all except the most feeble active nuclei.

For spectroscopically complete AGN samples it is then straightforward to construct an Hα luminosity function of low-redshift AGN. We have done this for two samples: (1) Our own HES dataset, using the spectra from the candidate conformation observations. (2) The 'Dwarf Seyfert' sample of Ho et al. (1995), using the deblended broad Hα luminosities from Ho et al. (1997). The combined LF is shown in Fig. 4. At brighter luminosities, the shape of the Hα LF is basically indistinguishable from the low-z QLF in the broad B_J band (Wisotzki (2000a)), in particular there is no evidence in the HES data for a turnover into a flatter low-luminosity tail. This is dramatically confirmed by the Dwarf Seyferts forming a seamless continuation, only slightly below a simple power law extrapolation from intermediate towards low luminosities.

At face value, this seems to be in strong conflict with the results of Londish et al. (2000), assuming that both soft X-rays and broad Hα luminosities represent similar primary emission components – essentially, both are used as estimators of the UV output. That latter assumption may be incorrect if the 'Dwarf Seyferts' are operating on ADAF-type accretion flows as suggested by Ho (1999); (also these proceedings). Furthermore, it is by no means obvious that X-ray selected AGN should yield the same *optical* LF as an optically selected sample.

6. Low-Luminosity AGN at High Redshifts

The space density of high-z low-luminosity AGN is still very poorly known, at $z \gtrsim 3$ even almost unconstrained. However, this quantity is extremely important also beyond AGN physics, as the AGN contribution to the metagalactic ionising UV background radiation field depends sensitively on the shape of the low-luminosity end of the QLF. Useful samples are small, and spectroscopic follow-up is expensive. The faintest currently available complete samples are:

- Marano field (Zitelli et al. 1992): 52 AGN, $z < 2.9$ and $B < 22$. Partly visual selection.

- CFRS (Schade et al. 1996): 6 AGN with $z < 3$ and $I < 22.5$; no QSO-specific selection.

- HDF (Jarvis & MacAlpine 1998; Conti et al. 1999; Sarajedini et al. 2000): Between 1 and 20 AGN candidates, but only 1 certain spectroscopic identification (of a type 2 AGN).

Most researchers simply *assume* that the faint-end slope derived at lower z can be extrapolated; given the presently existing data, it is futile to speculate about stringent tests of such assumptions. This sobering situation is hopefully to change over the next few years with more powerful multiplex spectroscopic facilities arriving at large telescopes. The new generation of wide-field imagers, hase made generating faint candidate samples a possibility, but spectroscopic follow-up is still the major hurdle. A possible approach to circumvent this bottleneck is to use photometric redshifts based on a multicolour survey database. To take an example, our recently started COMBO-17 survey (Wolf et al. (2001)) features deep images in 17 different photometric bands (5 broad- and 12 medium-band filters) and can yield redshift estimates with an accuracy fully sufficient for luminosity calculations. First results will be available shortly.

7. Conclusions

To a very qualitative level, the general picture of quasar evolution is no longer contentious, but several quantitative issues are still unsolved. We can probably say that the uncertainties at intermediate redshifts and intermediate luminosities are relatively small, and that the overall evolution properties of such objects can be considered known, but that the more extreme objects maintain their elusive status. Some of the central questions that are still open:

- When was the peak of QSO activity in the universe? Does its location or shape depend on luminosity, and if so, in what way? Does it make sense, after all, to speak of a well-defined 'quasar epoch', or can substantial AGN formation be traced out to very high redshifts?

- The interpretation of traditional single-band surveys hinges on the assumption of average spectral energy distributions. New multicolour surveys can measure individual SEDs and vastly reduce the K correction uncertainties, but combining low- and high-z surveys remains a problem. Which 'luminosity' is sought to be represented?

- The local AGN distribution properties are still very uncertain and difficult to disentangle from their host galaxies. If indeed ADAF-type solutions become important at very low L, then how does the transition regime manifest itself in the luminosity function? Intermediate luminosity scales need to be probed – here the new large galaxy surveys such as 2dFGRS, SDSS and VIMOS will yield important constraints.

The fundamental question of 'obscured AGN' has not been addressed here, as optical selection is known to systematically avoid such objects. There can be no doubt that a significant fraction of low-luminosity (Seyfert-type) AGN is removed from directly seeing the central engine through obscuration. On the other hand, it is not clear and still a matter of debate whether this is an important effect also for high-luminosity AGN, as their much stronger incident UV radiation fields and possibly also winds might have a clearing effect on any present obscuration screens. All these issues will be taken up by other contributors to these proceedings. As a final remark on the continuing usefulness of optical AGN surveys, let me note that the whole issue of what AGN fraction might be obscured is irrelevant for the cosmologically important task of estimating the AGN contribution to the metagalactic UV background; here the statement holds: *What You See Is What You Get.*

References

Boyle B. J., Shanks T., Croom S. M., Smith R. J., Miller L., Loaring N., Heymans C., 2000, MNRAS, 317, 1014
Boyle B. J., Shanks T., Peterson B. A., 1988, MNRAS, 235, 935
Cheng F.-z., Danese L., de Zotti G., Franceschini A., 1985, MNRAS, 212, 857
Dunlop J. S., Peacock J. A., 1990, MNRAS, 247, 19
Fan X., Strauss M. A., Schneider D. P., et al., 2001, AJ, 121, 54
Goldschmidt P., Miller L., 1998, MNRAS, 293, 107
Grazian A., Cristiani S., D'Odorico V., Omizzolo A., Pizzella A., 2000, AJ, 119, 2540
Hewett P. C., Foltz C. B., Chaffee F. H., 1993, ApJ, 406, L43
Ho L. C., 1999, ApJ, 516, 672
Ho L. C., Filippenko A. V., Sargent W. L., 1995, ApJS, 98, 477
Ho L. C., Filippenko A. V., Sargent W. L., Peng C. Y. T., 1997, ApJS, 112, 391
Huchra J., Burg R., 1992, ApJ, 393, 90

Jones L. R., McHardy I. M., Merryfield M. R., et al., 1997, MNRAS, 285, 547
Kellerman K. I., Sramek R., Schmidt M., Shaffer D. B., Green R., 1989, AJ, 98, 1195
Köhler T., Groote D., Reimers D., Wisotzki L., 1997, A&A, 325, 502
Londish D., Boyle B. J., Schade D. J., 2000, MNRAS, 318, 411
Marshall H. L., 1987, AJ, 94, 628
Miyaji T., Hasinger G. ., Schmidt M., 2000, A&A, 353, 25
Peacock J. A., Gull S. F., 1981, MNRAS, 196, 611
Wisotzki L., 1998, AN, 319, 257
Wisotzki L., 2000a, A&A, 353, 853
Wisotzki L., 2000b, A&A, 353, 861
Wolf C., Dye S., Kleinheinrich M., Meisenheimer K., Rix H.-W., Wisotzki L., 2001, A&A, 377, 442

QSOs from a Variability-and-Proper Motion Survey

Helmut Meusinger, Jens Brunzendorf

Thüringer Landessternwarte Tautenburg, D-07778 Tautenburg, Germany

Ralf-Dieter Scholz

Astrophysikalisches Institut Potsdam, An der Sternwarte 16, D-14482 Potsdam, Germany

Mike Irwin

Institute of Astronomy, Madingley Road, Cambridge CB3 1HA, UK

Abstract. We report on a search for QSOs via variability and zero-proper motion. Candidates were selected by means of indices for overall variability, long-term variability, proper motion and image structure measured on a large number of Schmidt plates with a time-baseline of three decades. Spectroscopic follow-up observations of the candidates brighter than the pre-estimated completeness limit, $B_{\text{lim}} \approx 19.7$, of the survey revealed about 200 QSOs and Seyfert 1s with redshifts $z \approx 0-3$. We describe the survey strategy and discuss the properties of the resulting QSO sample.

1. Introduction

Most criteria for the selection of QSO candidates are based on the differences of their spectral energy distribution (SED) compared to stars. All such search methods are known to be biased (e.g., Hewett & Foltz 1994). Alternative strategies are provided by fundamental properties of quasars that are not directly based on the SED, like variability of flux densities (e.g., Hawkins 1983; Véron & Hawkins 1995; Bershady et al. 1998) and stationarity of positions (e.g., Kron & Chiu 1981). The search for objects which are both variable and stationary is supposed to be a powerful technique for efficiently finding QSOs. The selection bias of such a search strategy is expected to be quite different from those of more conventional optical search techniques. A combination of both constraints was successfully applied by Majewski et al. (1991) in a small survey field. Here, we use a similar search strategy and present preliminary results from a new combined variability and proper motion QSO search (in the following "VPM survey") in two Tautenburg Schmidt fields.

For an efficient VPM quasar search, a large number of homogeneous observations of a large number of faint objects are needed, with high astrometric and photometric accuracy and covering a long time-baseline of a few decades. These requirements can be matched only by archival plates from large Schmidt telescopes. The basic observational material of our VPM survey is provided by a large number of plates from the Tautenburg Schmidt telescope. In several fields, the Tautenburg plate archive contains homogeneous material of more than a hundred plates with epoch differences of up to three decades or more. In addition, due to its relatively large focal length the Tautenburg Schmidt plates have fewer problems with plate bending and are well suited for astrometry. A plate covers an area of about 10 square degrees.

2. Survey Strategy and Candidate Selection

For the VPM survey, we selected the fields centred on the globular clusters M3 and M92, respectively, where large numbers of B plates are available with a time-baseline of about 30 years and with typical limiting magnitudes between about 19.5 and 21.5. Whereas the basic strategy is the same for both fields, there are several differences in the details (see Scholz et al. 1997; Brunzendorf & Meusinger 2001). For the M3 field, which is at high Galactic latitude, 57 B plates were selected and measured with the APM facility in Cambridge. In the M92 field ($b \approx 35°$), where the search faces the problem of stronger contamination by foreground stars, a larger number of 162 B plates were selected and digitised by means of the Tautenburg plate scanner TPS. In both fields, several U- and V-plates were digitised as well in order to derive colour indices (which are, however, not used for the candidate selection).

The basic object sample contains all objects which are measured on a given number ($n > 3$) of plates. The analysis of the objects from these samples comprises the following steps: *1.)* classification according to the image structure and selection of star-like objects, *2.)* computation of relative proper motions from the linear regression of the measured positions as a function of the epochs; the goodness of fit is taken as a measure of the proper motion error, *3.)* transformation of relative proper motions to absolute proper motions using several hundred faint galaxies per field as reference frame; the measured proper motion in units of the proper motion error is taken as a proper motion index, *4.)* computation of the variability indices for the star-like objects, and *5.)* specification of the thresholds for the selection of stationary and variable objects.

The frequency distributions of the measured mean B magnitudes for all star-like objects indicate that substantial incompleteness sets in at $B \approx 20$. We adopt $B \approx 19.7$ for the completeness limit of the survey in both fields. Typical proper motion errors are less than about 2 mas/yr up to B = 19 and about 4 mas/yr at the survey limit. The photometric errors, expressed by the standard deviation of the magnitude measurements of an object on all plates, are typically less than 0.1 mag for $B < 18$ and rise in the M3 field up to ≈ 0.2 mag at the survey limit. In a first attempt of the data reduction in the M92 field (Brunzendorf & Meusinger 2001), this increase was considerably stronger, obviously due to a lower accuracy of the procedures applied there for the determination of the object parameters. Meanwhile, we have repeated the photometry for the M92 field using the SExtractor package by Bertin & Arnouts (1996) which yields a substantial reduction of the photometric uncertainties (Brunzendorf & Meusinger, in preparation).

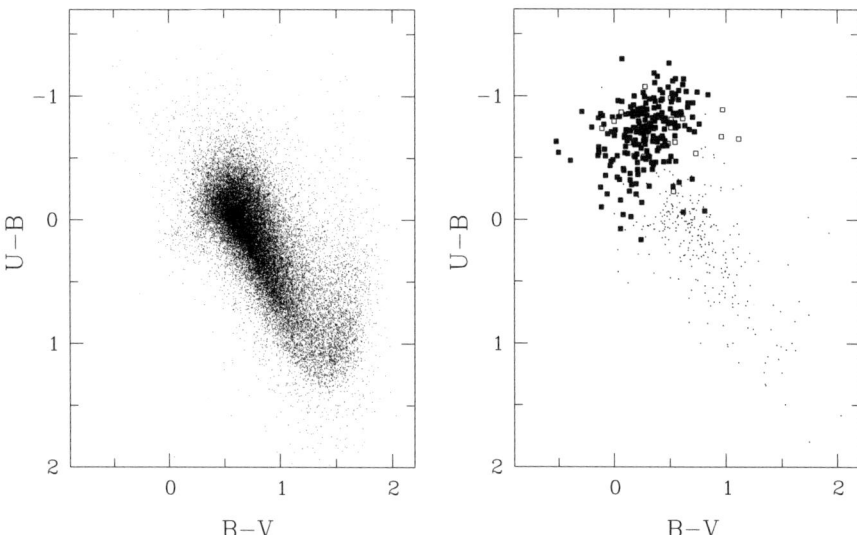

Figure 1. Colour-colour diagrams for all star-like objects in the two VPM search fields (*left*) and for the spectroscopically observed candidates *right*). On the right-hand side, QSOs are shown as filled squares, Seyfert 1s as open squares and stars as dots.

The variability of an object is assessed by measuring the deviation of the individual magnitudes about the mean magnitude, and is normalized by the average magnitude scatter for star-like objects in the same flux range. In the M 3 field, which overlaps with the blue grens CFHT survey (Crampton et al. 1989), the CFHT QSOs were used to define the variability selection thresholds in such a way that a success rate of at least 40% is expected in combination with a high completeness of 90%. In the M92 field, where no QSOs were known before, we had to define indices that are directly related to the probability of real variability: an object is considered a candidate if its probability for variability is larger than 95%. In addition, an index for long-term variability is computed which measures variability with time-scales longer than about a few months. The long-term variability index is defined either by means of structure function analysis (Scholz et al. 1997) or by the method of mean square successive differences (Brunzendorf & Meusinger 2001).

The basic samples contain 24,600 objects in the M 3 field and 35,000 objects in the M 92 field. After the exclusion of the objects near the plate margins and in the crowded cluster region and considering only star-like objects, the basic samples are reduced by a factor of 4 in the M 3 field and 2 in the M 92 field, respectively. About 70% of the objects from these reduced basic samples are rejected due to the zero-proper motion constraint. Finally, the variability selection strongly reduces the candidate sample to a manageable size of altogether about 300 candidates of high or medium priority.

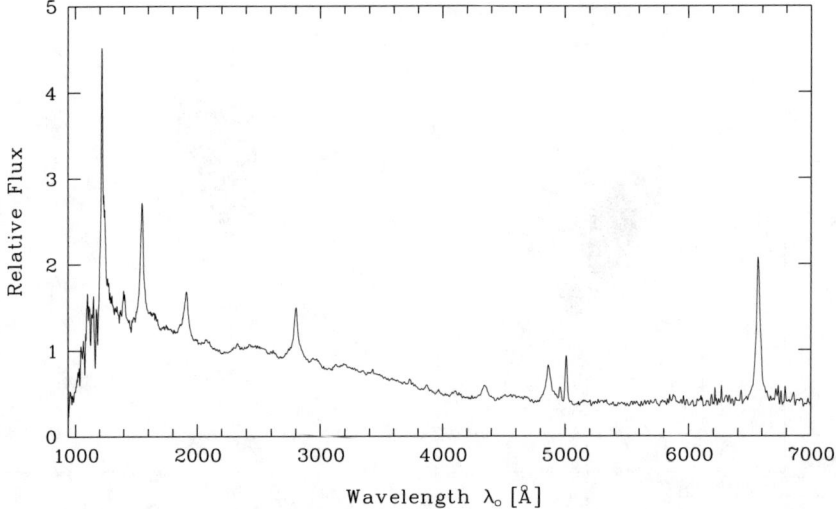

Figure 2. Composite spectrum (in the rest frame) obtained from the follow-up spectra of the QSOs and Seyfert 1s in the M 92 field.

3. Properties of the AGN Sample

Follow-up spectroscopy was performed for the brighter candidates with the Tautenburg multi-object spectrograph TAUMOK. The fainter candidates were observed with the faint-object spectrograph CAFOS at the 2.2m telescope of the DSAZ[1], Calar Alto, Spain. Up to now, follow-up spectra are available and reduced for altogether 412 candidates. 182 QSOs and Sy1s were found, corresponding to a total success rate of 44%. Preliminary results for the M 92 field were presented by Meusinger & Brunzendorf (2001). (The publications of the results for the M 3 field and from the improved photometry in the M 92 field, respectively, are in preparation.)

Despite a few BAL QSOs and a few objects with relatively weak broad emission lines, there is no indication of a substantial population of QSOs with anormalous SEDs. As can be seen from the colour-colour diagram in Fig. 1, all candidates with extremely red colours proved to be galactic foreground stars. Further, nearly all of the VPM QSOs populate the area of the colour-colour diagram which would be surveyed by a classical two-colour search. Up to the limit of the present survey, the QSO population detected by the VPM method is obviously not significantly different from the population detected by colour surveys. The same conclusion is reached from the colour-redshift diagrams and from the detailed analysis of the spectra. The composite spectrum of the VPM QSOs (Fig. 2) is very similar to those of quasars from other optical surveys.

---

[1]German-Spanish Astronomical Centre operated by the Max-Planck-Institute for Astronomy, jointly with the Spanish National Comission for Astronomy

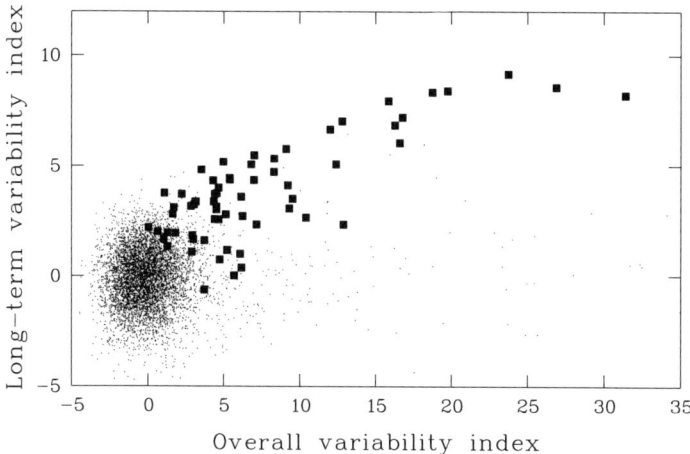

Figure 3. Long-term variability *versus* overall variability for the starlike objects with zero-proper motions (dots) and QSOs/Seyfert 1s (filled squares) in the M 92 field.

For practical reasons, we divided the candidate sample into three priority classes. In the high priority class, all 141 candidates were observed with a success rate as high as 96% in the M3 field and 71% in the M92 field. High priority means that both overall variability and long-term variability are highly significant. In addition, spectra were taken for a number of medium-priority candidates. QSO candidates of low priority were observed especially when only one of the selection criteria was slightly violated. Figure 3 shows the long-term variability index versus the index for total variability. Evidently, quasars show in many, yet not in all, cases strong long-term variability. This confirms the importance of the long-term variability index, but also the use of both variability indices for an efficient QSO variability search.

The redshift distributions are similar to those from colour surveys. A difference appears in the preliminary results from the M 92 field, where higher redshifts are slightly underabundant. This can be understood as due to the incompleteness of the variability selection at fainter magnitudes where the photometric accuracy is lower. A (preliminary) completeness of 50% was estimated in the M 92 field by the comparison of the magnitude-dependence of the variability selection threshold with the distribution of the variability indices computed for the (more complete) QSO sample investigated by Hook et al. (1994). For the M 3 field we estimate a completeness of 90% from the overlap with the CFHT QSOs. The cumulative surface density of VPM QSOs in the two search fields is remarkably high, especially for $B \leq 18.5$ (Fig. 4).

4. Conclusion

The combination of constraints for overall variability, long-term variability and zero-proper motion provides an efficient (yet expensive) AGN search method

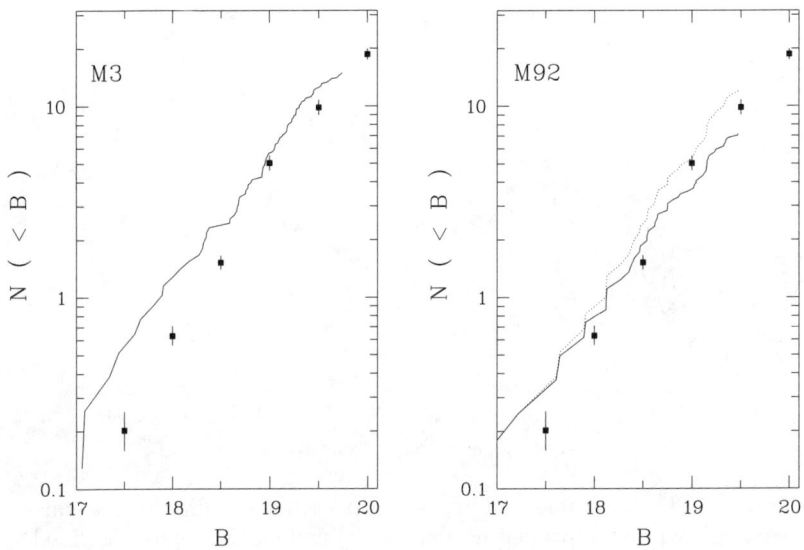

Figure 4. Cumulative surface densities of QSOs from the VPM survey (solid curves) compared with the data by Hartwick & Schade (1990) from conventional optical surveys (squares and error bars). The dotted curve for the M 92 field is the result of correction for incompleteness.

that yields sizeable quasar samples. Up to the limit of the present survey, the properties of the VPM-QSO sample do not significantly differ from those of the QSOs from more conventional optical searches. The resulting QSO sample will be well suited to investigate the long-term variability of QSOs.

References

Bershady, M.A., Trevese, D., & Kron, R.G. 1998, ApJ, 496, 103
Bertin, E., & Arnouts, S. 1996, A&AS, 117, 393
Brunzendorf, J., & Meusinger, H. 2001, A&A, 373, 38
Crampton, D., Cowley, A.P., & Hartwick, F. 1989, ApJ, 345, 59
Hartwick F.D.A., & Schade, D. 1990, ARA&A, 28, 437
Hawkins, M.R.S. 1983, MNRAS202, 571
Hewett, P.C., & Foltz, C.B. 1994, PASP, 106, 113
Hook, I.M., McMahon, R.G., Boyle, B.J. *et al.* 1994, MNRAS, 268, 305
Kron, R.G., & Chiu, L.-T. 1981, PASP, 93, 397
Majewski, S R., Munn, J.A., Kron, R.G., *et al.* 1991, ASP Conf. Ser. 21, 55
Meusinger, H., & Brunzendorf, J. 2001, A&A, 374, 878
Scholz, R.-D., Meusinger, H., & Irwin, M. 1997, A&A325, 457
Véron, P., & Hawkins, M.R.S. 1995, A&A296, 665

Spectroscopic and Variability Surveys for AGN in the Groth Survey Strip

Vicki Sarajedini

Department of Astronomy, University of Florida, Gainesville, FL 32611

Abstract. Preliminary results are presented for a spectroscopic survey of the Groth Survey Strip (GSS), a 30 by 3 arcminute region of the sky imaged with HST, for which several hundred galaxy spectra have been obtained as part of the DEEP project (http://deep.ucolick.org). At least 6 broad-line AGNs (primarily Seyfert 1s) have been detected as well as several narrow-line Seyfert 2 candidates. The seyfert galaxies detected in our survey have integrated absolute magnitudes extending to $M_B \sim -17.5$, probing fainter magnitudes and higher redshifts than existing optical spectroscopic surveys. We also discuss a variablity study of the GSS using the original HST images from 1994 and new images obtained in 2001. The high resolution obtained with HST allows us to isolate and measure variable galactic nuclei too faint to be detected from the ground, reaching nuclear magnitudes of $M_B \sim -16$ in galaxies to $z \sim 0.8$. The combination of these techniques provide a powerful probe of the population of low-luminosity AGNs at moderate redshifts.

1. Introduction

To better understand the nature of any class of extragalactic object, an accurate knowledge of the luminosity function (LF) over a wide range of absolute magnitudes and covering a range of redshifts is necessary. The AGN LF is populated by quasars at the brighter, primarily high redshift end and Seyfert galaxy nuclei, considered to be their intrinsically fainter counterparts, at the low luminosity, low redshift end (Cheng et al. 1985; Huchra & Burg 1992; Maiolino & Rieke 1995). See Wizotski (2002, this volume) for a discussion of the current state of the AGN LF. While bright QSOs are easily observable at all redshifts, fainter seyfert nuclei become increasingly difficult to detect at redshifts much beyond the local universe. Understanding how the faint end of the AGN LF evolves is of particular importance for determining the frequency and total space density of AGNs at earlier epochs.

The purpose of this paper is to show how the increased light gathering capabilities of large (i.e. 10-meter) telescopes and the high spatial resolution achieved with space-based telescopes allows for the detection of these lower luminosity AGN at higher redshifts. In Sections 2 and 3, we describe preliminary results of a spectroscopic survey for AGN using the Keck telescope. Section 4 outlines plans for a variability survey for AGN in this same region of the sky using Hubble Space Telescope images separated by 7 years. A summary of this research is presented in the final section.

2. The DEEP Project

The Deep Extragalactic Evolutionary Probe (DEEP; Koo et al. 1996; Simard et al. 2001) is a project designed to study the formation and evolution of distant field galaxies by combining images from HST with spectroscopic data from the Keck telescope. The Groth Survey Strip (Groth et al. 1994) is one of the fields targeted by DEEP for spectroscopic follow-up. The GSS is comprised of 28 contiguous WFPC2 fields located at 14h17m+52 imaged in the F606W and F814W filters. Optical spectra have been obtained through multi-slit masks with the Keck/Low Resolution Imaging Spectrograph for 775 objects in the GSS between 1995 and 1999. Of the 683 spectra with high enough S/N to identify spectral features and determine redshifts, 634 are galaxies and 49 are galactic stars. The galaxies extend to $I_{AB} \sim 24$ with a mean redshift of $z \simeq 0.8$. The typical exposure time is one hour with some of the fainter targets being exposed for up to a few hours. Since the majority of galaxies at these redshifts have sizes comparable to the seeing resolution ($\sim 1''$), spatial spectral information is not available in most cases and a one-dimensional spectrum has been produced for each object by summing several pixels along the spatial axis.

3. Spectroscopic Identification of AGN

Active Galactic Nuclei can be detected in galaxy spectra through the presence of broad, permitted lines covering a wide range of ionization. Optical spectra for the galaxies studied in this survey display broad H balmer series lines or singly ionized MgII emission (at rest $\lambda 2800 \text{\AA}$ for higher z objects) when a Seyfert 1 nucleus or QSO is present in the galaxy. Narrow-line Seyfert 2 or LINER galaxies are defined by the property that their forbidden lines and permitted lines have similar, narrow widths with line ratios indicating a more energetic ionizing source than hot stars alone can provide (although in the case of LINERs, this is still under discussion (see Ho 2002, this volume)).

At the present date, a subsample of 235 spectra from the GSS spectroscopic survey (those obtained between 1995 and 1997 with high enough S/N to discern spectral features) have been analyzed to search for AGN. Of these objects, 7 are galactic stars and 7 additional galaxies are high-z galaxies ($z \gtrsim 2.8$) not considered in this analysis for the sake of sample uniformity.

Inspection of the remaining 221 galaxy spectra has revealed only two objects with broad lines representing $\sim 1\%$ of the galaxies. These galaxies have redshifts of 1.15 and 1.22, each displaying broad MgII emission. Initial inspection of the most recently obtained spectra has revealed at least 4 more broad-lined galaxies with redshifts ranging between 0.6 and 1. The inclusion of the additional spectra are consistent with the finding that $\sim 1\%$ of the GSS galaxies appear to be Seyfert 1/QSOs. The integrated luminosities for these galaxies range from $-19.5 \gtrsim M_B \gtrsim -22.8$, just below the nominal dividing line between QSOs and seyferts. Since these are integrated magnitudes, the actual nuclear magnitudes are likely to be even fainter.

The majority of the galaxies display narrow emission lines. AGN can be differentiated from star-forming galaxies based on the emission-line ratios of the most prominent optical lines such as [OII]λ3727Å, [OIII]λ4959,5007Å, [NII] λ6548Å, [SII]λ6717,6730Å, Hα and Hβ (e.g. Veilleux & Osterbrock 1987). For the large fraction of galaxies in the GSS, however, many of these lines are redshifted out of the optical range. In addition, our spectra, like many large, multi-slit spectroscopic surveys, are not flux calibrated. For these reasons, traditional line ratio diagnostics are not applicable to our survey.

A new emission line diagnostic (Rola, Terlevich & Terlevich 1996) is being employed to differentiate AGN and star-forming galaxies in our survey spectra. This technique is based only on the equivalent widths of [OII] and Hβ, allowing for classification of galaxies to z$\simeq$0.8 with optical spectra and avoiding the necessity of flux calibration. Two distinct zones define the AGN region of the diagram, at EW(Hβ)<10 and EW(OII)/EW(Hβ)>3.5. Using a sample of local emission line galaxies, Rola et al. (1996) find that 87% of the AGN reside in these regions with 88% of the HII galaxies falling in the remaining region. Although this technique does not perfectly separate the two object classes, it does a fairly good job of identifying the majority of AGNs in a sample of emission line galaxies.

Of the 221 galaxy spectra in our subsample, 90 have both the [OII] and Hβ lines in the optical spectrum range. Out of those 90 galaxies, only 44 show both lines in emission. We plot these on the Rola diagram in Figure 1 with appropriate error bars. We note the importance of correcting the Hβ EW measurements for the underlying stellar continuum absorption. Without the detection of the Hα emission line for these galaxies, we can only estimate the amount of Hβ absorption in the continuum. We have chosen a moderate value of 3 Å for the underlying stellar absorption (Kennicutt et al. 1992, Tresse et al. 1996). This correction has the effect of pushing objects into the HII region of the diagram. The filled circles represent those galaxies clearly in the HII galaxy region. The open circles are those in the AGN region but with error bars extending into the HII region. The asterisks are those galaxies clearly in the AGN region of the diagram. If the probability that these galaxies are AGN is 88%, our lower-limit estimate on the total number of narrow-line AGN in the GSS out to z$\simeq$0.8 is 10%. If we include the additional 11 objects which lie in the AGN region but have error bars allowing them to be placed in either the AGN or HII regions, our fraction increases to 20%.

The AGN candidates in our survey have integrated absolute magnitudes extending to $M_B \simeq$–17.5, with nuclei that may be up to a magnitude fainter. This demonstrates the strength of the Keck telescope in probing activity in galaxies at much fainter luminosities and higher redshifts than previously possible. Our completed survey for AGN in the GSS will extend the AGN LF several magnitudes fainter at z$\sim$0.8.

A few other diagnostics are available to identify narrow-line AGN in our survey such as the strength of [NeIII]λ3869 and the presence of [NeV]λ3426. A few additional candidates have been detected through the presence of these emission lines. A full analysis of the complete sample of GSS galaxy spectra to detect AGN is presented in a future paper (Sarajedini et al. 2002).

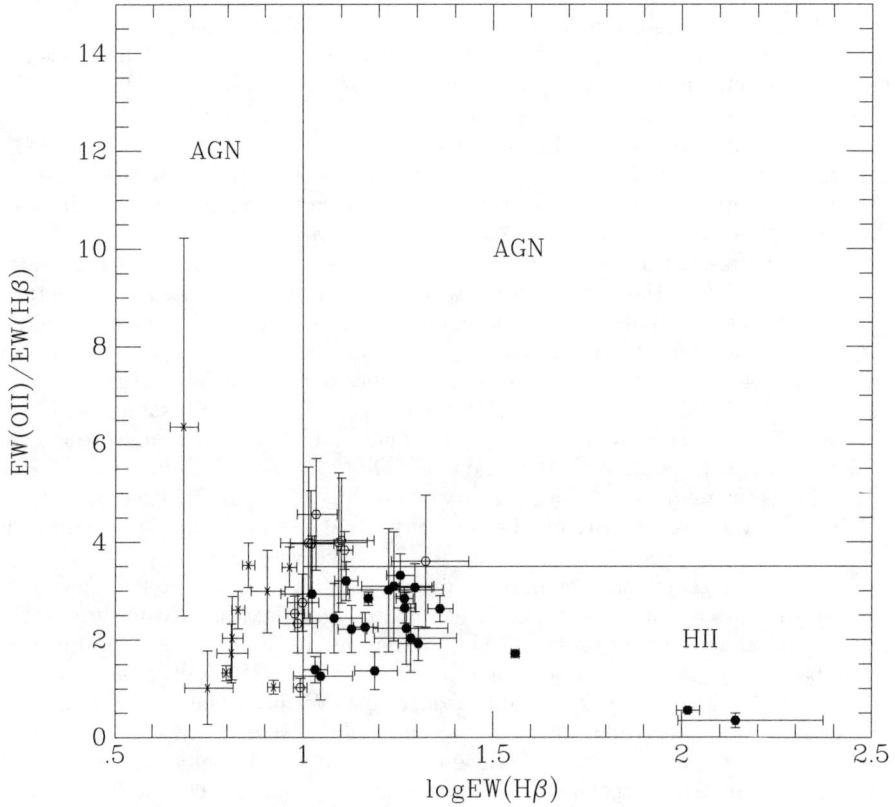

Figure 1. EW[OII]/EW(Hβ) ratio versus EW(Hβ). Solid symbols are those galaxies in the HII region of the diagram. Open symbols are those in the AGN region with error bars extending into the HII region and asterisks are galaxies in the AGN region of the diagram.

4. Detecting AGN through Variability Surveys

Obtaining spectra for as many galaxies as possible in a particular region of the sky is a robust way to find and classify the population of active galaxies. However, this technique requires many nights of observing on large telescopes. A less "expensive" way to detect AGN, and complement the spectroscopic survey, is through the use of multi-epoch images to identify variable sources.

Variability has long been known as an effective way to identify QSOs (e.g. Hawkins 1986) with Koo, Kron & Cudworth (1986) finding ~80% of their spectroscopic and color selected quasars in Selected Area 57 to be variable over an 11 year time period. In addition to quasars, Bershady et al. (1998) detected 14 extended variable objects in this region with Seyfert-like spectral characteristics. The variability amplitude for objects in SA57 was generally higher for active nuclei of lower luminosity, making this technique well suited for the selection of intrinsically faint QSOs and Seyfert-like nuclei.

We are conducting a survey to detect nuclear variabilty for galaxies in the GSS. In addition to the original HST images taken in 1994, a second epoch has been obtained for the entire region in the spring/summer of 2001 (J. Mould, PI). The unique high resolution capabilities of HST are necessary to isolate and measure faint, variable nuclei within brighter host galaxies. We can easily detect and measure structural parameters for galaxies to $V_{606} \simeq 25$ with redshifts extending to $z \simeq 1$ in the GSS (Simard et al. 1999). Many galaxies in this regime have shown evidence of central point source components in addition to a disk and/or bulge component (Sarajedini et al. 1999). Assuming a typical disk host galaxy with a scale length of 0.25", we estimate that nuclei comprising as little as 15% of the host galaxy light can be detected in galaxies down to $V_{606} \simeq 23.5$. The advantage of HST is the ability to do accurate photometry within smaller apertures, thus allowing us to probe much lower AGN/host galaxy luminosity ratios than can be done from the ground.

The success of this technique has been demonstrated with the Hubble Deep Field North. Based on observations of the HDF-N separated by two years, we have detected nuclear variability at or above the 3σ level in 8 of 633 galaxies at $I_{814} \leq 27.5$ (Sarajedini et al. 2000). Only 2 detections would be expected by chance in a normal distribution. At least one of these 8 has been spectroscopically confirmed as a Seyfert 1 galaxy. Based on the AGN structure function for variability (Trevese & Kron 1990), the estimated luminosities for the varying nuclear components extend to $M_B \simeq -16$ providing an interesting comparison with the population of local Seyfert galaxies at similar luminosities.

These results demonstrate the strength of this technique in probing faint AGN at $z \simeq 1$. Extending this search to the GSS should produce a much larger sample of faint AGN. Based on several different estimates for the number of AGN in the local universe (e.g. Huchra & Burg 1992; Maiolino & Rieke 1995; Ho et al. 1997) we expect to find at least ~ 45 Seyfert-like nuclei in the entire GSS (assuming no evolution to z=1) or ~ 120 if mild number density evolution has occurred. With this much larger sample, we will have the ability to not only determine number density evolution with statistical significance, but also study any changes with redshift in the shape of the LF for low-luminosity AGN.

5. Summary and Conclusions

With the aim of studying the evolution of AGN in the low-luminosity regime, we are conducting two different yet complementary surveys of galaxies in the GSS. We present the preliminary results of our spectroscopic search to identify broad and narrow-line AGN based on their spectral characteristics. Six broad-line AGN have currently been detected, representing about 1% of the galaxies for which spectra have been obtained as part of the DEEP project. A larger number (10 to ~ 20) of narrow-line AGN candidates have been detected using a new emission line diagnostic to identify Seyfert 2s and LINERs based on the equivalent widths of [OII] and Hβ. This represents between 10 and 20% of the galaxies to $z \simeq 0.8$ in which these lines could be detected. The absolute magnitudes of these galaxies extend to $M_B \simeq -17.5$ with a mean redshift of ~ 0.8.

We have also outlined a program to search for variable nuclei in GSS galaxies using HST images separated by 7 years. The high resolution capabilities of HST will allow us to detect and measure faint nuclei using small aperture photometry, probing lower nuclear/host galaxy ratios than possible from the ground. Using this technique on HDF-N images separated by 2 years, we have shown that varying nuclei as faint as $M_B \simeq -16$ may be detected for galaxies out to $z \simeq 0.8$.

The results of these surveys will be used in conjunction with GSS observations in other wavelengths such as X-ray (XMM; Griffiths et al. 2000), infrared (SIRTF), radio (VLA FIRST) and sub-millimeter (SCUBA) to better understand the nature of the AGN and their host galaxies.

Acknowledgments. VS would like to acknowledge the many members of the DEEP Team (UC Santa Cruz) for work in obtaining and reducing spectral data presented here. Financial support for part of this work comes from NASA grants through the Space Telescope Science Institute, operated by AURA, Inc. under NASA contract NAS5-26555.

References

Bershady, M. A., Trevese, D. & Kron, R. G. 1998, ApJ, 496, 103

Cheng, F. Z., Danese, L., De Zotti, G. & Franchesini, A. 1985, MNRAS, 212, 857

Griffiths R. E., Miyaji, T. & Ptak, A. 2000, BAAS 197, 5606

Groth, E. J., Kristian, J. A., Lynds, R., O'Neil, E. J., Balsano, R., Rhodes, J., & the WFPC-1 IDT. 1994, BAAS, 185, 5309

Hawkins, M. R. S. 1986, MNRAS, 219, 417

Ho, L. C., 2002, this volume

Ho, L. C., Filippenko, A. & Sargent, W. L. W. 1997, ApJS, 112, 315

Huchra, J. & Burg, R. 1992, ApJ, 393, 90

Kennicutt, R. C. 1992, ApJS, 79, 255

Koo, D. C. et al. 1996, ApJ, 469, 535

Koo, D. C., Kron, R. G. & Cudworth, K. M. 1986, PASP, 98, 285

Maiolino, R. & Rieke, G. H. 1995, ApJ, 454, 95

Rola, C. S., Terlevich, E. & Terlevich, R. J. 1996, MNRAS, 289, 419

Sarajedini, V. L., Green, R. F., Griffiths, R. E., & Ratnatunga, K. 1999, ApJ, 514, 746

Sarajedini, V. L., Gilliland, R. L. & Phillips, M. M. 2000, AJ, 120, 2825

Sarajedini, V. L. et al. 2002, in preparation

Simard, L., Koo, D. C., Faber, S. M., Sarajedini, V. L., Vogt, N. P., Phillips, A. C., Gebhardt, K., Illingworth, G. D., & Wu, K. L. 1999, ApJ, 519, 563

Simard, L. et al. 2001, ApJ, submitted

Tresse, L., Rola, C. S., Hammer, F., Stasinska, G., Le Fevre, O., Lilly, S. J. & Crampton, D. 1996, MNRAS, 281, 847

Trevese, D. & Kron, R. G. 1990, in Variability of Active Galactic Nuclei, ed. H.R. Miller & J.P. Witta (Cambridge: Cambridge University Press), 72

Veilleux, S. & Osterbrock, D. E. 1987, ApJS, 63, 295

Wisotzki, L. 2002, this volume

Spectral Variability of Quasars in the Optical Band

Dario Trèvese

Dipartimento di Fisica, Università di Roma "La Sapienza", Piazzale A. Moro 2, I-00185 Roma, Italy

Fausto Vagnetti

Dipartimento di Fisica, Università di Roma "Tor Vergata", Via della Ricerca Scientifica 1, I-00133 Roma, Italy

Abstract. We performed a new analysis of the B and R light curves of a sample of PG QSOs. We confirm a variability-redshift correlation and its interpretation in terms of spectral variability. We find an intra-QSO and an inter-QSO $\alpha - L$ correlation. The former can be explained neither by presence of the host galaxy nor by changes of the accretion rate. Hot spots due to accretion disk instabilities could explain the observations.

1. Introduction

The physical origin of variability is still substantially unknown. Even restricted to non-blazar objects, the most diverse mechanisms have been considered: gravitational lensing due to intervening matter (Hawkins 1996), supernova explosions (Aretxaga, Cid Fernandes & Terlevich 1997), instabilities in the accretion disk (Kawaguchi et al. 1998) and star collisions (Torricelli-Ciamponi et al. 2000). A small number of multi-wavelength studies of low redshift objects has provided several indications about different emission components (see Ulrich, Maraschi & Urry 1997). The most robust and general results come from statistical studies which indicate hardening of the spectrum in the bright phase (Cutri et al. 1985; Edelson, Krolik & Pike 1990; Kinney et al. 1991; Paltani & Courvoisier 1994). However, most of the statistical information on AGN variability has been derived, so far, from single-band light curves of magnitude limited samples of objects (Angione et al. 1972; Bonoli et al. 1979; Hawkins 1983; Trèvese et al. 1989; Cristiani, Vio & Andreani 1990; Trèvese et al. 1994 (T94); Hook et al. 1994; Bershady, Trèvese & Kron 1998). In this case the strong luminosity-redshift (L-z) correlation, caused by the crowding of objects towards the limiting flux makes it difficult to disentangle the intrinsic variability-luminosity (v-L) and variability-redshift (v-z) correlations. Correlation analyses depend on the specific variability index adopted. Giallongo, Trèvese & Vagnetti (1991) (GTV) found a positive v-z correlation, later confirmed by Cristiani et al. (1996), considering a variability index defined on the basis of the rest-frame structure function. The suggestion of GTV that QSOs at high redshift appear more variable since they are observed at a higher rest-frame frequency where the variability is stronger, is consistent with observations. Direct statistical evidence of a spectral hardening in the bright phase was found by Giveon et al. (1999) (G99) and by Trèvese, Kron & Bunone (2001). However, variability mechanisms proposed so far cannot be simply differentiated on qualitative grounds since most of them imply a hardening of the optical-UV spectrum in the bright phase. Here we report on preliminary results of a new analysis (Trèvese & Vagnetti 2002), based on the data made available to the community by the Wise Observatory group (G99), consisting of B and R light curves of 42 nearby, i.e. $z < 0.4$, and bright, i.e. $B < 16$ mag PG QSOs observed with the 1-m Wise Observatory telescope with a median time interval of 39 days for a total duration of 7 years, with r.m.s. photometric uncertainty $\sim 0.01, 0.02$ mag in B and R, respectively. G99 show a correlation between the color changes $\Delta(B-R)$ and the brightness variations ΔB and ΔR, corresponding to an average hardening of the spectrum in the bright phase. They do not find a correlation between variability and redshift, at variance with GTV, T94, and Cristiani et al. (1996), but they ascribe this to the difficulty of disentangling v-z, v-L and L-z correlations in the sample, which spans a small redshift interval. We adopt $H_o = 50$ km s^{-1} Mpc^{-1}, $q_o = 0.5$, unless otherwise stated.

2. Variability-Redshift Correlation

To measure the amplitude of variability we define, for each object, the first order structure function, as in Di Clemente et al. (1996) (D96): $S(\tau, \Delta\tau) = [(\frac{\pi}{2}\overline{|m(t+\tau)-m(t)|}^2 - \sigma_n^2)]^{1/2}$, where $m(t)$ is either the B or the R magnitude, t is the rest-frame time, τ is the time lag between the observations, σ_n is the relevant r.m.s. noise and the bar indicates the average taken over all the pairs of observations lying in the time interval $\tau \pm \Delta\tau$. We define four variability indices $S_i(\tau \pm \Delta\tau)$, with $i = B, R$, and $\tau = 0.3 \pm 0.09$ yr, 2.0 ± 0.6 yr. The subscripts B and R refer to the observing band and the values of τ and $\Delta\tau$ have been chosen for comparison with previous analyses (D96). None of these four indices shows a significant correlation with redshift when the whole sample is considered, confirming the result of G99. If we restrict the analysis to a magnitude bin $-23.5 < M_B < -22.5$, around the average absolute magnitude of the sample $< M_B > = -22.75$, to disentangle v-L and L-z correlations, we find a v-z correlation coefficient $r_{v,z} = 0.39$, which is marginally significant ($P(>r) = 0.09$) despite the small number of objects (19) in the bin.

We take the ensemble averages of the four variability indices defined above, over the same subsample of 19 objects, to examine the dependence of variability on redshift. For each observing band we compute the average rest-frame observing frequency of the sample. We must take into account the dependence of variability on magnitude for comparison with D96, reducing S by an amount $\Delta S = (\partial S/\partial M_B)\,\Delta M_B$, where for $S(M_B)$ we adopt model A of Cristiani et al. (1996) and ΔM_B is the difference between the average absolute magnitudes of the present sample and the sample of D96. The increase of variability with the observing frequency is shown in Figure 1a, where the new points appear consistent with the previous results. The general trend can be quantified as: $\partial S_i/\partial \log \nu_{rest} = \partial S_i/\partial \log(1+z) \simeq 0.25 - 0.3$ in agreement with GTV, Cristiani et al. (1996) and D96. Despite poor statistics, the present results confirm

3. Spectral Variability

Direct evidence of an average hardening of the SED in the bright phase was obtained by Trèvese, Kron, & Bunone (2001) from U,B_J,F,N photometry of a sample of 40 QSOs observed at two epochs. A statistical analysis of SED variations of individual objects is made possible by the B and R light curves of G99, containing on average 40 points per object and allowing the computation of the time dependent spectral slope for each QSO:

$\alpha(t) = -(0.4[(B - R) - (B_o - R_o)])/\log \frac{\lambda_R}{\lambda_B} - 2$,

where B_o, R_o are the zero points of B,R photometric bands. This is reported in Figure 1b as a function of the relevant intrinsic luminosity L_{ν_B}. Each small cloud of points represents an individual QSO at different epochs. The relevant regression lines are reported for each cloud. They show, with a few exceptions, a positive correlation between the spectral slope and the intrinsic luminosity, which we call the intra-QSO $\alpha - L$ correlation. The distribution of clouds in the α-L_{ν_B} plane also shows an inter-QSO $\alpha - L$ correlation, indicated by the thick regression lines, corresponding to a highly significant ($P(> r) = 6 \cdot 10^{-5}$) correlation coefficient. $r_{\alpha-L} = 0.58$.

We quantify the spectral variations of individual objects by a spectral variability parameter (SVP) which represents the spectral slope changes per unit log-luminosity change: $\beta(\tau) \equiv \frac{\alpha(t+\tau) - \alpha(t)}{\log L_B(t+\tau) - \log L_B(t)}$. where $L_B(t)$ is the luminosity in the B band and τ is an appropriate time delay. For the following analysis we define the SVP of each QSO as the mean value β_m in a single bin $0 < \tau < 1000$ d. In Figure 2, β_m values for the 42 QSOs are reported versus the relevant time average $\bar{\alpha}$. The dot-dashed line represents a black body of fixed area and varying temperature whose spectral slope is $\alpha_{BB}(x) = 3 - xe^x/(e^x - 1)$ and the SVP is $(d\alpha/dT)/(d\log B_\nu/dT) = (\ln 10)[1 - x/(e^x - 1)] \equiv \beta_{BB}(x)$, where $x \equiv h\nu/kT$, with T, h, k equal to the temperature, Planck, and Boltzmann constants respectively. Typical values $\langle \bar{\alpha} \rangle = -0.2 \pm 1.0$, $\langle \beta_m \rangle = 2.2 \pm 0.9$, correspond to temperature changes of a black body of $T \approx 10^4$ K, in approximate agreement with the result of Trèvese, Kron & Bunone (2001).

According to Cid Fernandes et al. (2000), color changes in G99 data imply the presence of a constant spectral component, redder than the variable one, which could be identified with either the non-flaring part of the QSO spectrum or with the host galaxy, also considered by Romano & Peterson (1998). We evaluate the latter effect through numerical simulations based on templates of the QSO and host galaxy SEDs from Elvis et al. (1994) (see also Trèvese & Vagnetti (2001)). QSO and host spectra are added with a relative weight $\eta \equiv \log(L_H^Q/L_H^g)$, where L_H^Q and L_H^g are the total H band luminosities of the QSO and the host galaxy, respectively. We explore a wider range $-3 < \eta < 3$. Variability is represented by small changes $\Delta \eta$ deriving from a luminosity change which corresponds to the r.m.s. variability $\sigma_B = 0.16$ mag. For each synthetic spectrum, representing the QSO plus host SED at a given time, we compute $\alpha(\bar{\nu}, t) \equiv (\partial \log L_\nu / \partial \log \nu)_{\nu=\bar{\nu}}$, $\bar{\nu} = \sqrt{\nu_B \nu_R}$, then we derive the SVP β. The

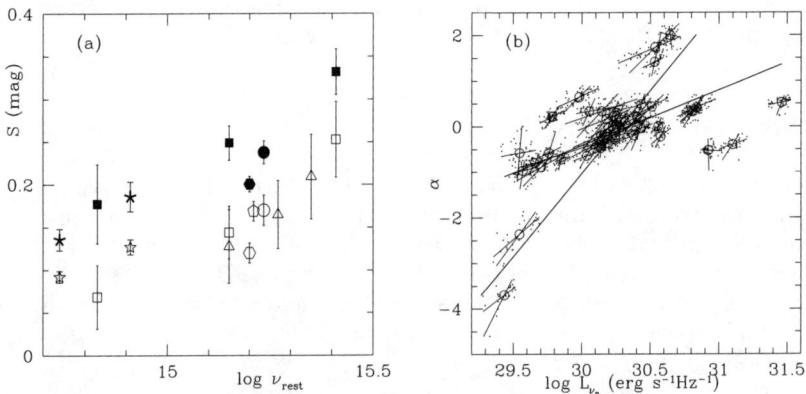

Figure 1. (a) Variability S versus rest-frame frequency for various QSO samples. Filled symbols correspond to $\tau = 0.3 \pm 0.09$ yr and open symbols correspond to $\tau = 2.0 \pm 0.3$ yr. Stars: Trevese & Vagnetti 2002; squares: D96; triangles: Cimatti et al. 1993; pentagons: Hook et al. 1994; hexagons: Cristiani et al. 1990; circles: T94. (b) Instantaneous spectral slope α versus monochromatic luminosity L_{ν_B}. Regression lines $\alpha - L_{\nu_B}$ are reported for each QSO (thin lines, intra-QSO correlation), and for the population (thick lines, inter-QSO correlation).

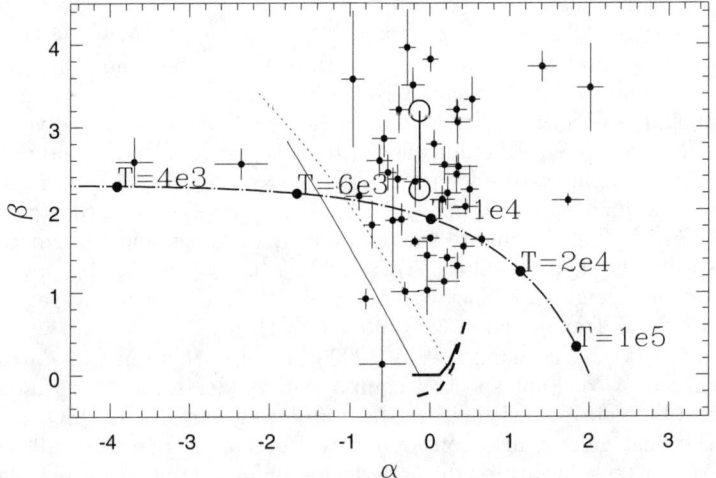

Figure 2. The spectral variability parameter β_m versus the average spectral slope for each QSO of the sample. Dot-dashed line: black bodies of different T; continuous line: the effect of the host galaxy, for z=0; dotted line: the same for z=0.4; thick continuous line: spectral variability caused by an increase of $\dot{m}$ from 0.1 to 0.3 from the Siemiginowska et al. (1995) model; thick dashed line: the same for $\dot{m}$ increasing from 0.3 to 0.8. Large open circles: hot spot model for $T = 2 \cdot 10^5$ K (upper) and $T = 2 \cdot 10^4$ K (lower).

result is shown in Figure 2. The same computation is repeated for the maximum redshift of the G99 sample, $z = 0.4$, to check the dependence of the result on QSO redshift. Although an appropriate choice of η can reproduce the observed β or α separately, the curves are clearly shifted with respect to the distribution of the observational points, indicating that the host galaxy is not sufficient to account for the observed changes of the spectral shape. This also implies that spectral variations are intrinsic to the active nucleus and that the constant red continuum, resulting from the analysis of Cid Fernandes et al. (2000), cannot be identified with the host galaxy but it must be, at least in part, due to the nucleus itself. Thus it could be identified with the spectrum of the non-flaring part of the accretion disk.

We can check whether a change of the accretion rate $\dot{M}$ can account for the observed variations of the spectral shape. We considered the accretion disk model of Siemiginowska et al. (1995) (S95), corresponding to a Kerr metric and modified black body SED. A grid of models has been considered for $\log M/M_\odot = 7.0, 8.0, 9.0, 10.0$, $\dot{m} \equiv \dot{M}c^2/L_E = 0.1, 0.3, 0.8$ (where L_E is the Eddington luminosity $L_E = 4\pi G c m_p/\sigma_e M$ with the usual meaning of symbols) and inclination $\mu \equiv \cos\theta = 1, 0.75, 0.5, 0.25, 0.1$. A change of $\dot{M}$ produces a variation of both luminosity and the SED shape. It is interesting to see how the spectral changes between two $\dot{M}$ states compare with the observed ones, as done by S95 and Tripp et al. (1994). The two curves on the bottom right of Figure 2 represent β and α changes obtained by varying M (from right to left), as computed for two different values of $\dot{m}$ and for $\mu = 1$, from the Kerr metric, modified blackbody model of S95 (their Table 4). The spectral variations are clearly smaller, on average, than the observed ones, meaning that a transition, e.g., from a lower to a higher $\dot{M}$ regime implies a larger luminosity change for a given slope variation.

This result suggests that transient phenomena, like hot spots produced on the accretion disk by instability (Kawaguchi et al. 1998), instead of a transition to a new equilibrium state, may better explain the relatively large changes of the local spectral slope. The available models of instability phenomena do not provide a spectrum of the hot spot. Thus we try a simple "model" based on the addition of a black body flare to the disk SED, represented by the average QSO SED of Elvis et al. (1994). $\Delta B = 0.16$ mag, which corresponds to the r.m.s. variability of the sample, can be obtained by a hot spot of $T_{BB} \approx 2 \cdot 10^5$ K and $A = 5 \cdot 10^{30}$ cm^2, producing $\beta = 3.2$, or $T_{BB} \approx 2 \cdot 10^4$ K, $A = 1, 3 \cdot 10^{32}$ cm^2, giving $\beta = 2.2$, as shown by the large circles in Figure 2. A sudden heating of a fraction of the disk surface is thus capable of producing the observed change of the SED in the B and R bands and the relevant intra-QSO α-L correlation.

References

Angione, R. J., & Smith, H. J. 1972, IAU Symp. 44, 171
Aretxaga, I., Cid Fernades, & Terlevich, R., 1997, MNRAS, 286, 271
Bershady, M. A., Trèvese, D., & Kron, R. G. 1998, ApJ, 496, 103
Bonoli, F., Braccesi, A., Federici, L., & Zitelli, V., 1979, A&AS, 35, 391
Cid Fernandes, R., Sodrè, L., Vieira da Silva, L., 2000 ApJ, 544, 123

Cimatti, A., Zamorani, G., & Marano, B. 1993, MNRAS, 263, 236
Cristiani, S., Vio, R., & Andreani, P. 1990, AJ, 100, 56
Cristiani, S., et al 1996, A&A, 306, 395
Cutri, R. M., et al. 1985, ApJ, 296, 423
Di Clemente, A., et al. 1996, ApJ, 463, 466
Edelson, R.A., Krolik, J. H.,& Pike, G. F. 1990, ApJ, 359, 86
Elvis, M., et al. 1994, ApJS, 95, 1
Giallongo E., Trèvese D., Vagnetti F.: 1991, ApJ, 377, 345
Giveon, U., et al. 1999, MNRAS, 306, 637
Hawkins, M. R. S., 1983, MNRAS, 202, 571
Hawkins, M. R. S., 1996, MNRAS, 278, 787
Hook, I. M., et al. 1994, MNRAS, 268, 305
Kawaguchi, T., et al. 1998, ApJ, 504, 671
Kinney, A. L., et al. 1991, ApJS, 75, 645
Paltani, S., & Courvoisier, T. J.-L. 1994, A&A, 291, 74
Romano, P., & Peterson, B. M., 1998, ASP Conf. Ser. 175, 55
Siemiginowska, et al. 1995, ApJ, 454, 77
Torricelli-Ciamponi, G., et al. 2000, A&A, 358, 57
Trèvese, D., Kron, R. G., & Bunone A., 2001, ApJ, 551, 103
Trèvese, D., et al. 1994, ApJ, 433, 494 (T94)
Trèvese, D., et al. 1989, AJ, 98, 108
Trèvese, D., & Vagnetti, F., 2001, in 'QSO Hosts and their Environments', I. Marquez et al. Eds., in press (astro-ph/0102252)
Trèvese, D., & Vagnetti, F., 2002, ApJ, (in press) (astro-ph/0110075)
Tripp, T. M., Bechtold, J., & Green, R. F. 1994, ApJ, 433, 533
Ulrich, M. H., Maraschi L., & Urry C. M. 1997, ARA&A, 35, 445

AGN SURVEYS
ASP Conference Series, Vol. 284, 2002
R.F. Green, E.Ye. Khachikian, D.B. Sanders

AGNs in Shakhbazian Compact Groups

A.S. Amirkhanian and A.G. Egikian

Byurakan Astrophysical Observatory, Byurakan, 378433, Armenia

H. Tiersch and D. Stoll

Sternwarte Koenigsleiten, Muenchen, Germany

The results of CCD spectroscopic observations of Shakhbazian compact groups of galaxies (SHCGs) with the 1.54-m (La Silla, Chile), 2.2-m (Calar Alto, Spain) and 2.6-m (Byurakan) telescopes are presented. According to these preliminary data, about 10% of member galaxies in SHCGs are emission-line galaxies (ELGs) including the broad-line AGNs (of classical Seyfert 1 type) and the narrow-emission-line galaxies.

A research program has been developed in the University of Potsdam, Potsdam Astrophysikalisches Institut in cooperation with other observatories (particularly with Byurakan Astrophysical Observatory) to perform photometric and spectroscopic investigations of galaxies in the SHCGs. Within the framework of this program the redshift (radial velocity) measurements have been carried out for more than 200 galaxies in 36 SHCGs. The MIDAS software package was used for processing and interpreting of the galaxy spectra. Most of these redshifts were measured for the first time. 180 member galaxies (90%) in these groups have absorption spectra typical of E and S0 galaxies. Twenty galaxies (10%) turn out to be ELGs. They are in the range $0.02 \leq z \leq 0.17$, i.e., the SHCGs lie in approximately the same redshift space as Abell clusters. These compact groups contain predominantly elliptical and lenticular galaxies (del Olmo 1988; Amirkhanian 1989) like the cores of rich, regular, centrally condensed clusters of galaxies. The fraction of spirals falls in the densest matter concentrations. On the other hand, it is a well-established fact that in the local universe the active objects tend to avoid the cores of dense clusters of galaxies (e.g. Green and Yee 1984). That is why the discovery of an emission-line population with broad-line AGNs in SHCGs (Tiersch et al. 1999) was unexpected. As shown by Dressler, Thompson and Shectman (1985) in their sample of 1268 galaxies in the feilds of 14 rich clusters the ELGs comprise 31% of the field galaxies but only 7% of the cluster galaxies. Similarly, according to their statistics AGNs make up 5% of the field sample, but only 1% of the cluster sample. They note that the difference in the distribution of morphological types can only partially explain this effect. Obviously, some sort of environmental influence is present.

A cursory investigation in the literature of the frequency of ELGs in the Hickson compact groups (HCGs) (Shimada et al. 2000) and the south compact groups (SCGs) (Coziol et al. 2000) has resulted in conflicting data. Shimada et al. (2000) present results of their optical spectroscopy of 69 galaxies belonging to 31 HCGs of galaxies. They found that for each morphological type there is no statistically significant difference in the frequencies of occurence of ELGs among the HCGs and in the field. The field galaxies data are taken from Ho, Filippenko and Sargent (1995; 1997). But according to their statistics (Shimada et al. 2000) the difference for the total sample is significant, in that the ELGs (AGNs + HII nucleus galaxies) and the absorption galaxies are found in 70% and 30% of the HCG galaxies, respectively. Meanwhile, in the field the ELGs make up 89% and the absorption galaxies comprise only 11% of the sample.

The ELGs and the absorption galaxies comprise 73% and 27% of the SCGs galaxies, respectively (Coziol et al. 2000). This result is similar to that for HCGs. The population of galaxies in Shakhbazian groups differs strongly by morphological content. The frequency of emission-line galaxies in the HCGs and SCGs is much higher than in Shakhbazian groups. This difference is likely to be due to the following explanation. The great majority of Shakhbazian groups are at larger distances with respect to the HCGs, most of which are nearby systems at redshifts below 0.05 (Hickson et al. 1992). Moreover, extensive spectroscopic studies are needed to obtain the spectral activity types for all the galaxies in the 377 groups of the Shakhbazian sample (Shakhbazian 1973; Baier and Tiersch 1979 and references therein).

Finally we note that Shk 355/4 is a classical Seyfert 1 galaxy at the same redshift as its host group with z=0.0943 (Tiersch et al. 1999). Moreover Shk 278/4 is also a broad-line AGN in an early-type galaxy (z=0.1205). This is the first active object in the SHCGs discovered by Spanish astronomers (del Olmo and Moles 1991). It has a remarkable resemblance to QSOs or QSO-like objects. At the same time, although more than 70% of the galaxies in SCGs probably have an active nucleus, there is no Seyfert 1 among 193 galaxies in 49 southern groups spectroscopically investigated by Coziol et al. (2000).

References

Amirkhanian, A. S. 1989, PhD. Thesis, Byurakan Astrophysical Observatory, Armenia

Baier, F. W., & Tiersch, H. 1979, Astrofizika, 15, 33

Coziol, R., Iovino, A., & Carvalho, R. R. 2000, AJ, 120,47

del Olmo, A. 1988, PhD. Thesis, University of Granada, Spain

del Olmo, A., & Moles, M. 1991, A & A, 245,27

Dressler, A., Thompson, I. B. , & Shectman, S. A. 1985, ApJ , 288,481

Green, R. F. , & Yee, H. K. C. 1984, Ap.JS, 54,495

Ho, L. C., Filippenko, A. V., & Sargent, W. L. W. 1995, ApJS, 98,477

Ho, L. C., Filippenko, A. V., & Sargent, W. L. W. 1997, ApJS, 112, 315

Shakhbazian, R. K. 1973, Astrofizika, 9, 495

Shimada, M., Ohyama Y., Nishiura, S., Hurayama, T., & Taniguichi Y. 2000, AJ, 119,2664

Tiersch, H., Stoll, D., Neizvestny, S., Amirkhanian, A. S., & Egikian, A. G. 1999, in Active Galactic Nuclei and Related Phenomena, IAU Symp. 194, eds. Y. Terzian, D. Weedman, E. Khachikian, Astron. Soc. Pacif., Chelsea, Michigan, p. 394

A Morphological Optical Survey of Nearby AGN

C.S. Boschetti

Dipartimento di Astronomia, Università di Padova, Italy

S. Ciroi[1], J. Funes[2], A. Omizzolo[2], P. Rafanelli[1], G.M. Richter[3], A. Rifatto[4], J. Vennik[5]

[1] *Dipartimento di Astronomia, Università di Padova, Italy;* [2] *Specola Vaticana, Città del Vaticano;* [3] *Astrophysikalisches Institut Potsdam, Germany;* [4] *Osservatorio Astronomico di Capodimonte, Napoli, Italy;* [5] *Tartu Observatory, Toravere, Estonia*

Abstract. We present preliminary results of an optical survey of a spectroscopically selected sample of nearby AGNs ($z < 0.1$). The aim of the program that we are currently carrying out is to demonstrate that the activity phenomenon is strongly related to galaxy interactions. In detail, we want to study the morphology of our sample of galaxies in order to reveal the presence of features like bars, rings, double nuclei, isophotal distortions and tidal tails, most of which are not detectable on the POSS plates. The work is based on the analysis of broad band B, V and R images, the study of their photometric parameters and the results of the application of a Laplacian adaptive filter, able to enhance hidden and faint structures.

1. Introduction

Seyfert galaxies are the nearest, and hence most widely studied AGNs, usually associated with spirals, but some are also found in ellipticals.

Whether Seyferts tend to be interacting with nearby companions (Dahari 1984) or not (Bushouse 1986) is still a matter of debate. Nevertheless, several statistical studies carried out within our group have shown an excess of Seyferts among galaxy pairs and an excess of physical companions in samples of Seyferts (Rafanelli et al. 1995, 1997; Salvato & Rafanelli 1997).

Other clues to the connection between interaction and activity come from morphological studies, which indicate that many Seyferts are amorphous or disturbed galaxies (MacKenty 1990), and from N-body simulations, which show how disk instabilities in colliding galaxies can lead rapidly to the formation of strong bars and to inflows of gas fueling starburst and/or AGN activity before galaxy merging (Mihos & Hernquist 1996; Barnes & Hernquist 1996; Mihos 1999).

We present preliminary results of an optical imaging survey of a spectroscopically selected sample of nearby AGN ($z<0.1$). Only isolated galaxies are extracted, following the criteria given by Rafanelli et al. (1995), and studied to reveal morphological signatures of recent interactions like bars, double or multiple nuclei, isophotal distortions and tidal tails, most of which are not detectable on the POSS plates.

2. Observations & Data Analysis

As examples of the AGN survey we are still currently carrying out, we present three galaxies: HE 1338-1423, ESO 362-G018 and NGC 2768. The first two objects were observed in April, 1995, at the ESO 2.2m telescope with EFOSC2 and with a typical seeing of $\sim 1''$. The 1k×1k CCD (19μ pixel) covered a field of 5.7' with a scale of 0.336''/px. The data of the third object were taken in October, 1998, at the VATT 1.8m telescope, with a 2k×2k CCD camera (15μ pixel) covering a field of 6.4' with a a scale of 0.4''/px (after binning). The seeing was $\sim 2''$.

Broad band B,V and R images were obtained for each galaxy and reduced within the IRAF environment then an Adaptive Filter procedure (Richter et al. 1991; Lorenz et al. 1993) was applied to strongly reduce the noise (smooth) and/or enhance hidden and faint structures (Laplacian).

The morphology of the galaxies was analyzed by fitting their isophotes with free ellipses (Jedrzejewski 1997) and plotting as a function of the equivalent radius (R_{eq}) the brightness profile, the ellipticity (e), position angle ($P.A.$) and center (x,y) of the ellipses, and the parameters which measure the deviation from perfect elliptical isophotes. In addition, a model of each object was built on the basis of the best-fitting ellipses and subtracted from the original image to study the residuals.

3. Results

HE 1338-1423 is a Seyfert galaxy (z=0.041) without any morphological classification. The V image and the isophotal analysis reveal a little bar with semi-major axis of $\sim 4''$ (~ 3.2 kpc, assuming H_0=75 km s^{-1} Mpc^{-1}) oriented at P.A. $\sim 160°$. The brightness profile was fitted trying to take into account the contributions of the main components: a point source for the AGN, an exponential bulge, a bar and a disk. The galaxy is clearly disk dominated and the residuals of the fit reveal a large structure with at least three maxima at about 7'', 10'' and 14'' (~ 5.6, 8 and 11.2 kpc), corresponding to different spiral arms. Such arms are very well enhanced by means of the Laplacian adaptive filter application.

ESO 362-G018 is a Seyfert 1.5 (z=0.012) classified in NED as S0/Sa. The B image shows an asymmetric structure, as indicated by the evident changing of the ellipses centers, a very bright and elliptical nucleus and an emission knot at $\sim 8''$ (~ 1.8 kpc) NE of it. Two faint tidal tails are visible, one of $\sim 10''$ (~ 2.3 kpc) in the NW direction and the other very extended (> 30 kpc) to the South. The first one is clearly connected to the knot in the residual image, after the model subtraction, which shows also two other structures in the SW and SE directions. We suggest that the knot is in fact a companion galaxy in an advanced merging phase with the Seyfert.

NGC 2768 is a LINER (z=0.004) galaxy classified as S0. The isophotal analysis of the R image confirms its early type and symmetric morphology: the brightness profile follows the $r^{1/4}$ De Vaucouleurs law typical of elliptical galaxies, the P.A. is almost constant around $\sim 95°$ and the ellipticity increases smoothly outwards. Notwithstanding these properties, the 4th order cosine coefficient C_4 (Bender et al. 1989) is negative up to $\sim 50''$ (~ 4 kpc), indicating that the inner isophotes are boxy. As expected, the residual image shows missing light along the principal axes and excess light along two diagonal lines. This feature is believed to be the result of a merger remnant. (Hernquist & Quinn 1989). These few examples extracted from the many data of our survey show that active galaxies with apparent regular morphology have systematically more complicated structures and hide very often traces of past or current interaction with other galaxies. In the future we plan to select and observe a control sample of non-active galaxies to put stronger constraints to the interaction-activity connection.

References

Barnes, J. & Hernquist, L. 1996, ApJ, 471, 115
Bender, R., Surma, P., Döbereiner, S. et al. 1989, A&A, 217, 35
Bushouse, H.A. 1986, AJ, 91, 255
Dahari, O. 1985, ApJS, 57, 643
Hernquist, L. & Quinn, P.J. 1989, ApJ, 342, 1
Jedrzejewski, R.I 1987, MNRAS, 226, 747
MacKenty, J.W. 1990, ApJS, 72, 231
Mihos, C. & Hernquist, L. 1996, ApJ, 464,641
Mihos, C 1999, preprint, astro-ph/9903115
Rafanelli, P., Violato, M. & Baruffolo, A. 1995, AJ, 109, 1546
Rafanelli, P., Temporin, S. & Baruffolo, A. 1997, AN, 318, 249
Salvato, M. & Rafanelli, P. 1997, AN, 318, 237

Byurakan Surveys: Density of Bright AGN

L.K. Erastova

Byurakan Astrophysical Observatory, Armenia, 378433

It is clear now that morphological criteria don't divide galaxies from QSOs. Many galaxies are stellar-like objects and, conversely, quasars have host galaxies of various luminosities. For instance, two objects - SBS 1120+586A and SBS 1123+598 - absolutely do not differ from stars on the charts of the POSS and were classified as BSO-type. But they turned out to be galaxies with moderate luminosities (Markarian et al. 1988). Erastova (2000) produced a list of stellar-like galaxies from the SBS. Out of 339 KUV objects, 107 (31.6%) are emission-line galaxies of various luminosities (Darling & Wegner 1996). A considerable number of active galaxies appear among CSO - Case stellar objects. Conversely, objects having extended images on low-dispersion spectral plates turned out to be QSOs. SBS 1520+530 is a gravitationally lensed QSO with z=1.855 and on our low-dispersion plates appears as a galaxy (Chavushyan et al., 1997).

It is claimed that the surface density of bright quasars from the Bright Quasar Survey (BQS - Schmidt & Green (1983)) is considerably low and that it departs from the relation of lg(N<B)-B established from fainter magnitudes (see Wampler & Ponz 1985, Markarian et al. 1987, Goldschmidt et al. 1992). There are many different reasons proposed (Köhler et al. 1997). We consider that these reasons are not enough. The general reason for the discrepancy is probably that QSO surveys include in their samples QSOs and Sy1-type galaxies in arbitrary proportion. On the other hand, we have another survey - the Markarian survey of UV-continuum galaxies, which covers the same interval of magnitudes. Some fraction of the Sy1-type Markarian galaxies are also in the BQS survey. It's clear, because both Markarian and Green used the same criterion, UV-excess, for discovering active objects. We decided to combine the Sy1 galaxies and QSOs from the Markarian sample with the QSOs from the BQS survey and create a new composite AGN sample for investigation of their surface density. We claim that we have one population of AGN found in these two surveys: Markarian Sy1's and QSOs from the BQS. We have taken the area of the composite sample to be 10714 sq. deg. We have chosen from Markarian et al. (1989) the Sy1-type galaxies and QSOs and added all QSOs from the BQS sample. So, a new sample of 101 AGN was created with a limiting magnitude $\sim 15^m.5$, which we used for constructing the lg(N<B)-B relation. Now it's easy to derive the lgN(<B)-B relation and determine the surface density of active extragalactic objects - QSOs and Sy1-type galaxies. This important diagnostic relation between blue magnitude and number of objects per sq. deg. is shown in Figure 1.

It may be approximated by the linear regression lgN=βB+const, where β=0.60 ±0.06 in the range of magnitudes $B = 12^m.5 - 15^m.5$. The same relation was fit for the HES Survey from Köhler et al. (1997). In both cases the magnitudes were not corrected for extinction in our Galaxy or for the contribution of the host galaxy to the total brightness of the active galaxy. It

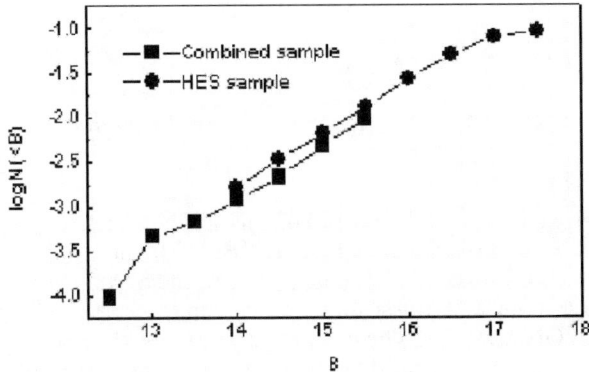

Figure 1. Squares represent lgN(<B)-B for the combined sample for 13^m5-15^m5 magnitude; dots show the same dependence for the HES survey. These two curves join smoothly without gaps between them.

matches well our relation with a slope of $\beta=0.59\pm0.04$ for the range of magnitudes $B = 14^m0 - 17^m5$. The surface density is 0.01 ± 0.01 per sq. degree for objects up to $B = 15^m5$ with redshifts less than z=2.0. We propose that this newly created combined sample of local objects may serve to determine the zero-point of surface and space densities as well as other properties of active extragalactic objects. Now we have a flux-limited and well-defined sample of AGN (QSOs and Sy1-type objects). One of the first experiments to join samples of Sy-type galaxies and QSOs was made by Stepanian et al. (1989) from the SBS objects.

References

Chavushyan V.H. et al.,1997, A&A, 318, L67.
Darling G.W.,& Wegner G.,1996, AJ, 111, 865.
Erastova L.K., 2000, Astrofizika, 43, 191.
Goldschmidt P., et al.,1992, MNRAS,255,65.
Köhler T. et al., 1997, A&A,325,502.
Markarian B.E. et al., 1987, Astrofizika, 26, 15.
Markarian B.E., Lipovetsky V.A., & Stepanian J.A.,1988, Astrofizika, 29, 548.
Markarian B.E. et al.,1989, SAO Communications., N62,117p.
Schmidt M., & Green R.F., 1983, ApJ,269,352.
Stepanian J.A. et al, 1989, IAU Symp.N185, Dordrecht:Reidel, 31.
Wampler E.J., & Ponz D., 1985, ApJ, 298, 448.

Comparison of ELG and UV Galaxies from the SBS Survey

M. V. Gyulzadian

BAO, Byurakan 378433, Armenia, E-mail:mgyulz@bao.sci.am

Abstract.
In this paper we compare the results of spectrophotometric observations for more then 40 emission line galaxies (ELG) without UV-excess and 70 galaxies with UV-excess (UV-galaxies) from the Second Byurakan Survey (SBS) carried out with the scanner on the 6-m telescope of the Special Astrophysical Observatory. The comparison was aimed to answer the questions: are there any essential differences between these two types of galaxies? And what is the fraction of AGN among ELG and UV-galaxies? To answer this question we have investigated some physical parameters. Great care was taken in classifying the spectra of these galaxies as "HII region-like" or "AGN-like" using a new diagnostic method for emission-line galaxies (Rola et al.). We find that 14% of emission line galaxies (ELG) have Sy2 and LINER-like spectra. The transition or ambiguous objects among them comprise 20%. The fraction of LINER-like and Sy2 objects among UV-galaxies is 21% while that of the transition or ambiguous objects in UV-galaxies is 28%. From SBS galaxies, 18% are Sy2s and LINER-like galaxies.

1. Introduction

In 1978 the Byurakan Observatory marked the completion of the First (FBS) and the beginning of the Second Byurakan Sky Survey (SBS). The SBS is an objective prism survey, which was carried out with the Byurakan Observatory 1-m Schmidt telescope in combination with a set of three low dispersion objective prisms ($1.5^o, 3^o, 46^o$). The prisms' dispersions were respectively 1800, 900 and $280 Å/mm$ at Hγ. The observations were done on well-sensitized plates, which gave an opportunity to observe objects down to 19-20 magnitude. In the observed fields with an area of about 1000 square degrees ((=8h-17h,(=$49^o - 61^o$), nearly 1300 nonstellar (galaxies) objects and 1700 star-like objects were discovered.

As in the case of the First Byurakan Survey (FBS, Markarian 1967) with the 1.5-degree prism, unwidened spectra were obtained and UV-excess objects were selected. Similarly to the FBS, the SBS classified objects as stellar ("s") or diffuse ("d") type. In addition, from observations with the 3- and 4-degree prisms a large number of emission-line galaxies without UV-excess were identified. They are called emission-line galaxies (ELG) and also classified as stellar ("s") or diffuse ("d") type, with "sd" and "ds" intermediate types without UV-excess numbers 1,2,or 3 (only s; sd; ds; d).

2. Observation and Reduction

Spectra (in the spectral range $3500\text{Å} - 5800\text{Å}$) of galaxies discovered at the observatory were obtained at the Nasmyth focus of the SAO 6-m telescope with the SP-124 spectrograph with a dispersion of $100 - 200 \text{Å}/mm$(Stepanian et al. 1990). In this paper the results for more then 110 galaxies from the SBS are presented. The galaxy spectral observations were carried out during 1992 on the 6-m telescope of the Special Astrophysical Observatory using the 1024-channel photon counting scanner (Stepanian et al. 1992a, Stepanian et al. 1992b). The data were reduced with the AIDA system in BAO. From these galaxies some galaxies with UV-excess (UV-galaxies) and emission line galaxies (ELG) (without UV-excess) were chosen (Gyulzadian 1995a,b).

3. Spectral Classification

A good segregation between starburst and active galactic nuclei, i.e. Sy 2s and LINERs, is obtained from diagnostic diagrams involving the $[OII]\lambda 3727\text{Å}$, $H\beta$ and $[OIII]\lambda 5007\text{Å}$ relative intensities or the $[OII]\lambda 3727\text{Å}$ and $H\beta$ equivalent widths. Furthermore, the color index of the continuum underlying $[OII]\lambda 3727\text{Å}$ and $H\beta$ provides an additional separation parameter between the two types of emission line galaxies. We have classified the spectra of these galaxies as "HII region-like" or "AGN-like" using a diagnostic method for emission-line galaxies. The first H II region-like object among Mrk galaxies was Mrk 94 (Arp & Khachikian 1974); forty objects were found subsequently by Sahakian & Khachikian (1975). Future work will include the comparison with other H II - or AGN-like galaxies (e.g., Burenkov 1991). 66% of ELG, 36% of UV-galaxies and 54% of our total SBS sample are HII galaxies. We find that 14% of emission line galaxies (ELG) have Sy2 and LINER-like spectra. The transition or ambiguous objects among them comprise 20%. The fraction of LINER-like objects and Sy2 among UV-galaxies is 21%; the transition or ambiguous objects comprise 28%. Among the whole sample of SBS galaxies 18% are Sy2 and LINER-like galaxies. In HII galaxies the emission-line spectrum is dominated by the emission originating in HII regions, where ultraviolet photons emitted by OB stars ionize the surrounding gas, while in the case of active galaxies, the gas-ionizing source is much harder and has the shape of a power law. Furthermore, star-forming regions are always associated with HII regions and notably with HII galaxies, while that is not necessarily the case for active galaxies. An active nucleus can disrupt the surrounding gas through its intense radiation field, X-ray heated wind, and possible radio jets (Begelman et al. 1983, Begelman et al. 1984).

The transition/ambiguous objects among SBS galaxies are located mainly with the H II galaxies in the new diagrams with little spillover into the AGN region. This could lead essentially to three possibilities: (1) These objects are in fact H II/AGN composites, with an AGN component weaker than the emission component produced from ionization by stellar sources; (2). They are H II galaxies with a peculiar hot stellar population; (3) these galaxies just have different properties in their ionized regions. It is difficult to determine what makes these objects different from H II galaxies.

We have done the SBS sample identification with IRAS objects. 30% of UV galaxies and 30% of ELG are IRAS objects.

Acknowledgments. I would like to thank J. Stepanian for the observational material.

References

Arp, H. & Khachikian, E. Ye., 1974, Astrofiz., 11, 207, 1975.
Begelman, Blandford, & Rees, 1984, Rev.Mod.Phys. **6** 255.
Begelman, McKee, & Shields, 1983,APJ.,**271** 70.
Burenkov, A. N., 1991, Preprint of the SAO, No. 67.
Gyulzadian, M., 1995a, Astrofiz., **32** 501.
Gyulzadian, M., 1995b, IAU 171 386.
Markarian, B.,1967, Astrofiz., **3** 55.
Rola, C.S., Terlevich, E., and Terlevich, R.,1997, MNRAS, **289**, 419.
Sahakian, K. A. & Khachikian, E. Ye., 1975, Astrofiz., 11, 207.
Stepanian, J. et al., 1990, Astrofiz., **3** 89.
Stepanian, J., Lipovetsky, V., Erastova, L. Shapavalova, A., Gyulzadian, M., 1992a, Astrofiz. Issled., **35** 15.
Stepanian, J., Lipovetsky, V., Erastova, L., Gyulzadian, M. Izotov, Yu., Guseva, N., 1992b, Astrofiz. Issled., **35** 32.

Seven Samples of SBS Galaxies in Selected Fields

S. A. Hakopian and S. K. Balayan

Byurakan Astrophysical Observatory (BAO), Byurakan 378433, Armenia. E-mail: susanaha@bao.sci.am

Abstract. The results of follow-up spectroscopy completed for seven samples of SBS galaxies in selected fields are briefly presented.

It's well known that the only criterion for selection of the FBS objects (First Byurakan Survey; Markarian 1967) was the UV excess. Together with this, one more criterion has been used for selection of SBS objects (Second Byurakan Survey; Markarian et al. 1983), that is, the presence of emission lines on the low dispersion spectra of the plates of the Second Survey. Almost 1300 objects showing such features have been included in the sample of galaxies of the SBS.

Spectra of Markarian (FBS) galaxies, obtained with higher, intermediate resolution confirmed their supposed nature for the vast majority of them. In contrast, follow-up spectroscopy of SBS galaxies revealed a number of absorption objects, especially at magnitudes close to the limit of the plates at about 19.5^m.

We studied SBS galaxies in seven selected fields (Hakopian & Balayan 1998). To complete follow-up spectroscopy for these galaxies, spectral observations of the faint "tails" of the seven samples were carried out with ByuFosc spectrograph (Movsessian 2001) at the 2.6m telescope at the Byurakan Observatory of Armenia, with the Long Slit Spectrograph (LSS, at http://www.sao.ru/~gafan/devices/LSS) at the 6m telescope of the SAO of Russia. 3-D spectroscopy obtained for a number of galaxies with the Multi Pupil Fiber Spectrograph (MPFS, at http://www.sao.ru/~gafan/devices/MPFS) at the 6m has been used for redshift determination as well as for more detailed study (Hakopian & Balayan 2001 and references therein).

As a result, seven samples of SBS galaxies in the selected fields have finally been constructed to include only objects with confirmed emission spectra with AGN properties (Sy types, LINERS and of Composite behaviour) and with features of star formation activity in different stages, including BCDG and so on.

Table 1. Data on Selected Samples of SBS Galaxies

Selected field	I	II	III	IV	V	VI	VII
N (%AGN)	70(7%)	42(14%)	53(6%)	79(10%)	58(7%)	70(10%)	44(20%)
Mark gal	3 (1)	3 (1)	5 (0)	3 (0)	4 (0)	4 (0)	2 (1)

Current results are illustrated in Fig.1 and summarized in Table 1, the rows of which give: (1) order number of the selected SBS field (see Figure 1); (2)

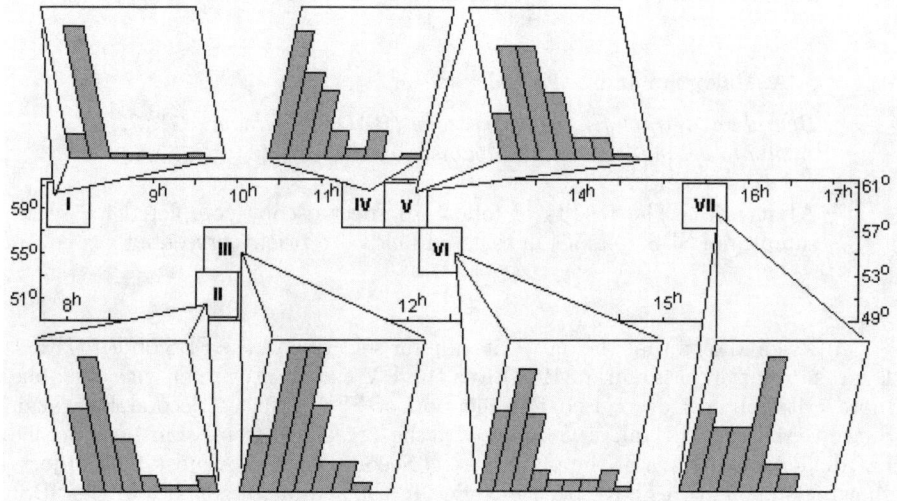

Figure 1. SBS sky region, location of the selected fields and corresponding redshift distribution of SBS galaxies in the fields.

number of galaxies in the sample with the percentage of AGNs in parentheses; (3) number of Markarian galaxies (included in the number given in the previous row) and number of Sy type among them in parentheses.

Figure 1 schematically gives the sky region of the Second Byurakan Survey and the location of the seven selected fields, each of 16 deg^2. Corresponding redshift distributions in these fields are presented by linked histograms. The range of z values on the x-axis $0 - 0.16$, the same in all seven cases, includes the numbers in the samples given in Table 1, except for one or two galaxies in each field with much greater redshifts.

References

Markarian, B.E. 1967, Astrofizika, 3, 55
Markarian, B.E., Lipovetsky, V.A. & Stepanian, J.A. 1983, Astrofizika, 19, 29
Hakopian, S. & Balayan, S. 1999, in "Active Galactic Nuclei and Related Phenomena", Proceedings of IAU Symp. 194, eds. Terzian Y., Weedman D. & Khachikian E., Astron. Soc. Pacific, Chelsea, Michigan, 162
Movsessian, T., Boulesteix, J., Gach, J.-L. et al. 2001, Baltic Astronomy, 9, 652
Hakopian, S. & Balayan, S. 2001, Astrophysics (in press)

Properties of the Low-z NELGs from the VPM Survey

Helmut Meusinger, Jens Brunzendorf

Thüringer Landessternwarte Tautenburg, D-07778 Tautenburg, Germany

Abstract. We discuss the properties of the low-redshift narrow-emission line galaxies detected in the framework of a variability-and-proper motion (VPM) search for AGNs. The VPM-NELG sample is obviously not a random selection of normal field galaxies.

1. Introduction

The variability-and-proper motion search (VPM) provides an unconventional, yet powerful technique for efficiently finding AGNs. Presently, we are performing a VPM-QSO survey in two fields, based upon a large number of fully-digitised Schmidt plates with a long time-baseline (see Meusinger et al, this conference). The search in the M 3 field was strictly confined to star-like objects. In the M 92 field, this constraint was relaxed allowing also for candidates with image profiles slightly different from star-like images. Hence, the spectroscopic follow-up observations in this field revealed a substantial number of low-redshift objects ($z < 0.3$), among them one QSO, five Seyfert 1s, 27 narrow-emission line galaxies (NELGs), and three early-type galaxies without emission lines (Meusinger & Brunzendorf 2001). The large fraction of NELGs is surprising.

2. Properties of the NELG Sample

All 27 NELGs were selected because of their high variability indices measured on the Schmidt plates. Redshifts of $z = 0.03 - 0.25$ were derived from the follow-up spectra taken with CAFOS at the 2.2 m telescope of the DSAZ[1] at Calar Alto, Spain. The distribution of the absolute B magnitudes is similar to those of galaxies with dwarf-Seyfert nuclei (Ho et al. 1997) or of galaxies selected for their compact nuclei (Sarajedini et al. 1999). The NELGs are blue ($\langle U - B \rangle \approx -0.3$ in the observer frame) and show strong Hα emission with equivalent widths of up to 1,000Å. Since the quality of the follow-up spectra was insufficient for a clear-cut discrimination between the principal ionisation sources (AGN versus young stars), spectra of higher resolution and higher signal-to-noise ratio were taken in a subsequent observation campaign. On the resulting diagnostic line ratio diagrams (Fig. 1a-c), a substantial fraction of these galaxies is located near the demarcation line between H II galaxies and AGNs.

---

[1]German-Spanish Astronomical Centre operated by the Max-Plank-Institute for Astronomy, Heidelberg, jointly with the Spanish National Commission for Astronomy

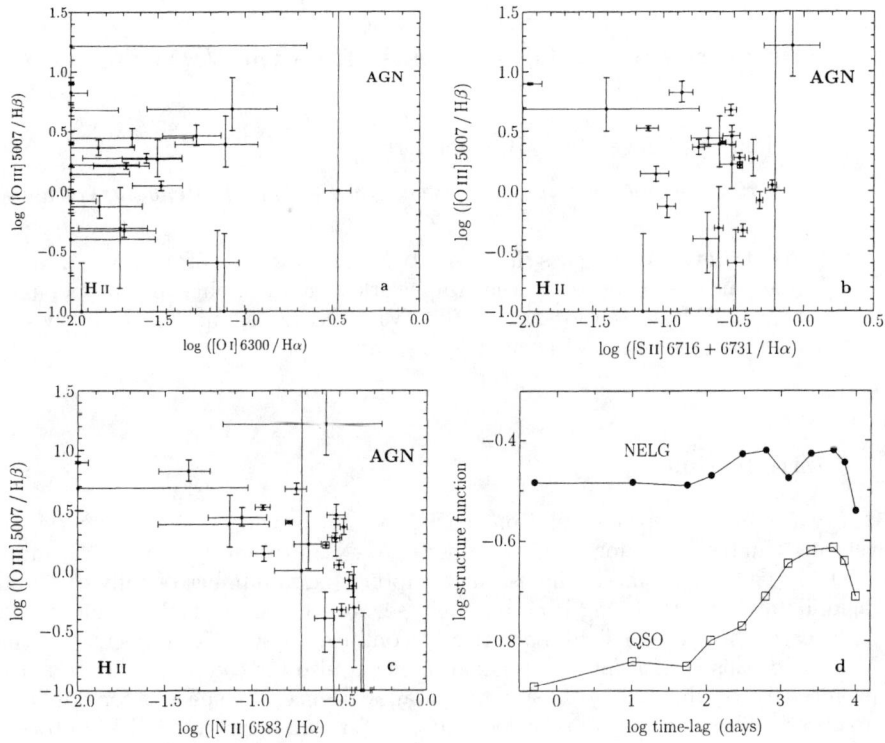

Figure 1. **a - c** Diagnostic line ratio diagrams of the VPM-NELGs. The demarcation curves were taken from Osterbrock (1989). **d** Sample-averaged structure function from the light-curves of NELGs (upper polygon) and QSOs (lower polygon) in one of the VPM fields.

The sample-averaged structure function (Fig. 1d) reflects the strong variability measured for the NELGs. The flatness of the structure function indicates a remarkable difference compared to the QSO sample from the same database: QSOs show pronounced long-term variability while NELGs do not. We are going to check the reality of this effect as well as the role of photometric errors for the measured variability by means of additional CCD time-series observations.

References

Ho, L.C., Filippenko, A.V., & Sargent, W.L.W. 1997, ApJ, 487, 568
Meusinger, H., & Brunzendorf, J. 2001, A&A, 374, 878
Osterbrock, D.E. 1989, Astrophysics of Gaseous Nebulae and Active Galactic Nuclei, Univ. Sci. Books, Mill Valley
Sarajedini, V.L., Green, R.F., Griffiths, R.E. *et al.* 1999, ApJ, 524, 746

The Blue Stellar Objects of the First Byurakan Survey

A.M. Mickaelian

Byurakan Astrophysical Observatory (BAO), Byurakan 378433, Armenia. E-mail: aregmick@bao.sci.am

Abstract. The Second part of the FBS is devoted to the discovery and study of blue stellar objects. It was carried out in an area of 4009 deg^2 and identified 1103 objects: hot subdwarfs and white dwarfs, cataclysmic variables, QSOs and Seyfert galaxies, etc. The discovery of new bright QSOs led to a re-evaluation of their surface density (0.012 deg^{-2}). Several interesting cataclysmic variables have been discovered as well.

1. The Second Part of the First Byurakan Survey

The second part of the First Byurakan Survey (FBS) was conducted in 1987-1996 for selection and further study of blue stellar objects (BSOs) on the basis of the FBS observational material (Mickaelian 2000). 278 FBS fields (4009 deg^2 surface area) have been inspected by eye with 7$^\times$ and 15$^\times$ lenses in the region $+33° < \delta < +45°$ and $+61° < \delta < +90°$ (at $|b| > 15°$). The main purpose of this work was discovery of new bright QSOs, Seyferts, other compact galaxies, cataclysmic variables (CV), white dwarfs (WD), hot subdwarfs, HBB stars, and other peculiar stellar objects. 1103 objects have been selected, including 716 new BSOs. In all, 11 lists have been published and the FBS BSOs catalog is available at the CDS (Abrahamian et al. 1999). The completeness of the sample for objects with B<16.5^m and U-B<-0.5 has been estimated as about 67%. Subsamples of candidate QSOs, WDs, CVs, and other objects have been constructed for further detailed studies.

2. Bright QSOs and Seyfert Galaxies

New QSOs have been discovered in the FBS. 15 have been found in the subarea of the FBS with 2250 deg^2 in common with the Palomar-Green Survey ($|b| > 30°$), substantially changing the existing numbers. In this area, we have constructed the most complete sample for bright QSOs (26 QSOs) (Mickaelian et al. 2001a). The surface density of bright QSOs (B<16.16^m) is estimated as 0.012 deg^{-2} and the completeness of the Bright Quasar Survey (BQS) (Schmidt & Green 1983) is re-estimated to be 53%. An interesting bright (14^m) very luminous (-24.6^m) NLS1 galaxy FBS 0732+396 has been found at z=0.118 with strong FeII emission lines.

In all, there are 108 bright (B<17.0^m) AGNs in the FBS subarea with $|b| > 30°$, including 40 objects discovered in the FBS (Véron-Cetty & Véron 2001). FBS QSOs and Seyferts have redshifts 0.063-2.00 (QSOs with larger redshifts are red enough that it is difficult to distinguish them in the FBS spectra) and absolute B magnitudes in the range $-20.3^m \div -29.9^m$. NVSS/USNO and ROSAT/USNO cross-correlations, and checking the low-dispersion spectra in FBS with further follow-up spectroscopy identified new QSOs on the FBS plates which had not been selected before, including a new ROSAT NLS1 galaxy RXS J170535+3340 with z=0.118 having unusual line ratios.

3. White Dwarfs

White dwarfs have broad absorption lines and are easily recognized on the FBS plates. Moreover, FBS spectra allow selection of all WDs (not only blue): this task may be completed in the very near future, after the digitization of the FBS. Some subtypes are rather interesting: pulsating white dwarfs (ZZ Ceti stars), magnetic WDs, polars, planetary nebula nuclei (DO stars, PG 1159 type objects), etc.: several objects of these types have been found and dozens are expected to be found. Polarimetric observations revealed 0.7-6.0% linear polarization for 5 objects, including FBS 1704+347 (a possible polar), and FBS 1815+381 (a variable magnetic WD). The total number of WDs is estimated to be 270 in the whole sample (24% of the whole sample). Lists of candidate pulsating and magnetic WDs have been compiled and a detailed study of the most interesting objects has been conducted.

4. Cataclysmic Variables

Seven cataclysmic variables have been revealed already from the FBS sample. Two new CVs, FBS 0019+348 and FBS 0306+333, have been observed with the Observatoire de Haute-Provence (OHP) 1.93m telescope in France (in collaboration with M.-P.Véron-Cetty and P.Véron).

A new bright (V=12.6) cataclysmic variable (Mickaelian et al. 2001b), RXS J16437+3402, was found by cross-correlation of ROSAT/USNO objects and further inspection of the FBS spectra. Spectroscopic and photometric observations show it to be a novalike cataclysmic variable of the SW Sex subclass. Its spectroscopic period is within the period "gap" for such objects. More observations are needed to describe the physics of this object. The total number of CVs in the FBS sample is estimated to be 35 (3% of the whole sample). More objects will be found in the total FBS area after the digitization of the survey.

References

Abrahamian, H.V. et al. 1999, at http://vizier.u-strasbg.fr/
Mickaelian, A.M. 2000, AATr, 18, 557
Mickaelian, A.M. et al. 2001a, Ap, 44, 14
Mickaelian, A.M. et al. 2001b, astro-ph/0108377, A&A (in press)
Schmidt, M., & Green, R.F. 1983, ApJ, 269, 352
Véron-Cetty, M.-P. & Véron, 2001, A&A, 374, 92

The New BL Lac Candidates from the FBS

O. Kh. Torosyan

BAO,Armenia 378433; E-mail: ofelia@bao.sci.am

Abstract. Six new possible BL Lac candidates from the FBS are presented.

1. Introduction

The optical spectrum of BL Lac objects is the most distinguishable characteristic for them. In spite of that, there are only 10 objects classified as BL Lacs from optical surveys. Most of these objects have been selected from radio or X-ray surveys and are called, respectively, RBLs and XBLs. The number of BL Lac's has now increased by tens or hundreds on account of X-ray sources (Padovani and Giommi, 1995; Kock et al., 1996; Wolter et al., 1996; etc.).

Some Markarian galaxies have characteristics of BL Lac's: the variability and nonthermal radiation. Three of them (Mkn 180, Mkn 421, Mkn 501) are the only known Lacertids from the Markarian galaxies today. They are the most famous RBLs and XBLs simultaneously. Mkn 421 and Mkn 501 are the first extragalactic objects detected at TeV energies by ground-based observations (there are only 4 other such objects now, and all of them are BL Lacs); and there isn't any foundation to think that they are the exceptional ones among the 1500 Mkn galaxies.

The completeness of these objects was estimated by the well known relation (Fig.1): $\lg N(m_v) = 0.6 m_v + const$.

The slope of this relation is about 0.57 for the relatively bright objects ($m_v < 16^m$) and only 0.3 for the objects with $16^m < m_v < 17^m$ in the regions of the FBS survey.

2. Selection Criteria

The absence of emission lines in the spectra of the Markarian galaxies with starlike nuclei and E or S0 morphological typeswith may be the common selection criteria for possible candidate BL Lacs: almost all of the 34 RBLs from the 1 Jy sample (Stickel et al., 1993); more then 94% of BL Lac host galaxies observed by the 3.6m CFHT (Wurtz et al., 1996); more then 92% of BL Lac host galaxies observed by HST (Urry et al., 2000) are galaxies with E or S0 morphology. Just by these principal criteria the possible candidate BL Lacs from the catalogue of Markarian galaxies have been selected (CCAO, 1989):

• They are classified in the catalogue (CCAO, 1989) as the galaxies without emission lines (information on the emission is absent or they are specified as

absorption galaxies as well);
- if there is information, the morphologies of the host galaxies are necessarily elliptical (E) or lenticular(S0), or any information is absent;
- the nuclei of the galaxies are classified (in Byurakan classification) as starlike (s) or semi-starlike objects (sd, ds);
- the lower bound of absolute magnitude is taken as $M_v = -18^m$ (the lower bound for the list of BL Lac objects);
- the values of visual angular sizes of the galaxies are conventionally taken as $D_{op} \geq 10''$

3. Results

There are 224 objects in the catalogue of 1500 Markarian galaxies (CCAO, 1989) without emission lines. 184 of them have star-like or semi-star-like nuclei. and are possible candidate BL Lacs. Information about IR emission is absent for almost all of them (180 of 184). In contrast, all of the Sy galaxies have IR emission above the IRAS threshold. Only Mkn 501 has IR emission above the IRAS threshold of the three well-known FBS Lacertids. For nine of them, the radio fluxes are more than 100 mJy for at least one of the four wavelengths in the 1995 catalogue of the radio survey of Markarian galaxies (Bicay, Kojoian et al., 1995) and may be classified as RBLs (Table 1). For 6 of these objects, the average radio spectral indexes are less then 0.6, which is one of the necessary characteristics of BL Lacs (Stocke, et al., 1991).

In Table 1 are presented, respectively, the redshifts, the visual magnitudes, the absolute magnitudes, the spectral types of the nuclei by Markarian classification, the galaxies morphological types and the presence of emission (e) or absorption (a) (CCAO, 1989), the radio fluxes at 1.415 GHz and 4.755 GHz in mJy (Bicay, Kojoian, et al., 1995) as well as the radio spectral indices (Sanamian, Kandalian, 1980).

Table 1. The Possible New RBL Candidates from Mkn Galaxies

Mkn No.	Z	m_v	M_v	Sp (core)	Mph Host	Sp Type	$F_{1.415}$ (mJy)	$F_{4.755}$ (mJy)	α_R
11	0.0133	14.4	-19.7	s3	-	e?	n/a	288	0.19
*180	0.0458	15.5	-21.1	s1e	-	BL	n/a	227	0.18
*421	0.0308	13.1	-22.6	s1e	E1	BL	...	584	0.15
422	0.0316	14.8	-21.0	sd3e:	-	...	758	-	-
452	0.0174	14.0	-20.4	sd3	S0	a	...	≤ 268	-
*501	0.0337	13.7	-22.3	s2e:	S0	BL	n/a	1199	0.16
514	0.0470	14.8	-21.1	s2	E	a	228	157	0.60
1007	0.0180	14.2	-20.4	s3e:	S0	a	...	172	0.44
1077	0.0229	16.0	-19.8	ds2	-	n/a	≤ 378	-	-

Acknowledgments. I am grateful to E. Ye. Khachikian, V. O. Chavushian and R. A. Kandalian for consultation and useful advice.

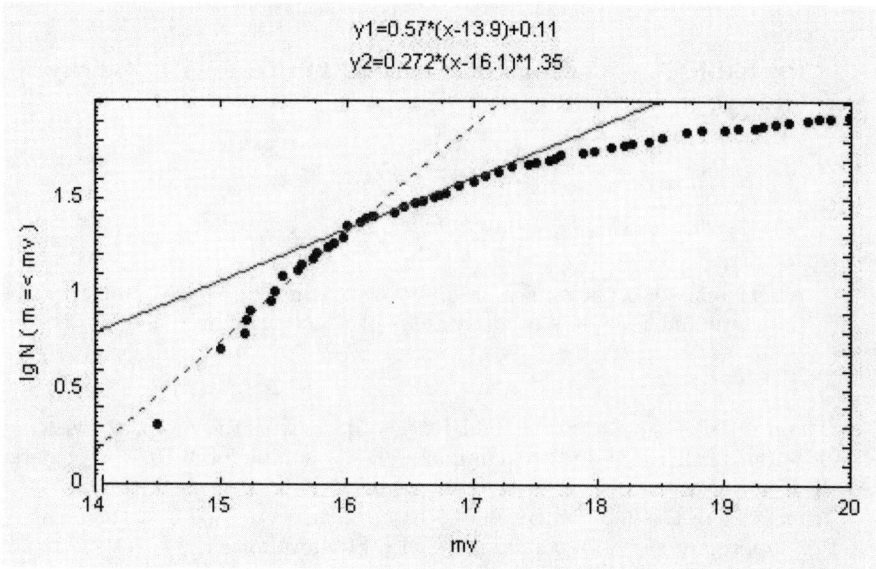

Figure 1. The relation $\lg N(m_v) - m_v$ for BL Lacs in the regions of the FBS. y_1 - dotted line; y_2 - solid line

References

Bicay, M. et al.,1995 AJ, 98, N2
FBS, 1989, CCAO, 62,3
Kock, A. et al., 1996, A&A, 307, 745
Lipovetsky, V., 1987, CCAO, 53, 47
Mazarella, J., 1986, ApJS, 62, 751
Padovani, P., Giommi, 1995, MNRAS, 277, 1477
Petrosian, A. R., Turato, M., 1986, A&A, 163, 26
Sanamian, V. A., Kandalian, R. A., 1980, Afz, 16, 3, 425
Stickel, M., Frid, J. W. and Kuhr, H., 1993, A&AS, 98, 39
Stocke, et al., 1998, ApJS, 76, 813
Torosyan, O. Kh.; -in prep. (Afz)
Urry, C. M., et al., 2000, AJ, 532, 816
Wurtz R., et al., 1996, ApJS, 103, 109

Probable Associations of BL LAC Objects with Zwicky and Abell Clusters

O. Kh. Torosyan

BAO,Armenia 378433; E-mail:ofelia@bao.sci.am

Abstract. BL Lac's may be found everywhere in space. But they are the dominant members in all systems of which they are part.

The probable associations of 266 BL Lac objects with the Zwicky (Zwicky et al. 1961-1968; Hill 1983) and the complete all-sky sample of 4076 Abell cluster (Abell, Corwin, & Olowin 1989) are discussed.

Almost all of the 266 real BL Lacs (the list of objects used in this work) are giant or supergiant radio-loud galaxies with E,S0 morphology, which usually are the central members of clusters of galaxies. There are many observational results (Xie et al. 1992; Wurtz et al. 1993; Falomo, Pesce, & Treves 1993a,b; Pesce, Falomo, & Treves 1995) to show that BL Lacs really are the central dominant members of clusters of galaxies with Abell richness 1,2. But there are facts also (Stickel, Fried, & Kuhr 1993; Ledlow & Owen 1995) which show that BL Lacs are the dominant members in small groups of galaxies, and are situated at the centers of poor (Abell richness class 0) clusters.

The primary neighborhood was found for each of the BL Lacs with respect to the centers of the clusters. The goal was to ascertain the fraction of BL Lacs in the central parts of Zwicky and Abell clusters.

The observed $(n(Zw.cont.))$ and the expected numbers with standard deviations $(N(Zw.c.)_{exp} \pm \sigma)$ of the projected BL Lac's in the Zwicky cluster contours, the observed $(n(1Ra))$ and the expected numbers of projected BL Lac's in the centers of Abell clusters within one Abell radius $(N(1Ra)_{exp} \pm \sigma)$ as well as in the field are presented in Table 1.

The observed (2 : 5 : 8) and the expected (2.4 : 4.95 : 7.65) distributions of BL Lacs are the same within the 95% confidance level by the χ^2 test, in the regions with respective distances from the centers of Abell clusters : $(R_1 \leq \frac{1}{3}R_a)$, $(R_2 = \frac{1}{3}R_a - \frac{2}{3}R_a)$, $(R_3 = \frac{2}{3}R_a - 1R_a)$.

Table 1. The BL Lac's in the Zwicky and Abell clusters

BL Lac	$n(Zw.cont.)$	$N(Zw.cont.) \pm \sigma$	$n(1Ra)$	$N(1Ra) \pm \sigma$
in the clust.	11	10.07 ± 2.42	1	0.96 ± 0.95
in the field	13	13.98 ± 2.41	22	23.04 ± 0.95

The maximum observed redshift for BL Lacs is 1.8; about 30% of them have redshifts $Z \geq 0.5$. And it is interesting to examine the distribution of BL Lacs with redshift from the perspective of their large-scale structure in space.

Table 2 presents the results of the known and expected counts of BL Lacs in intervals of redshift (ΔZ) and declination (δ): N_1, N_2, are the observed and (1), (2), (3) are the expected counts of BL Lacs with their standard deviations in the respective intervals of redshift (ΔZ) and declination (δ), N_{Zw} is the counts of the Lacertids that were projected in the Zwicky cluster contours (Zwicky et al., 1961-68).

Table 2. The Distribution of BL Lacs by Redshift and Declination

BL Lac	$\delta \leq -3.0°$		$\delta \geq -3.0°$		in the Zw. cont.	
ΔZ	N_1	$N_{1\,\exp} \pm \sigma$	N_2	$N_{2\,\exp} \pm \sigma$	N_{zw}	$N_{zw\exp} \pm \sigma$
< 0.3	3	9.1 ± 1.1	87	139.5 ± 6.9	21	34.8 ± 4.4
0.3 ÷ 0.5	6	18.3 ± 1.6	15	24.0 ± 4.3	6	9.9 ± 1.4
> 0.5	12	36.6 ± 2.2	24	38.5 ± 5.3	2	3.3 ± 0.8

It may be concluded that the BL Lacs with $Z \leq 0.3$ (the average distances for the Abell and Zwicky clusters): a) may be discovered not only in the centers of clusters as was expected. b) they are the dominant members in all the systems wherever they are found: clusters of galaxies or small groups of galaxies, at different distances from the centers of the clusters, as well as in the field.

References

Abell, G. O, Corwin, H. G., Olowin, R. P., 1989, ApJS, 70, 1, (Machine-Readable Version)
Falomo, R., Pesce, J. E. & Treves, A., 1993a, AJ, 105, 2031
Falomo, R., Pesce, J. E. & Treves, A., 1993b, ApJ, 411, L63
Hill, R. S., Document. for the Machine-Readable Version of the Cat. of Galaxies and of Clusters of Galaxies, S A S C, 1983, S S D-T-1-5069-022-82
Ledlow, M. J. & Owen, F N., 1995 AJ, 110
Pesce, J. E., Falomo, R. & Treves, A., 1995, ApJ, 438, L9
Stickel, M., Fried, J. W., & Kuhr, H.,1993, A&AS, 98, 393
Wurtz, R., Ellingson, E., Stocke, J. T. & Yee, H. K. C.,1993, AJ, 106, 869
Xie, G. Z., et al., 1992, ApJS, 80, 683
Zwicky, F., et al., Cat. of Gal. and Clust. of Gal.(CG CG),1961-68, vv1,2,3,4,5,6.

Part 2
Infrared and Submillimeter Surveys for AGN

Lutz Wisotzki, John Hutchings, Sylvain Veilleux

Zeljko Ivezic and daughter

Spectroscopic Diagnostics for AGNs

Sylvain Veilleux

Department of Astronomy, University of Maryland, College Park, MD 20742

Abstract. A review of the spectroscopic tools needed to characterize AGNs is presented. This review focusses on ultraviolet, optical and infrared emission-line diagnostics specifically designed to help differentiate AGNs from starburst-dominated galaxies. The strengths and weaknesses of these methods are discussed in the context of on-going and future AGN surveys.

1. Introduction

The first decade of the 21st Century promises to become the Golden Age of extragalactic astronomy. The 2dF and Sloan Digital Sky Surveys have already made significant contributions to our knowledge of the extragalactic universe. Ongoing and planned wide-field (pencil-beam) imaging and spectroscopic surveys with 4m (8m)-class telescopes from the ground and in space (e.g., SIRTF, SOFIA, Herschel, NGST) will nicely complement these large-scale surveys and should go a long way to answer some of the most fundamental questions in extragalactic astronomy: How do galaxies form? How do they evolve? How do supermassive black holes fit in this picture of galaxy formation? Which objects are the main contributors to the overall energy budget of the universe? To properly answer these questions, one will need to differentiate objects powered by nuclear fusion in stars (i.e. normal and starburst galaxies) from objects powered by mass accretion onto supermassive black holes (quasars and AGNs). A wide variety of diagnostic tools have been used in the past for this purpose with different degree of success.

Due to space limitations, the present discussion focusses on *emission-line* diagnostics. The fundamental principles behind these diagnostics are reviewed in §2. Next, the main diagnostic tools available in the ultraviolet, optical, and infrared domains are described in §3, §4, and §5, respectively. A table listing the main diagnostic lines is given in each of these sections. Additional factors which may complicate the use of these tools are discussed in §6. A summary is given in §7 along with an outlook on the future. Note that this review is not meant to be exhaustive; it is meant to emphasize the practical aspects of starburst/AGN spectral classification. Readers who are looking for a more detailed discussion of the physics behind these diagnostic tools should refer to the original papers listed in the text.

2. Basic Principles

Activity driven by mass accretion onto supermassive black holes differs in many ways from star-formation activity. The thermal and non-thermal processes associated with the accretion disk and its surroundings (e.g., corona) are at the origin of the "hard" ionizing continuum detected in quasars and AGNs (e.g., Krolik 1999). Material in the vicinity of the nucleus will bear the imprint of this strong radiation field. The deep gravitational potential at the center of these galaxies allows the presence of high-density ($\gtrsim 10^9$ cm^{-3}), high-velocity ($\gtrsim$ 2000 km s^{-1}) gas clouds in the inner parsec of quasars and AGNs. This so-called broad-line region or BLR is a powerful diagnostic of nuclear activity in galaxies. The main signatures of the BLRs are broad recombination lines which are unaffected by the effects of collisional de-excitation at high densities. Two general methods have been used in the past to detect BLRs in galaxies: direct spectroscopy and spectropolarimetry. This last method relies on the presence of dust or electrons ("mirrors") to scatter the BLR signature towards the line of sight (e.g., Antonucci 1993). Direct spectroscopy searches for the presence of the broad recombination lines at wavelengths where the effects of dust extinction are reduced. As shown in Table 1 for representative Galactic extinction (see, e.g., Cardelli, Clayton, & Mathis 1989; Draine & Lee 1984; Draine 1989; Lutz et al. 1996; Lutz 1999), great increase in sensitivity can in principle be obtained by observing at longer wavelengths.

Table 1.   Galactic Dust Extinction and Column Densities

λ	$\tau(\lambda)/\tau(H\alpha)$	N_H(cm^{-2}) @ $\tau(\lambda) = 1$
Lyα 1216 Å	2.0 – 4.5	$0.5 - 1.0 \times 10^{21}$
V band 5500 Å	1.2	1.7×10^{21}
Hα 6563 Å	1.0	2.2×10^{21}
J band 1.25 μm	1/3	6.1×10^{21}
H band 1.65 μm	1/4.5	9.8×10^{21}
K band 2.2 μm	1/7	1.6×10^{22}
L band 3.4 μm	1/15	3.4×10^{22}
M band 5.0 μm	1/30	6.4×10^{22}
N band 10 μm	1/15	3.2×10^{22}
12 μm	1/30	6.2×10^{22}
25 μm	1/60	1.3×10^{23}
60 μm	1/400	8.6×10^{23}
100 μm	1/700	1.5×10^{24}

In highly obscured objects with $N_H \gtrsim 10^{24}$ cm^{-2}, direct detection of the BLRs becomes very difficult and one has to rely on spectropolarimetry to search for the presence of a BLR. The obscuring screen may not be opaque in all directions, however. The ionizing radiation field may be able to escape in certain directions and ionize the surrounding material on scales beyond the obscuring material. Distributed in the shallower portion of the gravitational potential ($\sim$ 0.1 – 1 kpc), this "narrow-line region" or NLR is another excellent probe of nuclear activity. The ionizing spectra of all but the hottest O stars cut off near the He II edge (54.4 eV; Dopita et al. 1995). In contrast, the ionizing spectrum of AGNs contains a relatively large fraction of high-energy photons (e.g., Elvis et al. 1994). Optically thick gas clouds ionized by the hard continuum of AGNs will present a stratified ionization structure with (1) a highly ionized inner face (closest to the AGN), (2) a large partially zone with characteristic fraction of ionized hydrogen $H^+/H \sim 0.2 - 0.4$ produced by the deposition of keV X-rays (recall that the absorption cross sections of H^0, He^0, and all other ions decrease rapidly with increasing energy; Osterbrock 1989), and (3) a neutral zone facing away from the AGN. The fast free electrons in the partly ionized zone will have a positive effect on the strengths of low-ionization lines produced by collisional effects, while the highly ionized conditions in the inner face will favor the production of emission lines from ions with high ionization potentials (e.g., Ferland & Netzer 1983; Ferland & Osterbrock 1986, 1987; Binette, Wilson, & Storchi-Bergmann 1996).

Based on these physical principles, one should choose narrow emission line diagnostics following ten basic rules or "Commandments" (a reminder of the 1700th anniversary of the adoption of Christianity as a national religion in Armenia):

1. Thou shalt use lines which emphasize the differences between H II regions and AGNs; i.e., use high-ionization lines or low-ionization lines produced in the partially ionized zone.

2. Thou shalt use strong lines which are easy to measure in typical spectra.

3. Thou shalt avoid lines which are badly blended with other emission or absorption line features.

4. Thou shalt use lines with small wavelength separation to minimize sensitivity to reddening.

5. Thou shalt use line ratios from the same elements or involving hydrogen recombination lines to eliminate or reduce abundance dependence.

6. Thou shalt avoid lines from Mg, Si, Ca, Fe – depleted onto dust grains.

7. Thou shalt use lines easily accessible to current UV/optical/IR detectors.

8. Thou shalt avoid lines affected by strong stellar absorption features.

9. Thou shalt avoid lines affected by strong atmospheric features.

10. Thou shalt use lines at long wavelengths to reduce the effects of dust extinction.

3. Ultraviolet Emission-Line Diagnostics

When possible, ultraviolet diagnostic tools should be avoided because of their sensitivity to dust extinctions (see the Tenth Commandment and Table 1). However, investigators of the high-redshift universe often have very little choice but to study this region of the electromagnetic spectrum. The ultraviolet domain is potentially a rich source of diagnostic lines. The main emission lines are listed in Table 2. Among the most useful diagnostics to discriminate between AGNs and starbursts are the N V $\lambda1240$/He II $\lambda4686$, N V $\lambda1240$/C IV $\lambda1549$, N V $\lambda1240$/Lyα, and C IV $\lambda1549$/Lyα emission-line ratios. As shown in Figure 1, these ratios are sensitive functions of the shape of the ionizing continuum (harder spectra provide more heating per photoionization, therefore increasing the temperature). These line ratios have been used extensively in studies of high-z quasars (e.g., Hamann & Ferland 1999) and radio galaxies (e.g., Röttgering et al. 1997; Villar-Martin et al. 1996, 1999), and the analysis of low-z AGNs/LINERs (e.g., Ho et al. 1996; Barth et al. 1996, 1997; Maoz et al. 1998; Nicholson et al. 1998) and starburst galaxies (e.g., Robert, Leitherer, & Heckman 1993).

Table 2.   Ultraviolet Emission-Line Diagnostics

Low-to-Moderate Ionization Lines		High-Ionization Lines	
Line	χ(eV)	Line	χ(eV)
C III 977 Å	24.4	O VI 1032, 1038 Å	114
N III 991, 1750 Å	29.6	N V 1240 Å	77.4
Lyβ 1026 Å, Lyα 1216 Å	13.6	O IV] 1407 Å	54.9
Si IV 1394, 1403 Å	33.5	N IV] 1488 Å	47.4
O III] 1663 Å	35.1	C IV 1549 Å	47.9
N III] 1750 Å	29.6	He II 1085, 1640 Å	54.4
Si III 1895 Å	16.3		
C III] 1909 Å	24.4		
Fe II 2080, 2500, 3300 Å	7.9		
[O III] 2322 Å	35.1		
C II] 2326 Å	11.3		
Si II 2336 Å	8.2		
Mg II 2798 Å	7.6		

4. Optical Emission-Line Diagnostics

The excellent quantum efficiency of current CCDs combined with the high transparency and low emissivity of the Earth's atmosphere at optical wavelengths make optical spectroscopy the easiest way to identify AGNs. Table 3 lists the strongest diagnostic lines between 3000 Å and 1 μm. Classification schemes involving several line ratios which take full advantage of the physical distinction between the two types of objects and minimize the effects of reddening correction and errors in the flux calibration have proven very useful for the identification of galaxies as AGNs or starbursts (e.g., Phillips, Baldwin, & Terlevich 1981; Veilleux & Osterbrock 1987; Osterbrock, Tran, & Veilleux 1992; Dopita et al. 2000). Examples of emission-line diagrams are shown in Figure 2 for ultraluminous infrared galaxies from the 1-Jy sample (ULIGs; these are IRAS galaxies with infrared luminosities between 8 and 1000 μm larger than or equal

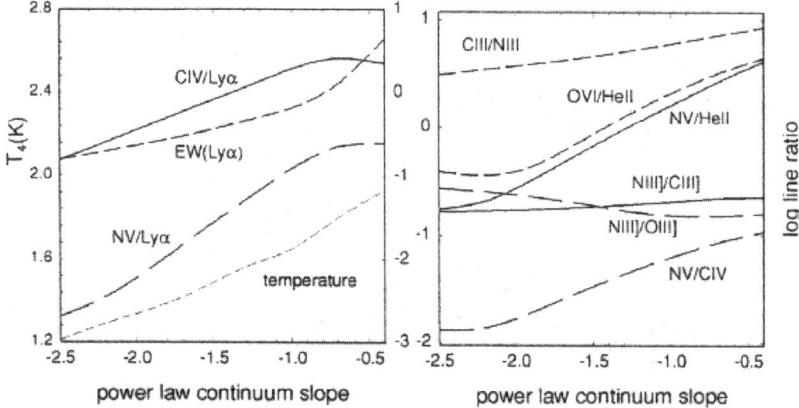

Figure 1. Predicted UV line flux ratios, gas temperatures, and dimensionless equivalent widths in Lyα for clouds photoionized by different power-law spectra (ν^α). The UV-to-X-ray slopes of QSOs are roughly consistent with $\alpha \approx -1.5$. From Hamann & Ferland (1999).

to 10^{12} L$_\odot$; Kim 1995). The results from this classification indicates that the fraction of Seyfert nuclei increases from $\sim 5\%$ at $\log[L_{\rm IR}/L_\odot] = 10 - 11$, to $\sim 50\%$ at $\log[L_{\rm IR}/L_\odot] > 12.3$ (Veilleux et al. 1995, Kim, Veilleux, & Sanders 1998; Veilleux, Kim, & Sanders 1999a).

5. Near-Infrared Emission-Line Diagnostics

Infrared-bright galaxies such as those discussed in the previous section are hosts to large quantities of molecular gas and dust (e.g., Solomon et al. 1997). The optical line ratios measured in these objects are undoubtedly affected by dust extinction. It is therefore important to also observe these objects at longer wavelengths to verify the results derived from the optical spectra. Near-infrared spectroscopy has had success finding obscured BLRs in several ULIGs (e.g., Hines 1991; Veilleux et al. 1997b, 1999b; spectropolarimetry has lent support to some of these findings: Hines 1991; Hough et al. 1991; Hines & Wills 1993; Hines et al. 1995; Young et al. 1993). This technique has also proven useful in the study of highly reddened BLRs in intermediate Seyferts (1.8's and 1.9's; Goodrich 1990; Rix et al. 1990) and in optically classified Seyfert 2 and radio galaxies (e.g., Blanco, Ward, & Wright 1990; Goodrich, Veilleux, & Hill 1994; Ruiz, Rieke, & Schmidt 1994; Hill, Goodrich, DePoy 1996; Veilleux, Goodrich, & Hill 1997a).

The line of choice for ground-based near-infrared searches of obscured BLRs in nearby galaxies is Paα at 1.8751 μm (Table 4). Under Case B recombination (Osterbrock 1989), this line is one-third the strength of Hα and is *twelve* times stronger than Brγ λ2.1655, the next best diagnostic line (e.g., Goldader et al. 1995). This huge gain in intensity more than compensates the slightly larger

Table 3. Optical Emission-Line Diagnostics

Low-to-Moderate Ionization Lines		High-Ionization Lines	
Line	χ(eV)	Line	χ(eV)
[O II] 3727, 7325 Å	13.6	[Ne V] 3346, 3426 Å	97.1
[Ne III] 3869, 3968 Å	41.0	[Fe V] 3840, 3893, 4071 Å	54.8
[O III] 4363, 5007 Å	35.1	[Fe VII] 3588, 3760, 4071, 5721, 6087 Å	99.0
Fe II 4500, 5190, 5300 Å	7.9	He II 4686 Å	54.4
Hβ 4861 Å, Hα 6563 Å	13.6	[Fe XIV] 5303 Å	344
He I 5876, 7065 Å	24.6	[Fe X] 6375 Å	235
[O I] 6300, 6363 Å	0.0	[Fe XI] 7892 Å	262
[N II] 5755, 6548, 6583 Å	14.5		
[S II] 6716, 6731 Å	10.4		
[S III] 6312, 9069, 9531 Å	23.3		

optical depth due to extinction at the shorter wavelength of Paα (see 4th column in Table 4).

Another important AGN diagnostic line in the K band is [Si VI] λ1.962. The existence of five-times ionized silicon ions requires energies larger than 167 eV (Table 4). This forbidden line has been detected in a number of optically selected Seyfert 2 galaxies with a strength comparable to that of [Fe VII] λ6087 (roughly a tenth the strength of Hβ), as expected from photoionization by a AGN power-law continuum (Oliva & Moorwood 1990; Greenhouse et al. 1993; Marconi et al. 1994; Oliva et al. 1994; Thompson 1995, 1996). Near-infrared spectroscopic surveys of ULIGs have confirmed the optical results: the fraction of objects with genuine AGNs (with a BLR or strong [Si VI] λ1.962 feature) is at least $\sim$ 20 – 25%, but reaches $\sim$ 35 – 50% for those objects with log[$L_{ir}/L_{\odot}$] > 12.3. Nevertheless, the presence of an AGN in ULIGs does not necessarily imply that AGN activity is the dominant source of energy in these objects. A more detailed look at the AGNs in these ULIGs is needed to answer this question.

Table 4. Hydrogen Recombination Lines and some High Ionization Lines in the Near-Infrared

Hydrogen Recombination Lines				High-Ionization Lines	
Line	$\lambda(\mu m)$	$F/F_{H\alpha}$	$A_\lambda/A_{H\alpha}$	Line	χ(eV)
Hβ	0.4861	1.00	1.48	[S IX] 1.252 μm	328
Hα	0.6563	2.85	1.00	[Si X] 1.430 μm	351
Paγ	1.0938	0.090	0.45	[Si XI] 1.932 μm	401
Paβ	1.2818	0.162	0.34	[Si VI] 1.962 μm	167
Paα	1.8751	0.332	0.18	[Ca VIII] 2.321 μm	128
Brγ	2.1655	0.0275	0.14	[Si VII] 2.483 μm	205
Brα	4.0512	0.0779	0.05	[Si IX] 3.935 μm	303

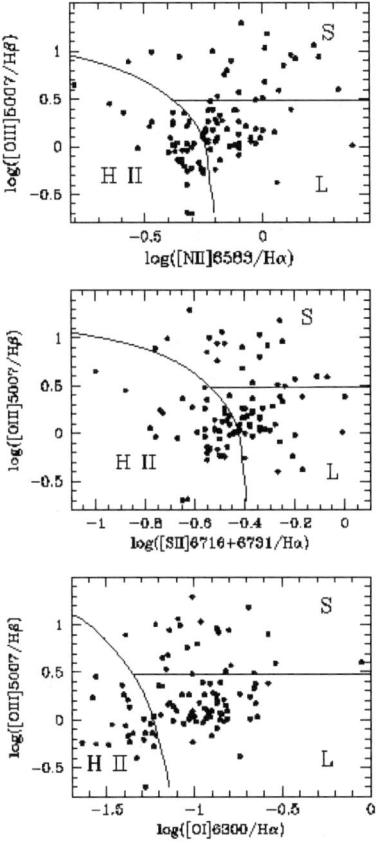

Figure 2. Examples of optical line ratio diagrams used for the classification of ultraluminous infrared galaxies. From Veilleux et al. (1999a).

A strong linear correlation has long been known to exist between the continuum (or, equivalently, bolometric) luminosities of broad-line AGN and their emission-line luminosities (e.g., Yee 1980; Shuder 1981; Osterbrock 1989). This correlation has often been used to argue that the broad-line regions in AGNs are photoionized by the nuclear continuum. If this is the case, the broad-line–to–bolometric luminosity ratio is a measure of the covering factor of the BLR (e.g., Osterbrock 1989). This correlation can be used to estimate the importance of the AGN in powering ultraluminous infrared galaxies (Veilleux et al. 1997b, 1999b). In ULIGs powered uniquely by an AGN, we expect the broad-line luminosities to fall along the correlation for AGNs. Any contribution from a starburst will increase the bolometric luminosity of the ULIG without a corresponding increase in the broad-line luminosity. Starburst-dominated ULIGs are therefore expected to fall below the "pure-AGN" correlation traced by the optical quasars in a diagram of $L_{H\beta}(BLR)$ plotted as a function of L_{bol}. The

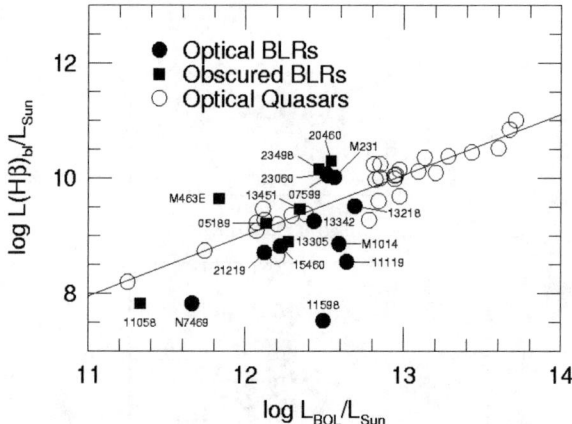

Figure 3. Dominant energy source of ultraluminous infrared galaxies based on their broad-line luminosities. The solid line is the best fit for the optical quasars. From Veilleux et al. (1999b).

data of ULIGs with optical and obscured BLRs are shown in Figure 3. A discussion of the methods and assumptions which were used to create this figure is presented in Veilleux et al. (1999a). Figure 3 strongly suggests that most ($\sim$ 80%) of the ULIGs with optical or near-infrared BLRs in the 1-Jy sample are powered predominantly by the quasar rather than by a powerful starburst. In other words, *the detection of an optical or near-infrared BLR in a ULIG (about 20% of the total 1-Jy sample) appears to be an excellent sign that the AGN is the dominant energy source in that ULIG.*

6. Mid-to-Far Infrared Emission-Line Diagnostics

The mid-to-far infrared region has long been known to be a rich source of emission-line diagnostics in active and starburst galaxies (e.g., Watson et al. 1984; Roche et al. 1984; Aitken & Roche 1985; Crawford et al. 1985; Lugten et al. 1986; Duffy et al. 1987; Roche et al. 1984, 1991, Spinoglio & Malkan 1992; Voit 1992). With the Infrared Space Observatory (ISO) much progress has been made in this area of research in recent years (see review by Genzel & Cesarsky 2000). One of the most important applications of ISO spectroscopy has been its use as a tool to distinguish between star formation and AGN activity in obscured environments. High ionization fine structure lines are strong in the NLR of AGNs but very weak in starbursts (Table 5). Fine-structure lines have smaller excitation energies than their optical counterparts, so they are less temperature sensitive and less model dependent. Unfortunately, they are also fainter than their optical counterparts, and are therefore difficult to detect even in genuine, optically-selected AGNs (e.g., Genzel et al. 1998).

Table 5. Strongest Fine-Structure Lines expected from AGNs

Line	χ(eV)
[Ar III] 9 μm	27.6
[S IV] 10.5 μm	34.8
[Ne II] 12.8 μm	21.6
[Ne V] 14.3, 24.2 μm	97.1
[Ne III] 15.6, 36.0 μm	41.0
[S III] 18, 34 μm	23.3
[O IV] 26 μm	54.9
[Si II] 35 μm	8.2
[O III] 52, 88 μm	35.1

This work has since been extended to larger samples, using the polycyclic aromatic hydrocarbon (PAH) diagnostic to reach fainter sources (e.g., Lutz et al. 1998; Rigopoulou et al. 1999; Tran et al. 2001). The PAH features at 3.3, 6.2, 7.7, 8.7, and 11.2 μm are ubiquitous in normal galaxies and starbursts but absent near an AGN (e.g., Roche et al. 1991). Obscured regions also show absorption features, the strongest ones at 9.7 and 18 μm being due to silicate dust, which complicate the placement of the continuum near the PAH features. An object-by-object comparison of the optical and ISO spectral types for ULIGs in the 1-Jy sample reveals a remarkably good agreement between the two classification schemes if optically classified LINERs are assigned to the starburst group (Fig. 4). These results indicate that strong AGN activity, once triggered, quickly breaks the obscuring screen at least in certain directions, thus becoming detectable over a wide wavelength range.

7. Complications

7.1. Contribution from Shocks

Violent gas motions associated with AGN-driven or starburst-driven outflows or galaxy mergers may cause shock waves with velocities of 100 – 500 km s^{-1} in the ISM of the host galaxies. The shocks may produce a strong flux of EUV and soft X-ray radiation which may be absorbed in the shock precursor H II region (e.g., Sutherland, Bicknell, & Dopita 1995). The combination of the low-ionization emission-line spectrum from the post-shock material and the high-ionization emission-line spectrum from the precursor H II region can reproduce many of the spectroscopic signatures of LINERs and narrow-line AGNs (e.g., Dopita & Sutherland 1995). Fortunately, there are important physical differences between shock ionization and photoionization by an AGN (e.g., Morse, Raymond, & Wilson 1996). First, the line ratios produced in photoionized objects should be independent of the gas kinematics, while they are expected to correlate with the kinematics of the shock-ionized material. This effect is seen in a few optically and infrared-selected LINERs (Veilleux et al. 1994, 1995). The ionizing ultraviolet and soft X-ray continuum in shock-ionized objects should be extended on the

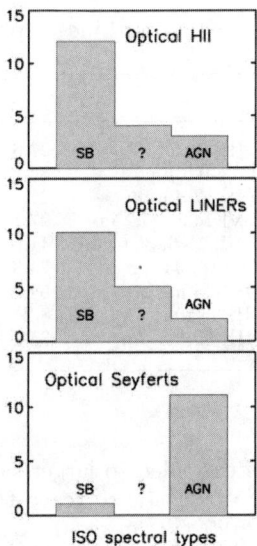

Figure 4. Optical versus mid-infrared classification of ultraluminous infrared galaxies. From Lutz, Veilleux, & Genzel (1999).

same scale as the shock structure, while it is expected to be a point source in the case of pure AGN photoionization. Finally, the electron temperature in shock ionized objects is expected to be considerably higher. Temperature-sensitive line ratios such as C III $\lambda 1909/\lambda 977$ and N III $\lambda 1750/\lambda 991$ in the ultraviolet and [O III] $\lambda 5007/\lambda 4363$ and [N II] $\lambda 5755/\lambda 6583$ in the optical range are the prime diagnostics of shock excitation. This method was used by Kriss et al. (1992) to deduce that shock excitation is likely to be important in the NLR of NGC 1068.

7.2. Aperture Effects

Circumnuclear starbursts often accompany AGNs (see, e.g., recent reviews by Veilleux 2000 and Gonzales Delgado 2001). The strength of the AGN signature is therefore a function of the size of the extraction aperture. This effect is particularly evident among infrared-selected galaxies where circumnuclear starbursts are nearly always present. Figure 5 shows the line ratios of luminous infrared galaxies as function of aperture size. The line ratios in some of these objects are seen to drift towards the H II region locus with increasing aperture size; large apertures dilute the AGN signature. Aperture effects will be particularly important in samples which cover a broad redshift range where a constant angular aperture corresponds to a wide range in linear scale. For a meaningful statistical analysis of the spectral classification one should use a fixed *linear* aperture for all objects in the sample (regardless of redshifts).

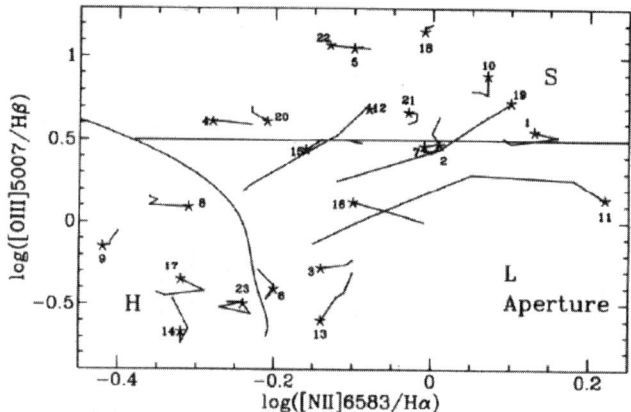

Figure 5. Aperture effects. Line ratios as a function of the size of the extraction aperture. The asterisks mark the nuclear values. The size of the extraction aperture increases (generally doubles) between each data point. From Veilleux et al. (1995).

7.3. Morphological Biases

Strong trends exist between the presence of an AGN, the mid-to-far infrared colors, and the host morphology. Objects with "warm" infrared colors (e.g., IRAS $f_{25}/f_{60} > 0.2$) often harbor an AGN at optical or near-infrared wavelengths or in polarized light (de Grijp et al. 1985; Veilleux et al. 1995, 1997b, 1999ab; Heisler, Lumsden, & Bailey 1997). Infrared-selected samples are often biased towards or against the presence of AGNs (but this is not the case for the 1-Jy sample; Kim & Sanders 1998). The same thing can be said about galaxy morphology. ULIGs often show signs of galaxy interactions. Most ULIGs are involved in the merger of two relatively large galaxies. Optically-classified Seyferts (especially those of type 1) are generally found in advanced mergers, while H II galaxies and LINERs are found in all merger phases (Fig. 6; see also Veilleux 2001). This means that surveys which specifically look for compact objects will be biased against starburst galaxies and are not statistically reliable for spectral classification purposes.

7.4. Metallicity Effects

The line ratio diagnostics discussed in this review often are a sensitive function of the metal contents in the ionized gas (see, e.g., Ferland & Netzer 1983; Veilleux & Osterbrock 1987 for early papers describing the effects of metallicity). The metallicity is well known to be correlated positively with the mass of the host galaxies (e.g., Bender, Burnstein, & Faber 1993), although this result has only been proven at low redshifts. In the early universe, one would expect declining metal abundances with increasing redshifts. The redshift dependence of the relative abundances of the elements involved in the emission-line ratios is a complex function of the star formation history and chemical evolution (including

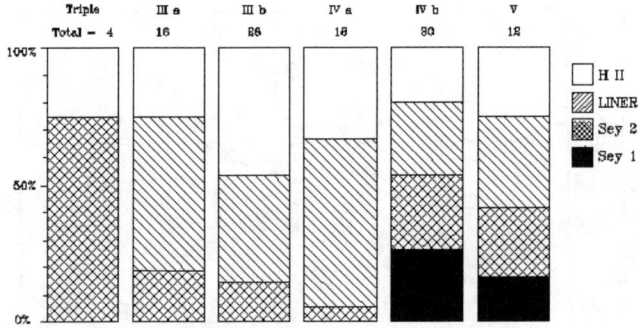

Figure 6. Morphological biases among infrared-selected AGNs. The hosts of ultraluminous infrared galaxies which are optically classified as Seyferts generally are advanced mergers (morphological classes IVa, IVb, or V). From Veilleux et al. (2002, in prep.). See also Veilleux (2001).

the effects of gas accretion and outflows) of the host galaxy environment (see, e.g., Hamann & Ferland 1999 for a discussion of QSO hosts). The usefulness of emission-line diagnostics at high redshifts will directly depend on the availability of accurate metallicity measurements and diagnostic tools properly calibrated in terms of metallicity.

8. Summary

UV–Optical–IR emission-line ratios are powerful diagnostics tools to discriminate between starbursts and AGNs. The following ratios have been shown to be the most reliable tools for this purpose.

1. Ultraviolet: N V $\lambda1240$/Lyα, N V $\lambda1240$/He II $\lambda1640$, C IV $\lambda1548$/Lyα.

2. Optical: [O III] $\lambda5007$//Hβ, [N II] $\lambda6583$/Hα, [S II] $\lambda\lambda6724$/Hα, [O I] $\lambda6300$/Hα, [O II] $\lambda\lambda7324$/Hα, [Fe VII] $\lambda6087$/Hα, [Ne V] $\lambda3426$/Hβ, He II $\lambda4686$/Hβ.

3. Near-infrared: Obscured broad Paα 1.875 μm, [Si VI] 1.962 μm/Paα.

4. Mid-Infrared: [Ne V] 14 μm/[Ne II] 12.8 μm, [O IV] 26 μm/[Ne II] 12.8 μm, EW(PAH 7.7 μm), overall SEDs especially 25 μm/60 μm colors.

A number of issues complicate the use of line ratios as discriminants between starburst and active galaxies, but additional measures can be used to clarify the situation:

1. Shock ionization: If shocks are important, one would generally expect correlations between the line ratios and gas kinematics, a UV continuum

extended on the same scale as the shock structure, and high gas temperatures.

2. Aperture effects: One should use a constant linear aperture to avoid variations in the contributions from circumnuclear starbursts.

3. Morphological bias: The spectral classification is likely to depend on the morphology of the host, especially the merger phase. Selection methods based on morphology will bias the sample.

4. Metallicity: Massive host galaxies in the local universe have larger metallicity, but high-redshift galaxies should be less dusty and less metal rich. One needs to use emission-line diagnostics which are properly calibrated as a function of metallicity and reddening.

Several new instruments will help refine the diagnostic tools discussed in this paper. The Cosmic Origins Spectrograph (COS), to be installed in 2003 on HST, will provide the high ultraviolet throughput needed to calibrate the UV diagnostic tools as a function of metallicity, evaluate the importance of shock ionization with the use of the C III and NIII temperature-sensitive line ratios, and to help resolve the circumnuclear starbursts and shock-excited winds around AGNs. The advent of SIRTF will help in the calibration of the infrared diagnostic tools as a function of metallicity *and dust extinction*. This spacecraft will also be a powerful instrument to search for infrared-bright AGNs. Ground-based work with adaptive optics and integral-field units will improve the sensitivity of searches for obscured AGNs by focussing on the inner regions of galaxies and avoiding the circumnuclear material associated with other phenomena. Spectroscopic follow-ups from the ground will help identify and classify AGN candidates in space-based and submm-selected samples.

Acknowledgments. The ground-based study on ultraluminous infrared galaxies discussed in this paper is done in collaboration with Drs. D. B. Sanders and D.-C. Kim. The author gratefully acknowledges the financial support of NASA through LTSA grant number NAG 56547.

References

Aitken, D. K., & Roche, P. F. 1985, MNRAS, 213, 777
Antonucci, R. 1993, ARA&A, 31, 473
Baldwin, J. A., Phillips, M. M., & Terlevich, R. 1981, PASP, 93, 5
Barth, A. J., et al. 1996, AJ, 112, 1829
Barth, A. J., et al. 1997, AJ, 114, 2313
Bender, R., Burstein, D., & Faber, S. M. 1993, ApJ, 411, 153
Binette, L., Wilson, A. S., & Storchi-Bergmann, T. 1996, A&A, 312, 365
Blanco, P. R., Ward, M. J., & Wright, G. S. 1990, MNRAS, 242, 4P
Cardelli, J. A., Clayton, G. C., & Mathis, J. S. 1989, ApJ, 345, 245
Crawford, M. K., et al. 1985, ApJ, 291, 755
De Grijp, M. H. K. Miley, G. K., Lub, J., & de Jong, T. 1985, Nature, 314, 240

Draine, B. T., & Lee, H. M. 1984, ApJ, 285, 89
Draine, B. T. 1989, in Infrared Spectroscopy in Astronomy, ed. B. H. Kaldeich (ESA-SP290; Noordwijk: ESA), 93
Dopita, M. A. & Sutherland, R. S. 1995, ApJ, 455, 468
Dopita, M. A., et al. 2000, ApJ, 542, 224
Duffy, P. B., et al. 1987, ApJ, 315, 68
Elvis, M., et al. 1994, ApJS, 95, 1
Ferland, G. J., & Netzer, H. 1983, ApJ, 264, 105
Ferland, G. J., & Osterbrock, D. E. 1986, ApJ, 300, 658
Ferland, G. J., & Osterbrock, D. E. 1987, ApJ, 318, 145
Genzel, R., & Cesarsky, C. J. 2000, ARA&A, 38, 761
Genzel, R., et al. 1998, ApJ, 498, 579
Goldader, J. D., et al. 1995, ApJ, 444, 97
Gonzales Delgado, R. M. 2001, in Issues in Unification of AGNs, in press (astro-ph/0109505)
Goodrich, R. W. 1990, ApJ, 355, 88
Goodrich, R. W., Veilleux, S., & Hill, G. J. 1994, ApJ, 422, 521
Greenhouse, M. A., et al. 1993, ApJS, 88, 23
Hamann, F., & Ferland, G. 1999, ARA&A, 37, 487
Heisler, C. A., Lumsden, S. L., & Bailey, J. A. 1997, Nature, 385, 700
Hill, G. J., Goodrich, R. W., & de Poy, D. L. 1996, ApJ, 462, 163
Hines, D. C. 1991, ApJ, 374, L9
Hines, D. C., et al. 1995, ApJ, 450, L1
Hines, D. C., & Wills, B. J. 1993, ApJ, 415, 82
Ho, L. C., Filippenko, A. V., & Sargent, W. L. W. 1996, ApJ, 462, 183
Hough, J. H., et al. 1991, ApJ, 372, 478
Kim, D.-C. 1995, Ph.D. Thesis, University of Hawaii
Kim, D.-C., & Sanders, D. B. 1998, ApJS, 119, 41
Kim, D.-C., Veilleux, S., & Sanders, D. B. 1998, ApJ, 508, 627
Kim, D.-C., et al. 1995, ApJS, 98, 129
Kriss, G. A., et al. 1992, ApJ, 394, L37
Krolik, J. H. 1999, Active Galactic Nuclei, Princeton University Press.
Lugten, J. B., et al. 1986, ApJ, 311, L51
Lutz, D. 1999, in The Universe as seen by ISO, ed. P. Cox and M. F. Kessler (ESA SP-427; Noordwijk: ESA, 623
Lutz, D., et al. 1996, A&A, 315, L269
Lutz, D., et al. 1998, ApJ, 505, L103
Lutz, D., Veilleux, S., & Genzel, R. 1999, ApJ, 517, L13
Maoz, D., et al. 1998, AJ, 116, 55
Marconi, A., et al. 1994, A&A, 291, 18
Morse, J. A., Raymond, J. C., & Wilson, A. S. 1996, PASP, 108, 426

Nicholson, K. L., et al. 1998, MNRAS, 300, 893
Oliva, E., & Moorwood, A. F. M. 1990, ApJ, 348, L5
Oliva, E., et al. 1994, A&A, 288, 457
Osterbrock, D. E. 1989, Astrophysics of Gaseous Nebulae and Active Galactic Nuclei, University Science Books
Osterbrock, D. E., Tran, H. D., & Veilleux, S. 1992, ApJ, 389, 196
Rigopoulou, D., et al. 1999, AJ, 118, 262
Rix, H.-W., et al. 1990, ApJ, 363, 480
Robert, C., Leitherer, C., & Heckman, T. M. 1993, ApJ, 418, 749
Roche, P. F., et al. 1984, MNRAS, 207, 35
Roche, P. F., et al. 1991, MNRAS, 248, 606
Röttgering, H.J.A., et al. 1997, A&A, 326, 505
Ruiz, M., Rieke, G. H., & Schmidt, G. D. 1994, ApJ, 423, 608
Shuder, J. M. 1981, ApJ, 244, 12
Solomon, P. M., Downes, D., Radford, S. J. E., Barrett, J. W. 1997, ApJ, 478, 144
Spinoglio, L., & Malkan, M. A. 1992, ApJ, 399, 504
Sutherland, R. S., Bicknell, G. V., & Dopita, M. A. 1993, ApJ, 414, 510
Thompson, R. I. 1995, ApJ, 445, 700
——. 1996, ApJ, 459, L61
Tran, Q. D., et al. 2001, ApJ, 552, 527
Veilleux, S. 2000, in the proceedings of the Ringberg meeting "Starbursts – Near and Far", September 2000, in press (astro-ph/0012121)
Veilleux, S. 2001, in the proceedings of the Granada meeting "QSO Hosts and their Environments", January 2001, in press (astro-ph/0104401)
Veilleux, S., et al. 1994, ApJ, 433, 48
Veilleux, S., Goodrich, R. W., & Hill, G. J. 1997a, ApJ, 477, 631
Veilleux, S., et al. 1995, ApJS, 98, 171
Veilleux, S., Kim, D.-C., & Sanders, D. B. 1999a, ApJ, 522, 113
Veilleux, S., et al. 2002, ApJ, in press
Veilleux, S., & Osterbrock, D. E. 1987, ApJS, 63, 295
Veilleux, S., Sanders, D. B., Kim, D.-C. 1997b, ApJ, 484, 92
Veilleux, S., Sanders, D. B., Kim, D.-C. 1999b, ApJ, 522, 139
Villar-Martin, M., Binette, L., & Fosbury, R. A. E. 1996, A&A, 312, 751
Villar-Martin, M., et al. 1999, A&A, 351, 47
Voit, G. M. 1992, ApJ, 399, 495
Watson, D. M., et al. 1984, ApJ, 279, L1
Yee, H. K. C. 1980, ApJ, 241, 894
Young, S., et al. 1993, MNRAS, 260, L1

The 2MASS Red AGN Survey

R.M. Cutri, B.O. Nelson

Infrared Processing and Analysis Center/California Institute of Technology, MS 100-22, Pasadena, CA 91125, email: roc@ipac.caltech.edu, nelson@ipac.caltech.edu

Paul J. Francis

The Australian National University, Canberra, ACT 0200, Australia, email: pfrancis@mso.anu.edu.au

Paul S. Smith

Univ. of Arizona, Steward Obs., Tucson, AZ 85721, USA, email: psmith@as.arizona.edu

Abstract. The Two Micron All Sky Survey (2MASS) provides an unprecedented, uniform photometric data set for large samples of AGN discovered at other wavelengths, and also forms the basis to search for previously unknown, obscured AGN. We present the results of a highly efficient near infrared color-based AGN survey using 2MASS that has already discovered 485 new, red AGN and QSOs. The extrapolated surface density of the 2MASS red AGN is $\sim$0.57 deg^{-2} for $K_s \leq 15.5$ mag. The ratio of Type 1 to Type 2 AGN among the newly discovered objects is 4:1, similar to proportions found in X-ray and deep ISO surveys, but the inverse of that found in IRAS surveys. The median redshift of the the new 2MASS red AGN sample is z=0.22, and all but three have z<0.7. The color distribution, polarization and X-ray properties data suggest that most of the 2MASS-discovered AGN are red because of obscuration by dust in and around their nuclei.

1. Introduction

Much of the current wisdom about the distribution, energetics and evolution of Active Galactic Nuclei (AGN) and Quasi-stellar Objects (QSOs) is based on samples of objects found in UV-excess or optical color surveys. While it has been known for at least 30 years that some AGN are partially obscured by dust in and around their nuclei (e.g. Mrk 231, NGC 1068), there has been increasing evidence in the past decade that large numbers of AGN have remained hidden from traditional UV/optical surveys because of dust obscuration. Infrared and radio surveys, less affected by dust extinction, suggest that between 50% and 80% of all AGN have been missed in the existing short wavelength surveys (Low et al. 1988; Webster et al. 1995; Francis et al 1999). The existence of a large population of obscured AGN has also been proposed to account for at least part of the hard X-ray background (e.g. Comastri et al. 1995; Gilli et al. 2001), and recent recent deep Chandra observations may be revealing this population (Mushotzky et al. 2000).

Seemingly at odds with the findings of the radio, infrared and X-ray surveys are the results of recent deeper optical-color surveys that claim to be no more than 10-20AGN (e.g. Meyer et al. 2001; Ivezic et al. this volume). We use data from the highly uniform Two Micron All Sky Survey (2MASS; Skrutskie et al. 1997) to address these apparently disparate results by conducting a large-scale search for previously unknown, red AGN. The 2MASS red AGN survey uses a simple, highly efficient color-selection criterion that is well-suited to identify relatively low redshift AGN. The objective of this search is not to find all AGN detected by 2MASS, but rather to test for the existence of a population of extremely red AGN. We present here the preliminary results of this Survey.

2. Red AGN Search Criteria

A characteristic of known AGN is that their near infrared colors are redder than most foreground stars and normal galaxies (e.g. Rieke 1978; Neugebauer et al. 1987). Obscured AGN will stand out from foreground stars and low redshift galaxies with even greater contrast. Because 2MASS provides uniform photometry in the J (1.25μm), H (1.65μm) and K_s (2.17μm) near infrared bandpasses over the full-sky, we tuned a color-selection criterion to best discriminate reddened AGN from UV/optically-selected ones using the measurements of large samples of known AGN. 2MASS detects 100%, 99% and 75% of the PG QSOs (Schmidt & Green 1986), the Hamburg/ESO QSOs (Wisotzki et al. 2000), and the LBQS QSOs (Hewitt et al. 1995), respectively, in all three near-infrared bands. 95% of the combined set of AGN from these optical/UV surveys have $J - K_s$<2.0. Therefore, we employ this color limit as the primary selection criterion to identify obscured AGN.

We selected candidates from the 2MASS Point Source Working Databases by requiring a detection in all three 2MASS bands and $J-K_s$>2.0. To minimize contamination by foreground AGB stars, which also occupy this part of color space, the search was limited to the $|b| > 30^o$ sky and excluded ~170deg^2 covering the Large and Small Magellanic Clouds. There are 16,977 unique candidates in an effective area of ~20,400deg^2 (0.83 candidates deg^{-2}) that satisfy these criteria. The brightness and color distribution of the color-selected candidates are shown in Figure 1. Less than 5% of the candidates are previously identified objects, and $\leq$ 1% are previously known AGN. Virtually all the candidates with K_s≤11 mag are high latitude Galactic AGB and carbon stars.

No requirement was imposed on the either the near-infrared or optical morphology of the candidates. Many sources appear slightly extended on either the 2MASS images or Palomar and ESO sky survey prints. Dilution of the nuclear light by starlight from the host galaxy may influence the observed infrared and optical color properties of the sample. The rest-frame 2MASS colors of galaxies (Jarrett et al. 2000) are significantly *bluer* than our color-selection criterion, so the lowest redshift, lowest luminosity AGN are missed by our sample. The galaxy k-correction shifts $J - K_s$ colors to the red with increasing redshift, though, and by redshift z~ 0.4 – 0.5, some "normal" galaxies may begin to shift into our

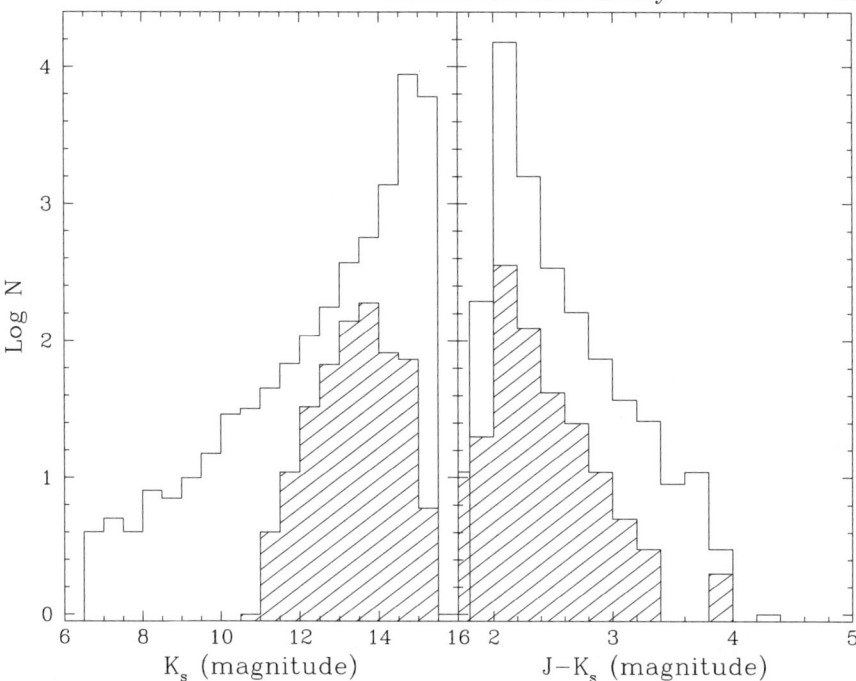

Figure 1. (*left*) The K_s brightness distribution of the 16,997 color-selected red AGN candidates selected from the 2MASS Point Source Working Databases. (*right*) The $J - K_s$ color distribution of 2MASS red AGN candidates. In both panels, the distribution of sources with follow-up optical spectroscopy is shown in the shaded regions.

color window. An L^* galaxy at $z = 0.4$ will be nearly 2 magnitudes below the 2MASS K_s detection limit, though, so only the brightest cluster galaxies will be marginally detectable at that distance ($K_s \sim 15.0$ mag; Brough et al. 2001). We expect the contamination from non-active galaxies to be relatively small in our sample, but it may increase towards the faintest K_s magnitudes.

Dilution by starlight from the host galaxy will make the optical colors of an AGN appear redder, and less AGN-like. Thus, it would not be unexpected to find galaxies harboring extinguished active nuclei that exhibit relatively normal galaxy-like optical colors. The active nucleus would be more apparent at longer wavelengths where the affect of obscuration is reduced.

3. Source Classification and Properties of the 2MASS AGN

Optical spectroscopy has now been carried out for 704 (4.1%) of the red AGN candidates at a variety of telescopes. Object classification and redshift measurements were made from the optical spectra: 384 (44%) of the objects are confirmed broad-line Type 1 AGN (Sy1/QSO), 101 (14%) are narrow-line Type 2 AGN (Sy2/LINER), 108 (15%) are weak emission line or normal galaxies, and 71 (10%) are carbon and L-dwarf stars. Approximately 5% of the candidates could not be identified because of insufficient signal-to-noise ratio in their optical spectra. Type 1 AGN are taken to be any object with broad emission-line components, and thus contain all of the intermediate AGN classes. We make no luminosity-based distinction between Sy 1 and QSO. The ratio of Type 1 and Type 2 (no broad component) AGN is $\sim 4:1$.

Figure 1 also shows the brightness and color distributions of the sources that have optical spectroscopy. The sampling is uniform in color space. However, relatively few of the faintest sources have been observed because most of the optical follow-up was done on modest-sized telescopes,

3.1. Redshift Distribution

The 2MASS red AGN discovered to-date span a redshift range of $0.03 \leq z \leq 2.52$, as illustrated in Figure 2. However, this search finds predominantly low redshift objects; the median redshift of the sample is 0.22, and only three AGN are found at z> 0.8. Near-infrared color-selection biases the 2MASS search to low redshift objects because of the k-correction of AGN spectral energy distribution (e.g. Hyland & Allen 1982). The predicted $J - K_s$ color for the radio-quiet QSO template of Elvis et al. (1994) is plotted versus redshift in Figure 2. QSOs from the optically-selected Hamburg/ESO survey (Wisotzki et al. 2000) and LBQS survey (Hewitt et al. 1995) are also shown in Figure 2, confirming the general trend and illustrating the typical scatter for an ensemble of objects. The predicted color of unobscured QSOs rapidly shifts to the blue with increasing redshift. The observed QSO colors exhibit a peak near z$\sim$2 because Hα enters the K_s bandpass. Although 2MASS has the sensitivity to detect QSOs out to redshift $z=3$ and even higher, they are rare in our color-selected sample.

At low redshifts ($z \leq 0.5$), 2MASS is revealing large numbers of previously unknown red AGN. Taken together with the optical/UV-selected objects, they show that the intrinsic spread in $J - K_s$ colors of AGN is at least a factor of two larger than previously known.

3.2. Luminosity Distribution

The bias against higher redshift objects means that the highest luminosity objects are underrepresented in the 2MASS red AGN sample. If the samples are limited to $z < 0.4$, the 2MASS Type 1 AGN and Hamburg/ESO QSOs have a similar range of K_s-band luminosity, as shown in Figure 3. The median and RMS values of log $h^{-2}L_K/L_\odot$ are 9.8 +/- 0.3 and 9.6 +/- 0.5 for the 2MASS Type 1 AGN and Hamburg/ESO QSOs, respectively. However, since $\leq 5\%$ of the Hamburg/ESO QSOs have $J - K_s > 2.0$, the samples are largely mutually exclusive.

3.3. Detection in Radio, Far Infrared and X-Ray Surveys

Approximately 50% of the confirmed 2MASS Type 1 and 2 AGN have counterparts within 2" in the FIRST radio survey (Becker et al. 1995). Virtually all of the radio-detected sources are radio-quiet or "moderate" with $S_{1.4GHz}/S_{K_s} \leq 1$. Only $\sim$20% of the new 2MASS AGN were detected in one or more bands by IRAS, indicating that the 2MASS red AGN search does not effectively sample the extreme IR QSO population found by Low et al. (1988). Only $\sim$10% of the 2MASS AGN are detected in the ROSAT Faint Source Survey.

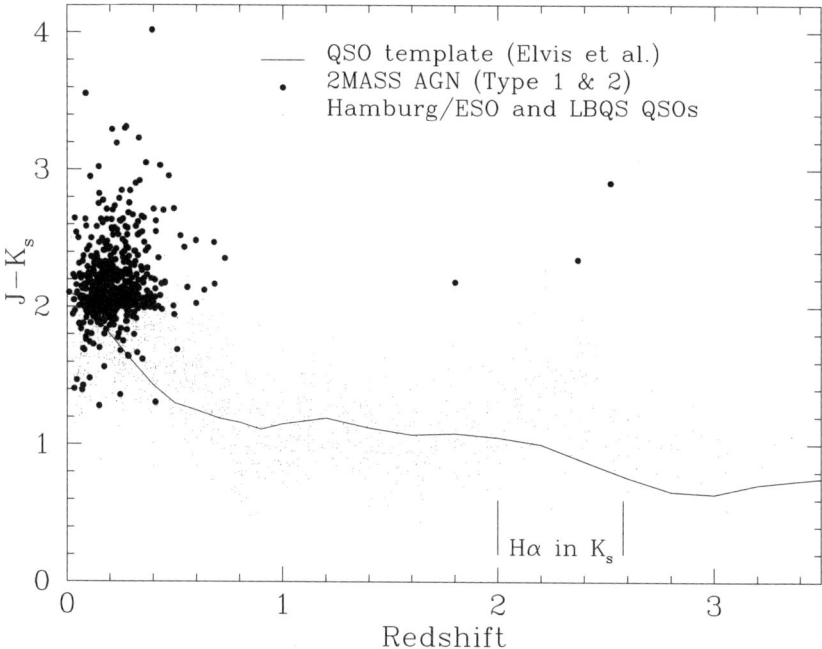

Figure 2. The 2MASS $J - K_s$ colors of AGN plotted a function of redshift. New 2MASS AGN are shown as large, filled circles, and optical-color selected QSO from the Hamburg/ESO and LBQS samples are shown as small points. The line shows the color predicted using the radio-quiet QSO template of Elvis et al. (1994). The redshift range in which the H-alpha emission line enters the K_s window and can contaminate the $J - K_s$ color is shown.

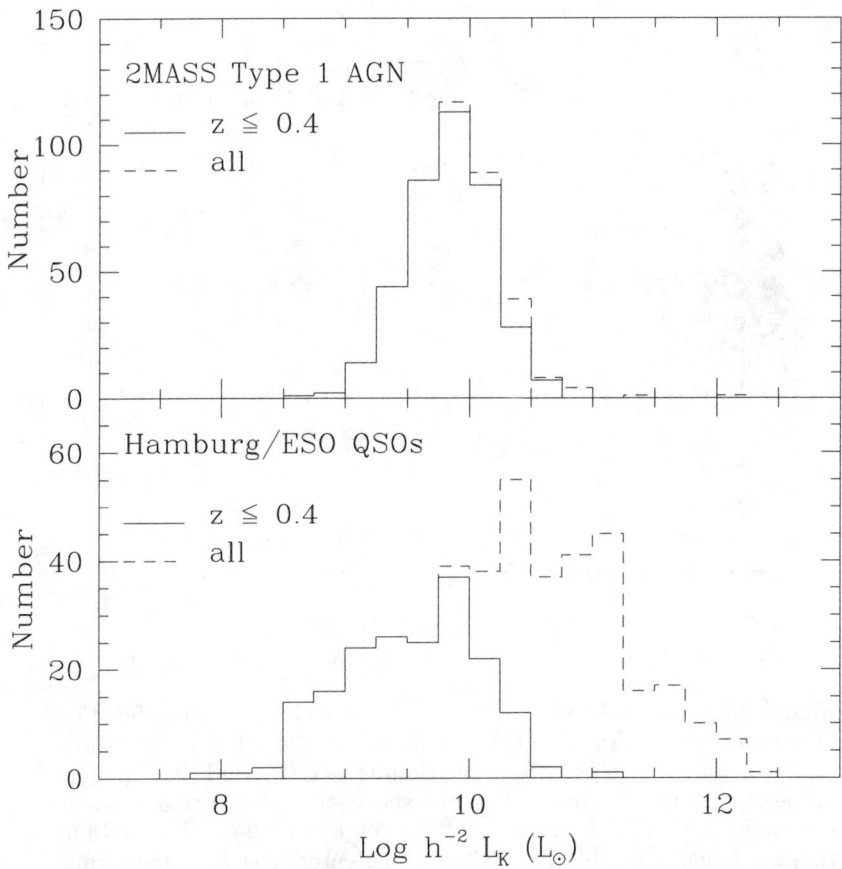

Figure 3. (*top*) K_s-band luminosity distributions for the 2MASS Type 1 AGN (*top*) and Hamburg/ESO QSOs (Wisotzki et al. 2000) (*bottom*). In each panel, the solid curves correspond to the subset of objects with $z \leq 0.4$, and the dashed curves to the full samples. The 2MASS color-selection finds AGN as luminous as the Hamburg/ESO survey in the local universe, but is biased against finding the most distant and luminous objects.

4. Discussion

If the 69% AGN detection rate achieved for the 2MASS red AGN sample holds for the entire sample of 16,977 candidates, there will be $\sim 0.57 deg^{-2}$ previously unknown Type 1 and 2 red AGN brighter than $K_s \leq 15.5$ mag (~ 0.04 mJy) over the sky. This is equivalent to the Hamburg/ESO QSO surface density extrapolated to $B_j \leq 17.8$ mag (Wisotzki et al. 2000), or that at $B_j \leq 18.2$ mag derived from the complete spectroscopic survey of the Fornax cluster field by Meyer et al. (2001). This extrapolation is tempered by the fact that so few of the candidates fainter than $K_s > 14.5$ have been observed to-date, though, so the behavior for the faint sources, which are in the majority, is largely unknown.

The modest sensitivity of 2MASS, and the color-selection technique we apply to the flux-limited 2MASS data set inherently biases us against finding the most heavily obscured and most distant objects. 2MASS is > 99% complete to $J \leq 15.8$ mag and $K_s \leq 14.3$ mag at high latitudes. Sources are detected up to ~ 1 mag fainter than the completeness limit, but with decreasing completeness. The red color limit selects against higher redshift objects because the k-correction shifts objects to bluer $J - K_s$ colors with increasing redshift (Figure 2). The three-band detection requirement also biases against finding the most extreme red and possibly distant objects which are turning up in other searches. Gregg et al. (2002) and Lacy et al. (2002) discovered a number extremely red, high redshift and high luminosity QSOs by matching FIRST radio and 2MASS infrared sources. Most of these were not detected in all three 2MASS bands. Thus, the density of red, obscured AGN inferred from this search may actually underestimates the true density.

The Fornax survey field can be used to compare the volume densities of AGN from optical- and infrared-selected surveys. Meyer et al. (2001) and Drinkwater et al. (2001) found 71 QSOs in the 3.1 sq.deg. field, and seven (2.2 deg^{-2}) have $z \leq 0.5$. There are eight red 2MASS AGN candidates in that region (2.5 deg^{-2}). This is much higher than the surface density of the full 2MASS red AGN sample, and it may be a result of the cluster in this field. Two of the red AGN candidates were identified by Meyer et al. as low redshift QSOs (FCSS J033720.9-345123, z=0.325, $J - K_s$=2.03 and FCSS J033736.6-353335, z=0.469, $J - K_s$=2.19). The remaining six 2MASS candidates are fainter than the nominal optical completeness limit of survey. If 69% of the 2MASS candidates are AGN, the total would be 5 – 6, bringing the combined number of QSOs in the $z \leq 0.5$ volume to 10 – 11. Optical color-selection finds seven of them, infrared color-selection finds 5 – 6, and two objects are in common to both techniques.

4.1. IR and Optical Colors

The near infrared color distribution of the 2MASS red AGN is consistent with $\sim$1-5 magnitudes of visual extinction being applied to the optical/UV-selected population. However, the optical-to-infrared colors are not consistent with a simple screen extinction model. Most of the 2MASS red AGN are too bright in the blue for the amount of extinction needed to produce the observed $J - K_s$ colors. As discussed earlier, though, host galaxy contamination of the optical nuclear light might pump up the blue brightness especially of the active nucleus is obscured. In addition, the USNOA-2.0 magnitudes tend to overestimate the brightness of extended sources since magnitudes were estimated using a brightness-diameter relation.

2MASS red AGN exhibit a range in $B - K_s$ color that spans over 7 magnitudes. The blue optical magnitudes range from $B_j \sim 15$ to > 21 mag. This is similar to the range observed in the radio-loud AGN samples of of Webster et al. (1995) and Francis et al. (1999). Flux-limited optical/UV searches will be biased against finding the most heavily obscured, nearby objects. Near-infrared and radio selection removes this bias and is sensitive to AGN over a much broader range of optical brightness. This may very well be the origin of the disparity in the claimed completeness of shallower optical/UV and infrared/radio surveys. The optical/UV surveys are highly complete, but down to specific optical brightness limits. IR and radio-selection finds the optically-faint AGN in the same volume.

4.2. Polarization

Smith et al. (2002) carried out a survey of optical broadband polarization for a representative subset of 70 new 2MASS red AGN. The 2MASS red AGN constitute one of the most highly polarized samples of radio-quiet AGN. Nine of the observed 2MASS AGN have polarization $P > 3\%$. For comparison, none of the PG QSOs have $P > 3\%$ (Berriman et al. 1990). The polarized fraction of the 2MASS AGN is comparable to the IRAS-discovered AGN (e.g. Hines et al. 1995). The intermediate type AGN (1.5-1.9) have the highest polarization fraction, and the degree of polarization correlates weakly with $J - K_s$ color and K_s luminosity. However, not all of the reddest or most luminous objects are highly polarized. Galactic starlight likely dilutes the polarized emission in these objects, so the measured degrees of polarization are probably lower limits.

Spectropolarimetry has been obtained for two of the most highly polarized 2MASS red AGN by Smith et al. (2000). Both objects show strong broad lines and *blue* optical continua in the polarized light spectra, indicating that we are observing the broad line regions in scattered light in these objects. At least for these two objects, obscuration by dust affects the optical continua and spectra, and likely contributes to the observed red broadband colors.

4.3. X-ray Emission

Chandra observations of a subset of 23 (17 Type 1 and 6 Type 2) 2MASS red AGN were conducted by Wilkes et al. (2002). They found that the 2MASS AGN are generally weak X-ray emitters, with X-ray-to-K_s flux ratios ranging from just below to over a factor of 100 lower than the low-redshift, broad-line QSO sample of Elvis et al. (1995). The reddest $(J - K_s)$ sources are also the weakest X-ray sources. The hardness-ratios derived from the X-ray measurements indicate that these objects have intrinsic neutral hydrogen absorbing column densities of $N_H \sim 10^{21-23} cm^{-2}$, larger than that of PG QSOs, but smaller than those found for samples of Seyfert 2 galaxies. However, the softest spectrum sources do not show evidence for intrinsic absorption, so at least some of the 2MASS AGN may be underluminous in X-rays.

A comparison between the amount of dust needed to explain the near-infrared colors by reddening a "normal" QSO spectrum and that derived from the X-ray spectral hardness showed that the values of E_{J-K}/N_H are between a factor of a few and ~ 100 lower than the Galactic value. This suggests that either the X-ray absorbing material is relatively dust free, or has unusual dust properties, or that the lines of sight towards the X-ray and infrared continuum emitting regions are different.

5. Summary

A simple color-based search of the 2MASS Point Source Database is revealing large numbers of previously unknown, predominantly low redshift AGN. Four times more Type 1 AGN than Type 2 AGN are found. Even though this search grows increasingly incomplete towards faint K_s flux levels and for increasingly red colors, extrapolating the early AGN identification rate to the entire sample suggests that 2MASS will find $> 10,000$ new AGN in the $|b| > 30°$ sky, $> 7,000$ of which will be Type 1.

The redshift and color distributions, source counts, and polarization and X-ray properties of the 2MASS red AGN sample all suggest that we preferentially detect objects with obscured nuclei. Flux-limited optical/UV surveys will be biased against detection of such objects because they are often below the optical brightness limits. However, deeper optical surveys do begin to detect them (e.g. Meyer et al. 2001). Given the known incompleteness of this search, and the results of recent far-infrared, radio and X-ray surveys for such objects, it is likely that the obscured AGN population comprise a significant fraction of the all AGN in the universe.

6. Acknowledgements

This publication makes use of data products from the Two Micron All Sky Survey, which is a joint project of the University of Massachusetts and the Infrared Processing and Analysis Center/California Institute of Technology, funded by the National Aeronautics and Space Administration and the National Science Foundation. RMC and BON acknowledge the support of the Jet Propulsion Lab which is operated by the California Institute of Technology under contract to NASA.

References

Becker, R.H., Helfand, D.J., White, R.L. & McMahon, R.A. 1995, ApJ, 450, 559
Berriman, G.B., Schmidt, G.D., West, S.C. & Stockman, H.S. 1990, ApJS, 74, 869
Brough, S., Collins, C.A., Burke, D.J., Mann, R.G. & Lynam, P.D. 2001, MNRAS.
Comastri, A., Setti, G., Zamorani, G., & Hasinger, G. 1995, A&A, 296, 1
Drinkwater, M., Engel, C., Phillips, S., Jones, S. & Meyer, M. 2001, AAO Newlsetter, 97, 4
Elvis, M., Wilkes, B.J., McDowell, J.C., Green, R.F., Bechtold, J., Willner, S.P., Oey, S.P., Polomski, E. & Cutri, R. 1994, ApJS, 95, 1
Francis, P.J., Whiting, M.T. & Webster, R.L. 1999, PASA, 17, 56

Gilli, R., Salvati, M. & Hasinger, G. 2001, A&A, 366, 407

Gregg, M.D., Lacy, M, White, R.L., Glikman, E., Helfand, D., Becker, R.H. & Brotherton, M.S. 2002, ApJ, 564, 133

Hewitt, P.C., Foltz, C.B. & Chaffee, F.H. 1995, AJ, 109, 1498

Hines, D.C., Schmidt, G.D., Smith, P.S., Cutri, R.M. & Low, F.J. 1995, ApJ, 450, L1

Hyland, A.R. & Allen, D.A. 1982, MNRAS, 199, 943

Jarrett, T.H., Chester, T., Cutri, R.M., Schneider, S., Skrutskie, M.F. & Huchra, J.P. 2000, AJ, 119, 2498

Lacy, M., Gregg, M.D., Becker, R.H., White, R.L., Glikman, E. and Helfand, D.J. 2002, BAAS, 200, 503

Low, F.J., Cutri, R.M., Huchra, J.P. & Kleinmann, S.G. 1988, ApJ, 327, L41

Meyer, M.J., Drinkwater, M.J., Phillips, S. & Couch, W.J. 2001 MNRAS, 324, 343

Mushotzky, R.F., Cowie, L.L., Barger, A.J. & Arnaud, K.A. 2000, Nature, 404, 459

Neugebauer, G., Green, R.F., Matthews, K., Schmidt, M., Soifer, B.T. & Bennet, J. 1987, ApJS, 63, 615

Rieke, G. 1978, ApJ, 226, 550

Schmidt M. & Green, R.F. 1986, ApJ, 305, 68

Skrutskie, M.F. et al. 1997, in "The Impact of Large Scale Near-IR Sky Surveys," eds. F. Garzon et al. (Kluwer (Netherlands), 25

Smith, P.S., Schmidt, G.D., Hines, D.C., Cutri, R.M. & Nelson, B.O. 2000, ApJ, 545, L19.

Smith, P.S., Schmidt, G.D., Hines, D.C., Cutri, R.M. & Nelson, B.O. 2002, ApJ, 569, 23

Webster, R.L., Francis, P.J., Peterson, B.A., Drinkwater, M.J. & Masci, F.J. 1995, Nature, 375, 469

Wisotzki, L., Christlieb, N., Bade, N., Beckmann, V., Khler, T., Vanelle, C. & Reimers, D. 2000, A&A, 358, 77

The Optical, Infrared and Radio Properties of Extragalactic Sources Observed by SDSS, 2MASS and FIRST Surveys

Ž. Ivezić[1], R.H. Becker[2], M. Blanton[3], X. Fan[1], K. Finlator[1], J.E. Gunn[1], P. Hall[1], R.S.J. Kim[4], G.R. Knapp[1], J. Loveday[5], R.H. Lupton[1], K. Menou[1], V. Narayanan[1], G.R. Richards[6], C.M. Rockosi[7], D. Schlegel[1], D.P. Schneider[6], I. Strateva[1], M.A. Strauss[1], D. Vanden Berk[8], W. Voges[9], B. Yanny[8], for the SDSS Collaboration

[1]*Princeton University,* [2]*University of California,* [3]*The New York University,* [4]*The John Hopkins University,* [5]*University of Sussex,* [6]*Pennsylvania State University,* [7]*University of Washington,* [8]*Fermi National Accelerator Laboratory,* [9]*Max-Planck-Institute für Extraterrestrische Physik*

Abstract. We positionally match sources observed by the Sloan Digital Sky Survey (SDSS), the Two Micron All Sky Survey (2MASS), and the Faint Images of the Radio Sky at Twenty-cm (FIRST) survey. Practically all 2MASS sources are matched to an SDSS source within 2 arcsec; $\sim$11% of them are optically resolved galaxies and the rest are dominated by stars. About 1/3 of FIRST sources are matched to an SDSS source within 2 arcsec; $\sim$80% of these are galaxies and the rest are dominated by quasars. Based on these results, we project that by the completion of these surveys the matched samples will include about 10^7 stars and 10^6 galaxies observed by both SDSS and 2MASS, and about 250,000 galaxies and 50,000 quasars observed by both SDSS and FIRST. Here we present a preliminary analysis of the optical, infrared and radio properties for the extragalactic sources from the matched samples. In particular, we find that the fraction of quasars with stellar colors missed by the SDSS spectroscopic survey is probably not larger than $\sim$10%, and that the optical colors of radio-loud quasars are $\sim$0.05 mag. redder (with 4σ significance) than the colors of radio-quiet quasars.

1. Introduction

The increasing availability of large scale digital sky surveys spanning many wavelengths offers an unprecedented view of the Universe. The positional matching of such surveys is of obvious scientific interest. Not only can wide wavelength coverage provide a comprehensive description of the various classes of astrophysical object, but characterizing the most populous families can help isolating more peculiar, and usually more interesting, objects. In this contribution we discuss the matching of early SDSS data with the 2MASS and FIRST surveys. The details of this work are presented elsewhere (Finlator *et al.* 2000, Menou *et*

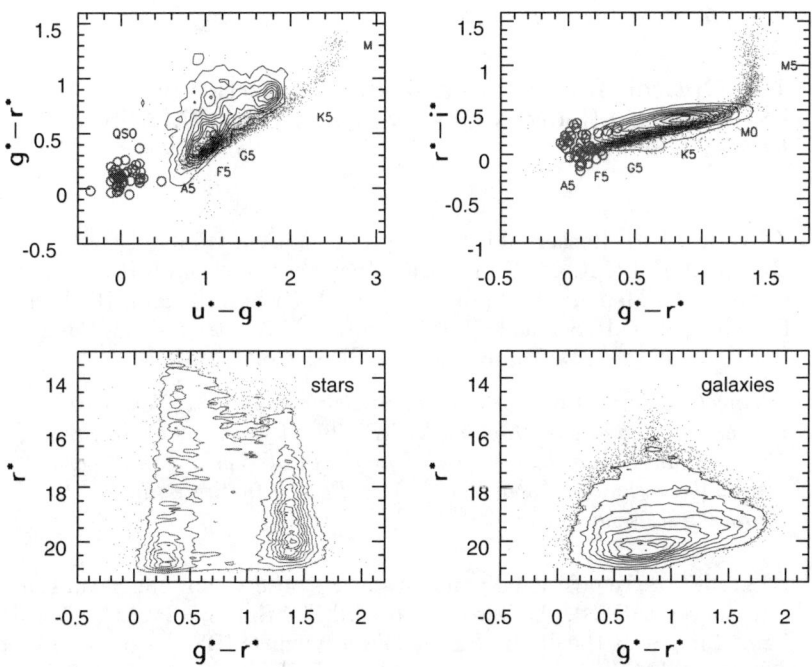

Figure 1. The color-color and color-magnitude diagrams which summarize photometric properties of SDSS sources.

al. 2001, Knapp et al. 2001, Ivezić et al. 2001, Ivezić et al. 2002) and here we summarize the most important results.

The SDSS (York et al. 2000, Stoughton et al. 2002, and references therein) is a digital photometric and spectroscopic survey which will cover one quarter of the Celestial Sphere in the North Galactic cap and produce a smaller area ($\sim$225 deg^2) but much deeper survey in the Southern Galactic hemisphere. It utilizes five broad bands (u', g', r', i', z') with central wavelengths ranging from 3550 r^*A to 8930 r^*A, and will detect about 10^8 stars and a similar number of galaxies brighter than $\sim 22^m$. For about one million of the brightest galaxies and 100,000 quasar candidates the SDSS will also obtain high-quality spectra. 2MASS (Skrutskie et al. 1997) surveyed the entire sky in near-infrared light (J, H, and K_s bands) and catalogued $\sim$300 million stars, as well as several million galaxies. The FIRST survey (Becker et al. 1995) provides the most comprehensive view of the Universe at 20 cm and will detect about a million radio galaxies and quasars.

2. SDSS Color-Color and Color-Magnitude Diagrams

Of the three surveys discussed here, SDSS provides the most detailed information about the detected objects due to its multi-color optical photometry, the highest angular resolution, and the spectroscopic data. The position of an object in SDSS color-color and color-magnitude diagrams can be efficiently used to constrain its nature. The color-color and color-magnitude diagrams which summarize photometric properties of SDSS sources are shown in Figure 1 (magnitudes are marked as m^* because the calibration is still uncertain within $\sim 5\%$). We use the "model" magnitudes, as computed by the photometric pipeline ("photo", Lupton et al. 2002). The model magnitudes are measured by fitting an exponential and a de Vacouleurs profile, and using the formally better model in r to evaluate the magnitude. Photometric errors are typically 0.03^m at the bright end ($r^* < 20^m$), and increase to about 0.1^m at $r^* \sim 21^m$, the faint limit relevant in this work (for more details see Ivezić et al. 2000 and Strateva et al. 2001, hereafter S01).

The top two panels in Figure 1 display the $g^* - r^*$ vs. $u^* - g^*$ and $r^* - i^*$ vs. $g^* - r^*$ color-color diagrams for $\sim$ 300,000 objects observed in 20 deg^2 of sky. The unresolved sources are shown as dots, and the distribution of resolved sources is shown by linearly spaced density contours. The low-redshift quasars (z $\lesssim$ 2.5), selected by their blue $u^* - g^*$ colors indicating UV excess ($0.6 < u^* - g^* <$ 0.6, $-0.2 < g^* - r^* < 0.6$), are shown as circles. Most of the unresolved sources marked as dots are stars. For a more detailed discussion of the stellar properties in the SDSS photometric system see Finlator et al. (2000, hereafter F00) and references therein. Here we only briefly mention that the position of a star in color-color diagrams is mainly determined by its spectral type, as marked, and that the modeling of the stellar populations observed by SDSS (F00) indicates that the majority of these stars ($\sim 99\%$) are on the main sequence. The lower two panels in Figure 1 display the color-magnitude diagrams for unresolved (left) and resolved (right) sources, with the distributions shown by linearly spaced density contours. The distribution of galaxies in the SDSS color-color diagrams has been studied by Shimasaku et al. (2001) and S01. S01 found that galaxies show a strongly bimodal distribution of the $u^* - r^*$ color, also visible in the upper left panel in Figure 1, and demonstrated that the two components can be associated with the spiral (blue component) and elliptical (red component) galaxies.

The aim of this work is to find out where in the diagrams shown in Figure 1 2MASS and FIRST sources are found, and how many 2MASS and FIRST sources are not detected by SDSS.

3. The Positional Matching of SDSS and 2MASS Sources

The positional matching of SDSS and 2MASS sources is described by F00, who also discussed the optical and infrared properties of stars. The analysis of extragalactic sources observed by both SDSS and 2MASS will be described in detail by Ivezić et al. (2002). Practically all 2MASS sources ($\sim 98\%$ for point sources from the PSC and $\sim 97\%$ for extended sources from the XSC) are matched to an SDSS source within 2 arcsec. About $\sim 11\%$ of the 2MASS PSC sources are optically resolved galaxies and the rest are dominated by stars. Practically all ($\sim 98\%$) sources from the 2MASS XSC are associated with an optically resolved source.

Figure 2 shows the color-magnitude diagrams for sources observed by both SDSS and 2MASS. The top two panels show representative optical and near-IR diagrams for $\sim$10,000 stars, and the bottom two panels show analogous diagrams for a similar number of galaxies. The vertical dashed lines at $J - K_s = 1$ in

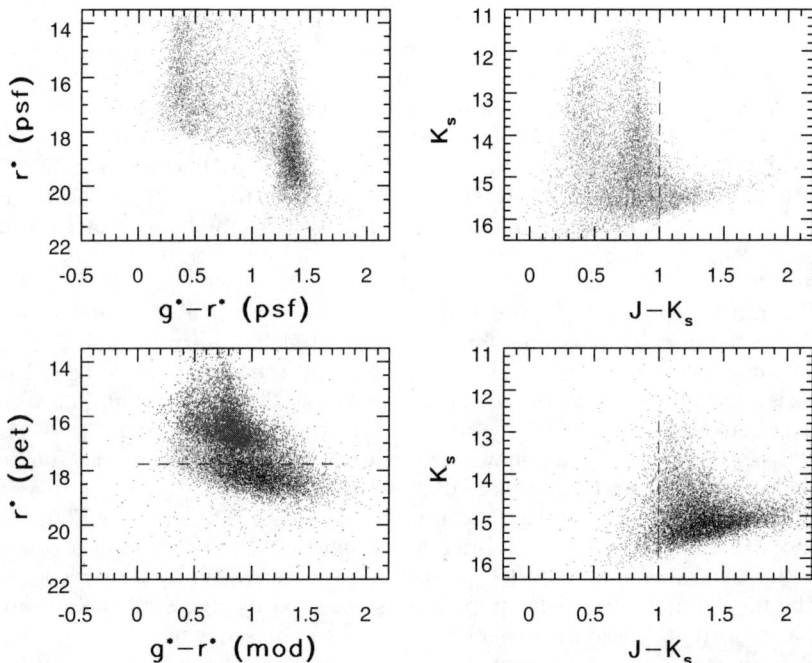

Figure 2. The color-magnitude diagrams for stars (top) and galaxies (bottom) observed by both SDSS and 2MASS.

the two right panels roughly separate stars from galaxies in 2MASS data (only ~22% of these galaxies are resolved in 2MASS images). The horizontal dashed line in the lower left panel shows the magnitude limit for the SDSS spectroscopic galaxy survey. Practically all sources from the 2MASS XSC are brighter than that limit, and are shown as dark dots (on top of gray dots that mark sources from the 2MASS PSC). The fraction of 2MASS galaxies in each of the regions in the r^* vs. $g^* - r^*$ color-magnitude diagram that track different morphological types (shown by dashed lines in Figure 6, for details see Ivezić et al. 2002) is listed in Table 1. Based on the matched source density, we project that by the completion of the SDSS the matched SDSS-2MASS sample will include about 10^7 stars and 10^6 galaxies. About half of these galaxies will be part of the SDSS spectroscopic survey, and about 10^5 will have photometry better than 0.1 mag. in all 8 bands.

Figure 3 shows the color-color diagrams for stars and galaxies from Figure 2, which also have better than 10σ detections in all three 2MASS bands. Stars and galaxies are shown as dots, and the unresolved sources with optical colors indicative of low-redshift quasars are shown as circles (open for sources with better than 5σ detections in all three 2MASS bands and solid for better than 7σ). Note that it is not possible to separate galaxies from stars using only optical colors (see upper left panel). However, the addition of IR data allows nearly perfect separation, and the lines shown in the two right panels outline the regions populated by stars, galaxies and low-redshift quasars (the i^*-K_s color

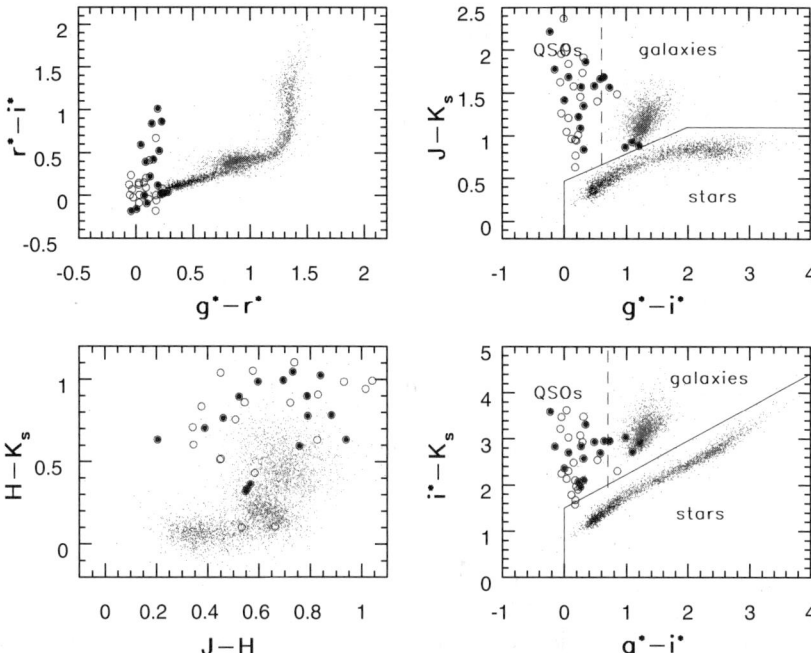

Figure 3. The color-color diagrams for sources from Figure 2 with high signal-to-noise 2MASS detections.

is based on the "psf" i^* magnitude). This clean separation can be used to gauge the success of the SDSS star-galaxy separation at the bright end. We find that 99.3% of the sources deemed as resolved by the SDSS photometric pipeline are found in the region marked as "galaxies". The bottom left panel shows an infrared color-color diagram constructed with 2MASS data. Using SDSS data we find that sources with $H - K_s > 0.3$ are predominantly extragalactic, while the bluer sources are stars.

3.1. Search for Reddened Quasars Using SDSS and 2MASS

Dust-reddened quasars are hard to distinguish in optical color-color diagrams because the plausible reddening vectors are roughly parallel to the stellar locus. However, such objects may be more easily distinguishable from stars in optical-infrared color-color diagrams because the reddening moves objects from the region marked "QSOs" to the region marked "galaxies", thus missing the stellar locus (because quasars have redder $J - K_s$ colors than stars). Since SDSS imaging data can easily separate optically unresolved and resolved sources, selecting sources with e.g. $J - K_s > 1$ and $g^* - i^* > 0.5$ can in principle reveal dust-obscured quasars. In a pilot study, we have obtained spectra for 30 optically unresolved sources from the region marked "galaxies". Most of them are M stars, and not a single one was confirmed to be a quasar. This null result places a strong upper limit on the fraction of reddened quasars.

Table 1. The Galaxy Distribution^a in the r^* vs. $g^* - r^*$ Diagram.

Region	Counts	Blue	Red	PSC^b	XSC^b	FIRST
Ia	24.6±0.5	95.0	5.0	38.6	17.9	3.9
Ib	58.0±0.7	16.0	84.0	77.9	38.8	4.4
Ic	13.2±0.3	2.2	97.8	79.1	20.9	8.8
all I	95.8±0.9	34.5	65.5	67.9	31.0	4.9
IIa	317±1.8	97.4	2.6	1.9	0.0	0.1
IIb	517±2.2	54.3	45.7	2.2	0.0	0.2
IIc	304±1.8	7.4	92.6	14.4	0.0	0.8
IId	84.0±0.9	2.4	97.6	21.0	0.1	3.3
all II	1222±3.5	50.2	49.8	6.5	0.0	0.5
I + II	1318±3.7	49.1	50.9	10.9	2.3	0.9

a) All entries are percentages, except counts which are deg^{-2}.
b) Refers to 2MASS catalogs.

Cutri et al. 2001 discuss the selection of red AGNs from the 2MASS PSC database using the condition that $J - K_s > 2$. These sources may be dust-reddened quasars with optical colors indistinguishable from those of stars and thus missed by SDSS spectroscopic targeting of quasars. We have analyzed a sample of 241 such candidates selected from a 220 deg^2 large region (Ivezić, Cutri, Nelson, et al. 2002). The majority of these sources are optically resolved (226), and are dominated by red (elliptical) galaxies (194) with redshifts up to ∼0.4. Their very red $J - K_s$ colors seem to be a consequence of the K correction. The majority of optically unresolved matches (13 out of 15) show UV excess indicative of quasars with redshifts ≲2.5 and thus are easily distinguishable from stars using optical colors alone. This is confirmed for 9 sources for which SDSS spectra are available. One of the remaining two unresolved sources is a confirmed L dwarf, and the other one has optical-infared colors consistent with also being an L dwarf. Consequently, this preliminary analysis indicates that the fraction of moderately reddened quasars ($A_V \lesssim 5$) which are missed by the SDSS quasar survey is small. While the precise value of this fraction is somewhat model-dependent, it seems to be less than ∼10%.

4. The Positional Matching of SDSS and FIRST Sources

The positional matching of SDSS and FIRST sources is described in detail by Ivezić et al. 2001 (see also Menou et al. 2001 and Knapp et al. 2001). Optical identifications can be made for ∼31% of FIRST sources. The majority of the FIRST sources identified with an SDSS source are optically resolved, and their fraction among the matched sources is a function of the radio flux, increasing from ∼50% at the bright end to ∼90% at the FIRST faint limit (1 mJy). The cumulative fraction of the optically unresolved sources among the SDSS-FIRST matches is ∼16%. The colors of optically unresolved sources, as well as SDSS spectra for objects brighter than the flux cutoff for the spectroscopic

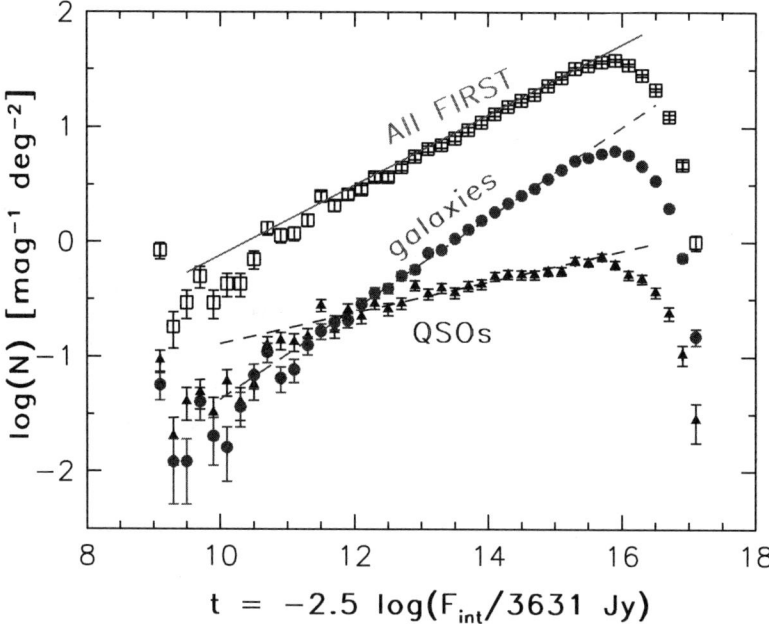

Figure 4. The comparison of the radio counts for SDSS-detected quasars and galaxies.

targeting, indicate that they are dominated by quasars. Figure 4 compares the differential radio counts for SDSS-detected quasars, marked by dots, and galaxies, marked by triangles (we introduced an AB radio magnitude defined as $t = -2.5 \log(F_{int}/3631 \mathrm{Jy})$, where F_{int} is the total radio flux). The dashed lines show the best fits in the $11.5 < t < 15.5$ range: for quasars

$$\log(N) = -2.24 + 0.14\,t, \qquad (1)$$

and for galaxies

$$\log(N) = -5.33 + 0.40\,t. \qquad (2)$$

We find no significant differences in the counts of FIRST sources with and without an optical identification: they both follow $\log(N) = C + 0.3\,t$ relations. To illustrate this point, we compare the sum of counts for quasars and galaxies (renormalized to account for the matching fraction), shown as open squares, to the counts of all FIRST sources, shown by the solid line. As evident, the two distributions are similar.

Based on the matched source density, by the completion of SDSS and FIRST the matched sample will include ∼250,000 galaxies and ∼50,000 quasars. As discussed in detail by Ivezić et al. 2001, the majority of these quasars are radio-loud.

4.1. The Properties of Quasars Observed by SDSS and FIRST

One of the most important advantages of a radio-selected sample of quasars is that it can be used to estimate the fraction of quasars with stellar colors that are

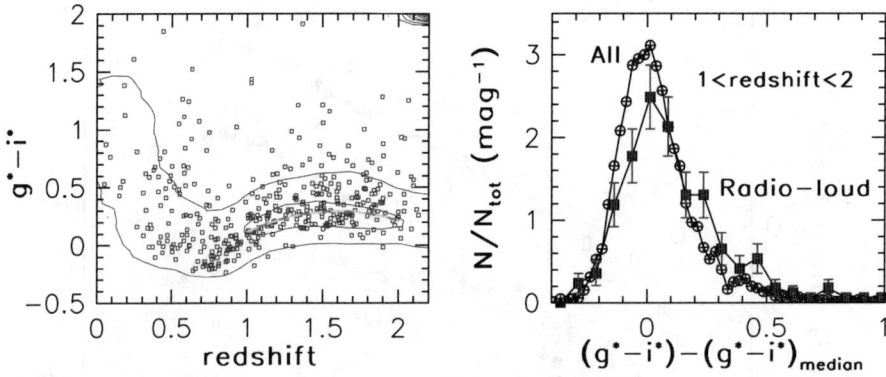

Figure 5. The comparison of optical colors for radio-loud and optically selected quasars (see Section 4.1).

missed by optical surveys, such as SDSS. The majority of optically unresolved SDSS-FIRST sources have non-stellar colors. We find that the fraction of sources with colors indistinguishable from stellar is $\sim$10%, assuming that the colors of radio-loud ($F_\nu^{radio} > 10\, F_\nu^{optical}$, for more details see Ivezić et al. 2001) quasars are similar to the colors of radio-quiet quasars. We do, however, detect a small but statistically significant difference between the colors of radio-loud quasars and radio-quiet quasars. The left panel in Figure 5 shows the dependence of QSO g^*-i^* color on redshift (for a similar dependence of other SDSS colors on redshift see Richards et al. 2001). The distribution of optically selected and unresolved quasars with $i^* < 19$ is shown by contours. The 464 radio-loud quasars from the matched sample are shown as squares. The thick line shows the median g^*-i^* color of all optically selected quasars in the redshift range 1–2, which is subtracted from the g^*-i^* color to obtain a color excess. The right panel compares the distribution of this color excess for all optically selected quasars in that redshift range, shown by circles, to the distribution for 225 radio-loud quasars, shown by squares. The g^*-i^* color-excess distribution for radio quasars appears to be different from the distribution for the whole sample. First, the median excess for the radio subsample is redder by $\sim$0.05 mag, with about 4σ significance. Second, the fraction of objects with very large color excess is larger for the radio subsample. Adopting 0.5 mag for the minimum value of the color-excess, we find that 2.4$\pm$0.2% of quasars have such extreme g^*-i^* colors, while this fraction is 7.1$\pm$1.8% for the radio-loud quasars.

4.2. The Properties of Galaxies Observed by SDSS and FIRST

Figure 6 displays the r^* vs. $g^* - r^*$ color-magnitude diagram for SDSS-FIRST galaxies, shown by dots, compared to the distribution of all SDSS galaxies, shown by linearly-spaced contours. The left panel shows the distribution of the SDSS-FIRST galaxies brighter than r^*=21.5, and the right panel shows the distribution of galaxies for which SDSS spectra are available. For details about the spectroscopic targeting of SDSS galaxies see Strauss et al. 2001 and Eisenstein et al. 2001. The dashed lines outline regions with different galaxy morphology and

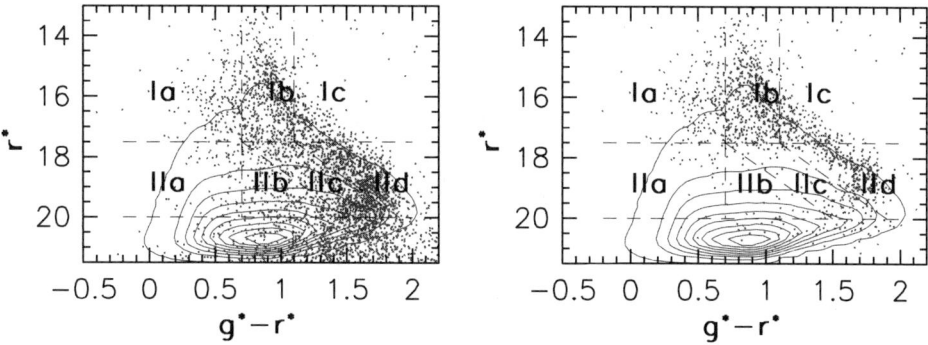

Figure 6. The r^* vs. $g^* - r^*$ color-magnitude diagram for SDSS-FIRST galaxies. The left panel shows all SDSS-FIRST galaxies, and the right panel those for which SDSS spectra are available.

fraction of radio galaxies, as listed in Table 1. At the bright optical end ($r^* <$ 17.5) these radio-galaxies represent ∼5% of all SDSS galaxies, with the radio fraction for red galaxies ∼2 times higher than for blue galaxies. In the magnitude range 17.5 < r^* < 20, ∼0.5% of SDSS galaxies are detected by FIRST, and the radio fraction of the reddest galaxies, dominated by giant ellipticals, is ∼40 times larger than the radio fraction of the bluest galaxies. We find that radio galaxies in a redshift-limited sample have statistically indistinguishable colors and luminosity distribution from other galaxies from the same volume. Nevertheless, the preliminary analysis of spectra indicates that the fraction of active galaxies is higher for the FIRST-detected galaxies than for all SDSS galaxies selected in the same regions of the optical color-magnitude diagrams. A more quantitative analysis of this effect will be presented in a future publication.

5. Discussion

The preliminary analysis of sources in common to the SDSS, 2MASS and FIRST surveys indicates the enormous potential of combining modern digital sky surveys. The eight-color highly accurate photometry, the morphological information, radio properties, and redshift information for samples several orders of magnitude larger than previously available is bound to place the studies of extragalactic sources at an entirely new level. A good example of a result made possible by both a large sample and accurate photometry is the small, yet statistically significant, difference between the optical colors of radio-loud and radio-quiet quasars. Another result with potentially large astrophysical significance is the upper limit of 10% on the fraction of quasars with stellar colors, and finding that optical colors of FIRST-detected galaxies are not significantly different from other galaxies in a volume-limited sample.

Acknowledgments

The Sloan Digital Sky Survey (SDSS) is a joint project of The University of Chicago, Fermilab, the Institute for Advanced Study, the Japan Participation Group,

The Johns Hopkins University, the Max-Planck-Institute for Astronomy (MPIA), the Max-Planck-Institute for Astrophysics (MPA), New Mexico State University, Princeton University, the United States Naval Observatory, and the University of Washington. Apache Point Observatory, site of the SDSS telescopes, is operated by the Astrophysical Research Consortium (ARC). Funding for the project has been provided by the Alfred P. Sloan Foundation, the SDSS member institutions, the National Aeronautics and Space Administration, the National Science Foundation, the U.S. Department of Energy, the Japanese Monbukagakusho, and the Max Planck Society. The SDSS Web site is http://www.sdss.org/.

References

Becker, R.H., White, R.L., & Helfand, D.J. 1995, ApJ, 450, 559

Cutri, R.M., 2001, in *The New Era of Wide Field Astronomy*, ASP Conference Series, Vol. 232. Eds. R. Clowes, A.Adamson, and G. Bromage. San Francisco: Astronomical Society of the Pacific. ISBN: 1-58381-065-X, p.78

Eisenstein, D., et al. 2001, in prep.

Finlator, K., et al. 2000, AJ, 120, 2615 (F00)

Ivezić, Ž., et al. 2000, AJ, 120, 963

Ivezić, Ž., et al. 2001, *Optical and Radio Properties of Sources Observed by the FIRST Survey and the Sloan Digital Sky Survey*, in prep.

Ivezić, Ž., et al. 2002, *Optical and Infrared Colors of Extragalactic Objects Observed by the Two Micron All Sky Survey and the Sloan Digital Sky Survey*, in prep.

Ivezić, Ž., Cutri, R.M., Nelson, B., et al. 2002, *The Properties of the 2MASS AGN Candidates Observed by the Sloan Digital Sky Survey*, in prep.

Knapp, G.R., et al. 2001, in prep.

Lupton, R.H. et al. 2002, *The SDSS Photometric Pipeline*, in prep.

Menou, K., et al. 2001, ApJ, *Broad Absorption Line Quasars in the Sloan Digital Sky Survey with VLA-FIRST Radio Detections*, in press (November issue), also astro-ph/0102410

Richards, G.T., et al. 2001, AJ, 121, 2308

Skrutskie, M.F. et al. 1997, The Impact of Large-Scale Near-IR Sky Surveys, ed. F. Garzon et al. (Dordrecht: Kluwer), 25

Stoughton, C., et al. 2002, *SDSS: Early Data Release*, AJ, in press

Strateva, I., et al. 2001, AJ, 122, 1861 (S01)

Strauss, M., et al. 2001, in prep.

York, D.G., et al. 2000, AJ, 120, 1579

Dubious Deductions from AGN Survey Data

Robert Antonucci
University of California, Physics Department, Santa Barbara, CA 93106, USA

Abstract. The participants in this meeting are almost all carrying out the hard work of making many different types of AGN surveys. Since it's so much easier to criticize other people's work than to do actual work myself, I'll just present some demurs regarding recent papers drawing conclusions from various AGN survey data. In particular I'll mention some questionable interpretations of surveys of Seyfert 2 near-UV polarization; interpreting the results of searches for polarized broad H-alpha lines in Seyfert 2s; testing the beam model for radio galaxies and quasars; testing the unification of Seyfert spectral types with a torus; and finally testing the energy sources for ULIRGs, especially those with LINER optical spectra. Only the polarized broad H-α results are examined in detail here.

1. Levels of Polarization of the Nuclear Light in Seyfert 2 Galaxies

Koski (1978) showed that Seyfert 2s and Narrow Line Radio Galaxies generally have UV excesses relative to the SEDs for the Pop II stars which dominate the optical light in arcsec apertures. He showed the spectra could be parameterized approximately by a normal Pop II SED, plus a suitably normalized power law which dominates in the near-UV. The brightest powerful Seyfert 2 nucleus, that of NGC1068, can indeed be fit quite well that way (Miller and Antonucci 1983, McLean et al 1983, Antonucci et al. 1994), and furthermore, the power-law component has a wavelength-independent polarization of $\sim 16\%$ in arcsec apertures. It is crucial to note that the broad permitted lines show the same percent P as the power-law continuum component accounting for the "UV excess". This strongly suggests that the scattering paths of the broad lines and the power law are very similar geometrically. Finally, the permitted lines have normal equivalent width in polarized flux. This confirms that the hidden source is simply a normal Seyfert 1 nucleus, whose light is scattered into the line of sight.

The polarization angles of NGC1068 and other Seyfert 2s lie perpendicular to the radio axes, so that the photons' last flights before scattering into the line of sight were *along* the axes. Thus the equatorial directions must be opaque, and the torus model was born. In principle the polar scattering could result from intrinsically anisotropic emission, but the anisotropy would need to be the same for the broad emission lines as for the power law, which would be difficult to arrange by some mechanism other than shadowing. In fact if these two components differ substantially in their isotropy in any significant number of objects, we'd expect to see one component without much of the other fairly frequently in Type 1 objects. The broad line equivalent width dispersion would then be very large and this is not seen.

It was evident from the beginning of the era of high SNR spectropolarimetry that most other objects behave somewhat differently (Antonucci 1984 and Tran et al. 1995 on 3C234; Miller and Goodrich 1990; Tran 1995). It's quite general that the broad line equivalent widths are normal in the polarized flux spectra of narrow line objects, that is, they match those of broad line objects. Thus the basic picture is the same as for NGC1068: a normal broad line nucleus is seen in reflection. But in general the UV "featureless continuum" has a polarization much lower than that of the broad emission lines (Kay 1994). Thus there is another, unpolarized UV continuum component, now known to derive from hot stars in many cases (Heckman et al. 1995, Gonzalez-Delgado et al. 1998). But the key from the unified model point of view is that the broad lines are highly polarized (at 16%, NGC1068 has about the *lowest* P), so that the scattering is in one or two polar cones.

The point here is that several authors have taken the percent P from the UV continuum—which is often low—and rejected polar scattering on that basis. For example, see the key role of this erroneous argument in the discussion of Malkan et al. 1998.[1]

2. Polarized Broad H-α and the Generality of Seyfert Unification

Since the early days of the geometrical unified models of various kinds, attempts have been made to assess the generality of the models. For example, Tran (2001) has recently discussed whether or not all Seyfert 2s have hidden Type 1 nuclei of any significant luminosity. It's usually difficult to prove that something is *not present*, and people haven't yet converged on the answer to this.[2]

Tran (2001) has given a variety of arguments and concludes that there are in fact many "real" Seyfert 2 galaxies without hidden Type 1 nuclei. His paper may be correct but I want to take this opportunity to review some places where

---

[1] The way to measure the broad line polarization, and hence the polarization of the scattered light alone, is to divide the polarized flux by the total flux. However, it is generally impossible to see the line clearly in total flux, so one just derives high lower limits to the broad line polarization in most cases. Tran (1995) claimed to measure the broad lines in total flux in some objects. However, M. Kishimoto and I looked carefully at the case of Mrk 477, in preparing a paper on polarization imaging. We examined the total-flux plots, overlaying the permitted Ba lines on various forbidden lines; we saw little or no evidence for broad wings in the total flux. We thus placed an upper limit on the total flux of the broad components, leading to a conservative lower limit on broad H-α and H-β polarization of 10%. Thus we disagree with the intrinsic value of only 2–3% quoted in Tran 1995. We used the same data.

[2] One way to do it is to see whether there's any significant "waste heat"—dust reradiation from the matter obscuring the putative hidden nucleus. D. Whysong and I are studying 3C radio galaxies in the mid-IR with this in mind, and have shown, for example, that M87 can have no such hidden nucleus at any remotely relevant luminosity level: astro-ph 0106381. Also, Meisenheimer et al. (2001) present a very good study of ISO data, and reach a similar conclusion that many lower-luminosity radio galaxies lack significant waste heat.

his arguments seem quite uncertain—most of these are already discussed in his paper.

Tran (2001) presents spectropolarimetric data on two Seyfert 2 samples: the 12μ-selected sample of Rush et al. 1993, and the CfA sample of Huchra and Burg 1992. The main finding seems to be that those with detected hidden broad emission line regions ("HBLRs") have higher radio power relative to the far-IR power than the non-HBLRs on average, and that they also have warmer 25μ–60μ colors on average. These differences are taken to show that the two types (HBLRs vs. non-HBLRs[3]) of Seyfert 2 are intrinsically different, and that in particular there is a large subset with no hidden Type 1 nucleus.

I've often preached about the great improvement in interpretability of tests of this type when the samples are selected by a property which is thought to be fairly isotropic, such as far-IR, and H. Schmitt explores this in detail in his article in this volume. (See also Schmitt et al. 2001.) That's certainly not the case with the 12μ sample, since the very high column densities of the tori result in substantial optical depths at that wavelength, though this sample is much better in this regard than those based on say, UV excess. Aside from all theoretical considerations, and aside from reference to the X-ray columns, it seems obvious that the 12μ emission is quite anisotropic simply because the HBLR Seyfert 2s have much steeper mid-IR spectra than bare Type 1 nuclei (see e.g., Edelson & Malkan 1986). This wouldn't be the case if the mid-IR were emitted isotropically. The X-ray columns reinforce this: they are *much* larger in the Type 2s of all kinds than in the Type 1s, and this huge extra column density is probably dusty molecular gas. That last point is confirmed semi-quantitatively at least from molecular line maps.[4] See for example Planesas et al. 1991. Thus I can't agree with the characterization of the 12μ sample as complete and unbiased with respect to orientation.

For a few important related fine points of Tran's paper, he asserts that the contradictory conclusions of Heisler et al. 1997 may be due to sample "incompleteness" of the latter. As discussed below and in Schmitt's contribution to these proceedings, I believe the selection of the latter is in fact better, based largely on the far-IR which is thought to be relatively isotropic. The Heisler paper was recently updated as Lumsden et al. 2001, and the interpretation given there is more consistent with Tran's interpretation.

Other caveats about Tran's paper:

1) The polarimetry data come from Lick, Palomar and Keck Observatories. No indications are made that the observations are of uniform depth, and no upper limits are given for broad H-alpha for the putative non-HBLRs. It would not be simple to rectify this situation. However, I already see from a footnote

---

[3]Note that "non-HBLR" in Seyfert 2s means there is no BLR present at all, since the Seyfert 2 classification precludes a visible BLR.

[4]The X-ray columns in Type 1 nuclei, while much smaller in general, are still usually larger than those expected from the optical reddening/extinction estimates, for Galactic dust size distributions and dust/gas ratios (Maiolino et al. 2001 and references therein). It's possible that *this* gas contains only large grains, or perhaps is dust-free, but it's quite a different issue from the enormous excess columns of Type 2s relative to Type 1s.

to Moran et al. 2001 that *two of Tran's "non-HBLR" objects do show the broad H-α in polarized flux* with better data.

2) The warm mid-IR average colors of the HBLR objects have been discussed before (e.g. Heisler et al. 1997, Lumsden et al. 2001; Alexander 2001, and references therein; see also Miller and Goodrich 1990), with arguments as to whether the warmer colors might derive from lower inclinations as expected for tori theoretically. That could explain the lack of polarized broad Hα if the objects with the highest inclinations have partially obscured scattering regions. There is also evidence given in these papers that at least part of the effect (the correlation between detectability of a scattered BLR and mid-IR "warmth") is a contrast issue related to the relative dominance of the AGN in a particular object, and this seems natural and unavoidable too.

Tran argues (following Alexander 2001) that the cooler mid-IR spectra of the non-HBLRs cannot be due to higher extinction to the warm regions (and thus presumably to torus inclination) because the X-ray columns for the samples under consideration are statistically indistinguishable. To me that's suggestive at best because of the sample mismatches, and more importantly, because it's a purely qualitative argument. No theory predicts the magnitude of the expected difference between the columns of the HBLRs and non-HBLRs if the difference is due to inclination. One would need to have such a robust prediction to compare with any observational *upper limit* on the statistical column density difference, in order to prove a discrepancy.

3) A point that I find at least strongly suggestive (given the problematic selection criteria), is that the HBLR Seyfert 2s have a larger ratio of radio/far-IR than the others.[5] This seems to be a modest effect though (see Tran's Fig. 1; no statistical significance is given in the paper, but a significance of 0.8% is given by a K-S test: H.T., pc). If it is accepted to be statistically significant for the two populations, a next step would be to examine whether free-free absorption in the radio would be significant for the objects of highest inclination. Many recent papers on Seyfert radio properties have concluded free-free optical depths can be substantial in the cm region (e.g., Gallimore et al. 1997, Ulvestad 1999, and several others). Also a starburst component might have a different radio/far-IR ratio. But a different starburst contribution doesn't constitute strong evidence against a hidden BLR.

Tran and also Thean et al. 2001 find that the absolute radio luminosities of the HBLR Seyfert 2s are significantly greater than those of the non-HLBRs for the 12μ sample. However, with imperfect selection criteria, I'm more comfortable with comparing ratios of nearly isotropic properties such as discussed in the previous paragraph.

Also a general caveat should be kept in mind. The fact that a statistical difference in the populations of the putative types of 2 may exist is only suggestive of a qualitative difference in the physics. If all have hidden BLRs, the

---

[5]The paper also cites Moran et al. 1992 as providing evidence for intrinsically greater absolute radio luminosities in HBLR Seyfert 2s; that result was subject to severe selection effects, and was essentially retracted in Moran et al. 2000. Certainly given the anisotropic selection criteria for the Tran objects, I'd hesitate to make much of the radio power difference in the Tran paper.

two types would necessarily differ in average torus covering factor, so needn't be identical in other intrinsic properties.

4) The point is made that the [OIII] 5007/ Hβ ratio median is 9.9 ± 1.3 for the HBLR nuclei but only 6.8 ± 1.5 for the non-HBLRs. This brings up the vexing question of definitions for Seyfert 2. Well known examples of bright Seyfert 2s have ratios near 10. Half of the "non-HBLR Seyfert 2s" as used in the paper have ratios less than 6.8. The ratios aren't tabulated in the paper, but I'd interpret this as meaning that many so-called non-HBLRs are composites, with important LINER or starburst contributions to many of them. Certainly these components will reduce the integrated dust temperatures in the sense observed. But again that doesn't constitute evidence against a hidden BLR commensurate with the amount of high-ionization narrow line gas present.

5) Tran asserts that the non-HBLR Seyfert 2s simply have no hidden BLR. But the BLR is always accompanied by a "power-law" continuum, and this continuum is the only viable explanation for some of the narrow line ratios. Thus the question arises: how are the non-HBLR narrow line regions ionized? He speculates that the latter are "dominated by other nuclear and circumnuclear processes such as starbursts." This seems consistent with the statement that they have lower excitation, but also seems to reduce the question to semantics. No one doubts that there are objects dominated by e.g. starbursts, which thus have different narrow line ratios than Seyferts... it's just that the starbursts are relatively more important in some objects than in others, and the excitation is a measure of that. Again I don't see any implication that there's no hidden BLR commensurate with the requirements for narrow line ionization level in a particular object.[6]

6) The paper of Pappa et al. (2001) is cited as showing the existence of low intrinsic X-ray absorption in two Seyfert 2s of very low luminosity. Many Seyfert 2's are Compton thick and thus show no absorption turnover at low X-ray energies, but for these two, other arguments are given that suggest that this isn't likely. As Pappa et al. point out, the lack of an observed BLR could be intrinsic, or else it could be due to a dusty warm absorber as documented in several more luminous objects. I'm not aware that either of these two, NGC3147 or NGC 4698, has been checked for broad polarized Hα, but that would certainly be worthwhile.

My personal conclusion from all this is that the existence of objects without the Big Blue Bump and accompanying broad emission lines—commensurate with the luminosity of highly excited narrow line gas—has not been shown robustly. This is actually pretty similar to Tran's conclusion that "it is the strength of the AGN engine that seems to be the dominant factor in determining the visibility of the HBLR." Undoubtedly if the nuclear continuum and BLR strength can be turned down in a particular object, everything else being the same, the

---

[6]Nuclear activity seems manifest mainly as the Big Blue Bump and the resulting broad and narrow emission lines in radio quiet AGN. There is no question that some radio galaxies lack significant visible *or* hidden BBBs, based on the mid-IR argument mentioned in Footnote 2, but they are still AGN because of the radio activity. But what AGN activity exists in a non-HBLR non-radio-loud Seyfert 2? In finance, the analogy would be a bond which is a "zero coupon perpetuity," which makes no interest payments, and never pays back the principle. At that point it's not much of a bond, and a non-HBLR Seyfert 2 might be similarly ill-defined.

scattered light signal would be harder to see. The question is, does anything happen *qualitatively* at low AGN luminosities? Do these components decrease faster than say the strength of the high-excitation narrow line gas? It seems to me we're as far as ever from answering that question.

3. Testing the Generality of Seyfert Unification with Isotropic Properties

It would be great to have a dollar for every paper that concluded that Seyfert 1s and 2s are intrinsically different by showing that samples differ in some way, without selecting the sample by an isotropic property. No unified model says that any old batch of objects of one type is equivalent to any old batch of objects of another type. As an extreme example, with UV selection, an object whose UV excess is just scattered light must be much more powerful intrinsically than one whose UV excess is seen directly. For objects like NGC1068, only a percent or so of the nuclear UV is scattered into the line of sight, so Seyfert 2s found in this way come from five magnitudes higher on the luminosity function than Seyfert 1s found this way. No wonder they have more CO, L(far-IR), radio emission, etc, etc! I'll just mention one recent example. The Malkan et al 1998 paper, based on the Seyferts in the HST archives, concludes that Seyfert 2s are more likely to have nuclear dust structures. According to the verbal contribution of Maiolino et al. to the Guillermo Haro Workshop held at UNAM in 2000, and in a pc, this effect disappears if the Seyfert types are matched for [OIII] luminosity. (Other demurs regarding Malkan et al. 1998 can be found in Antonucci 1999a.)

A must-read paper in this context is Keel et al. 1994, which analyzes properties of Seyferts selected by 60μ flux and 25μ–60μ color. That sample should be pretty good—maybe the best that is available right now. Several old saws about Seyfert 2s having greater narrow line luminosities, ratios of narrow line luminosities to radio, etc, are disproven there. Not everything in Keel et al. can be easily explained by orientation though!

H. Schmitt is leading a major program of study of the sample from the Keel et al. paper, and he gives a report in these proceedings. (See also Schmitt et al. 2001.) He shows that several claimed differences between Seyfert 1s and 2s go away with this type of selection. To make the sample even better, we hope to add in the Seyferts dropped from the Keel et al list because of the 25μ–60μ color criterion. The result should be quite good.

There is a crucial limitation of this type of test, however. As emphasized in the past by e.g., A. Lawrence, there *must* be a range of covering factors for the dusty tori, and so those classified as Type 2 *must* have higher average covering factors. Therefore as populations, the 1s and 2s *must* be intrinsically different in their statistical properties at some level. This has *no direct implication* that some objects lack hidden BLRs!

To reiterate, even a perfect study, using isotropic selection and real upper limits, is expected to show differences between 1s and 2s in their intrinsic statistical properties. One could only confidently expect broad overlap in the properties of the 1s vs. the 2s even if every 2 has a hidden Type 1 nucleus. Such a difference would by itself have no direct implication that some of the Type 2s lack a BLR.

4. Testing the Beam Model for Radio Galaxies and Quasars with Isotropic Properties

This is an old issue, pretty much settled long ago I think. It still carries a good lesson, very closely analogous to that of the previous section. Blandford and Rees (1978) and Blandford and Konigl (1979) proposed that superluminal radio sources were simply normal double sources seen from very low inclinations, i.e., seen from along the axes of the radio jets. Early tests of the idea using the associated radio projected linear sizes (Browne[7] et al. 1982) and the narrow emission lines (Heckman 1983) found major discrepancies with it. However the double-lobed objects used in the comparison samples to the superluminals came from orders of magnitude higher on the luminosity function, because they satisfied survey flux limits with their nearly isotropic diffuse radio emission, whereas the superluminals satisfied survey flux limits because of their beamed core emission. It was exactly like the Seyfert 1/2 sample problems referred to in the last section. This was discussed in detail in Antonucci and Ulvestad 1985.

5. Testing the Energy Sources for Ultraluminous Infrared Galaxies with Mid-IR Spectroscopy

As for the previous section, I'll give just a brief sketch of this important issue, with a reference to a recent fuller discussion: Antonucci 2001.

The issue is the energy sources in Ultraluminous Infrared Galaxies, and whether recent ISO mid-IR spectral surveys provide breakthroughs in this area.

A large subset of the ULIRGs have properties *exactly* as expected for ultraluminous "Quasar 2s," that is to say, powerful high-ionization narrow line emission, powerful mid-far infrared emission, no optical point sources, but instead diffuse "mirror" regions revealing light from a hidden quasar. Note that the optical continuum is not expected to scale with nuclear luminosity because in low luminosity Seyfert 2s, it's almost always dominated by the underlying old stellar population.

The narrow emission lines have much lower equivalents widths in quasars than in Seyfert 1s, so their luminosity doesn't scale proportionately to the nuclear optical/UV either (e.g., Boroson & Green 1992, and Wills et al. 1993). It's a big effect.

As with nearby Seyfert 2s, it's difficult to be very precise about any starburst contribution to ULIRG energetics, but some well-studied cases reveal that it is often substantial. However, it is almost universally true that the objects characterized by Seyfert-2 like narrow line ratios in the optical show the same type of emission line spectrum in the mid-IR. Thus, the latter spectral region just reveals the same emission component as the optical for this group of ULIRGs.

At lower luminosities especially, many ULIRGs have starburst optical spectra, and also starburst mid-IR spectra. Again the mid-IR observations reveal just the same emission region as the optical, at least in a qualitative sense.

---

[7]Browne was, however, an early advocate for matching diffuse radio power.

Where the mid-IR data show something new is among the many ULIRGs with *LINER* optical spectra: in almost all of these cases the mid-IR spectra show lines from HII regions (e.g., Lutz et al. 1999)! This is interpreted most simply (and in the published papers, essentially solely) as proving that the true nuclei, and by implication the dominant energy sources, are compact starbursts. However, in general the star formation uncovered this way cannot be shown to dominate (or otherwise) the galaxy energetics because there is no accurate conversion from any particular starburst spectral feature to the bolometric luminosity of the associated starburst population.

Thus the ISO data show the presence of substantial but poorly determined starburst luminosity somewhere in the galaxies, generally *at the obscuration depth penetrated by light of these wavelengths*. Unfortunately, obscured AGN are known to possess column densities of $\gtrsim 10^{24}$ cm^{-2} in most cases, where $A(V) \gtrsim 1000$ and the mid-IR from the nucleus is quite obscured. There is a moderate range of parameter space such that the mid-IR can't penetrate the gas/dust columns, but the hard X-rays can. Above $> 10^{25}$ cm^{-2} or so however, the dusty gas becomes "Compton thick" and even the hard X-rays are blocked.

The point is that recently many ULIRGs classified as LINERS in the optical and starbursts in the mid-IR have since revealed powerful $\sim$ 10keV X-ray sources, often with luminosities suggestive of hidden AGN which are capable of providing the entire observed bolometric power. In other LINER ULIRGs a hidden powerful AGN is revealed by spectropolarimetry, though with that technique we have no good way of estimating the contribution of the AGN to the bolometric luminosity. These cases show that published conclusions that $> 50\%$ of the energy in a ULIRG derives from a hidden starburst are invalid. The same applies to the SCUBA sources, which were placed directly on the "Madau" diagram of stellar luminosity density evolution, without the slightest evidence that they're powered by stars!

The details and references are given in the review article cited at the top of this section (as well as in papers by experts such as Sanders and Mirabel, and Veilleux and several others). I just mention two cases here because they are really amusing to me. The well-studied ULIRG NGC6240 is a *LINER* in the optical, but a starburst in the mid-IR. In fact it's a "template" starburst in the mid-IR according to Genzel et al. 1998. The Genzel team claims to have shown that this and many other examples are starburst- dominated on this basis. Only trouble is, NGC6240 and many others have powerful AGN X-ray sources coming directly through the obscuring matter at 10keV (Vignati et al. 1999).

I don't know of any starburst spectral feature that can be accurately used to determine the starburst bolometric luminosity, so the starburst features can only say that there is such a component, and put a relatively low minimum value on its luminosity. The situation is a little better for AGN in that the X-ray luminosity is empirically a pretty good ($\sim$ factor of 3?) estimator of the bolometric value, based on the unobscured cases. Vignati et al. (1999) point out that that relation indicates that the bolometric luminosity of the hidden AGN in NGC6240 is consistent with the entire observed bolometric luminosity. Conservatively, I conclude from that that the AGN makes a significant contribution, though it's not certain that it's $> 50\%$.

The situation for NGC4945 is similar. Spoon et al. (2000) have analyzed the mid-IR spectrum finding that the starburst "may well power the entire bolometric luminosity, ... [but] are also consistent with an up to 50% contribution from an embedded AGN." Immediately afterward, Madejski et al. (2000) published a spectacular wide-band SED showing the few times 10^{24} X-ray column, and deriving the unabsorbed nuclear X-ray luminosity alone at $\sim 10^{43}$ erg/s, a substantial fraction of that in the mid-far-IR, and indicative of a bolometric luminosity consistent with that observed. It is very important to note that there are no indications of activity in NGC4945 other than the hard X-ray source. This means that active galaxies can look perfectly normal at other wavelengths! They do not need to show even the LINER-like emission lines!

Acknowledgements

For comments on an earlier draft I thank L. Kay, T. Heckman, B. Wills, M. Kishimoto, H. Schmitt, J. Ulvestad, and R. Barvainis. Detailed comments from B. Wills led to many changes. Support came from NSF grants NSF AST96-17160 and NSF AST00-98719.

References

Alexander, D. M. 2001, MNRAS, 320, L15

Antonucci, R. R. J. 1984, ApJ, 278, 499

Antonucci, R. R. J. & Ulvestad 1985, ApJ, 294, 158A

Antonucci, R., Hurt, T., & Miller, J. 1994, ApJ, 430, 210

Antonucci, R. 1999, ASP Conf. Ser. 161: High Energy Processes in Accreting Black Holes, 193 -also known as astro-ph/9810067

Antonucci, R 2001 - astro-ph /0103048 - to be pub in AGN Surveys, Proceedings of IAU Colloquium 184. Edited by R.F. Green, E.Ye. Khachikian, and D.B. Sanders. Publisher: ASP, Dates: June 18-22, 2001, Location: Byurakan, Armenia

Blandford, R. D. & Rees, M. J. 1978, Pittsburgh Conference on BL Lac Objects, Pittsburgh, Pa., April 24-26, 1978, Proceedings. (A79-30026 11-90) Pittsburgh, Pa., University of Pittsburgh, 1978, p. 328-341

Blandford, R. D. & Konigl, A. 1979, ApJ, 232, 34

Boroson, T. A. & Green, R. F. 1992, ApJS, 80, 109

Browne, I. W. A., Clark, R. R., Moore, P. K., Muxlow, T. W. B., Wilkinson, P. N., Cohen, M. H., & Porcas, R. W. 1982, Nature, 299, 788

Edelson, R. A. & Malkan, M. A. 1986, ApJ, 308, 59

Gallimore, J. F., Baum, S. A., & O'Dea, C. P. 1997, Nature, 388, 852

Genzel, R. et al. 1998, ApJ, 498, 579

González Delgado, R. M., Heckman, T., Leitherer, C., Meurer, G., Krolik, J., Wilson, A. S., Kinney, A., & Koratkar, A. 1998, ApJ, 505, 174

Heckman, T. M. 1983, ApJ, 271, L5

Heckman, T. et al. 1995, ApJ, 452, 549

Heisler, C. A., Lumsden, S. L., & Bailey, J. A. 1997, Nature, 385, 700
Huchra, J. & Burg, R. 1992, ApJ, 393, 90
Kay, L. E. 1994, ApJ, 430, 196
Keel, W. C., de Grijp, M. H. K., Miley, G. K., & Zheng, W. 1994, A&A, 283, 791
Koski, A. T. 1978, ApJ, 223, 56
S.L. Lumsden (1,2), C.A. Heisler (3), J.A. Bailey (2), J.H. Hough (4), S. Young, MNRAS, in press. Also known as astro-ph/0106263.
Lutz, D., Veilleux, S., & Genzel, R. 1999, ApJ, 517, L13
Maiolino, R., Marconi, A., & Oliva, E. 2001, A&A, 365, 37
Malkan, M. A., Gorjian, V., & Tam, R. 1998, ApJS, 117, 25
McLean, I. S., Aspin, C., Heathcote, S. R., & McCaughrean, M. J. 1983, Nature, 304, 609
Meisenheimer, K., Haas, M., Müller, S. A. H., Chini, R., Klaas, U., & Lemke, D. 2001, A&A, 372, 719
Miller, J. S. & Antonucci, R. R. J. 1983, ApJ, 271, L7
Miller, J. S. & Goodrich, R. W. 1990, ApJ, 355, 456
Moran, E. C., Halpern, J. P., Bothun, G. D., & Becker, R. H. 1992, AJ, 104, 990
Moran, E. C., Barth, A. J., Kay, L. E., & Filippenko, A. V. 2000, ApJ, 540, L73
Moran, E. C., Kay, L. E., Davis, M., Filippenko, A. V., & Barth, A. J. 2001, ApJ, 556, L75
Pappa, A., Georgantopoulos, I. Stewart, G.C., Zezas, A.L. astro-ph/0104061, to be published in MNRAS
Planesas, P., Scoville, N., & Myers, S. T. 1991, ApJ, 369, 364
Rush, B., Malkan, M. A., & Spinoglio, L. 1993, ApJS, 89, 1
Schmitt, H. R., Antonucci, R. R. J., Ulvestad, J. S., Kinney, A. L., Clarke, C. J., & Pringle, J. E. 2001, ApJ, 555, 663
Spoon, H. W. W., Koornneef, J., Moorwood, A. F. M., Lutz, D., & Tielens, A. G. G. M. 2000, A&A, 357, 898
Tran, H. D. 1995, ApJ, 440, 578
Tran, H. D. 2001, ApJ, 554, L19
Ulvestad, J. S., Wrobel, J. M., Roy, A. L., Wilson, A. S., Falcke, H., & Krichbaum, T. P. 1999, ApJ, 517, L81
Vignati, P. et al. 1999, A&A, 349, L57

Infrared Surveys for AGN

Harding E. Smith[1]

Center for Astrophysics & Space Sciences and Department of Physics, University of California, San Diego, La Jolla, CA 92093-0424, USA

Abstract. From the earliest extragalactic infrared studies AGN have shown themselves to be strong infrared sources, and IR surveys have revealed new populations of AGN. I briefly review current motivations for AGN surveys in the infrared and results from previous IR surveys. The Luminous Infrared Galaxies, which in some cases house dust-enshrouded AGN, submillimeter surveys, and recent studies of the cosmic x-ray and infrared backgrounds suggest that there is a population of highly-obscured AGN at high redshift. ISO Surveys have begun to resolve the infrared background and may have detected this obscured AGN population. New infrared surveys, particularly the SIRTF Wide-area Infrared Extragalactic Legacy Survey (*SWIRE*) will detect this population and provide a platform for understanding the evolution of AGN, Starbursts and passively evolving galaxies in the context of large-scale structure and environment.

1. Motivation: Active Galaxies in the Infrared

There are a number of motivations for infrared studies of AGN and for carrying out surveys for AGN at infrared wavelengths:

1. AGN unification models require a dusty, molecular obscuring screen (torus?) which is expected to emit in the infrared.

2. Cosmic X-Ray Background models require a population of highly obscured AGN which may be detectable in the infrared.

3. Infrared/Starburst emission is a frequent companion to AGN activity; IR emission originates from dust which is either associated with the active nucleus (*e.g.* the PG QSO sample; Sanders *et al.* 1989) or with circumnuclear starburst emission as in the classical Seyfert galaxies NGC 1068 & NGC 7469.

4. Luminous Infrared Galaxies frequently harbor AGN cores and may be a stage in the development of AGN from the merger of gas-rich galaxies.

---

[1] also, Infrared Processing and Analysis Center, Caltech/JPL, Pasadena, CA 91125

1.1. AGN Unification Models

The AGN Unification models (Antonucci 1993) which have been popular for some time, and which have garnered considerable observational support, require an optically-thick, dusty obscuring screen, frequently assumed to be toroidal (Pier & Krolik 1993; Granato & Danese 1994), which obscures the central engine and broad emission-line region from equatorial lines of sight. In this scenario broad-line objects (Type 1 AGN – Sy1 galaxies and classical QSOs) are objects viewed near the pole of the torus, while narrow-line objects (Type 2 AGN – Sy 2 galaxies) are viewed edge-on. Direct evidence for the existence of this obscuring screen is largely from studies of scattered broad-line emission in Sy 2 systems, but claims have been made that the mid-infrared spectra of some nearby galaxies match various torus models (Pier & Krolik 1993; Granato & Danese 1994). A critical point regarding this torus is that *it will be optically thick at all wavelengths from soft x-rays through the mid-infrared.* Even mid-infrared diagnostics may not reveal a highly obscured AGN. Hard X-rays and VLBI radio imaging are currently required to peer through the veil of obscuration.

1.2. The Cosmic X-Ray and Infrared Backgrounds

Recent models for the Cosmic X-Ray Background (CXB) require a population of highly obscured, perhaps even Compton-thick, AGN which emit primarily in the hard X-ray region in order to fit the X-ray background spectrum (Comastri *et al.* 1995; Gilli *et al.* 2001). This population increases in density with redshift. These AGN are expected to re-emit the absorbed radiation at infrared wavelengths providing a link between the CXB and the Cosmic InfraRed Background (CIRB). Recent measures of the CIRB with COBE (Hauser *et al.* 1998; Puget *et al.* 1996) show that more than half of the cosmic energy density (excluding the CMB) comes out in the infrared. An important question is then the relative importance of accretion energy due to AGN compared with stellar nucleosynthesis. Deep surveys with Chandra and XMM-Newton are resolving the CXB (Hornschemeier *et al.* 2001; Hasinger *et al.* 2001; Rosati *et al.* 2001) and ISO (Elbaz *et al.* 2002) has begun to resolve CIRB. New Surveys with SIRTF will convincingly determine the relationship between the faint X-ray and infrared populations revealing a great deal about the history of our Universe.

1.3. Luminous Infrared Galaxies

The most luminous galaxies in the Local Universe are Luminous Infrared Galaxies (LIGs) which emit the vast majority of their radiant power in the far-infrared between about 40–120μm. These are gas-rich systems which are in the late stages of collisions or mergers. Extrapolation from the properties of lower luminosity Starburst galaxies suggests that the LIGs should be active star-forming systems (see Sanders & Mirabel 1996 for a review). The LIGs also show many charactersiatics of AGN and their luminosities reach values comparable to those of luminous QSOs. Much effort has been focused on whether LIGs are powered principally by Starburst or AGN activity, although both types of activity are almost certainly present. The discussion has been framed around a scenario proposed by Sanders *et al.* (1988) in which a merger of gas-rich disk galaxies stimulates a massive nuclear Starburst which in turn feeds a coalescing AGN core in the galaxy nucleus. As the AGN turns on, radiation pressure drives out

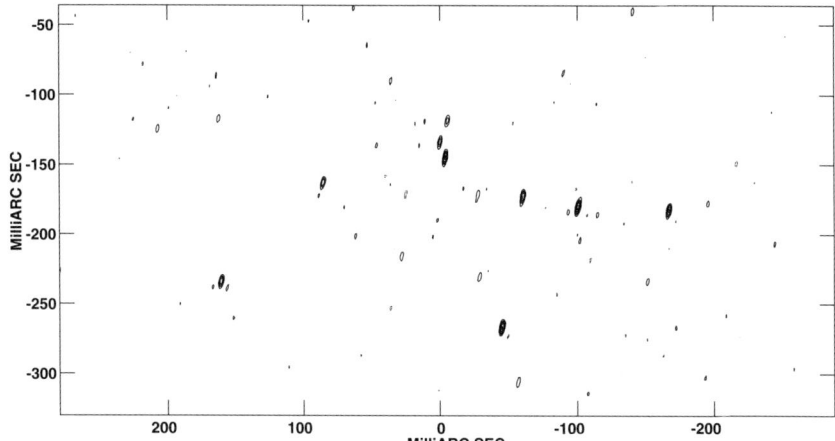

Figure 1. 18 cm VLBI image of the W Nucleus of Arp 220 from Smith et al. (1998). Over a dozen unresolved sources are interpreted as luminous radio supernovae in an intense starburst.

the shroud of dust, revealing a nascent quasar. The goal must be not only to understand the dominant source of energy in LIGs, but to understand the relationship between Starburst and AGN activity and other galaxy characteristics, and to place them into an evolutionary context.

We have for some time been using VLBI techniques to attempt to understand the power sources of LIGs and to place them in an evolutionary context. Analysis of a complete LIG sample (Smith, Lonsdale & Lonsdale 1998) suggests that many LIGs may be interpreted as intense Starbursts like Arp 220 (Figure 1), but a nearly equal number must house AGN cores, as in the case of Mrk 231, whose nuclear structure implies a recent ignition ($t << 10^6 yr$) for AGN activity. Evidence is strong that a significant number of LIGs house obscured AGN activity and schematics are consistent with this Starburst-to-AGN scenario.

Of particular interest are the extremely luminous high-z galaxies detected in recent SCUBA and other submillimeter surveys (Ivison et al. 2000). The SEDs of LIGs in the submillimeter have the unique characteristic that the positive K-correction offsets cosmological dimming for redshifts from $z \sim 1 - 10$ such that it is equally easy (or difficult) to detect infrared galaxies at $850\mu m$ over a range of high redshifts. Although only a handful of redshifts are available for submm sources owing to the large error circles for submm sources and the optical faintness of the small number of identified galaxies, existing redshifts confirm that the submm population lies at redshifts $1 < z < 3$. The existence of these sources implies a population of very luminous galaxies at very early epochs, with concomitant rapid evolution. The star-formation rates inferred from the luminosities of these systems are not easily produced in CDM galaxy formation models. In analogy with local LIGs, the submillimeter sources may be candidates for nascent QSOs at high-redshift.

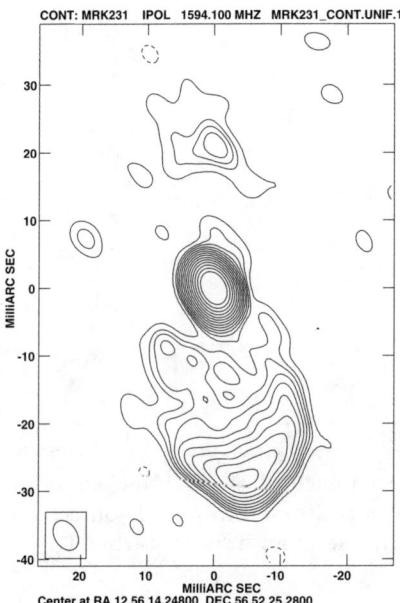

Figure 2. 18 cm VLBI image of the central 100pc of Mrk 231, interpreted as a nascent QSO; age $<< 10^6$yr.

2. Active Galaxies with IRAS

One of the most striking results from IRAS is that the infrared galaxy population evolves more rapidly than the optical galaxy population (Lonsdale et al. 1990) with a form similar to that for QSOs. Selection of Luminous IR Galaxies associated with FIRST radio sources (Stanford et al. 2001) or by FIR/Optical ratio (Smith et al. 2002; Lonsdale et al. 2002) have led to the detection of IRAS Faint Sources up to $z \sim 1$ with a small number of "hyperluminous" or lensed sources with redshifts in excess of 1 (e.g. Rowan-Robinson et al. 1991).

IRAS showed that the infrared spectra of QSOs are essentially similar, with a broad 'infrared bump' from about 2μm to 1 mm, with a significant fraction (10-50%) of the bolometric luminosity emitted at infrared wavelengths. This infrared emission is almost certainly thermal dust emission, but the location of this warm dust remains uncertain (Sanders et al. 1989) possibly associated with the outer edges of the accretion disk, the torus, or circumnuclear star formation.

Most of the known Sy 2 galaxies were discovered by IRAS, but the sensitivity of IRAS has limited our census of "Type 2 AGN" to $z < 0.2$. As Padovani (1998) has previously stressed, at $z < 0.2$ AGN Type 2 outnumber Type 1, suggesting that there are a large number of Sy 2 galaxies and Type 2 QSOs yet to be discovered at higher redshift. The expectation from our local census and from the CXB models suggests that an important task of future IR AGN Surveys will be detecting the high-redshift AGN 2 population, or explaining its absence in terms of AGN evolution.

3. 2MASS Quasars

The 2MASS AGN surveys have been reviewed by R. Cutri (2002; this volume) and the characteristics of that population will be discussed only briefly. The surface density of 2MASS AGN is of the order of 0.5 deg^{-2} with a space density comparable to that found in low-redshift optical-to-X-ray selected samples. The colors and polarization of 2MASS QSOs suggests that these are *reddened* rather than intrinsically red QSOs and the ratio of Type 1 AGN to Type 2 is about 2:1 — the obscured QSO 2's are not being found in 2MASS.

The color selection applied to detect 2MASS QSOs is $(J-K) > 2$ and Cutri *et al.* argue that these QSOs are sufficiently red to have been missed in previous optical surveys. Analysis by the SDSS and 2MASS teams (Ivezic *et al.* 2002; this volume) suggests, however, that the incompleteness of the SDSS to such red QSOs is less than 10%. The 2MASS AGN are faint in X-rays with a wide range in hardness-ratio (Wilkes *et al.* 2002) suggesting that the faintest and hardest sources may be highly obscured and could provide the missing X-ray population required to explain the CXB.

4. ISO AGN studies and Surveys

4.1. Mid-Infrared Diagnostics for AGN

ISO has shown that mid-infrared spectroscopy is a powerful tool for source classification. In a series of papers (Lutz *et al.* 1998, Tran *et al.* 2001) the midinfrared molecular band strengths have been used to discriminate between AGN-dominated (7.7μm line to continuum < 1) and Starburst-dominated (7.7μm to continuum > 1) LIGs. These results suggest that the majority of LIGs with $\log L_{fir} < 12(L_\odot)$, the "ULIGs", are Starburst-dominated, with an increasing AGN-fraction with increasing L_{fir}. Comparison with optical spectroscopy (Lutz, Veilleux & Genzel 1999) shows that the infrared classifications are consistent with classical optical excitation methods and suggests that infrared galaxies with *LINER* spectra are Starbursts (Veilleux 2002; this volume).

A somewhat different direction has been taken by Clavel *et al.* (2000) who compared the strengths of the mid-infrared bands among AGN types. In their sample there is a clear distinction between Sy 1 galaxies, showing low $W_\lambda(7.7\mu m)$, and Sy 1.5-2 galaxies with higher values of $W_\lambda(7.7\mu m)$. The 7.7μm *luminosities* of the Sy galaxy types are comparable, however. They interpret this result in terms of the classical AGN-torus model, suggesting that the warm AGN continuum from the inner torus is extinguished in the edge-on Sy 2 systems but visible in the Sy 1 systems, viewed face-on, whereas the extended, Starburst-related mid-infrared features are visible in both types of galaxies.

Taken together these results underscore the caveat discussed above, that AGN activity in compact, highly-obscured infrared galaxies may remain hidden even at mid-infrared wavelengths.

4.2. ISO Deep Surveys

A number of deep surveys were undertaken with the ISO satellite at wavelengths of 7, 15, 90, and 170μm. These are reviewed in detail by Taniguchi (2002; this volume) and the results will only be summarized here. The principal result from these surveys is the continuing high surface densities of infrared galaxies as ISO pushed to lower flux-densities — as low as $\sim 10\mu$Jy at the lower wavelengths — requiring continued steep evolution with redshift. There are a number of models with varying prescriptions for the evolving population (Franceschini et al. 2001; Rowan-Robinson 2001; Xu et al. 2001). The models of Xu et al., based upon a local complete 24μm sample, require luminosity evolution rates as high as $L \propto (1+z)^{4.2}$ combined with density evolution, $\rho \propto (1+z)^2$, up to $z \sim 1$ for the Starburst population to match the $log\,N - log\,S$ relation; the number of LIGs at $z \sim 1$ is thus estimated to be approximately 40 times higher than in the local Universe (Franceschini et al. 2001). The Xu et al. model has pure luminosity evolution for the AGN population, $L \propto (1+z)^{3.5}$, whereas the "normal" galaxy population evolves as $L \propto (1+z)^{1.5}$.

With about 2000+ galaxies resolved by ISO and over 400 spectroscopic observations, AGN account for about 10% of the identified sources in the ISO Surveys. The ratio of AGN type 1 to type 2 in these surveys is about unity, at variance with the IRAS results at higher flux-density.

Elbaz et al. (2002) have analyzed the deep ISO 15μm counts and estimated that the IR galaxies detected by ISO to a 15μm flux-density, $S_{15\mu m} > 50\mu$Jy, contribute over half of the CIRB and that AGN may contribute, at most about 20% of the IR background. Fadda et al. (2002) have combined the deep ISOCAM/XMM-Newton data from the HDFN and Lockman Hole with the brighter ELAIS S1/BeppoSAX data to estimate the AGN contribution to mid-IR surveys and the CIRB. Using X-ray emission as an indicator of AGN activity, Fadda et al. estimate that 15-20% of the mid-infrared emission in Lockman and HDFN originates from AGN. The detection (AGN) fraction of 15μm sources increases with X-ray energy from 30% below 2 keV to over 60% above 5 keV, as might be expected if the X-ray background is produced by obscured, high-column sources. Again, the fraction of the CIRB attributed to AGN-accretion energy is estimated to be less than 20%.

Further analysis of red X-ray luminous galaxies ($L_x \sim 10^{43}$-$10^{45} erg/s$) in the same sample (Franceschini et al. 2002) suggests that these may be highly-obscured AGN as predicted by CXB background models. The mid-infrared SEDs of these sources are well reproduced by model spectra of obscured QSOs with $\tau_{0.3\mu m} \sim 30$-40 and the ratio of Type 1-to-Type 2 AGN is 1:3 in agreement with predictions.

The areas surveyed remain small, less than a few hundred square arcminutes to $S_{15\mu m} < 100\mu$Jy, and statistics are restricted to small numbers with a few tens of confirmed AGN. The lower sensitivities at the longer wavelengths require the above analyses to employ template SEDs for estimating the contributions of ISO sources at the peak of CIRB near 140μm. With the wide range of mid-to-far-IR SEDS observed in the local Universe (e.g. the Far-IR/Mid-IR ratio may differ by an order of magnitude between a "typical Starburst" like M82 and a LIG such as Arp 220) these intriguing results remain tentative.

5. SWIRE: The SIRTF Wide-area InfraRed Extragalactic Survey

For *SIRTF*, the last of its *Great Observatories*, NASA has selected a set of Legacy Programs, designed to be major surveys of general interest to the Astronomical community and to be carried out in the first year of the SIRTF mission. The Survey data will be distributed to the community in time for use in preparation of *General Observer Proposals* with no proprietary period for Legacy data.

The SIRTF Wide-area InfraRed Extragalactic Survey (*SWIRE*, Dr. Carol Lonsdale, P.I.) is the largest of the six SIRTF Legacy Surveys (851 hours), surveying approximately 67 square degrees in all 7 SIRTF imaging bands. A current description of the SWIRE Survey is given on the SWIRE WebPages: http://www.ipac.caltech.edu/SWIRE. Table 1 lists the Survey sensitivities.

Table 1. SWIRE Sensitivity Limits (est. 5σ)

	IRAC			MIPS	
λ	Sensitivity	Resolution	λ	Sensitivity	Resolution
3.6μm	7.3μJy	0.9″	24μm	0.45mJy	5.5″
4.5μm	9.7μJy	1.2″	70μm	2.75mJy	16″
5.8μm	27.5μJy	1.5″	160μm	17.5mJy	36″
8.0μm	32.5μJy	1.8″			

The Survey will cover seven high-latitude fields, selected to be the most transparent, lowest background fields in the sky. The fields, covering between 5 and 15 sq. deg. include previously well-known IR extragalactic survey fields (e.g. Lockman and the ELAIS ISO Survey Fields) and x-ray fields (Chandra Deep South and XMM Large Scale Survey) are shown in Table 2.

Table 2. SWIRE Survey Fields

Field	Center (J2000) RA	Dec	Area (sq deg)	Background (MJy/Sr)
ELAIS S1	$00^h 38^m\ 30^s$	$-44°\ 00'$	14.8	0.42
XMM-LSS	$02^h 21^m\ 00^s$	$-05°\ 00'$	9.3	1.3
Chandra-S	$03^h 32^m\ 00^s$	$-28°\ 16'$	7.2	0.46
Lockman	$10^h 45^m\ 00^s$	$+58°\ 00'$	14.8	0.38
Lonsdale	$14^h 41^m\ 00^s$	$+59°\ 25'$	6.9	0.47
ELAIS N1	$16^h 11^m\ 00^s$	$+55°\ 00'$	9.3	0.44
ELAIS N2	$16^h 36^m\ 48^s$	$+41°\ 02'$	4.5	0.42

The SWIRE science goal is to enable fundamental studies of galaxy evolution in the infrared for $0.5 < z < 3$:

- Evolution of star-forming and passively evolving galaxies in the context of structure formation and environment.
- Spatial distribution and clustering of evolved galaxies, Starbursts, & AGN.
- The evolutionary relationship between galaxies and AGN and the contribution of AGN accretion energy to the cosmic backgrounds.

Galaxy evolution models which match the IRAS/ISO galaxy counts at all wavelengths from 7–100µm as well as the CIRB (Xu et al. 2001) predict that SWIRE will detect of the order of 2 million galaxies — spheroids and evolved stellar systems with IRAC, and active star-forming systems with MIPS. SWIRE will also detect about 25,000 classical AGN, and an unknown number, perhaps several times as many, dust-enshrouded AGN.

Recent estimates of the "Universal Star-formation History (*SFH*)" (Steidel et al. 1999) suggest that the bulk of cosmic evolution occurs between redshifts, $0.5 < z < 3$, the redshift interval for which SWIRE is optimized. The median redshift is predicted to be, $\langle z \rangle \sim 1$, where many estimates find a peak in the *SFH*; luminous infrared galaxies will be detected by *SWIRE* out to $z \sim 3$. Previous estimates of the *SFH* have varying, frequently large and uncertain corrections for extinction. *SWIRE* will directly measure the total star-formation rates as a function of redshift and environment over this critical range of time and redshift.

A key element in the *SWIRE* Survey design is to enable galaxy evolution studies in the context of large-scale structure/environment. One of the *SWIRE* Survey fields covers the deep survey areas of the XMM-LSS Survey (Pierre 2001) so that the infrared galaxy census may be directly tied to the presence of rich X-ray clusters to $z > 1$. *SWIRE* will sample several hundred, 100 Mpc scale co-moving volume cells enabling a variety of large-scale structure measures from correlation functions, power spectra, and counts-in-cells to direct comparison with model calculations. SWIRE's measures of the star-formation as a function of environment will be important input for CDM simulations which have been exceedingly successful in simulating the development of structure in the early Universe, but perhaps less so in simulating galaxy evolution within that structure owing to the complexity of the physics of star formation (*e.g.* Kay et al. 2002).

Of more direct importance to this Conference, the similarities of AGN SEDs in the mid-far Infrared suggests that *SWIRE* will be unbiased with respect to AGN types and ages, enabling a complete census of AGN out to redshifts greater than 1. Although the detection rates should be unbiased, the similarity between the SEDs of obscured AGN to those of Starbursts, and the extreme optical depths will make *identifying* the obscured AGN population very challenging. Low-frequency radio surveys will, of course, identify radio-loud AGN, but these make up only 10–15% of the AGN population. For this reason the XMM-LSS Survey, along with current and planned deeper surveys in hard X-rays will be vital to identifying *SWIRE* AGN.

5.1. Supporting Observations

An aggressive program of ground-based optical, near-infrared and radio observations is planned in support of the SWIRE Survey and we are actively pursuing other programs with HST, Chandra, XMM and Galex. As already described Chandra and XMM Surveys will be important for discovering the obscured AGN population, if it exists. SWIRE has entered into cooperation with the Galex team so that the SWIRE fields will be included in the Galex Deep Survey.

The SWIRE Optical-Near Infrared goal is to obtain moderate-depth optical multi-band ($g' \sim 25.7$, $r' \sim 25$, $i' \sim 24$; Vega magnitudes, 5σ detection for a 2" galaxy) data for the entire Survey area. At these limits we expect to detect approximately 2/3 of SWIRE sources detected by both MIPS and IRAC. The

ELAIS N1, N2 fields have already been imaged to somewhat shallower limits ($r' \sim 24$) as part of the INT Wide Field Survey and efforts continue to push deeper in the optical and into the near infrared as part of the UK SWIRE Program (ISLES and UKIDSS projects respectively). An extensive program for observations of ELAIS S1 is being undertaken at ESO. Optical and near-infrared imaging of the Lockman, Lonsdale and CDFS fields are being undertaken at KPNO and CTIO with the Mosaic cameras and FLAMINGOS infrared imager.

Two major SWIRE radio surveys are planned. The median 20cm flux density predicted for SWIRE Starburst galaxies is $\sim 43\mu$Jy — to faint to survey the entire area to this depth. We have therefore planned a deep pencil-beam VLA Survey (F. Owen, PI) and an extended shallow VLA survey (J. Condon, PI):

- SWIRE Lockman Deep Survey — 3μJy rms @ 20cm; $\alpha = 10^h 46^m$ $\delta = +59° 01'$; $30'$ VLA primary beam. The Deep VLA Survey is nearly completed and data analysis is just beginning.

- Cosmic Windows VLA Survey — 50μJy rms @ 20cm in the combined fields of SWIRE, Galex and XMM-LSS which are accessible to the VLA. This Survey is being proposed for the next VLA large survey program.

6. Benediction

The *SWIRE* Legacy Survey is a community Survey; the large dataset which is being accumulated reflects the synergies which between the Legacy program and other community surveys. With a couple of million galaxies and several tens of thousands of AGN, many with redshift estimates and SEDs from x-ray to radio, the SWIRE database will be released to the community through IPAC's Infrared Science Archive. We hope that *SWIRE* will provide a rich datamine for the entire community and will provide answers to many of the questions posed here. If you have projects that you would like to do with *SWIRE* data, please visit the *SWIRE* WebPages and/or contact one of the team members.

It is a pleasure to thank the local organizers, especially Areg Mickaelian and Ed Khachikian, and the scientific organizing committee for my second stimulating visit to beautiful Armenia. This research was supported by the US NASA.

References

Antonucci, R. 1993, ARA&A, 31, 473.
Clavel et al. 2000, A&A, **357**, 839.
Comastri, A., Setti, G., Zamorani, G., & Hasinger, G. 1995, A&A, **296**, 1.
Elbaz, D., et al. 2002, A&A, in press. (astro-ph/0201328)
Fadda, D., et al. 2002, A&A, in press. (astro-ph/0111412)
Franceschini, A., et al. 2002, A&A, in press. (astro-ph/0111413).
Franceschini, A., et al. 2001, A&A, **378**, 1.
Gilli, R., Salvati, M. & Hasinger, G. 2001, A&A, **366**, 407.
Granato, G. & Danese L., 1994, MNRAS, **268**, 235.
Hasinger, G., et al. 2001 A&A, **365**, 45.

Hauser, M., et al. 1998, ApJ, **508**, 25.
Hornschemeier, A., et al. 2001, ApJ, **554**, 742.
Ivison, R., et al., 2000, MNRAS, **315**, 209.
Kay, S., Pearce, F., Frenk, C., & Jenkins, A. 2002, MNRAS, **330**, 113.
Lonsdale, C. J, et al. 1990, ApJ, **358**, 60.
Lonsdale, C., Hurt, R., & Smith, H. E., & Xu, C. 2002, ApJ, *in preparation*.
Lutz, D., et al. 1998, ApJ, **505**, L103.
Lutz, D., Veilleux, S. & Genzel, R. 1999, ApJ, **517**, L13.
Padovani, P. 1998 in *New Horizons from Multi-Wavelength Sky Surveys*, ed. B. McLean, D. Golombek, J. Hayes, & H. Payne, (Kluwer), p. 257.
Pier, E. & Krolik, J. 1993, ApJ, **418**, 673.
Pierre, M. 2001 in *Where's the Matter?*, eds. L. Tresse & M. Treyer, *in press*. (astro-ph/0111242)
Puget, J-L., et al. 1996, A&A, **308**, 5.
Rosati, P. et al. 2002, ApJ, **566**, 667.
Rowan-Robinson. M., et al. 1991, *Nature*, **351**, 719.
Rowan-Robinson, M. 2001, ApJ, **549**, 745.
Sanders, D. B. & Mirabel, I. F. 1996, ARA&A, **34**, 749.
Sanders, D., et al. 1988, ApJ, **325**, 74.
Sanders, D., et al. 1989, ApJ, **347**, 29.
Smith, H. E., Lonsdale, C., & Lonsdale, C. 1998, ApJ, **492**, 137.
Smith, H. E., Lonsdale, C., Lonsdale, C. & Diamond, P. 1998, ApJ, **493**, L17.
Smith, H. E., Lonsdale, C., Hurt, R. & Siana, B. 2002, ApJ, *in preparation*.
Stanford, S., et al. 2000, ApJS, **131**, 185.
Steidel, C., et al. 1999, ApJ, **519**, 1.
Tran, Q., et al. 2001, ApJ, **552**, 527.
Wilkes, B., et al. 2002, ApJ, *in press*. (astro-ph/0112433)
Xu, C., Lonsdale, C., Shupe, D., O'Linger, J., & Masci, F. 2001, ApJ, **562**, 179.

Unveiling the Evolution of Type I AGNs in the IR (15μm) — As Seen by ISO in the ELAIS-S1 Region

Israel Matute, Fabio La Franca

Dipartimento di Fisica, Università degli studi "Roma Tre", Via della Vasca Navale 84, I-00146 Roma, Italy

Carlotta Gruppioni, Francesca Pozzi, Carlo Lari

Osservatorio Astronomico di Bologna, Via Ranzani 1, I-40127 Bologna, Italy

Abstract. We present the first estimate of the evolution of type 1 AGNs in the IR (15 μm) obtained from the ELAIS survey in the S1 region. We find that the luminosity function (LF) of Type 1 AGNs at 15μm is fairly well represented by a double power-law function with a bright slope of 2.9 and a faint slope of 1.1. There is evidence for significant cosmological evolution according to a pure luminosity evolution model $L_{15}(z) \propto (1+z)^k$, with $k=3.00^{+0.16}_{-0.20}$ in a $(\Omega_m, \Omega_\Lambda)=(1.0, 0.0)$ cosmology. This evolution is similar to what is observed at other wavebands. From the luminosity function and its evolution, we estimate a contribution of $\sim 2\%$ from Type 1 AGN to the total Cosmic Infrared Background (CIRB) at 15 μm.

1. Introduction

In the past, AGN samples have been mainly selected in the optical and more recently in the soft-X with ROSAT. The first statistically significant samples of AGNs in the Infrared come from the IRAS mission. The lack of enough sensitivity for the IRAS satellite has limited these samples to objects in the local Universe (z≤0.1). The evolution of the AGN population in the infrared could only be extrapolated from what was already known in the optical and X-ray.

The European Large Area ISO Survey (ELAIS) was carried out in order to extend to deeper fluxes the coverage of the luminosity-redshift space in the IR. This has given us the possibility to study the high redshift population (z≥0.2), and allowed us to have the first insights on the evolution of IR selected sources.

2. The Sample: ELAIS-S1

The European Large Area ISO Survey (ELAIS) is the largest single open time project conducted by ISO (Oliver et al. 2000), mapping an area of 12 deg² at 15μm with ISOCAM and at 90μm with ISOPHOT. Four main fields were chosen

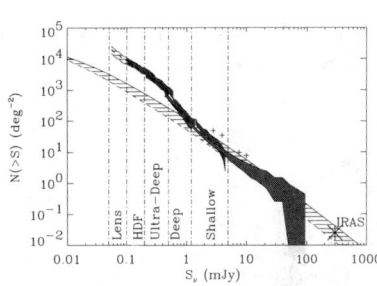
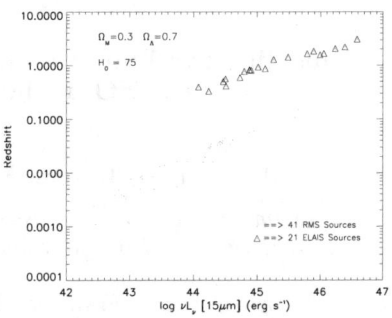

Figure 1. a) Integral Counts at 15μm from the Final Analysis of the ELAIS-S1 Region (light grey shaded area). Also plotted are the faintest IRAS surveys and the deep ISOCAM surveys (dark grey shaded areas). b) The luminosity/redshift distribution of our 21 ELAIS-S1 and 41 RMS Type I AGNs that entered into the computation.

(N1, N2, N3 in the northern hemisphere and S1 in the south) due to their high Ecliptic latitudes ($|\beta| > 40°$) and low cirrus emission.

An initial catalog for S1 (J2000, α : $00^h34^m44^s$, δ : $-43°34'44''$), covering an area of 3.96 deg^2, was produced using the Imperial College data reduction technique ("Preliminary Analysis", Serjeant et al., 2000). Optical identifications were possible thanks to an extensive R-band CCD survey, performed with the ESO 1.5m/Danish telescope. The spectroscopic follow-up program, carried out at the AAT at the AAO and the 3.6m/NTT at ESO/La Silla, of 114 sources in S1 fainter than R∼17.0 provided the sample presented here.

The ELAIS-S1 field has also been completely covered in the radio at 1.4 GHz down to 0.3 mJy (Gruppioni et al., 1999), and 50% covered in the X-rays with BeppoSAX (Alexander et al., 2001). Now the Final Analysis has been completed by the ELAIS team in Bologna. The resulting complete catalog includes more that 450 sources, with fluxes at 15μm down to 0.5 mJy and selected with S/N>5 (Lari et al., 2001).

For many years there has been a gap in flux between the brighter samples (>300 mJy), coming from the IRAS surveys and covering large sky areas, and the deep/pencil-beam surveys carried out by ISO at much fainter fluxes (<1 mJy). In the integral counts derived from the final analysis (Figure 4a, Gruppioni et al., in prep) we see how the ELAIS survey fills the whole flux range between these two regimes.

The nature of the objects identified by the spectroscopic follow-up is:
- High star-forming galaxies (45%);
- AGNs (Type 1 & 2) represent 30%;
- 15% of galaxies dominated by absorption lines;
- 4% of late-type stars;
- 6% of unclear classification (Type 2 AGN, Starburst, LINER).

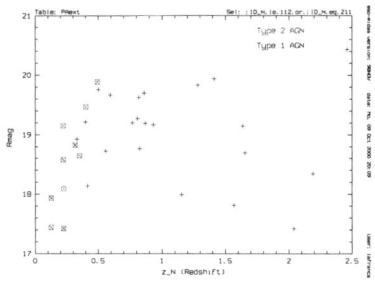 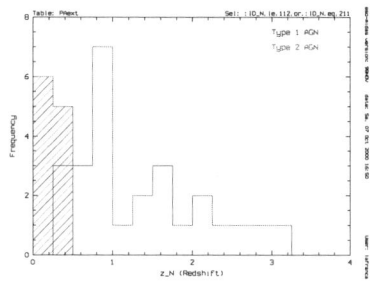

Figure 2. a) Distribution of Type I (plus symbols) and Type 2 (crossed squares) AGN in redshift-Rmag space. While the R-magnitude distribution is constant with a large spread for Type I AGN, Type 2 AGN show a trend with redshift, becoming fainter than R=20.0 at $z \sim 0.5$. b) Distribution in redshift space of Type I AGN and AGN2 (shaded area). If we consider the redshift bin $z=[0.0,0.5]$ the ratio of AGN2/Type I AGN becomes $11/3 \sim 4$.

Table 1. The percentages of all objects with emission lines

		type 1 AGN	type 2	ELG	LINERS
ELAIS-S1	%	20	10	50	2
15μm	\<z\>	1.4	0.3	0.2	0.3
IRAS	%	6	17	54	23
12μm	\<z\>	0.04	0.015	0.014	0.007

Note - IRAS subsample from Alexander & Aussel (2000)

The observed ratio of the number of Type 2 with respect to Type 1+2 AGNs is 1/3. But this result is mainly an artifact due to the different selection functions of the sample for Type 1 AGN and Type 2 objects. The optical selection introduced in the spectroscopic follow-up ($17.0 < R < 20.0$) seems to be the origin of the lack of Type 2 AGNs beyond $z \sim 0.5$. As a matter of fact, in Figure 2a a trend is observed for Type 2 with optical magnitudes being fainter at larger redshifts. In Figure 2b the redshift distributions of Type 1 and Type 2 AGN are shown. In the redshift bin where Type 2 AGNs are observed ($z=0.0$-0.5), the ratio Type2/Type1 becomes ~ 4, similar to the predictions of the standard unification model. The follow-up campaign carried out in September/November 2001 with the ESO telescopes, and currently under study, will verify if this trend is seen at fainter and brighter optical fluxes.

3. The Evolution of Type I AGN

We have assumed $H_0 = 75$ km s^{-1} Mpc^{-1} and a $(\Omega_m, \Omega_\Lambda) = (1.0, 0.0)$ cosmology. Our ELAIS preliminary sample of 21 Type 1 AGN is statistically significant enough to compute a first estimate of the evolution of these objects in the Mid-IR. The mean SED from Elvis et al. (1994) for radio quiet QSOs was adopted as a good representation of our sources. K-corrections in the IR were computed following Lang (1980) for each of the two different filters: ISOCAM-

LW3 at 15μm and IRAS 12μm. K-corrections in the R band were taken from Natali et al.(1998).

A subsample of Type 1 AGN was extracted from the catalog of Rush, Malkan & Spinoglio (1993) (RMS hereafter), as representative in the local universe of this type of object. The catalog consists of a sample of galaxies selected at 12μm from the IRAS Point Source Catalog PSCv2 (Moshir et al. 1991), and is complete down to 0.3 Jy. With the computed K-correction, 15μm νL_ν luminosities (L_{15}) were derived. Figure 1b represents, in luminosity-redshift space, all Type 1 AGN coming from ELAIS and RMS that have been used in this analysis.

Similar to what was found in the optical (La Franca & Cristiani 1997; Boyle et al. 2000) and in the X-rays (Miyaji et al. 2000; La Franca et al. 2001 submitted), we adopted a smooth double power-law for the space density distribution of QSOs and Seyfert 1s in the local universe ($z=0$):

$$\frac{d\Phi(L_{IR}, z=0)}{d\mathrm{Log}L_{IR}} = \frac{\Phi^*}{\left[(L_{IR}/L_*)^\alpha + (L_{IR}/L_*)^\beta\right]}$$

A standard pure luminosity evolution (PLE) has been adopted of the form $L_{15}(z) = L_{15}(0)(1+z)^k$. A parametric, unbinned maximum likelihood method was used to fit the evolution and luminosity function parameters simultaneously (Marshall et al., 1983) at 15μm. Since ELAIS identifications were not only flux limited at 15μm, but also in their R-band magnitude (17.0 < R < 20.0), a factor $\Theta(\mathbf{z,L})$ was introduced in the function 'S' to be minimized

$$S = -2\sum_{i=1}^{N} \ln[\Phi(z_i, L_i)] + \iint \Phi(z, L)\Omega(z, L)\Theta(z, L)\frac{dV}{dz}dzdL,$$

to correct for incompleteness, and applied only to the ELAIS sample. This factor Θ represents the probability that a source with a given luminosity at 15μm (L_{15}) has an R-magnitude (L_R) between the limits of the sample (17.0<R<20.0),

$$\Theta(z, L)\,(17.0 < R < 20.0 \mid L_{15}).$$

and was derived taking into account the 1σ internal spread in the assumed SED.

From the total available list of Type 1 AGN of the sample, the number of objects that entered the computation was defined as:

- RMS: sources with $F_{12\mu m} \geq 300$ mJy, 41 sources

- ELAIS: sources with $F_{15\mu m} \geq 1$ mJy, and 17.0 < R < 20.0, 21 sources

The resulting estimate of the LF of Type I AGN at 15μm with its parameters is shown in Figure 3. The probability that our data have been drawn from the fitted PLE model is 0.28, as given by the 2D KS test. The PLE model adopted here could not be sufficient to represent the space density of our sources; some density evolution may be required. Currently we are studying different parameterizations taking into account some degree of density evolution.

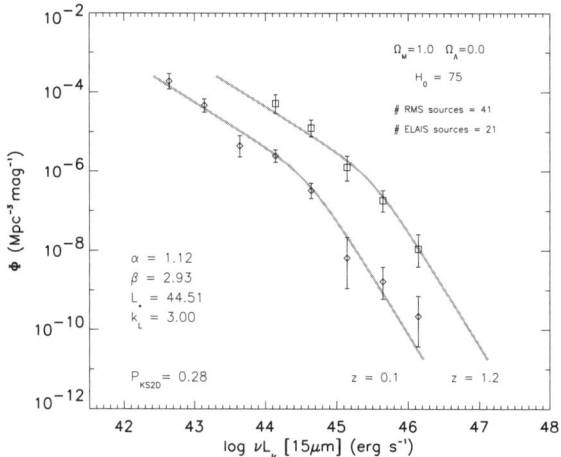

Figure 3. PLE fit to our 62 total sources (RMS + ELAIS-S1). The points correspond to the space densities of the observed sources, corrected for evolution within the redshift intervals. Sources with z=[0.0,0.2] are represented by diamonds and are mainly RMS sources. Sources with z=[0.2,2.2] are represented by squares, the ELAIS-S1 population. Lines plotted are the mid-LF at the central redshift of the interval considered. Also plotted are confidence levels at 1σ.

For the derived PLE model the contribution of Type 1 AGN to the CIRB at $15\mu m$ is $\nu I_\nu = 5.2 \times 10^{-11}$ Wm^{-2}sr^{-1} which corresponds $\sim 2\%$ of the lower limit to the CIRB calculated by Altieri et al. 1999 ($\nu I_\nu = 3.3 \times 10^{-9}$ W m^{-2} sr^{-1}) in this band (see also Hauser & Dwek 2001; Hauser 2001). At maximum, the total contribution of AGNs to the background at $15\mu m$ can be as high as $\nu I_\nu = 2.6 \times 10^{-9}$ Wm^{-2}sr^{-1} ($\sim 10\%$ of the background measured by Altieri et al. under the extreme assumptions: 1) that Type 2 AGN are as bright as Type 1 AGN at 15 μm, 2) that the ratio of Type 2 to Type 1 AGN is 4 at all redshifts, and 2) that Type 2 AGNs evolved with the same LF as Type I AGN.

4. Conclusions

Thanks to the spectroscopic campaigns carried out on the ELAIS preliminary catalog in S1, we have been able to build up statistically significant samples of Type 1 AGNs and estimate the evolution of their LF in the IR.

Our sample of 21 Type 1 AGN has allowed the first estimate of this evolution at $15\mu m$, and their contribution to the Cosmic Infrared Background at that wavelegth.

The LF is fairly well represented by a double power-law function with a significant cosmological evolution according to a PLE model with $L(z) \propto (1+z)^k$ and $k=3.00$.

From the evolution fitted we can derive a total contribution, for Type 1 AGN, of around 2% to the total IR background as measured by Altieri et al 1999. The contribution of Type 1 + Type 2 AGNs could be as high as 10% if Type 2 AGNs evolve in a similar way to Type 1, and if we take the ratio Type2/Type1 ~ 4 and to be constant with z.

Acknowledgments. Based on observations collected at the European Southern Observatory, Chile, ESO N°: 62.P-0783, 63.O-0117(A), 64.O-0595(A), 65.O-0541(A). This research has made use of the NASA/IPAC Extragalactic Database (NED) which is operated by the Jet Propulsion Laboratory, California Institute of Technology, under contract with the National Aeronautics and Space Administration. This research has been partially supported by ASI contracts ARS-99-75, ASI 00/IR/103/AS, ASI I/R/107/00, MURST grants Cofin-98-02-32, Cofin-99-034, Cofin-00-02-36, and a 1999 CNAA grant.

References

Alexander, D., Aussel H. 2000, in the Springer Lecture Notes of Physics Series, ISO Surveys of a Dusty Universe, astro-ph/0002200
Alexander, D., La Franca, F., Fiore, F. et al. 2001, ApJ, 554, 18
Altieri, B., Metcalfe, L., Kneib, J.P. et al. 1999, A&A, 343, L65
Boyle, B.J, Shanks T., Croom, S.M., Smith R.J., Miller L., Loaring N., Heymans C. 2000, MNRAS, 317, 1014
Brandt, W.N., Alexander, D.M., Hornschemeier, A.E., et al., 2001., AJ, in press, astro-ph/0108404
Elvis, M., Wilkes, B.J. et al. 1994, ApJS, 95, 1
Grossan, B.A. 1992, PhD thesis, MIT
Gruppioni, C. et al. 1999, MNRAS, 304, 199
Hauser, M. G. 2001, in Proc. IAU Symposium 204, The Extragalactic Infrared Background and its Cosmological Implications, Astron. Soc. Pac. Conf. Ser., vol. 204, p. 101, astro-ph/0105550
Hauser, M. G., Dwek, E. 2001, Annual Reviews of Astronomy and Astrophysics, 2001, Vol. 39, in press, astro-ph/0105539
La Franca F., Cristiani S. 1997, AJ, 113, 1517
Lari, C., Pozzi, F., Gruppioni, C. et al. 2001, MNRAS , 325, 1173
Lang, K.R. 1980, Astrophysical Formulae (Berlin:Springer)
Marshall, H.L., Avni, Y., Tananbaum, H., Zamorani, G. 1983, ApJ, 269, 35
Miyaji T., Hasinger G., Schmidt M. 2000, A&A, 353, 25
Moshir, M., et al. 1991, Explanatory Supplement to the IRAS Faint Source Survey, Version 2 (Pasadena:JPL)
Natali, F., Giallongo, E., Cristiani, S. & La Franca, F. 1998, AJ, 115, 397
Oliver, S., Rowan-Robinson, M., Alexander, D.M., et al. 2000, MNRAS, 316, 749
Rush, B., Malkan, M.A., Spinoglio, L. 1993, ApJS, 89, 1 (RMS)
Serjeant, S. et al. 2000, MNRAS, 316, 768

Testing the Unified Model with an Infrared Selected Sample of Seyferts

H. R. Schmitt and J. S. Ulvestad

National Radio Astronomy Observatory, P.O. Box 0, Socorro, NM87801

R. R. J. Antonucci

University of California, Santa Barbara, Santa Barbara, CA93106

C. J. Clarke and J. E. Pringle

Institute of Astronomy, Madingley Road, Cambridge CB3 0HA, England

A. L. Kinney

NASA Headquarters, 300 E St., Washington, DC20546

Abstract.
We present a series of statistical tests using homogeneous data and measurements for a sample of Seyfert galaxies. These galaxies were selected from mostly isotropic properties, their far infrared fluxes and warm infrared colors, which provide a considerable advantage over the criteria used by most investigators in the past, like ultraviolet excess. Our results provide strong support for a Unified Model in which Seyferts 2's contain a torus seen more edge-on than in Seyferts 1's and show that previous results showing the opposite were most likely due to selection effects.

1. Introduction

The Unified Scheme is based on the idea that the nucleus is surrounded by a dusty molecular torus, with orientation angle being the parameter which determines whether an AGN is perceived by observers as a Seyfert 1 or as a Seyfert 2 (Antonucci 1993). This scenario is supported by the detection of polarized broad emission lines in Seyfert 2's (Antonucci & Miller 1985) and the collimated escape of radiation from nuclear region, detected as Narrow Line Regions with conical shapes in Seyfert 2 galaxies (Pogge 1989) and jet like radio emission (Ulvestad & Wilson 1989).

It is now accepted that the Unified Model applies to a large fraction of Seyfert galaxies and is correct to first order. However, some observational results claim intrinsic statistical differences between Seyfert 1's and Seyfert 2's. Malkan et al. (1998) found that Seyfert 1's usually reside in earlier type host galaxies compared to Seyfert 2's, while Laurikainen & Salo (1995) and Dultzin-Hacian et al. (1999) found a higher percentage of companions around Seyfert 2's than in Seyfert 1's.

We believe that these differences between Seyfert 1's and Seyfert 2's are mostly due to the way these papers selected their samples. In order to be able to make a fair comparison between Seyfert 1's and Seyfert 2's and correctly address problems related to the Unified Model, it is necessary to use a sample selected from an isotropic property, believed to be independent of the orientation of the torus relative to the line of sight.

2. Sample and Data

One of the best ways to select an isotropic sample is based on the far infrared properties of the galaxies. According to the Pier & Krolik (1992) torus models, the circumnuclear torus radiates nearly isotropically at 60μm, so a sample selected in this way should be relatively free from selection effects.

Our sample was extracted from the survey of warm Seyfert galaxies defined by de Grijp et al. (1992), which was selected based on the strength of their IRAS 25μm and 60μm fluxes and warm infrared colors. We selected all Seyfert galaxies with z≤0.031, which gives a total of 29 Seyfert 1's and 59 Seyfert 2's. We have to point out, however, that this sample is not complete, since any Seyfert galaxy with infrared colors cooler than our criteria or for which there is no good quality IRAS data, are missed.

Another important point in this study is the use of high quality radio and optical images, obtained and measured homogeneously. We use broadband B and I images obtained for all the galaxies by Schmitt & Kinney (2000), which will be used to determine host galaxy inclinations and, in some cases, morphological types. We also used VLA A-configuration 3.6cm continuum images for 75 of the galaxies (Schmitt et al. 2001a), which will be used to determine the 3.6cm radio powers and the extent of the radio emission in these galaxies.

3. Infrared and Radio Luminosities

The left panel of Figure 1 presents the distribution of 60μm luminosities for Seyfert 1's and Seyfert 2's. We can see that they have very similar distributions, with the Kolmogorov-Smirnov test (KS-test) giving a 45% chance that two samples drawn from the same parent population would differ this much. This result is expected since the galaxies were selected by their infrared properties and shows that this measurement is isotropic.

The distribution of 3.6cm radio powers is presented on the right panel of Figure 1. Both Seyfert types have similar radio power distributions, as expected from the Unified Model, with the KS test showing that two samples selected from the same parent population would differ this much 11% of the time.

4. Host Galaxy Properties

The left panel of Figure 2 shows the observed distribution of the ratio between the host galaxies minor to major axis lengths (b/a). The observed distribution

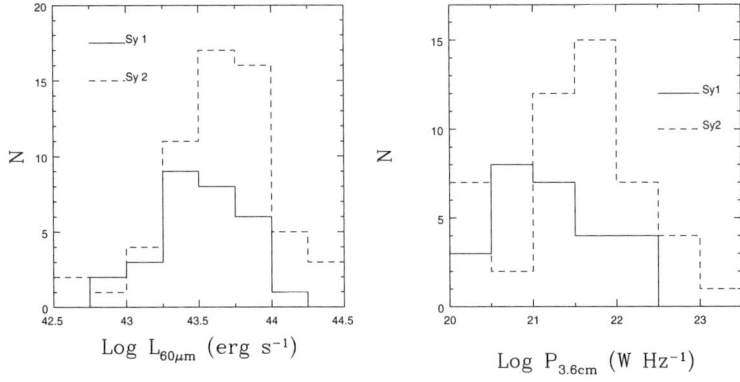

Figure 1. Left: histogram of the 60μm luminosity; right: histogram of the radio continuum 3.6cm power. Seyfert 1's and Seyfert 2's are represented by solid and dashed lines, respectively.

has a deficit of Seyfert 1 galaxies with b/a<0.5, while in Seyfert 2's this does not happen. The lack of galaxies with b/a<0.2 is due to the thickness of the disk. Comparing the b/a distribution for Seyfert 1's and Seyfert 2's we find that they are significantly different, with the KS test giving the probability that two samples drawn from the same parent population would differ this much only 4.7% of the time.

This result is similar to the one found by Keel (1980), who was the first to discover a deficiency of edge-on Seyfert 1 galaxies (see also Lawrence & Elvis 1982 and Maiolino & Rieke 1995). Although this result is in principle not expected from the Unified Model, it does not necessarily contradict it. The papers cited above suggested that, in the case of edge-on Seyfert galaxies, the gas and dust along the host galaxy disk can block the direct view of the Broad Line Region, thus leading to a classification as a Seyfert 2 galaxy.

The comparison between the Morphological Types of the host galaxies of Seyfert 1's and Seyfert 2's is shown on the right panel of Figure 2. The two distributions are very similar, with the KS test showing that two samples selected from the same parent population would differ this much 80% of the time.

5. Radio Sizes

The distributions of the logarithm of the extension of the 3.6cm radio emission in Seyfert 1's and Seyfert 2's is presented in Figure 3. Only 9 out of 26 Seyfert 1's (35%) show extended emission, and the rest are unresolved, as indicated by arrows in the figure. In the case of Seyfert 2's, 28 out of 48 galaxies (58%) have extended radio emission.

Given the fact that 50% of the galaxies in our sample are unresolved, we had to use survival analysis to compare the two distributions. The mean and standard deviations of the extension of the radio emission were calculated using the Kaplan-Meier estimator, which gave 148±65 pc and 348±97 pc for Seyfert 1's

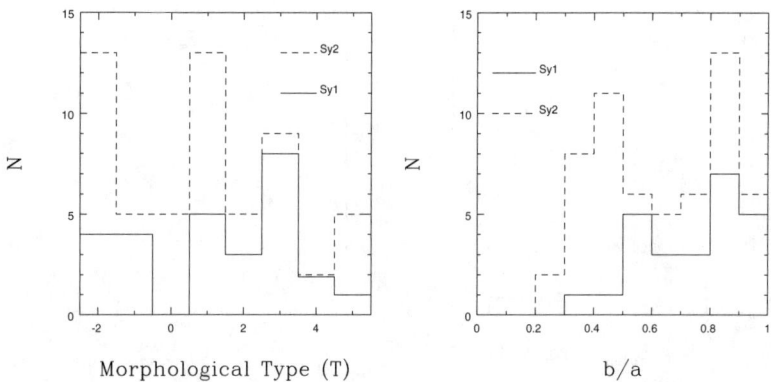

Figure 2. Histograms of morphological types, left, and the ratio between the semiminor and semimajor axes of the host galaxies, right.

and Seyfert 2's, respectively. We also compared whether Seyfert 1's and Seyfert 2's have similar distributions of the extension of the radio emission, using the Gehan-Wilcoxon test. We obtained a probability of 4.2% that the two samples are drawn from the same parent population. This confirms that Seyfert 1's have smaller extended emission, as predicted by the Model.

6. Frequency of Companions

Several mechanisms have been suggested to explain how to transport gas from the disk of a spiral galaxy to its nucleus, like interactions (Gunn 1979; Hernquist 1989) or bars (Schwartz 1981). The influence of interactions on the fueling of AGN has been the topic of several papers, but so far there is no consensus about this subject. Dahari (1984) and Rafanelli et al. (1995), among others, found Seyferts to have an excess of companions relative to normal galaxies. On the other hand, Fuentes-Williams & Stocke (1988) and Bushouse (1986) found that there is no detectable difference in the environments of Seyfert and normal galaxies. An intriguing result was obtained by Laurikainen & Salo (1995) and Dultzin-Hacyan et al. (1999), who showed that Seyfert 2's have a larger number of companions when compared to normal galaxies, while Seyfert 1's do not.

We used our broad band images, NED and the Digitized Sky Survey plates (DSS) to search for companions around our galaxies, and adopted the parameters used by Rafanelli et al. (1995) to determine if a galaxy has a companions. A galaxy is considered a companion if its distance to the galaxy of interest is smaller than 3 times the diameter of that galaxy (3D), the difference in brightness between them is smaller than 3 magnitudes ($|\Delta m| \leq 3$ mag) and the radial velocities difference is smaller than $|c\Delta z| \leq$1000 km s^{-1}.

According to these criteria, a total of 25 out of the 88 Seyfert galaxies in our sample have galaxies with $|\Delta I|$ or $|\Delta B| \leq 3$ mag and closer than 3D from them, which puts an upper limit of $< 28\pm6\%$ of possible interacting galaxies in this sample (the uncertainty is given by Poisson statistics). Of these 25 galaxies,

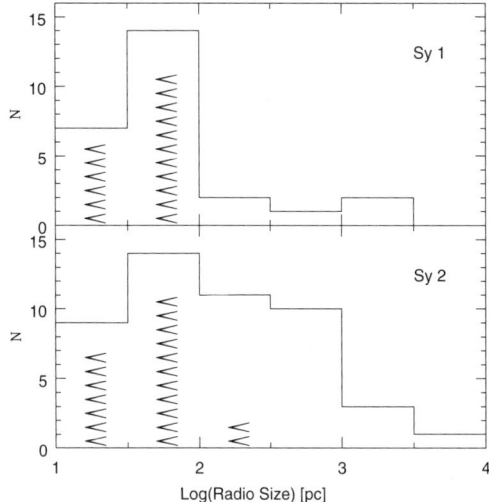

Figure 3. The distribution of the logarithm of the extension of the radio emission in Seyfert 1's and Seyfert 2's. The histograms represent the total number of galaxies in each bin, adding those with detected extended emission and upper limits, represent by arrows.

9 are Seyfert 1's and 16 are Seyfert 2's, which gives an upper limit of possible companions of $< 31\pm10\%$ and $< 27\pm7\%$, respectively. When we consider only the galaxies which satisfy the brightness, distance and velocity criteria, we find 17 Seyferts with confirmed companions. Since there is no information on the radial velocities for 8 of the possible companion galaxies, we assume that 17 is a lower limit, which gives that the percentage of Seyferts with companions is $> 19 \pm 5\%$. Of these 17 galaxies, 7 are Seyfert 1's and 10 are Seyfert 2's, corresponding to a lower limit of $> 24 \pm 9\%$ and $> 17 \pm 5\%$ galaxies with companions, respectively.

The percentage of confirmed companions in our sample is similar to the one found by Rafanelli et al. (1995) for the CfA sample, and also to the ones obtained by Schmitt (2001) for the Palomar sample, which is the same as for galaxies with other activity types. We should also notice that there is no apparent difference in the upper and lower percentage of companion galaxies in Seyfert 1's and Seyfert 2's, contradicting the results obtained by Laurikainen & Salo (1995) and Dultzin-Hacyan et al. (1999). An explanation why these paper got to their results is given by Schmitt et al. (2001b).

7. Summary

We presented a series of test to the Unified Model of Seyfert galaxies based on a sample of galaxies selected by their infrared properties, which presents several advantages relative to other samples. The far infrared is a mostly isotropic property, which is essential for testing Unified Models, since the sample is unbiased relative to the orientation of the torus. We also used homogeneous data and measurements. The detailed results are presented by Schmitt et al. (2001b).

We found that both Seyfert 1's and Seyfert 2's have similar 60μm luminosities and 3.6cm radio powers, as well as similar distributions of morphological types. The comparison between the host galaxy inclinations shows that there is a deficiency of Seyfert 1's in edge-on galaxies, which was known from previous studies and apparently contradicts the Unified Model. However, the model can be reconciled with the observations if we assume that Seyfert 1's observed edge-on will have their nucleus hidden by gas and dust in the galaxy disk. The extension of the radio emission in Seyfert 1's is, on average, smaller than in Seyfert 2's, as expected from the model. We also show that there is a similar percentage of Seyfert 1's and Seyfert 2's with companions.

These results, taken together, give strong support to the Unified Model. This indicates that previous results, which found differences in isotropic properties of Seyfert 1's and Seyfert 2's, were most likely due to selection effects.

Acknowledgments. Support for this work was provided by NASA grant AR-8383.01-97A. The National Radio Astronomy Observatory is a facility of the National Science Foundation operated under cooperative agreement by Associated Universities, Inc.

References

Antonucci, R. R. J. 1993, ARA&A, 31, 473
Antonucci, R. R. J. & Miller, J. S. 1985, ApJ, 297, 621
Bushouse, H. A. 1986, AJ, 91, 255
Dahari, O. 1984, AJ, 89, 966
de Grijp, M. H. K. et al. 1992, A&AS, 96, 389
Dultzin-Hacyan, D. et al. 1999, ApJ, 513, L111
Fuentes-Williams, T. & Stocke, J. T. 1988, AJ, 96, 1235
Gunn, J. 1979, in Active Galactic Nuclei, edited by C. Hazard & S. Mitton, (Cambridge University Press, Cambridge), p.213
Hernquist, L. 1989, Nature, 640, 687
Keel, W. C. 1980, AJ, 85, 198
Kinney, A. L. et al. 2000, ApJ, 537, 152
Laurikainen, E. & Salo, H. 1995, A&A, 293, 683
Lawrence, A. & Elvis, M. 1982, ApJ, 256, 410
Maiolino, R. & Rieke, G. H. 1995, ApJ, 454, 95
Malkan, M. A., Gorjian, V. & Tam, R. 1998, ApJS, 117, 25
Pier, E. A. & Krolik, J. H. 1992, ApJ, 401, 99
Pogge, R. W. 1989, ApJ, 345, 730
Rafanelli, P., Violato, M. & Baruffolo, A. 1995, AJ, 109, 1546
Schmitt, H. R. 2001, AJ, in press
Schmitt, H. R. & Kinney, A. L. 1996, ApJ, 463, 498
Schmitt, H. R. et al. 2001a, ApJS, 132, 199
Schmitt, H. R. et al. 2001b, ApJ, 555, 663
Schwartz, M. 1981, ApJ, 247, 77
Ulvestad, J. S., & Wilson, A. S. 1989, ApJ, 343, 659

Results from ISOCAM Deep Surveys: An Answer on the AGN Contribution to the Cosmic Infrared Background

Hervé Aussel

Institute for Astronomy, University of Hawaii, 2680 Woodlawn Drive, Honolulu, Hawaii, 96822, USA

Abstract. The amount of energy contributed by type-2 AGNs to the Cosmic Infrared Background (CIB) has recently been the subject of some debate, since models have shown it could be responsible for up to half. If this were the case, this contribution should be taken carefully into account before using the combined CIB and COB to derive the density of metals in the universe. I argue here that an observational answer to this problem comes from ISOCAM deep surveys performed at 15 μm, that do resolve into discrete sources the bulk of the CIB at 140 μm. Studies of the X-ray properties of these sources allow us to assess whether or not they are dominated by AGNs, and to derive that the type-2 AGNs contribution to the CIB is not greater than 20%.

Introduction

The Extragalactic Background Light (EBL) is the cumulative emission of all extragalactic objects along the line of sight. It is a very powerful tool for the study of galaxy evolution (Partridge & Peebles 1967): *if it is produced by star light*, it can be related to the present content of metals in the universe almost independently of the assumed cosmology (Cowie 1988; Bernstein, Weedman & Madore 2002). Starlight clearly dominates in the optical region (between 0.1 and 5 μm) where the EBL is often referred to as the Cosmic Optical Background (COB). However, measurements of the EBL in the X-ray and infrared/sub-millimeter domain have prompted new questions and generated some debate.

First, a far-infrared/sub-millimeter EBL has been discovered in the FIRAS data by Puget et al. (1996), at a very significant level. The values of this Cosmic Infrared Background (CIB) have been subsequently refined and extended to the DIRBE data down to 100 μm by Hauser et al. (1998) and Lagache et al. (2000) (see Figure 1). While in the local universe, about one third of the star light is absorbed by dust and reemited in the infrared (Soifer & Neugebauer 1991), this fraction is inverted in the EBL: between half and 2/3 of its integrated energy is emitted above 10 μm. This indicates that the amount of star light reprocessed by dust must increase with redshift, and that the infrared part of the EBL has to be taken into account to derive the metal content of the universe.

Second, the EBL has been measured in the X-ray domain (Cosmic X-ray Background: CXB) at a much lower level than the COB and the CIB (Giacconi et al. 1962). Deep ROSAT surveys have resolved 80% of the CXB in the range

0.5–2 keV, and subsequent follow-up studies have shown that it is produced by type-1 AGNs. However, such sources cannot explain the peak of the CXB at 30 keV. Comastri et al. (1995) have suggested that the high energy part of the CXB is due to a population of heavily obscured type-2 AGNs. Such a population has been confirmed by recent XMM/Newton (Hasinger et al. 2001) and Chandra surveys (Mushotsky et al. 2000) in the 2–10 keV range. While type-1 AGNs do not emit a significant fraction of their bolometric energy above $10\,\mu m$, such a population of type-2 AGNs has important implications on the nature of the CIB. Indeed, models of the CXB predict that an important fraction of the CIB is due to accretion in AGNs, from 20% up to 50% (Almaini, Lawrence & Boyle 1999). If this is the case, the contribution of this accretion-dominated emission has to be subtracted from the CIB before using it in conjunction with the COB to derive any conclusion on galaxy evolution and metal production in the universe.

It is therefore crucial to determine precisely the nature of the sources producing the bulk of the CIB at $200\,\mu m$. This can only be done by resolving the CIB into discrete sources and determining whether they are dominated by star formation or AGN emission. Unfortunately, neither IRAS nor ISOPHOT were sensitive enough to resolve a significant fraction of the CIB directly (see lower limits from counts on Figure 1). However, I will show we now have compelling evidence that mid-IR deep extragalactic surveys performed with ISOCAM at $15\,\mu m$ have indeed resolved the CIB between $100\,\mu m$ and $200\,\mu m$, and that we now have observational clues on the fraction of its energy originating from type-2 AGNs.

I will first review some basic properties of the EBL and discuss briefly the clues obtained as to its nature from sub-millimeter surveys. I will then review the results obtained in the mid-infrared (*i.e.* about $10\,\mu m$) that lead to the resolution of the bulk of the CIB: that mid-infrared extragalactic deep surveys performed by ISOCAM have resolved the CIB into discrete sources not only at the wavelength they were performed, but also in the far-infrared. I will then discuss the nature of the sources producing the CIB and show that it is mostly emitted by star formation.

1. The Extragalactic Background Light (EBL)

Predicted by Partridge & Peebles (1967), the EBL is the cumulative emission of all extragalactic objects. At frequency ν, its intensity is given by:

$$I(\nu) = \int_0^{+\infty} S(\nu) dN \qquad (1)$$

where dN is the number of extragalactic sources observed by unit areas with flux between S and $S+dS$ at frequency ν. We can rewrite (1) by assuming that the differential number count dN follows a power law in S, *i.e* $dN \propto S^{-\gamma}$ and we have

$$I(\nu) = \int_0^{+\infty} S(\nu)^{1-\gamma} dS \qquad (2)$$

It is clear from (2) that the slope γ of the differential number counts has to be lower than 2 in order for the background integral to converge, and that its intensity will be dominated by the region where the counts have a slope of $\gamma = 2$.

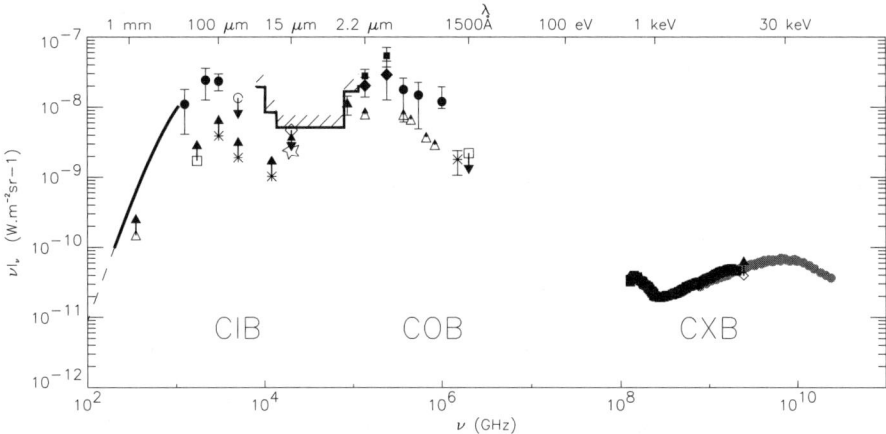

Figure 1. Spectral energy distribution of the Extragalactic Background Light (EBL). Measurements are in solid symbols, upper and lower limits in open symbols. Cosmic Infrared Background (CIB): solid line: best fit of the FIRAS detection by Lagache et al. (1999) - circles: DIRBE detections from Lagache et al. (2000) - open circle: upper limit from $60\,\mu$m IRAS maps (Mirville-Deschenes, Lagache & Puget 2002) - hatched region: upper limit from γ-rays absorption (Stanev & Franceschini 1998) - diamond: upper limit γ-rays absorption at $15\,\mu$m (Renault et al. 2000) - open star: ISOCAM number counts lower limit (Elbaz et al. 2002) - stars: IRAS number counts lower limits (Lonsdale et al. 1990) - square : ISOPHOT $170\,\mu$m lower limit (Dole et al. 2001) - Triangle : SCUBA $850\,\mu$m counts lower limit (Barger, Cowie & Sanders 1999). For the Cosmic Optical Background (COB): triangle: $3.5\,\mu$m DIRBE detection (Gordjian, Wright & Chary 2000) - diamond: (DIRBE/2MASS) detection (Wright 2001) - square: (DIRBE/2MASS) detection (Cambrésy et al. 2001) - circles: Bernstein, Freedman & Madore (2002) detection with HST- open triangle: HDF number counts (Pozzetti et al. 1998) - star: FOCA detection (Armand, Milliard & Deharveng 1994). open square : Martin, Hurwitz & Bowyer (1991) upper limit. Cosmic X-ray Background (CXB) - black squares: ASCA/ROSAT measurements (Gilli, Salvati & Hasinger 2001) - grey circles: HEAO1 measurements (Gruber, Matteson & Peterson 1999) - open diamond: Chandra number counts (Mushotsky et al. 2000)

Figure 1 displays the current estimates for the spectral energy distribution (SED) of the EBL. Such estimates can be direct measurements of an isotropic diffuse signal by experiments lacking angular resolution. This is the case in the far-infrared and sub-millimeter where the EBL was detected in the FIRAS and DIRBE data by Puget et al. (1996), and later confirmed and updated by Hauser et al. (1998) and Lagache et al. (2000). But the value of the EBL can also be obtained by integrating the extragalactic number counts until they converge as shown by equations (1) and (2). Technically, the value obtained is often a lower limit only, because one never reaches the zero bound of the integral, but provided upper limits from direct measurements, one can obtain a very good estimate of the EBL value. When the energy emitted by individual sources accounts for a significant fraction of the EBL, the background is said to have been resolved.

2. Clues from the sub-millimeter

The CIB has been partially resolved by deep surveys obtained at $850\,\mu$m. Indeed, Barger, Cowie & Sanders (1999) have shown that SCUBA counts down to 2 mJy account for 20–30% of the value measured by FIRAS and that it is most likely produced at this wavelength by sources of about 1 mJy. Barger et al. (2001) have compared Chandra and SCUBA observations of SSA13. Their X-ray sample consists of 20 sources and is complete at a level of 3×10^{-15} erg s^{-1}cm^{-2}, where between 58% and 89% of the 2–10 keV CXB is resolved. Only one source is recovered in their $850\,\mu$m map with an average sensitivity of 1.5 mJy. They deduce that the contribution of type-2 AGNs producing the CXB to the CIB at $850\,\mu$m is the range 9-13%.

A similar conclusion is reached by Severgini et al. (2000) who have studied a sample of hard X-ray sources observed by both Beppo-SAX and SCUBA. They also point out that most of the SCUBA population is dominated by star formation or Compton thick AGNs that contribute less than 6% of the CXB.

Both these studies point toward a low contribution of type-2 AGNs to the CIB. However, they can only make conclusions about the CIB at $850\,\mu$m. Unfortunately, the bulk of the CIB is observed between $100\,\mu$m and $250\,\mu$m: while SCUBA counts are dominated by luminous sources at redshift above 2, the CIB is produced by sources at lower redshifts where type-2 AGNs may have a different contribution. The definitive answer can only be obtained by resolving the bulk of the CIB.

3. ISOCAM resolves the EBL at $15\,\mu$m

ISOCAM (Cesarsky et al. 1999), the mid-infrared imager on board the ISO satellite has conducted many extragalactic surveys at $15\,\mu$m, at various depth and area coverage. The result of these observations are shown in Figure 2 from Elbaz et al. (2000), where the differential extragalactic number counts, normalized to the euclidean slope ($\gamma = 2.5$ in equation 2), are plotted against the observed flux. Five important remarks can be made on Figure 2:

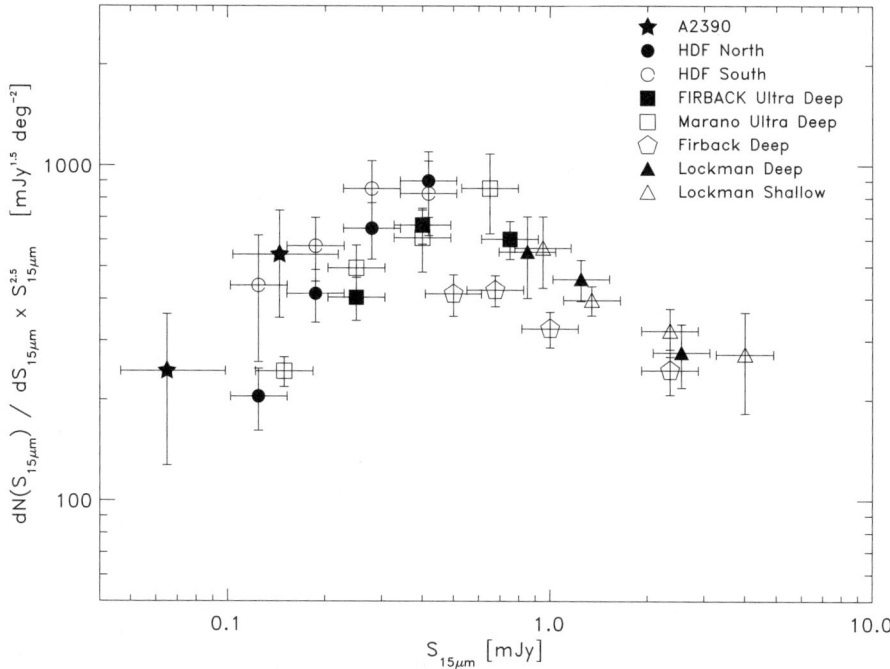

Figure 2. Extragalactic number counts at 15 μm from ISOCAM deep surveys, normalized to the euclidean slope, after Elbaz et al. (2000).

1) With respect to IRAS, ISOCAM has extended the depth of mid–infrared extragalactic surveys by more than 3 orders of magnitude, from 200 mJy in the IRAS Faint Source Survey (Lonsdale et al. 1990) to 50 μJy in the deepest areas.

2) All the number counts obtained on different Northern and Southern fields agree.

3) At 400 μJy, the number of sources observed per square degree is ∼3000, a factor of 10 in excess of the predictions of models without evolution (Franceschini et al. 2002).

4) The slope of the counts at high flux levels is super-euclidean with a slope of the differential counts of $\gamma = 3 \pm 0.1$. This cannot be explained alone by k-correction effects in the 15 μm filter, and strong evolution has to be advocated to reproduce the counts.

5) The slope of the counts at $S_\nu < 400\,\mu$Jy is under-euclidean, with $dN(S) \propto S^{-1.6}$.

Point 5 is especially important, because it means that the number counts are converging below 400 μJy. This means that we can obtain a reliable estimate of the value of the CIB at 15 μm by integrating the ISOCAM and IRAS counts, using equations 1 and 2. This exercise has been done by Elbaz et al. (2002) where a value of $2.4 \pm 0.5\,\text{nW}\,\text{m}^{-2}\text{sr}^{-1}$ has been derived.

As noted in § 2, this result would only constitute a lower value of the CIB at 15 μm if the counts were to steepen again at lower flux. Fortunately, we can obtain reliable upper limits for the value mid–IR CIB by studying the absorption of γ-ray photons emitted by blazars by the mid-IR photon radiation field between the source and the observer through the reaction $\gamma+\gamma \to e^++e^-$. Renault et al. (2001) derive a very good upper limit for the CIB at 15 μm of $4.7\,nWm^{-2}sr^{-1}$. This upper limit is very close to the lower limit derived from ISOCAM counts.

Both upper and lower limits are also in good agreement with the predicted value by Franceschini et al. (2001) of $3.3\,\mathrm{nW\,m^{-2}sr^{-1}}$. Adopting this figure, we deduce that ISOCAM deep surveys have resolved 73% of the 15 μm CIB.

4. ISOCAM resolves the bulk of the CIB

Admittedly, the title of this section is provocative. Indeed, ISOCAM observations were performed at 15 μm, not between 100 μm and 250 μm where the bulk of the CIB is observed. I will, however, argue that provided reasonable assumptions on the SED of the galaxies detected in ISOCAM surveys, it is possible to compute their contributions in the FIR, and show that they account for a large fraction of the CIB.

A first computation was performed by Aussel, Elbaz & Cesarsky (1999), assuming that all ISOCAM detected sources were Ultra Luminous Infrared Galaxies (ULIRGs) with a SED similar to the one of Arp 220 and with a median redshift of $z \sim 0.75$. With such simple assumptions, the CIB at 140 μm produced by the ISOCAM detected galaxies would be of the order of $250\,\mathrm{nW\,m^{-2}sr^{-1}}$, far above the actual measurement of $24.2\,\mathrm{nW\,m^{-2}sr^{-1}}$.

A more precise determination has been made by Elbaz et al. (2002), using the ISOCAM observations of the Hubble Deep Field (HDF) (Aussel et al. 1999). Let us note first from Figure 2 that the HDF samples the flux region where the counts flatten. Hence, according to Equation 2, the 15 μm EBL should be dominated by the sources observed at such a flux. This is indeed the case, with 48% of the 15 μm CIB beeing resolved between 100 μJy and 500 μJy, the flux range sampled by the ISO-HDF.

The HDF region is extremely interesting, because a nearly complete redshift survey has been obtained down to $R = 23.5$ by Cohen et al. (2000), and because it has also been observed at radio wavelengths (Richards 2000; Garret et al. 2000) and in X-rays (Brandt et al. 2000). In this region, the ISOCAM sample consists of 41 galaxies, complete down to 100 μJy. All galaxies but one have a redshift in the Cohen et al. (2000) catalog. It is therefore possible to compute the mid-IR luminosity of the ISOCAM galaxies: the main difficulty is the k–correction of the observed 15 μm flux, given the complicated shape of the SED of galaxies in the mid-IR region and its variation from object to object.

To overcome this problem, Elbaz et al. (2002) use a library of templates for the whole infrared spectrum of galaxies calibrated with local samples, together with the fact that the mid-IR luminosity is an excellent tracer of the total bolometric luminosity of galaxies (Spinoglio et al. 1995). Given the observed luminosity without the k–correction term (*i.e.* the luminosity at $\lambda_{obs} = \lambda_{rest}/(1+z) = 15\mu m$), they compute a range of possible SEDs for the galaxies, their IR luminosities and their contributions at 140 μm. They find that the contribution of the ISOCAM galaxies detected in the HDF at 15 μm contribute to the 140 μm CIB is $16.0 \pm 4.6\,\mathrm{nW\,m^{-2}sr^{-1}}$, where the quoted error comes from the spread in possible SEDs for each galaxy. This constitutes at least 30% of the measured value, and possibly for all of it. Hence, the ISOCAM deep surveys do resolve the bulk of the CIB. In comparison, ISOPHOT deep surveys only resolve 4% of the CIB at 170 μm (Dole et al. 2001), due to the lack of sensitivity of the instrument.

It is interesting to note that in the Elbaz et al. (2002) computations, ULIRGs only contribute 16% of the 15 μm CIB, but 36% of the 140 μm CIB, while LIRGs ($L \in [10^{11} L_\odot, 10^{12} L_\odot]$) represent respectively 44% and 48% of the 15 μm and 140 μm CIB. This is a first indication that the contribution of type-2 AGNs to the bulk of the CIB might be low, since Genzel et al. (1998) show that the central black hole is the dominant source of power only in the most luminous sources (but see Antonucci in these proceedings for a critique of this work). Indeed, Elbaz et al. (2002) estimate the contribution of AGNs in the HDF to ∼19% of the 15 μm EBL and only 4% to the 140 μm EBL due to the flatter SED of these objects.

Therefore, it appears from this work that the CIB is largely dominated by star formation. An independent test of this assumption can be made : 21 of the 41 ISOCAM galaxies are detected at 1.4 GHz in the HDF by Richards (2000) and Garrett et al. (2000). Using the model of Condon (1992), we can compute the star formation rate (SFR) in these galaxies from their radio flux, and compare it to the one derived from the infrared luminosities: an excellent agreement is found, with a slope of 1 and a dispersion compatible with the photometric errors (Aussel 2002). Moreover, the ISOCAM galaxies that are *not* detected in the radio have, from their IR-inferred SFR, 1.4 GHz fluxes below the detection limits of both the VLA and Westerbrok observations.

5. The AGN contribution to the EBL from ISOCAM surveys

It has been pointed out many times during this conference that the only means to detect a dust embedded AGN is by observing in the hard X-ray domain. Hence, a good constraint on the contribution of type-2 AGNs to the CIB can be obtained by following up the ISOCAM sources with XMM/Newton and Chandra. Fadda et al. (2002) present such a study in the HDF and the Lockman Hole fields.

The Lockman Hole field has been observed by both XMM/Newton and ISOCAM in a region of 218 square arc minutes. The X-ray observation reach a sensitivity of $1.4 \times 10^{-15}\,\mathrm{erg\,s^{-1}cm^{-2}}$ in the 2–10 keV band and $2.4 \times 10^{-15}\,\mathrm{erg\,s^{-1}cm^{-2}}$ in the 2–10 keV band, respectively resolving 80% and 60% of the CXB in these bands. At 15 μm, the ISOCAM observation probes the 500 μJy–2 mJy flux range (see Figure 2) and resolves 23% of the CIB at this wavelength (Elbaz et al. 2002). Twenty-four of the 76 X-ray sources are detected in the mid-IR, with the fraction of 15 μm detection rising with X spectrum hardness: from 30% in the 0.5–2 keV band to 63% in the 5–10 keV band. This is a confirmation that the hard CXB is produced by type-2 AGNs. On the other hand, only 10% of the ISOCAM sources are detected in the X-ray.

Fadda et al. (2002) combine these results fromon the Lockman Hole with the cross-correlation of the Chandra 1 Ms exposure (Brandt et al. 2001) and the ISOCAM 15 μm catalog of the HDF (Aussel et al. 1999). Using the median spectral index between X-ray and mid-IR of the sources detected in both fields, and assuming that *all the sources detected in X-ray are AGN-dominated*, they infer a contribution of type-2 AGNs to the 15 μm CIB of $17 \pm 2\%$ and note that this figure is also an upper limit to their contribution to the whole CIB, since the SED of these objects peaks at 20 μm.

This conclusion has been refined by Alexander et al. (2002) using the 1 Ms Chandra and ISOCAM data on the HDF, combined with the spectroscopic catalog of Cohen et al. (2000). They note that up to 100% of the X-ray sources that have been spectroscopically classified as emission line galaxies (ELGs) are detected at 15 μm, while the mid-IR detection rate of absorption line galaxies and AGN-dominated spectra are at the 20% level. Though none of these ELGs are detected in the hard Chandra band (2–8 keV) (probably because Chandra is less sensitive than XMM in the hard band), they perform a statistical analysis of their spectral slope by stacking all of the ELG X-ray detections and deduce a photon index $\Gamma \simeq 2$. This is an higher value than the photon index derived for the population of optically faint ($I \geq 24$) X-ray detections thought to be type-2 AGNs for which $\Gamma \simeq 1.3$. Also, the range of soft X-ray band luminosities of these sources is compatible with those of local starbursts like M 82 and NGC 3256.

This led Alexander et al. (2002) to the conclusion that most of the X-ray emission of the 15 μm detected sources does not come from an AGN, but from star formation. Therefore, the assumption used by Fadda et al. (2002) that all X-Ray sources are dominated by an AGN is much too strong, and the numbers they derive for the contribution of AGNs to the 15 μm and 140 μm CIB are upper limits.

Conclusion

The ISOCAM deep surveys at 15 μm have led to the resolution into discrete sources of the CIB both in the mid- and far-IR. From the infrared, radio and X-ray properties of the sources producing the CIB, it appears that at least 85% of it is due to star formation, with an upper limit of type-2 AGN's contribution of 15% at 100 μm to 250 μm where the bulk of the CIB is emitted. In comparison, type-1 AGNs represent only about 2% of the CIB at 15 μm and do not contribute noticeably to the whole CIB because of their very warm SED (Matute et al. 2002).

It should be noted that this conclusion is reached mainly by using the HDF data set that is composed of a flux-limited complete sample of 41 15 μm detected sources with measured redshifts. Hopefully, the statistics will improve as more spectroscopic followups of ISOCAM deep fields are undertaken, in the Lockman Hole, the Marano Field and the HDF-South (see Figure 2). Together, these fields represent a sample of more than 500 sources that will allow improved measurements. From early unpublished results on these fields, it appears that the conclusions presented here will not change drastically, and that the high value predicted by some models (Almaini, Lawrence & Boyle 1999) is ruled out.

In the near future, SIRTF with its MIPS instrument will perform surveys at 24 μm that will resolve the CIB between 200 μm and 400 μm in the same way that ISOCAM did. Both CAM and MIPS results will be tested by FIRST/Hershel that will be the first satellite to be in a position to resolve directly the CIB between 60 μm and 500 μm.

Acknowledgments

I thank the organizers for their invitation to this extremely interesting conference that has prompted the further investigations of the IR/X-ray properties of ISOCAM galaxies presented in § 5. I thank my collaborators D. Alexander, D. Elbaz and D. Fadda for their inputs. This research is supported by a Fellowship of the James Clerk Maxwell Telescope.

References

Alexander, D. M. et al. 2002, ApJ, 568, L85
Almaini, O., Lawrence, A. & Boyle, B. 1999, MNRAS, 305, 59
Armand, N. A., Milliard, B. & Deharveng, J. M. 1994, A&A, 284, 12
Aussel, H. 2002, Ap&SS, 281, 441
Aussel, H. et al. 1999, A&A, 342, 313
Aussel, H., Elbaz, D. & Cesarsky C. J. 1999, Ap&SS, 266, 307
Barger, A. J., Cowie, L. L. & Sanders, D. B. 1999, ApJ, 518, L5
Barger, A. J. et al. 2001, AJ, 121, 662
Bernstein, R. A., Freedman, W. L. & Madore, B. F. 2002, ApJ, 571, 107
Brandt, W. N. et al. 2001, AJ, 122, 1
Cambrésy, L. et al. 2001, A&A, 555, 563
Cesarsky, C. J. et al. 1996, A&A, 315, L32
Comastri, A. et al. 1995, A&A, 296, 1
Condon, J. J. 1992, ARA&A, 30, 575
Cohen, J. G. et al. 2000, ApJ, 538, 29
Cowie, L. L. 1988, in Proc. of the NATO ASI Vol. 240, The Post-Recombination Universe, ed. N. Kaiser & A. N. Lasenby (Dortrecht: Kluwer), 1
Dole, H. et al. 2001, A&A, 372, 364
Elbaz, D. et al. 1999, A&A, 351, L37
Elbaz, D. et al. 2002, A&A, 384, 848
Garrett, M. A. et al. 2000, A&A, 361, L41
Giacconi R. et al. 1962, Phys.Rev.Lett, 9, 439
Gilli, R., Salvati, M. & Hasinger, G. 2001, A&A, 366, 407
Gordjian, V., Wright, E. L. & Chary, R. R. 2000, ApJ, 536, 550
Gruber, D. E., Matteson, J. L. & Peterson, L. E. 1999, ApJ, 520, 124
Fadda, D. et al. 2002, A&A, 383, 838
Franceschini, A. et al. 2001, A&A, 378, 1
Genzel, R. et al., 1998 ApJ, 498, 579
Hauser, M. G. et al. 1998, ApJ, 508, 25

Lagache, G. et al. 1999, A&A, 344, 322
Lagache, G. et al. 2000, A&A, 354, 247
Lonsdale, C. J. et al. 1990, ApJ, 358, 60
Martin, C, Hurwitz, M. & Bowyer, S. 1991, A&A, 379, 549
Matute, I. et al. 2002, MNRAS, 332, L11
Mirville-Deschene, M. A. & Lagache, G. & Puget J.-L., A&A, in press, (astro-ph/0207312)
Mushotsky, R. F. et al. 2000, Nature, 404, 459
Partridge, R. B. & Peebles, P. J. E. 1967, ApJ, 148, 377
Pozzetti, L. et al. 1998, MNRAS, 298, 1133
Puget, J.-L., et al. 1996, A&A, 308, L5
Renault, C. et al. 2001, A&A, 371, 771
Richards, E. A. 2000, ApJ, 533, 611
Severgnini, P. et al. 2000, A&A, 360, 457
Spinoglio, L. et al. 1995, ApJ, 453, 616
Soifer, B. T. & Neugebauer, G. 1991, AJ, 101, 354
Stanev, T & Franceschini, A. 1998, ApJ, 494, L59
Wright, E. L. 2001, ApJ, 553, 538

Blazars from the CLASS Survey

M. J. M. Marchã

CAAUL, Observatório Astronómico de Lisboa, Tapada da Ajuda, 1349-018 Lisboa, Portugal

A. Caccianiga

CAAUL, Observatório Astronómico de Lisboa, Tapada da Ajuda, 1349-018 Lisboa, Portugal

Abstract.
This paper presents preliminary results from the CLASS blazar sample; the aim is to study the cosmological properties of radio selected BL Lacs and their closest relatives, the other low radio luminosity flat spectrum sources.

1. Introduction

The term *blazar* is typically used to group together two types of Active Galactic Nuclei (AGNs): the Flat Spectrum Radio Quasars (FSRQs) and the BL Lac objects. Despite the dichotomy in the type of optical spectra and luminosity range observed, the 2 types of sources show a Spectral Energy Distribution (SED) which is dominated by non-thermal processes. In this work we are primarily interested in the low-power tail of the blazar class, typically made up of BL Lac objects, where many important issues related to the unified models are expected to find a solution. In particular, the beaming model, which unifies BL Lacs and FR I, predicts some important features of the BL Lac luminosity function (LF) that should be observed in the low-power range ($P<10^{24}$ W Hz^{-1}). We have thus created a new radio survey of flat-spectrum sources, the CLASS blazar survey, with the specific aim of exploiting the low-power tail of the blazar class. In this paper we present some preliminary results about the cosmological properties of BL Lacs as derived from this survey.

2. The Sample

The CLASS blazar survey consists of 325 flat-spectrum radio sources selected according to the following criteria:
1) $35° \leq \delta \leq 75°$ and $|b^{II}| \geq 20°$
2) $S_{5GHz} \geq 30$ mJy
3) flat radio spectrum, i.e. $\alpha_{1.4}^{4.8} \leq 0.5$ ($S_\nu \propto \nu^{-\alpha}$)
4) red magnitude (corrected) ≤ 17.5

The details of the selection, as well as the radio properties of the sample, are discussed in Marchã et al. (2001), whereas the spectroscopic classification obtained thus far is presented in Caccianiga et al. (2001). In particular, 70% of the sources have been spectroscopically classified into 3 broad categories:
1) **Type 0** - objects with weak or no emission lines (39%).
2) **Type 1** - objects with broad emission lines (46%).
3) **Type 2** - objects with narrow emission lines (15%).

This paper concentrates on the preliminary results based on the 88 sources of Type 0 identified so far in the CLASS blazar survey.

3. The Type 0 Classification

An object is classified as a Type 0 if its optical spectrum shows only emission lines with Equivalent Width (EW) $\leq$ 5Å. This class includes both BL Lacs and PEGs (Passive Elliptical Galaxies). The distinction between the 2 depends on the value of the CaII break (Δ) at 4000 Å, which is defined in Dressler & Shectman (1987).

Since a typical elliptical galaxy shows values of Δ between 40% and 60%, a significantly lower value is usually interpreted as an indication of the presence of an extra source of continuum, i.e., a BL Lac nucleus. A limiting value of $\Delta \leq$ 25% was proposed by Stocke et al. (1991) in order to separate BL Lacs from the remaining type 0 sources. However, it has since been proposed that such a restrictive value may indeed miss weak BL Lacs simply because the ratio of optical emission between the BL Lac and its host galaxy would be too small to be recognized. As a consequence, an "expanded" classification of BL Lac has been used in the selection of new samples of BL Lacs (Marchã et al. 1996, Laurent-Muehleisen et al. 1998, Caccianiga et al. 1999, Perlman et al., 1998), Here we adopt this 'expanded' classification and divide the type 0 sources into 3 categories according to 3 regimes of Δ:
- **"Classical" BL Lacs: $\Delta \leq$ 25%**
- **BL Lac candidates: 25% < Δ < 40%**
- **PEGs: $\Delta \geq$ 40%**

The imposition of a sharp separation between BL Lacs (and BL Lac candidates) and PEGs is somewhat arbitrary and unlikely to be related to the intrinsic properties of the active nucleus. In fact, the measured contrast of a source is inversely proportional to the strength of the non-thermal contribution from the AGN, which means that as the AGN gets weaker, the measured contrast will get larger. Hence, the deeper the survey, the more important is the choice of contrast used in the selection of BL Lacs. A value for this quantity that is too limiting can introduce significant incompleteness in the final sample, as is widely discussed in Browne & Marchã (1993) and Marchã & Browne (1995). More quantitatively, by using the formalism discussed in these papers, we predict that a high fraction (43%) of the low-power (down to 10^{23} W Hz^{-1}) BL Lacs are not expected to produce a significant reduction of Δ (i.e. leaving $\Delta \geq$ 40%). This value is in good agreement with the observed fraction of PEGs found in the survey (45%).

It is clear that with the currently available data it is not possible to prove or disprove that all (or a fraction) of the PEGs found in the CLASS blazar sample actually contain a hidden BL Lac nucleus in their cores. However, the alternative to this scenario is to invoke a new population of low luminosity, core-dominated, flat radio spectrum sources which so far has no place in the current unified scheme for radio-loud objects. We are thus faced with two scenarios: one where PEGs hide weak BL Lac nuclei, and another where they are an entirely different population. The consequences for the statistical properties of BL Lacs will be different in both scenarios, and for that reason, the next Section dealing with the cosmological properties will treat both situations separately.

4. Cosmological Properties

The cosmological evolution of BL Lacs has been a matter of debate over the last decade, and it is still considered an open problem. At the origin of the debate are the opposing results obtained about ten years ago when using the only 2 statistically complete samples of BL Lacs available at the time, one selected at radio frequencies (1 Jy sample; Stickel et al. 1991) and the other in the X-ray (the EMSS-C sample; Morris et al. 1991). In the case of the radio-selected sample, the result was consistent with a population that suffered no evolution, or mildly positive, whereas in the case of the X-ray selected sample, the result was clearly indicative of a population that evolves negatively. More recent results (Bade et al. 1998; Giommi, Menna & Padovani 1999) based on ROSAT data seem to confirm this discrepancy showing negative evolution for High-energy peaked BL Lacs (HBL), and no evolution for the Low-energy peaked BL Lacs (LBL). On the other hand, the samples selected by combining radio and X-ray data (e.g. the REX survey, Caccianiga et al. 1999 and the DXRBS survey, Perlman et al. 1998) are not finding strong evidence for either negative or positive evolution (Padovani 2001, Caccianiga et al. 2002 and Caccianiga et al, this conference). Even if some attempts at explaining these results in terms of different selection effects have been done (Giommi et al. 2001), the general picture is still very unclear.

The CLASS blazar survey offers an independent tool to asses the evolutionary properties of BL Lacs although, to date, the missing 30% of identifications prevent us from drawing firm conclusions. Nevertheless, assuming that the missing 30% are distributed like the remaining objects, we have applied the V_e/V_a test (Avni & Bahcall 1980) to estimate the cosmological evolution of the CLASS BL Lacs. The results can be found in Table 1.

	BL Lacs (48)	PEGs (40)	All type 0 (88)
$< V_e/V_a >$	0.61	0.48	0.55
	±0.04	±0.05	±0.03

Table 1. Mean V_e/V_a values for the entire sample of type 0 sources; the numbers in brackets correspond to the number of sources in each category

The value of $< V_e/V_a >= 0.61$ for BL Lacs (including the BL Lac candidates) was obtained by assigning a redshift of $z = 0.26$ for the 25 featureless objects. This redshift corresponds to the mean of the remaining 23 BL Lacs (with redshift) in the sample. It is important to note, however, that the value

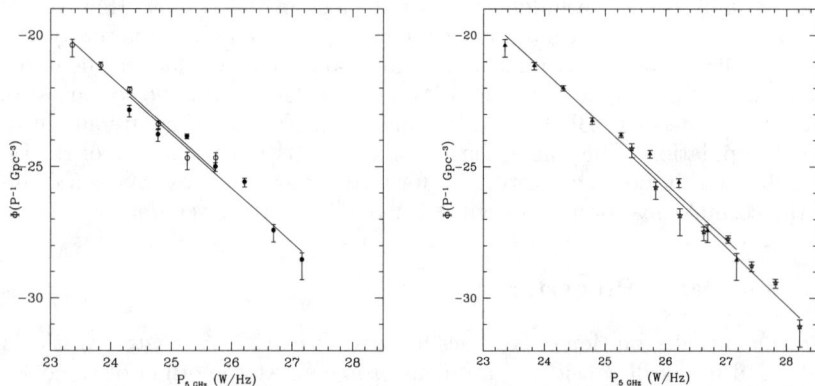

Figure 1. Left: The RLF of BL Lacs (filled circles) and PEGs (open circles). Right: The RLF of all the type 0 (triangles) compared to the BL Lac RLF from the 1 Jy sample (stars)

of $< V_e/V_a >$ does not change very much by changing this assumption. For instance, using a much lower or higher mean redshift of $z = 0.1$ or $z = 0.5$ the $< V_e/V_a >$ varies from 0.59 to 0.62, respectively. For the remainder of the analysis, the BL Lacs without a measured redshift are assigned $z = 0.26$.

The values of $< V_e/V_a >$ presented in Table 1 show that PEGs are consistent with a scenario of no evolution, while the $< V_e/V_a >$ of BL Lacs (without PEGs) is significantly (2.8σ) away from the case of no evolution. This result has interesting consequences for the discussion of the evolution of BL Lacs. If BL Lacs (including BL Lac candidates) alone are considered, then the above results give statistical evidence for mildly positive cosmological evolution for this class of objects. However, if PEGs are the population of 'diluted' BL Lacs, as discussed in the previous section, then V/V_{max} becomes consistent with a non-evolving population.

4.1. The Radio Luminosity Function (RLF)

The study of the RLF of radio-loud objects is particularly interesting because, according to the unified schemes, there should be a relationship between the LF of beamed and unbeamed (parent) populations. The effect of beaming on the LF was studied in detail by Urry & co-workers (see Urry & Padovani, 1995 and references therein for a review). In summary, if the unbeamed LF is well represented by a power-law with a given slope within a luminosity range, then the effect of beaming is to produce a break in the LF at a luminosity L_{break} that depends on the Doppler factor. The slope of the beamed LF will be the same as that of the parent population above this L_{break} but it will flatten below this luminosity.

The left panel of Fig. 1 shows the RLF for the BL Lacs (including BL Lac candidates as filled circles), and PEGs (open circles) of the CLASS blazar sample. (The LF discussed here assumes a no-evolution scenario, a 70% normalization of the surveyed area in order to account for the 70% spectroscopic identification obtained so far, and a bin size $\Delta logL$=0.5.) The 2 LFs cover $\sim$ 4 orders of magnitude in radio luminosity, with the PEGs extending to as low as $\sim 10^{23}$ W Hz^{-1}. There is good agreement between both LFs over their overlapping region. A least squares fit to each data set gives a slope of -2.00($\pm$0.19) for the PEGs, and -1.96($\pm$0.20) for BL Lacs. Since the beaming model of Urry & Padovani (1995), based on the data of the 1 Jy BL Lac data, predicts a flattening of the beamed LF at about $L \sim \times 10^{25}$ W Hz^{-1}, we also investigate this possibility using our data. A possible flattening below $L_{break} \sim 2 \times 10^{25}$ W Hz^{-1} is consistent with our BL Lac data. In this case, a least squares fit gives a slope of -1.06($\pm$0.52) below and -2.47($\pm$0.22) above L_{break}. However, even if a flattening of the BL Lac LF is consistent with our data, it is certainly not required.

The right panel of Figure 1 shows the LF for the entire type 0 population (BL Lacs + PEGs) represented by triangles, against that of the 1 Jy sample of BL Lacs which is represented by stars. The concordance between the 2 LFs is remarkable. A least squares fit applied to each set of data gives a slope of -2.09($\pm$0.11) for the type 0 and -2.18($\pm$0.18) for the 1 Jy data.

The 2 LFs overlap over 2 orders of magnitude in luminosity and as mentioned before, the LF for CLASS type 0 extends to 2 orders of magnitude below the luminosities of the 1 Jy BL Lacs. Nevertheless, the slopes found for both LFs are very similar, and in this case there is little evidence for a flattening of the RLF below $\sim 10^{25}$ W/Hz. Hence, if weak BL Lac nuclei do exist in the core of PEGs, then the flattening of the RLF is either very mild, or it occurs at lower luminosities. In either case, there are consequences for the beaming parameters to be considered in the BL Lac population.

5. Conclusion

In this paper the preliminary results concerning the cosmological properties of the 88 type 0 objects discovered so far in the CLASS blazar survey were briefly discussed. This group of sources contains both BL Lacs and PEGs, and it consists of all the objects without strong emission lines found in the 70% complete spectroscopic identification process. The analysis considered 2 possible scenarios: one considering that PEGs hide BL Lac nuclei, and another where they do not. When BL Lacs were considered alone, then mild positive evolution was found. However, this trend disappeared under the hypothesis that PEGs harbor BL Lacs. The RLF was also investigated under the 2 possible scenarios, and when kept separately, the data yielded RLFs of similar slopes for BL Lacs and PEGs. Even though there was no need to invoke a break in the LF for BL Lacs, a flattening at low radio luminosities could not be excluded. The position of this eventual flattening and the respective slopes below and above the L_{break} are consistent with those found by Urry & Padovani (1995). Nevertheless, if BL Lacs and PEGs are considered together, the resulting RLF extends to 2 orders of magnitude below the luminosities sampled by the 1 Jy sample of BL Lacs, and the 2 LFs are well represented by simple power-laws of similar slopes. In this scenario no flattening is observed in the LF for the type 0 objects of the CLASS sample.

The fact that the spectroscopic identification level is only 70% complete means that the discussion presented here is preliminary and that further data are needed. Apart from the completion of the identification process, further optical, radio and X-ray data are being considered in order to establish on a firmer basis whether PEGs do indeed harbor BL Lacs, or whether they constitute a new population of flat spectrum radio sources. The answer to this question will undoubtedly have repercussions on the statistical properties of the BL Lac class.

References

Avni, Y. & Bahcall, J.N. 1980, ApJ, 235, 694
Bade, N., Beckmann, V., Douglas, N. G., Barthel, P. D., Engels, D., Cordis, L., Nass, P., Voges, W. 1998, å, 334, 459
Browne, I.W.A. & Marchã, M.J.M. 1993, MNRAS, 261, 795
Caccianiga, A., Maccacaro, T., Wolter, A., della Ceca, R., Gioia, I. 1999, ApJ, 513, 51
Caccianiga, A., Marchã, M. J. M., Antón, S., K.-H, M., Neeser, M. J. 2001, MNRAS, in press
Caccianiga, A., Maccacaro, T., Wolter, A., della Ceca, R., Gioia, I. 2002, ApJ, in press
Giommi, P., Menna, M. T., Padovani, P. 1999, MNRAS, 310, 465
Giommi, P., Pellizoni, A., Perri, M., Padovani, P. 2001, in ASP Conf. Ser. Vol. 227, Blazar Demographics and Physics, ed. P. Padovani & C. M. Urry, (San Francisco; ASP), 227
Laurent-Muehleisen, S. A., Kollgaard, R. I., Ciardullo, R., Feigelson, E. D., Brinkmann, W., Siebert, J. 1998, ApJS, 118, 127
Marchã, M. J. M. & Browne, I. W. A. 1995, MNRAS, 275, 954
Marchã, M. J. M., Browne, I. W. A., Impey, C. D., Smith, P. S. 1996, MNRAS, 281, 425
Marchã, M. J., Caccianiga, A., Browne, I. W. A., Jackson, N. 2001, MNRAS, 326, 1455
Morris, S. L., Stocke, J. T., Gioia, I. M., Schild, R. E., Wolter, A., Maccacaro, M., della Ceca, R. D. 1991, ApJ, 380, 49
Padovani, P., 2001, in ASP Conf. Ser. Vol. 227, Blazar Demographics and Physics, ed. P. Padovani & C. M. Urry, (San Francisco; ASP), 163
Perlman, E., Padovani, P., Giommi, P., Sambruna, R., Jones, L. R., Tzioumis, A., Reynolds, J. 1998, AJ, 115, 1253
Stickel, M., Padovani, P., Urry, C. M., Fried, J. W., Kühr, H. 1991, ApJ, 374, 431
Stocke, J. T., Morris, S. L., Gioia, I., Maccacaro, T., Schild, R. E., Wolter, A., Fleming, T. A., Henry, J. P. 1991, ApJS, 76, 813
Urry, C.M. & Padovani, P., 1995 PASP, 107, 803

Discovery of Active Galactic Nuclei in Mid- and Far-Infrared Deep Surveys with ISO

Yoshiaki Taniguchi

Astronomical Institute, Graduate School of Science, Tohoku University, Aramaki, Aoba, Sendai 980-8578, Japan

Abstract. We present a summary on the discovery of active galactic nuclei in mid- and far-infrared deep surveys with use of the Infrared Space Observatory.

1. Introduction

Active galactic nuclei (AGNs) such as quasars have been one of the important issues of modern astrophysics since the discovery of the quasar 3C 273 in 1963 (Schmidt 1963). Their central engines have been thought to be mass-accreting, supermassive single black holes around which gravitational energy is transformed into huge kinetic and radiation energy with the help of gaseous accretion disks (e.g., Rees 1984). Since AGNs are intrinsically bright in any frequency regime, various kinds of AGN surveys have been conducted by using radio through optical to X-ray telescope facilities (e.g., Peterson 1997).

Since current unified models for AGNs introduce a dusty torus around the central engine (see for reviews, Antonucci 1993; Urry & Padovani 1995), any infrared information is absolutely necessary to understand AGN phenomena because such dusty tori radiate anisotropic infrared emission (e.g., Pier & Krolik 1992; Clavel et al. 2000; Laurent et al. 2000; Murayama, Mouri, & Taniguchi 2000 and references therein). Another important issue related to infrared observations of AGNs seems to be possible starburst-AGN connections (Weedman 1983; Sanders et al. 1988; Heckman et al. 1989; Mouri & Taniguchi 1992, 2001; Cid Fernandes et al. 2001; Storchi-Bergmann et al. 2001). In particular, infrared emission properties are highly useful in investigating circumnuclear star formation and its relation to the nuclear activity in AGNs (e.g., Genzel et al. 1998; Tran et al. 2001).

Most Seyfert galaxies and a part of quasars were indeed detected at a wavelength range between 12 μm and 100 μm by the Infrared Astronomical Satellite, *IRAS*, launched in 1983 (see for a review, Soifer, Houck, & Neugebauer 1987). Although the *IRAS* all-sky survey covered over 96% of the sky, its completeness flux limit was 0.5 Jy at 12 μm, 25 μm, and 60 μm, and 1.5 Jy at 100 μm. Therefore, *IRAS* could probe infrared galaxies and AGNs up to redshift $\sim 0.2 - 0.3$ (e.g., Veilleux et al. 1999; Borne et al. 2000) except for some unusually bright (or gravitationally amplified) sources beyond redshift $z \sim 0.5$, e.g., IRAS F10214+4724 at $z = 2.28$ (Rowan-Robinson et al. 1991) and IRAS F15307+3252 at $z = 0.93$ (Cutri et al. 1994); see also Rowan-Robinson (2000).

More than 10 years after the launch of *IRAS*, the Infrared Space Observatory, *ISO*, was launched in 1994 (Kessler et al. 1996). The two infrared array detectors, ISOCAM (Cesarsky et al. 1996) and ISOPHOT (Lemke et al. 1996), were used to carry out mid-infrared (MIR) and far-infrared (FIR) deep surveys, respectively. In this paper, we will give a summary of the MIR and FIR deep surveys with *ISO* and then describe what kinds of AGNs are newly found in these surveys (see for a review, Genzel & Cesarsky 2000).

2. MIR Deep Surveys with ISO

ISOCAM has two independent channels each containing a 32×32 pixel detector; 1) the short wavelength channel (2.5 μm to 5.5 μm) and the long wavelength one (4 μm to 18 μm) (Cesarsky et al. 1996). Since the short wavelengths can be accessible from ground-based telescope facilities, the long wavelength camera was used to carry out MIR deep surveys; i) at 7 μm (Taniguchi et al. 1997; Sato et al. 1999, 2001; Rowan-Robinson et al. 1997; Aussel et al. 1999; Oliver et al. 2000; Lémonon et al. 1998; Altieri et al. 1999; Fadda et al. 2000), ii) at 12 μm (Clements et al. 1999), and iii) 15 μm (Rowan-Robinson et al. 1997; Aussel et al. 1999; Oliver et al. 2000; Lémonon et al. 1998; Altieri et al. 1999; Fadda et al. 2000). A summary of these surveys is given in Table 1; note that the 12 μm survey by Clements et al. (1999) is not included in this table.

Among the MIR imaging surveys tabulated, follow-up spectroscopy has been done for five surveys given in Table 2. Eleven AGNs (4 type 1 and 7 type 2 AGNs) were found among 78 sources in the CFRS-ISOCAM survey (Flores et al. 1999). However, only a few AGN were found in the remaining four surveys (Aussel et al. 1999; Taniguchi et al. 1997; Taniguchi 1999, 2000; Altieri et al. 1999; Fadda et al. 2000). In total, only 14 AGNs (7 type 1s and 7 type 2s) were found among 217 sources in the five surveys. This gives an AGN fraction, $f_{\rm AGN} = N_{\rm AGN}/N_{\rm gal} \approx 7\%$. In Figure 1, we show an optical spectrum of a quasar at $z = 1.025$ found by Taniguchi et al. (1997) as an example.

Although the number of AGNs found in the MIR surveys with *ISO* is not so large, it seems interesting to compare the above observational result with some model predictions. Oliver et al. (1997) estimated expected numbers of AGNs for their MIR survey of the HDF (see also Rowan-Robinson et al. 1997; Aussel et al. 1999), adopting the following two models; 1) PRR models (Pearson & Rowan-Robinson 1996), and 2) AF models (Franceschini et al. 1994); in this article, we do not give the details of the PRR or AF models.

a) AGNs expected in the 7 μm survey: In the case of the PRR models, the expected number of AGNs in the ISO/HDF field is 1.26 (0.66 type 1 and 0.6 type 2 AGNs) if the 7 μm limiting flux is 38.6 μJy and the AGN fraction is $f_{\rm AGN} \approx 22\%$. On the other hand, in the case of the AF models, they are 0.34 and 9%, respectively.

b) AGNs expected in the 15 μm survey: In the case of the PRR models, the expected number of AGNs in the ISO/HDF field is 1.51 (0.73 type 1 and 0.78 type 2 AGNs) if the 15 μm limiting flux is 255 μJy and the AGN fraction is $f_{\rm AGN} \approx 21\%$. On the other hand, in the case of the AF models, they are 0.25 and 3.6%, respectively.

Table 1. A Summary of the MIR Surveys with ISO

Field[a]	Area[b] (arcmin^2)	$F_{\lim}(7\mu m)$[c] (μJy)	$N(7\mu m)$[d]	$F_{\lim}(15\mu m)$[e] (μJy)	$N(15\mu m)$[f]
HDF-N[1]	5/9	65	7	200	45
LHNW[2]	9/—	32	27	—	—
SSA 13[3]	16/—	6	65	—	—
CFRS[4]	—/100	—	—	250	78
ELAIS[5]	11.7/22.8[g]	1000	~ 700	2000	~ 800
A2390C[6]	5.76/5.76	65	4	65	4
A2390[7]	6.76/6.76	25	31	40	34
A1689[8]	36/36	150	41	300	18

[a] HDF-N = Hubble Deep Field-North, LHNW = the NW field in the Lockman Hole, SSA 13 = Hawaii Small Selected Area No. 13, CFRS = Canada-France Redshift Survey Field, ELAIS = European Large Area ISO Survey field, A2390C = the central region of Abell 2390, A2390 = Abell 2390, & A1689 = Abell 1689. Their references are; 1. Aussel et al. 1999, 2. Taniguchi et al. 1997, 3. Sato et al. 1999, 2001, 4. Flores et al. 1999, 5. Oliver et al. 2000, & Serjeant et al. 2000, 6. Lémonon et al. 1998, 7. Altieri et al. 1999, & Fadda et al. 2000.
[b] Sky coverage. The first and second numbers means the sky coverage at 7μm and 15 μm, respectively.
[c] The 3 σ_{rms} flux limit at 7 μm in the survey.
[d] The number of objects detected at 7 μm in the survey.
[e] The 3 σ_{rms} flux limit at 15 μm in the survey.
[f] The number of objects detected at 15 μm in the survey.
[g] Note that the area for ELAIS is given in units of square degree.

Table 2. AGNs Found in the MIR Surveys with ISO

Field[a]	N_{gal}[b]	N_{AGN}[c]	f_{AGN}[d] (%)	N_{Type1}[e]	N_{Type2}[f]
HDF-N	49	1	2	1	0
LHNW	13	1	8	1	0
CFRS	78	11	14	4	7
A2390	32	0	0	0	0
A1689	45	1	2	1	0
Total	217	14	7	7	7

[a] The same as those in Table 1.
[b] The number of galaxies detected in the survey.
[c] The number of AGNs detected in the survey.
[d] The fraction of AGNs = N_{AGN}/N_{gal}.
[e] The number of Type 1 AGNs detected in the survey.
[f] The number of Type 2 AGNs detected in the survey.

It seems difficult to draw any firm conclusions from the above comparisons between the observations and models because of small-number statistics, although we do not see significant inconsistency between them. Future MIR surveys will give us firmer answers.

Among the *ISO* MIR deep surveys, it has been found that there are a number of MIR (either 7 μm or 15 μm) sources without optical/NIR counterparts (Taniguchi et al. 1997; Aussel et al. 1999; Flores et al. 1999). These results are summarized in Table 3. It is shown that $\approx 18\%$ of the MIR sources found in the three MIR surveys have no optical/NIR counterparts and $\approx 10\%$ of them above 5σ detection have no counterparts.

We cannot exclude the possibility that some of such sources may be attributed to unexpected noise. However, if a starburst galaxy with a mass of $10^{11} M_\odot$ at $z \sim 3$ is heavily reddened (e.g., $A_V \sim 10$), this galaxy could be detected in the MIR but not at 2 μm (Taniguchi 2000). Recent submillimeter deep surveys have been finding such dust-enshrouded high-z galaxies (Hughes et al 1998; Barger et al. 1998; Smail et al. 1999). Therefore, we also cannot exclude the possibility that deep NIR surveys may miss a certain part of reddened populations, causing an underestimate of either star formation density or nonthermal energy density or both in the universe. It seems important to keep in mind the presence of such MIR sources without counterparts in future investigations.

Table 3. MIR sources without optical/NIR counterparts

Field[a]	S/N	N_{gal}[b]	N_{no}[c]	f_{no}[d] (%)
HDF-N	$> 5\sigma$	49	7	14
	$> 4\sigma$	51	16	31
LHNW	$> 5\sigma$	13	2	15
CFRS	$> 4\sigma$	40	4	10
	$> 3\sigma$	34	4	15
Total		187	34	18

[a]The same as those in Table 1.
[b]The number of galaxies detected in the survey.
[c]The number of galaxies without optical/NIR counterparts.
[d]The fraction of no-counterparts = $N_{\text{no}}/N_{\text{gal}}$.

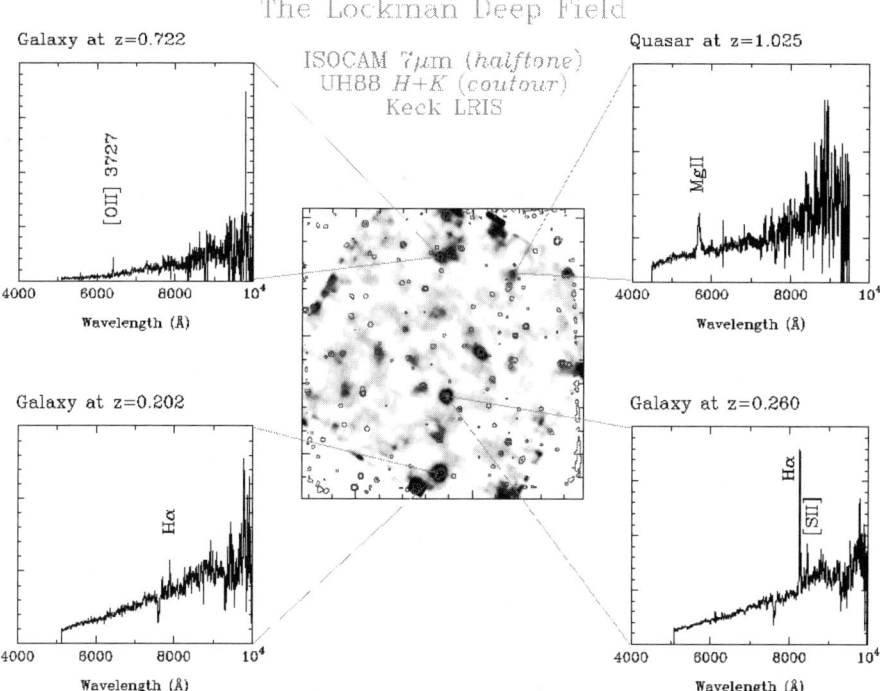

Figure 1. Optical spectra of four 7 μm sources found in the MIR deep survey by Taniguchi et al. (1997). The spectroscopic observations were made with LRIS on the W. M. Keck 1 telescope (Cowie et al. in preparation). One quasar at $z = 1.025$ was found in their survey (see also Taniguchi 1999). The middle panel shows the 7 μm image of the LHNW field (halftone; darker is brighter). The contours show the NIR (HK) image taken with the University of Hawaii 2.2 m telescope. Although the seeing size in the NIR image is good (FWHM≃0.8 arcsec), the image is blurred in order to make the comparison between the 7 μm and NIR images easier.

3. FIR Deep Surveys with ISO

ISOPHOT is an imaging photopolarimeter covering a wavelength range between 2.5 μm and 240 μm (Lemke et al. 1996). Among the several detectors, two detectors, C100 and C200, were used to carry out FIR deep surveys; i) FIRINDIV-DEEP&DEEP[1] (Kawara et al. 1998; Matsuhara et al. 2000), ii) FIRBACK-ELAIS; the 90 μm survey (Efstathiou et al. 2000; Serjeant et al. 2001), and the 170 μm survey (Puget et al. 1999; Lagache & Dole 2001; Dole et al. 2001). A summary of these survey is given in Table 4; note that we do not include another ISOPHOT survey at 60μm and 90 μm by Linden-Vornle et al. (2000) because no follow-up observation is available.

Although optical follow-up spectroscopy has not yet been completed for the two surveys, some preliminary results were reported recently. Serjeant et al. (2001) made optical spectroscopy for 20 sources detected in the FIRBACK-ELAIS survey with $f(90\mu m) > 100$ mJy, which are also detected either at 15 μm or at 1.4 GHz. Among them, they found two Seyfert galaxies at $z = 0.149$ and $z = 0.225$; the remaining 18 sources are either starbursts (16 sources) or early-type galaxies (2 sources). This gives an AGN fraction, $f_{AGN} = 2/20 = 10\%$. On the other hand, Kakazu et al. (2001a, 2001b; see also Murayama et al. 2001) made optical identification of 35 sources found in the FIRINDIV-DEEP&DEEP survey with $f(170\mu m) > 100$ mJy, using the W. M. Keck, Subaru, UH88, and VLA facilities. They identified two AGNs: a Seyfert 1.5 galaxy at $z = 0.206$ and a quasar at $z = 1.60$. The remaining 33 sources are classified as starbursts (21 sources), LINERs (10 sources), and early-type galaxies (2 sources). Since infrared-selected LINERs may be shock-heated galaxies (Taniguchi et al. 1999; Lutz, Veilleux, & Genzel 1999; cf. Imanishi, Dudley, & Maloney 2001), they are not genuine AGNs but starburst-related (i.e., superwind) galaxies. Therefore, their survey gives an AGN fraction, $f_{AGN} = 2/36 \approx 6\%$.

The above AGN fractions are considered to be tentative values because the number of observed galaxies are still small. We hope that future follow-up observations will give us firmer statistics. It seems worthwhile noting that Franceschini et al. (1989) estimated an AGN fraction in such FIR deep surveys; i.e., $f_{AGN} \sim 10\%$, being similar to the observed values.

Finally, we mention that 10 ultraluminous infrared galaxies (ULIGs) and 1 hyperluminous infrared galaxy (HyLIG) were identified as counterparts of 170 μm sources found in Kawara et al.'s (1998) survey (Kakazu et al. 2001a, 2001b; Murayama et al. 2001). Although the HyLIG[2] is a quasar at $z = 1.6$, the ULIGs are located at $z \simeq 0.3 - 0.8$. Such intermediate-z ULIGs had not yet been found by *IRAS*. It is also remarkable that the observed surface density of ULIGs is $\simeq 10$ degree^{-2}. Note that only ~ 120 ULIGs with $z < 0.3$ were found by *IRAS*

---

[1]In contrast to the FIRBACK-ELAIS program (acronym for Far Infrared Background), in this article, we call the Japanese-IfA/UH program FIRINDIV-DEEP&DEEP where INDIV means individual sources found in the FIR survey and DEEP&DEEP means a *deep* survey for *dust-enshrouded extragalactic populations (deep)*. Please make sure to note that DEEP&DEEP is different from the famous DEEP (Deep Extragalactic Evolutionary Probe) survey program promoted by David Koo at UCO/Lick Observatory, University of California (http://www/ucolick.org/ deep/home.html).

[2]See for discovery of a HyLIG in the ELAIS survey, Morel et al. (2001).

(Veilleux et al. 1999; Borne et al. 2000; see for a review Sanders & Mirabel 1996), giving a surface density of ~ 0.003 degree^{-2}. Therefore, it is suggested that the surface density of intermediate-z ULIGs is much higher by a factor of 3000 than that of low-z ULIGs. Since it has been often argued that ULIGs may be precursors of quasars (Sanders et al. 1988; Sanders & Mirabel 1996; Taniguchi, Ikeuchi, & Shioya 1999), it is interesting to investigate the nature of intermediate-z ULIGs found in the *ISO* FIR deep surveys. Future wide-field FIR deep surveys and their follow-up observations will be also very important to find new populations of intermediate-z and high-z ULIGs and HyLIGs.

Table 4. A Summary of the FIR Surveys with ISO

Program	Field[a]	$F_{\lim}(90\mu m)$[b] (mJy)	$N(90\mu m)$[c]	$F_{\lim}(170\mu m)$[d] (mJy)	$N(170\mu m)$[e]
FIRINDIV-DEEP&DEEP	LH 1.1 deg^2/ 1.1 deg^2	45	36	45	45
FIRBACK-ELAIS	Several 11.6 deg^2/ 3.89 deg^2	100(?)	120(?)	135	106

[a]LH = Lockman H<small>I</small> Hole. As for the FIRBACK survey fields, see http://wwwfirback.ias.u-psud.fr. The first sky coverage is for the 90 μm survey and the second one is for the 170 μm one.
[b]$3\sigma_{rms}$.
[c]The number of galaxies detected in the survey. The galaxies detected in FIRINDIV have $F(90\mu m) > 150$ mJy.
[d]$3\sigma_{rms}$.
[e]The number of galaxies detected in the survey. The numbers given in this column are that of galaxies with $F(170\mu m) > 150$ mJy for FIRINDIV and that of galaxies with $F(170\mu m) > 180$ mJy for FIRBACK.

4. Discussion and Summary

As summarized in this article, *ISO* enabled us to perform a number of MIR and FIR deep surveys (see also Genzel & Cesarsky 2000). We give a summary of the main points of this article below.

1) As for the MIR surveys, the *ISO* deep surveys at 7 μm and 15 μm are sensitive enough to detect sources down to 10 μJy (Taniguchi et al. 1997; Aussel et al. 1999; Sato et al. 2001b). Approximately, 10% of the detected sources appear to be AGNs, from low-z Seyfert galaxies to a high-z quasar at z 1 – 2 (Taniguchi 1999; Aussel 1999; Flores et al. 1999). Since the sky areas observed by the surveys are so small, it seems hard to make statistical arguments about AGN populations. However, the number counts of the

MIR sources show significantly stronger evolution than what is expected from no-evolution models (e.g., Elbaz et al. 1999). This may be attributed mainly to intense star formation in galaxies at intermediate- and/or high-z galaxies (e.g., Takeuchi et al. 2000; Chary & Elbaz 2001). Since dusty starburst galaxies sometimes harbor hidden AGNs (e.g., Sanders et al. 1988; Ivison et al. 2000; Willott et al. 2001), future follow-up observations will be important to understand what faint MIR sources are.

2) The *ISO* FIR deep surveys have also shown that approximately, 10% of the detected sources appear to be AGNs (Serjeant et al. 2001; Kakazu et al. 2001a, 2001b; Murayama et al. 2001). Although the total sky area surveyed by the FIR surveys exceeds 10 deg^2, optical follow-up observations have not yet been fully done. Therefore, future follow-up observations will be important to understand the nature of faint FIR sources. The most remarkable finding of the *ISO* FIR deep surveys is the discovery of numerous ULIGs at intermediate redshift between $z \simeq 0.3$ and $z \simeq 0.8$ (Kakazu et al. 2001a), because such populations had not yet been known from the *IRAS* survey. It seems quite likely that these populations contribute to the observed excess number count in the FIR (e.g., Elbaz et al. 1999; Takeuchi et al. 2000).

3) The exciting results above urge us to conduct new MIR and FIR surveys. Indeed, new space infrared telescope facilities will be launched soon; SIRTF[3] and IRIS (ASTRO-F[4]). We hope that these facilities will give us much more information on AGN populations in the infrared universe.

Acknowledgments. The author would like to thank the organizers of this nice IAU colloquium in the beautiful country, Armenia, E. Khachikian, Areg Mickaelian, Dave Sanders, and Richard Green. He would also like to thank his nice colleagues who have been working together on the MIR/FIR deep surveys with *ISO* and their follow-up observations, Len Cowie, Dave Sanders, Bob Joseph, Haruyuki Okuda, Kimiaki Kawara, Yasunori Sato, Toshio Matsumoto, Hideo Matsuhara, Yoshiaki Sofue, Ken-ichi Wakamatsu, Youichi Ohyama, Sylvain Veilleux, Min Yun, Takashi Murayama, Yuko Kakazu, and Tohru Nagao.

References

Altieri, B., et al. 1999, A&A, 343, L65
Antonucci, R. R. J. 1993, ARA&A, 33, 19
Aussel, H., Cesarsky, C. J., Elbaz, D., & Starck, J. L. 1999, A&A, 342, 313
Barger, A., et al. 1998, Nature, 394, 248

---

[3] http://sirtf.caltech.edu

[4] http://www.ir.isas.ac.jp/ASTRO-F/index-j.html

Borne, K. D., Bushouse, H., Lucas, R. A., & Colina, L., 2000, ApJ, 529, L77
Cesarsky, C. J., et al. 1996, A&A, 315, L32
Chary, R., & Elbaz, D. 2001, ApJ, 556, 562
Cid Fernandes, R., Heckman, T., Schmitt, H., Gonzáres Delgado, R. M. Gonzáres; Storchi-Bergmann, T. 2001, ApJ, 558, 81
Clavel, J., et al. 2000, A&A, 357, 839
Clements, D., Desert, F.-X., Franceschini, A., Reach, W. T., Baker, A. C., Davies, J. K., & Cesarsky, C. 1999, A&A, 346, 383
Cutri, R. M., Huchra, J. P., Low, F. J., Brown, R. L., vanden Bout, P. A. 1994, ApJ, 424, L65
Dole, H., et al. 2001, A&A, 372, 364
Efstathiou, G., et al. 2000, MNRAS, 319, 1169
Elbaz, D., et al. 1999, A&A, 351, L37
Fadda, D. et al. 2000, A&A, 361, 827
Flores, H., et al. 1999, ApJ, 517, 148
Franceschini, A., et al. 1991, A&AS, 89, 285
Franceschini, A., et al. 1994, ApJ, 427, 140
Genzel, R., & Cesarsky, C. J. 2000, ARA&A, 38, 761
Genzel, R., et al. 1998, ApJ, 498, 579
Heckman, T. M., Blitz, L., Wilson, A. S., Armus, L., & Miley, G. K. 1989, ApJ, 342, 735
Hughes, D. H., et al. 1998, Nature, 394, 241
Imanishi, M., Dudley, C. C., & Maloney, P. R. 2001, ApJ, 558, L93
Ivison, R. J., et al. 2000, MNRAS, 315, 209
Kakazu, Y., et al. 2001a, in this volume
Kakazu, Y., et al. 2001b, The Japanese-German Seminar on Studies of Galaxies in the Young Universe with New Generation Telescopes, edited by N. Arimoto, & W. Duschl, in press
Kawara, K., et al. 1998, A&A, 336, L9
Kessler, M., et al. 1996, A&A, 315, L27
Laurent, O., et al. 2000, A&A, 359, 887
Lagache, G., & Dole, H. 2001, A&A, 372, 702
Lemke, D., et al. 1996, A&A, 315, L64
Lémonon, L., Pierre, M., Cesarsky, C. J., Elbaz, D., Pelló, R., Soucail, G., & Vigroux, L. 1998, A&A, 334, L21
Linden-Vornle, M. J. D., et al. 2000, A&A, 359, 51
Lutz, D., Veilleux, S., & Genzel, R. 1999, ApJ, 517, L13
Matsuhara, H., et al. 2000, A&A, 361, 407
Morel, T., et al. 2001, MNRAS, 327, 1187
Mouri, H., & Taniguchi, Y. 1992, ApJ, 386, 68
Mouri, H., & Taniguchi, Y. 2001, ApJ, in press (astro-ph/0106155)
Murayama, T., Mouri, H., & Taniguchi, Y. 2000, ApJ, 528, 179

Murayama, T., et al. 2001, in preparation
Oliver, S., et al. 1997, MNRAS, 289, 471
Oliver, S., et al. 2000, MNRAS, 316, 749
Pearson, C., & Rowan-Robinson, M. 1996, MNRAS, 283, 174
Peterson, B. M. 1997, An Introduction to Active Galactic Nuclei (University of Cambridge Press)
Phillips, A. C., et al. 1997, ApJ, 489, 543
Pier, E. A., & Krolik, J. H. 1992, ApJ, 401, 99
Puget, J. -L., et al. 1999, A&A, 345, 29
Rees, M. J. 1984, ARA&A, 22, 471
Rowan-Robinson, M. 2000, MNRAS, 316, 885
Rowan-Robinson, M., et al. 1991, Nature, 351, 719
Rowan-Robinson, M., et al. 1997, MNRAS, 289, 490
Sanders, D. B., et al. 1988, ApJ, 325, 74
Sanders, D. B., & Mirabel, I. F. 1996, ARA&A, 34, 749
Sato, Y., et al. 1999, The Universe as Seen by ISO, edited by P. Cox, & M. F. Kessler (ESA-SP), 427
Sato, Y., et al. 2001a, ApJ, submitted
Sato, Y., et al. 2001b, A&A, submitted
Schmidt, M. 1963, Nature, 197, 1040
Serjeant, S., et al. 2001, MNRAS, 322, 262
Smail, I., et al. 1999, MNRAS, 308, 1061
Soifer, B. T., Houck, J. R., & Neugebauer, G. 1987, ARA&A, 25, 187
Storchi-Bergmann, T., Gonzáres Delgado, R. M., Schmitt, H. R., Cid Fernandes, R., & Heckman, T. M. 2001, ApJ, 559, 147
Takeuchi, T., et al. 2001, PASJ, 53, 37
Taniguchi, Y. 1999, Astrophysics with Infrared Surveys: A Prelude to SIRTF, ASP Conference Ser., 177, edited by M. D. Bicay, R. M. Cutri, & B. F. Madore, 89
Taniguchi, Y. 2000, Advances in Space Research, 25, 2233
Taniguchi, Y., et al. 1997, A&A, 328, L9
Taniguchi, Y., Ikeuchi, S., & Shioya, Y. 1999, ApJ, 514, L9
Taniguchi, Y., Yoshino, A., Ohyama, Y., & Nishiura, S. 1999, ApJ, 514, 660
Tran, Q. D., et al. 2001, ApJ, 552, 557
Urry, C. M., & Padovani, P. 1995, PASP, 107, 803
Veilleux, S., Kim, D. -C., & Sanders, D. B. 1999, ApJ, 522, 113
Weedman, D. W. 1983, ApJ, 266, 479
Willott, C. J., et al. 2001, in Proceedings of XXI Moriond Conference: "Galaxy Clusters and the High Redshift Universe Observed in X-rays", edited by D. Neumann, F.Durret, & J. Tran Thanh Van, in press (astro-ph/0105560)

A New MIR/submm Diagnostic for Dust-Enshrouded AGN

Martin Haas

Max-Planck-Institut für Astronomie, D-69117 Heidelberg, Germany

Abstract. We show that the PAH 7.7μm to continuum 850μm flux ratio can be used to reveal high mid-infrared extinction in ultraluminous infrared galaxies (ULIRGs). While the submm radiation is optically thin and represents the emission from essentially all dust grains, the PAH strength (measured by the peak height of the Polycyclic Aromatic Hydocarbonates at 7.7μm) is sensitive to dust extinction in the mid-infrared (MIR). As an application of the new diagnostic, after dereddening of the central MIR continuum and with the assumption of a disk-like dust distribution seen under a tilted angle, we find increasing evidence for a hidden quasar in the archetypal ULIRG Arp220.

1. Introduction

The IR 1-1000μm spectral energy distributions (SEDs) of ULIRGs show two basic types as illustrated in Fig. 1:

1. *Warm* ULIRGs: They have a steep flux rise from the NIR (around 2-5μm) to the MIR (around 10-25μm), a high $F_{25\mu m}$ to $F_{60\mu m}$ ratio (>0.2) and most of them show optical Seyfert spectra.

2. *Cool* ULIRGs: They have a flat NIR flux plateau followed by a sudden jump-like flux rise around 10μm, a low $F_{25\mu m}$ to $F_{60\mu m}$ ratio (<0.2) and most of them show LINER or HII type optical spectra.

The open question is whether the NIR-MIR continuum in cool ULIRGs could be suppressed by extinction on the order of $A_V > 100$. If "yes", then cool ULIRGs could also house a powerful AGN.

Standard methods for the MIR extinction use the Si 9.7μm absorption feature or the S[III] 18.7μm/33.5μm line ratios. These methods are based exclusively on a wavelength regime, the extinction of which is a priori unknown. The limitations could be that they probe only the shallow surface, but not the full dust column.

Therefore a new method is proposed which uses not only the MIR range alone, but also the submm regime at 850μm which is optically thin and traces the full dust column. The PAH (Polycyclic Aromatic Hydrocarbonates) are presumably mixed with the submm emitting grains. The submm emission serves as normalizer, while the PAH 7.7μm line strength is sensitive to extinction.

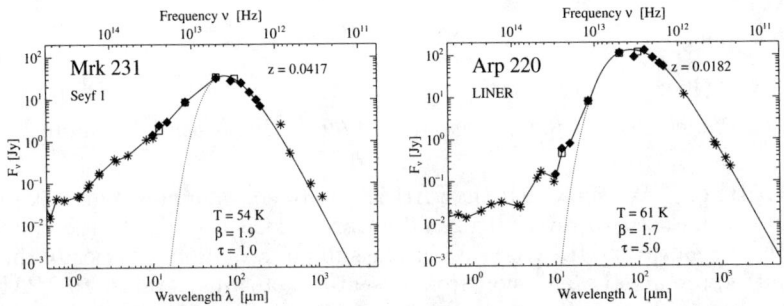

Figure 1. IR SED of Mrk 231 and Arp 220, a warm and a cool ULIRG (from Klaas et al. 2001)

2. Brief Outline of the Diagnosis

First results on the PAH 7.7μm/850μm ratio in ULIRGs have been published by Haas, Klaas, Müller et al. (2001). Also, a comprehensive study of this topic is presented by Haas (2001), and the reader can obtain a copy of this manuscript (just send an email to haas@mpia.de). Here, we restrict the discussion to a brief outline of the diagnosis:

2.1. The Data

We analyse a sample of 15 ULIRGs for which both the PAH spectra (from Rigopoulou et al. 1999) and the submm fluxes are available (from Rigopoulou, Lawrence & Rowan-Robinson 1996, Lisenfeld, Isaak & Hills 2000, Klaas et al. 2001). Also the FIR fluxes at 100μm are considered (from IRAS and from Klaas et al. 2001). These data are compared with those of a sample of 20 "normal" starforming galaxies (submm fluxes from Dunne et al. 2000, PAH spectra from the ISO archive). The PAH spectra refer to an aperture of 24" (with FWHM of the ISOPHOT-S point spread function $\approx$ 3"); the submm fluxes are derived assuming unresolved sources (FWHM of SCUBA beam at 850μm $\approx$ 15", FWHM of SEST beam at 1.3mm $\approx$ 24").

Fig. 2 shows the PAH spectra for two examples, and how the PAH 7.7μm line strength is derived.

2.2. The PAH/Submm Flux Ratio: A Tracer for High MIR Extinction

Fig. 3 shows the PAH 7.7μm to 850μm flux versus PAH 7.7μm to 100μm flux:

1. The y-axis shows the PAH 7.7μm/850μm distribution of the ULIRGs and the reference sample. Strikingly, all ULIRGs except Arp220 and UGC5101 lie in a confined range (around 5 $\pm$ 2), which is also the same as for the

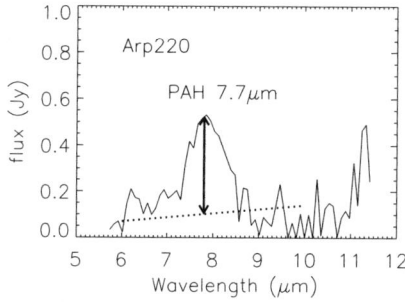 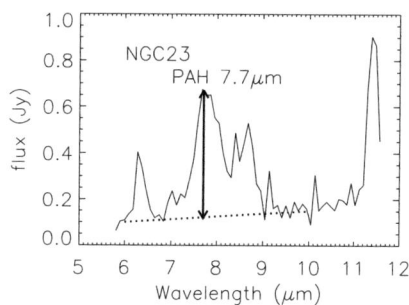

Figure 2. Examples of ISOPHOT-S 5-12 μm spectra (24″ aperture) for the ULIRG Arp220, and the comparison galaxy NGC 23. The dotted line indicates the continuum subtracted for the estimate of the PAH 7.7 μm strength. Note that adopting a lower continuum can increase the determined PAH strength by at most 20%.

normal galaxies. This suggests that for both samples the PAH and the sub-mm emission are related, and that no extraordinary excitation conditions are needed.

UGC5101 has a high mixed case extinction $A_V \approx 50$ derived from NIR-MIR spectroscopy (Genzel, Lutz, Sturm et al. 1998). Dereddening shifts it clearly into the range of the other ULIRGs.

2. The PAH 7.7μm/100μm ratio (x-axis) is lower for the ULIRGs than for the reference sample by a factor of about three. ULIRGs have warmer dust (30K<T<50K) than normal galaxies (20K<T<30K), and therefore their 100μm flux relative to that at 850μm is higher. Again, along the PAH 7.7μm/100μm distribution Arp220 lies below the other obviously more "typical ULIRGs" which populate a confined range.

In principle, the exceptional position of Arp220 could be due to PAH destruction (either by a quasar-like hard UV radiation field or strong collisions) or to an unusually high submm flux excess. Since both of these explanations remain unsatisfactory, as a most likely scenario we conclude that the low PAH 7.7μm/850μm flux ratio of Arp220 is mainly caused by extinction. A comparison with UGC5101 immediately suggests that the MIR extinction is about a factor of three higher in Arp 220.

Note that the precise value for the extinction at 7.7μm depends on the dust properties and extinction curves assumed. Here we consider two "extreme" cases: the interstellar one by Mathis, Rumpl & Nordsieck (MRN 1977, see Mathis, Mezger & Panagia 1983), and the one of dense protostellar clouds by Krügel & Siebenmorgen (1994).

It is reasonable to assume that the PAH carriers are not an isolated component in the interstellar medium (ISM), rather they are mixed with other constituents of the ISM. Thus we consider an ensemble of dust clouds which emit

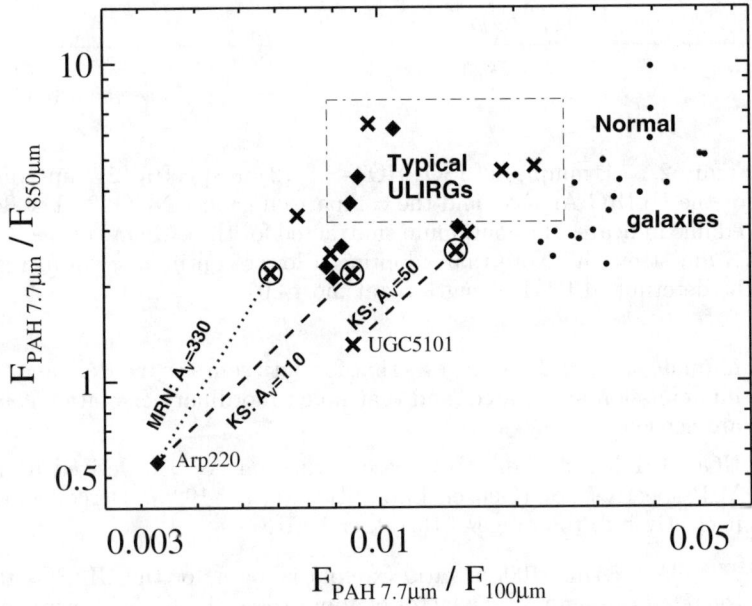

Figure 3. Two-colour diagram PAH 7.7μm to 850μm flux versus PAH 7.7μm to 100μm. The symbols are: crosses = Seyfert-ULIRGs, diamonds = LINER- and HII/SB-ULIRGs, points = normal galaxies. The errors are less than 30%. The positions of Arp220 and UGC5101 after dereddening are marked with the encircled crosses. For Arp220 two mixed case extinction curves are considered: interstellar dust (dotted line: MRN, Mathis, Rumpl & Nordsieck 1977), and protostellar dust (dashed line: KS, Krügel & Siebenmorgen 1994). The dash-dotted box indicates the min-to-max range of the typical ULIRGs after dereddenening the A_V values from Genzel, Lutz, Sturm et al. (1998) with the KS model; note that dereddening reduces the PAH 7.7μm/850μm dispersion.

the FIR and submm luminosity, and the PAH carriers are located preferentially at the borders of these clouds in the photodissociation regions. In this picture the PAH carriers are mixed with the dust clouds – at least on the spatial scale of several pc, which applies to our data. Consequently, for the PAHs we adopt not the commonly used "screen case" extinction, rather the "mixed case" extinction, which results in higher dust column densities.

To summarize, 13 out of 15 ULIRGs as well as 20 comparison galaxies populate the same confined range of the PAH 7.7μm/850μm flux ratio. Their MIR extinction may be moderate ($A_V \lesssim 3\text{-}10$), so that NIR-MIR spectroscopy can yield proper results. Also, it is not likely that they contain a hidden powerful AGN which has not yet been identified as such. Two ULIRGs, however, lie significantly below this range and their offsets are mainly due to extinction. The new PAH 7.7μm/850μm diagnostics is a promising tool to reveal high MIR extinction in ULIRGs.

2.3. Signatures Favoring a Hidden Quasar in Arp 220

The best signatures for an AGN are broad emission lines. Actually in Arp 220 a broad Br$_\alpha$ line has been discovered (Depoy, Becklin & Geballe 1987). The strength of this NIR line, however, was found to be too faint to account for the ultrahigh bolometric (FIR) luminosity (Depoy, Becklin & Geballe 1987). In order to do so, the extinction towards the region emitting this line would be required to exceed the incredibly high amount of $A_V \approx 130$, but at that time, other extinction estimates were by far lower ($A_V \approx 50$). An extinction of $A_V \approx 110$ is so high that even hard X-rays might be blocked (see also the excellent contribution by Iwasawa in this volumne). The supernova remnants resolved using VLBI at 18 cm (Smith et al. 1998), provide evidence for powerful starbursts whose extent is larger than the 100 pc diameter found for the bulk of the MIR emission (Soifer et al. 1999). They do not exclude the presence of a deeply hidden AGN.

As outlined in the introduction, an alternative tracer for an AGN is a powerful MIR continuum as well as a high MIR/FIR luminosity ratio. Thus our task is to find out the extinction for the MIR continuum, which may differ from that derived for the PAHs, depending on the geometry of the absorbing/reemitting dust. For example, imagine a central AGN as strong MIR continuum source, which (along our line of sight) is hidden by dust clouds emitting in the PAH 7.7μm line as well as the submm continuum. Then the PAH/submm diagnosis yields a *mixed* case extinction for the dust clouds, but the central source suffers from *screen* extinction by these dust clouds. (Actually the screen consists of half of the dust column, if the central source lies in its middle). The crucial point is that the dust column derived for the mixed case extinction is huge, and that the screen extinction of even half of this dust column is still very high. (In order to provide a given reddening curve by mixed case extinction with a dust column of homogeneous density, the emitters located deeply inside need to be extremely reddened, since the emitters at the shallow surface towards the observer are nearly unextincted.)

The presence of CO disks already indicates a *non-spherical axisymmetric geometry* tilted about 45° with respect to our line of sight (Downes & Solomon 1998, Sakamoto et al. 1999). The question is to decide between the two alternatives:

1. the line of sight towards the two nuclei of Arp 220 is relatively free of extinction, or

2. the two nuclei of Arp 220 are hidden by the disks.

High spatial resolution Keck telescope images between 3.4 and 24.5μm reveal that the colors of the two nuclear regions are similar to those for the extranuclear areas and do not show much lower extinction (Soifer et al. 1999). Furthermore, on NICMOS images between 1.1 and 2.2μm the NIR maximum is offset from the actual western nuclear position as traced by the radio continuum peak, and the eastern nucleus belongs to the most reddened regions (Scoville et al. 1998). These findings argue against the first alternative, hence the nuclei might be hidden by the disks.

It is reasonable to assume that the observed MIR continuum is a mixture of contributions from from (extranuclear) starbursts in the disks and from the nuclei. But we do not know their relative contributions. In Tab. 1 we list the values derived assuming purely mixed case extinction for the disks (rows 2 and 3) and purely screen extinction for the nuclei (row 4). The actual dereddened MIR continuum likely lies between the two extremes. Note that for the high extinction values also the FIR emission has to be dereddened, but the effect is moderate, since probably the mixed case extinction applies to the FIR.

Table 1. Observed and dereddened luminosities of Arp220.

	deredd. factor	L_{MIR} $10^{11} L_\odot$	deredd. factor	L_{FIR} $10^{11} L_\odot$	L_{MIR}/L_{FIR}
observed	1.0	2.1	1.0	9.0	0.23
MRN:A_V=330	10 (4...20)	$\approx$ 21.0	3.2	28.8	0.72
KS: A_V=110	4.0	8.4	1.7 (1.3 ... 2.0)	15.3	0.54
KS: $A_V$screen=90*	20.0	42.0	"	"	2.75

* with screen case on central MIR continuum using the half of the dust column derived from the PAH mixed case extinction, whereby the mixed case extinction probably still applies to the FIR emission

Fig. 4 shows the resulting MIR/FIR luminosity ratios for the ULIRGs, and how the dereddening moves Arp220 from the range observed for cool ULIRGs towards that range which is exclusively populated by PG quasars and typical warm ULIRGs housing a strong AGN. In addition, the nuclear regions are very compact (d < 100 pc), hence the MIR luminosity density exceeds that of known starburst (e.g. in M82) by a factor of about 1000. Since the dust (and the gas) is more dissipative than the stars, it tends to be distributed in a more compact area than the stars. Bearing this in mind, it is difficult to imagine, how starbursts alone can create such a high luminosity density, and *simultaneously* how the dust hides them entirely.

Therefore, a more natural explanation would be that – in addition to the prominent starbursts mainly responsible for the FIR luminosity – one (or both) of the nuclei of Arp 220 contains a powerful AGN providing the quasar-like MIR luminosity which is hidden to us.

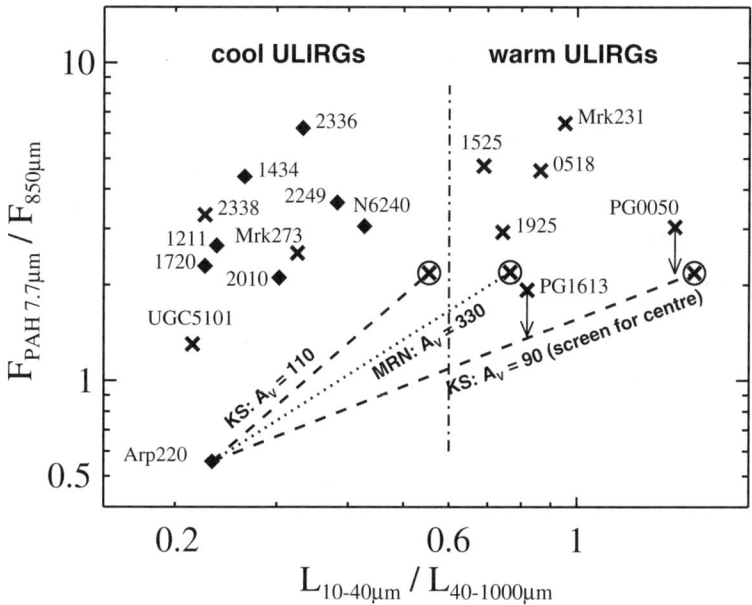

Figure 4. Two-"colour" diagram PAH 7.7μm to 850μm flux versus L_{MIR}/L_{FIR} (= $L_{10-40\mu m}/L_{40-1000\mu m}$). Symbols as in Fig.1. In addition, two PG quasars, where PAH spectra are available, are plotted. The dereddened positions of Arp220 correspond to the KS and MRN cases listed in Tab.1. The vertical dash-dotted line marks the division between warm and cool ULIRGs according to $F_{25\mu m}/F_{60\mu m} \approx 0.2$.

Acknowledgments. It is a pleasure for me to thank the organizers of this wonderful, friendly and fruitful meeting for the invitation to give this talk.

References

Depoy D.L., Becklin E.E., Geballe T.R., 1987, ApJ, 316, L63
Downes D., Solomon P.M., 1998, ApJ, 507, 615
Dunne L., Eales St., Edmunds M. et al., 2000, MNRAS, 315, 115
Genzel R., Lutz D., Sturm E. et al., 1998, ApJ, 498, 579
Haas M., 2001, Habilitationsschrift at Universität Heidelberg
Haas M., Klaas U., Müller S.A.H., Chini R, Coulson I., 2001, A&A, 367, L9
Klaas U., Haas M., Müller S.A.H., et al., 2001, A&A, submitted
Krügel E., Siebenmorgen R., 1994, A&A, 288, 929
Lisenfeld U., Isaak K.G., Hills R., 2000, MNRAS, 312, 433
Mathis J.S., Mezger P.G., Panagia N., 1983, A&A, 128, 212
Rigopoulou D., Lawrence A., Rowan-Robinson M., 1996, MNRAS, 278, 1049
Rigopoulou D., Spoon H.W.W., Genzel R., et al., 1999, AJ, 118, 2625
Sakamoto K., Scoville N.Z., Yun M.S., et al., 1999, ApJ, 514, 68
Scoville N.Z., Evans A.S., Dinshaw N. et al., 1998, ApJ, 492, L107
Smith H.E., Lonsdale C.J., Lonsdale C.J. Diamond P.J, 1998, ApJ, 493, L21
Soifer B.T., Neugebauer G., Matthews K. et al., 1999, ApJ, 513, 207

AGN SURVEYS
ASP Conference Series, Vol. 284, 2002
R.F. Green, E.Ye. Khachikian, D.B. Sanders

The Nature of the Faint Far-Infrared Extragalactic Source Population: Optical/NIR and Radio Follow-up Observations of ISOPHOT Deep-Field Sources using the Keck, Subaru, and VLA Telescopes

Yuko Kakazu[1], D. B. Sanders[1], R. D. Joseph[1], L. L. Cowie[1], T. Murayama[2], Y. Taniguchi[2], S. Veilleux[3], M. S. Yun[4], K. Kawara[5], Y. Sofue[5], Y. Sato[5], H. Okuda[6], K. Wakamatsu[7], T. Matsumoto[8], and H. Matsuhara[8]

[1]Institute for Astronomy, University of Hawaii. [2]Astronomical Institute, Tohoku Univesity. [3]Deptartment of Astronomy, University of Maryland. [4]FCRAO, University of Massachusetts. [5]Institute of Astronomy, University of Tokyo. [6]Gumma Observatory. [7]Gifu University. [8]ISAS.

Abstract. We report on optical and near-infrared (NIR) follow-up spectroscopy of faint far-infrared (FIR) sources found in our deep FIR survey by Kawara et al.

1. Introduction

Deep surveys at FIR and submillimeter wavelengths have been carried out in order to investigate the nature of dust-enshrouded galaxies at high redshift. As a contribution to this field, our group made a deep FIR survey using the ISOPHOT camera on board the *Infrared Space Observatory* (ISO) satellite (Kawara et al. 1998; Matsuhara et al. 2000). Mapping at 90μm and 170μm of two 44′ × 44′ fields in the Lockman Hole (LH_EX and LH_NW), a region exhibiting the lowest H I column density in the sky (Lockman et al. 1986), resulted in the detection of 36 sources with $f_{90} > 150$ mJy and 45 sources with $f_{170} > 150$ mJy. Given the relatively large size of the ISOPHOT beam at 170μm (∼90″), we have obtained opt/NIR images and spectra using telescopes on Mauna Kea and 6cm radio continuum maps using the VLA (Yun et al. 2002) to identify the most likely source of the 170μm emission. Here we report our initial identifications of the brightest of the ISOPHOT 170μm sources.

2. Results and Discussion

Redshifts of 35 FIR source candidates were determined using optical spectra obtained with ESI on Keck II during three observing runs in 2000 March and 2001 January. Infrared luminosities, $L_{\text{ir}}(8-1000\mu m)$, were then estimated by using the ISOPHOT fluxes and assuming an SED similar to that of Arp 220. We found one hyperluminous infrared galaxy (HyLIG: $L_{\text{ir}} > 10^{13} L_\odot$) at $z = 1.6$, 11 ultraluminous infrared galaxies (ULIGs: $L_{\text{ir}} > 10^{12} L_\odot$) at $0.3 < z < 1$, 12 luminous infrared galaxies (LIGs: $L_{\text{ir}} > 10^{11} L_\odot$), and 11 galaxies with $L_{\text{ir}} <$

213

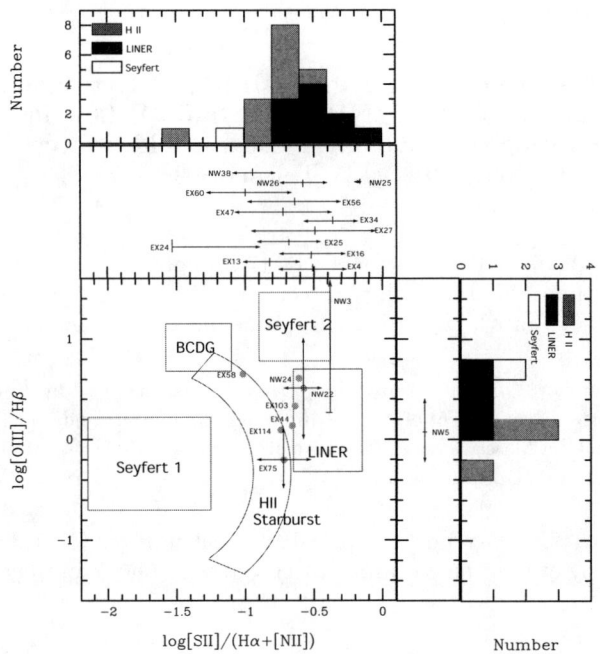

Figure 1. Low-resolution emission line diagnostics of ISOPHOT source candidates.

$10^{11}L_\odot$. Except for one LIG at $z = 0.365$, all of the galaxies with $L_{\rm ir} < 10^{12}L_\odot$) are at $z < 0.3$. The mean redshift for all sources is 0.31 ± 0.31.

The low-resolution ESI spectra were used to determine the optical spectral-type of the candidate ISOPHOT sources. Following procedures used by Murayama & Taniguchi (1998), the spectra were classified into four types – AGNs, LINERs, H II-type, and early-type (without emission lines). The HyLIG at $z = 1.6$ was found to be a quasar. One ULIG had an early-type spectrum and 10 ULIGs are H II galaxies. Among the remaining 23 lower-luminosity sources, there was one early-type galaxy, one Seyfert 2, 10 LINERs and 11 H II galaxies. Thus, based on our low-resolution ESI optical spectra most of the ISOPHOT 175μm sources appear to be powered primarily by star formation, consistent with the conclusion reached from an analysis of ISOPHOT number counts by Matsuhara et al. (2000) that most of the ISOPHOT sources are star-forming galaxies at $z < 1$.

References

Kawara, K., et al. 1998, A&A, 336, L9
Lockman, F. J., Jahoda, K. & McCammon, D. 1986, ApJ, 302, 432
Matsuhara, H., et al. 2000, A&A, 361, 407
Murayama, T. & Taniguchi, Y. 1998, PASJ, 50, 241
Yun, M. S., et al. 2002, in preparation

IRAS 03158+4227 – a ULIRG in a Widely Separated Pair of Galaxies

Helmut Meusinger, Bringfried Stecklum, Jens Brunzendorf

Thüringer Landessternwarte Tautenburg, D-07778 Tautenburg, Germany

Abstract. We present new deep optical images, optical spectroscopy, and high-resolution NIR images of IRAS 03158+4227, one of the most luminous ULIRGs from the IRAS 2 Jy sample. The data are best explained either by a multiple merger or by the assumption of a ULIRG triggered in an early phase of galaxy interaction.

1. Introduction

Ultra-luminous infrared galaxies (ULIRGs) are an important class of extragalactic objects which are probably related to AGNs. The standard picture of activity in nuclear regions of galaxies invokes dissipative gas infall toward the centre induced by galaxy-galaxy interactions. It is claimed that the ULIRG phenomenon is triggered during the final stage of galaxy mergers. Although ULIRGs show in general strong evidence for tidal distortions, IRAS 03158+4227 has been described as an apparently single and undisturbed system (Murphy et al. 1996).

2. Observations and Discussion

We performed observations of IRAS 03158+4227 at the DSAZ[1], Calar Alto, Spain. Spectra and deep optical images (BRI bands and unfiltered, respectively) were taken with the 2.2 m-telescope equipped with CAFOS at a seeing of typically $1''$. High-resolution imaging in the JHK' bands was performed using the adaptive optics system ALFA at the 3.5 m-telescope. In addition, wide field images were taken with the Tautenburg Schmidt CCD camera in the R and I band. Based on these new observations, IRAS 03158+4227 is found to be a member of a binary of two giant galaxies (G1 and G2, Fig. 1) with a projected nuclei separation of 47 kpc ($H_0 = 75 \,\mathrm{km\,s^{-1}\,Mpc^{-1}}$) and a radial velocity difference of less than about $200\,\mathrm{km\,s^{-1}}$. Strong emission lines are detected from the nuclear regions of both galaxies. There is no evidence for a double nucleus in the ULIRG's host G1 down to a separation of $0\rlap{.}''4$ (corresponding to about 1 kpc at the distance of the galaxy). The deep optical images (Fig. 1) clearly reveal that G2 has a faint but very extended material arm on the side opposite to G1. This structure is the most prominent peculiar morphological feature of the binary.

---

[1]German-Spanish Astronomical Centre operated by the Max-Planck-Institute for Astronomy, Heidelberg, jointly with the Spanish National Comission for Astronomy

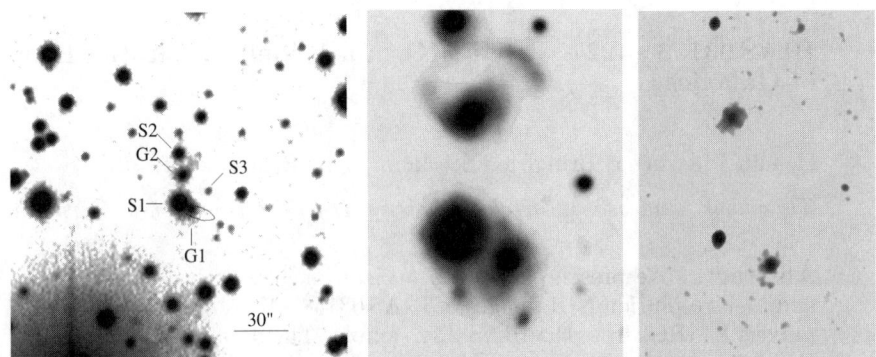

Figure 1. *Left:* the $3\rlap{.}''3 \times 3\rlap{.}''3$ field around the IRAS error ellipse of IRAS 03158+4227 (coadded R and I band images taken at a seeing of about $2''$); G1 and G2 are galaxies, S1 to S3 stars. *Middle:* optical composite image (BRI bands plus unfiltered image) of the G1-G2 pair at a seeing of about $1''$ after Lucy-Richardson deconvolution. *Right:* composite NIR image (JHK' bands) taken with the adaptive optics system ALFA. The size of the field is $30'' \times 42''$, N is up, E is left.

3. Conclusions

The long, faint tail emanating from G2 indicates strong tidal forces. The results of numerical simulations admit the interpretation of this structure as due to tidal interaction with G1 (Meusinger et al. 2001). It is tempting to speculate that the activities in both galaxies were triggered by the same process, namely the gravitational interaction of G1 and G2. In this case, IRAS 03158+4227 has to be interpreted as an early stage of merger like IRAS 23327+2913 (Dinh-V-Trung et al. 2001). Alternatively, IRAS 03158+4227 may be the result of a multiple merger (e.g., Borne et al. 2000) in a compact group: even though we do not find evidence for a close double nucleus, the host G1 might be a merger in a very advanced stage with a projected nuclei separation of less than 1 kpc. Then, however, the huge star formation rate derived from the infrared-luminosity of IRAS 03158+4227 seems to be be surprising (cf. Bekki 2001).

References

Bekki, K. 2001, ApJ, 546, 189
Borne, K.D., Bushouse, H., Lucas, R.A. et al. 2000, ApJ, 529, L77
Dinh-V-Trung, Lo, K.Y., & Kim, D.-C. 2001, ApJ, 556, 141
Meusinger, H., Stecklum, B., Theis, C. et al. 2001, A&A, in press
Murphy, T.W., Armus, L., Matthews, K. et al. 1996, AJ, 111, 1045

The Byurakan-*IRAS* Galaxy (*BIG*) Sample: The Redshift Survey

A.M. Mickaelian, S.K. Balayan and S.A. Hakopian

Byurakan Astrophysical Observatory (BAO), Byurakan 378433, Armenia. E-mail: aregmick@bao.sci.am

Abstract. The Byurakan-IRAS Galaxy (BIG) sample (1967 galaxies) is based on optical identifications of IRAS PSC sources at $\delta > +61°$ and $|b| > 15°$ (FBS area). A redshift survey for brighter objects (B<18^m) is being carried out with the Byurakan Observatory 2.6m, Special Observatory 6m, and Observatoire de Haute-Provence 1.93m telescopes. 213 objects have been observed, and redshifts in the range 0.008-0.173 have been measured. For this subsample, 15% of the objects are AGNs, and 15% are LIGs and ULIGs.

1. The *BIG* Sample

A program of optical identifications of all IRAS PSC (IRAS 1988) sources at high galactic latitudes was conducted in the Byurakan Observatory in 1995 (Mickaelian 1995). The First Byurakan Survey (FBS) (Markarian et al. 1989), served as the basis for this work. The area of the FBS with $+61° < \delta < +90°$ at galactic latitudes $|b| > 15°$ was included with a total surface area of 1487 deg^2. The identifications were made on the basis of the Digital Sky Survey (DSS) images, the First Byurakan Survey (FBS) low-dispersion spectra, and the IRAS infrared fluxes at 12μm, 25μm, 60μm and 100μm wavelengths. 1577 sources have been optically identified, with 1178 sources corresponding to galaxies. The BIG sample (Byurakan-IRAS Galaxies) was constructed of 1178 newly identified galaxies and 789 other IRAS galaxies in the same area, known before (Mickaelian 2001, Mickaelian et al. 2001). The sample contains compact galaxies, interacting pairs and groups, "mergers", radio and X-ray sources, etc. Study of the sample is important for a better understanding of star-formation, nuclear activity, interactions and connections between these phenomena. 350 newly identified galaxies were bright enough (B<18^m) to undertake a quick redshift survey with the available 2-6 m size telescopes. This was one of the subtasks of the whole Byurakan-IRAS program.

2. Spectroscopic Follow-up and the Redshift Survey

Medium-dispersion spectroscopic follow-up observations for the BIG objects with B<18^m in 1997-2000 were aimed at obtaining their redshifts and their classifications (Mickaelian et al. 1998; 1999; Balayan et al. 2001). New AGNs, composite spectrum objects and high-luminosity IR galaxies have been discovered too; study of starburst/AGN/interaction phenomena and their interrelationship also became possible for some objects, and subsamples of interesting objects have been defined for further detailed studies. The redshift survey is the first task. After these observations, the IR luminosity function will be constructed and the space distribution of BIG objects will be investigated.

The observations have been carried out with 3 telescopes: 1) the Byurakan Astrophysical Observatory (BAO, Armenia) 2.6m telescope with the ByuFOSC focal reducer and the TM 1060×514 CCD; 2) the Special Astrophysical Observatory (SAO, Russia) 6m telescope (in collaboration with A.N.Burenkov, S.N.Dodonov, V.L.Afanasiev and A.V.Moiseev) with the UAGS spectrograph and K 585×530 or TK 1024×1024 CCDs, and the Multi-Pupil Fibre Spectrograph (MPFS, Afanasiev et al. 1995) with the TK 1024×1024 CCD; 3) the Observatoire de Haute-Provence (OHP, France) 1.93m telescope (in collaboration with P.Véron and M.-P.Véron-Cetty) with the CARELEC spectrograph and TK 1024×1024 or EEV 2048×2048 CCDs.

Different spectral ranges have been observed with different equipment, from 3600 Å to 8000 Å. The dispersions are 1.8-5.8 Å/pix and spectral resolutions of 5-14 Å have been obtained. The S/N ratios vary in the range from 5:1 to 50:1, except for a few spectra. Study of the objects with different telescopes and observational methods is more efficient both for quick completion of the program and better quality of classifications. The most interesting cases of AGN containing interacting pairs are being studied by means of 2D spectroscopy with MPFS. The statistics of the observations of the BIG objects is given in Table 1: telescopes, equipment, number of IRAS sources with observed counterparts, number of observed BIG objects, number of spectra obtained.

Table 1.   Statistics of Observations of the BIG Objects in 1997-2000

Telescope	Equipment	IRAS sources	BIG objects	Spectra
BAO 2.6m	ByuFOSC	45	56	75
SAO 6m	UAGS	48	64	84
SAO 6m	MPFS	39	43	96
OHP 1.93m	CARELEC	64	69	74
All observations		172*	213*	329

* Some objects have been observed several times with different telescopes

3. Results Obtained

In all, 213 galaxies associated with 172 IRAS sources have been observed spectroscopically (some sources have more than one association, and all components have been observed to check their nature and find out which of them is responsible for the IR). Spectral observations revealed new AGNs and high-luminosity infrared galaxies. Redshifts for all observed galaxies have been measured (z=0.008–0.173), the distances are 80–1041 Mpc, the absolute magnitudes are in the range -17.5^m– -23^m and the calculated infrared luminosity is in the range 3×10^9 <L_{fir} <7.5×10^{12} (for H_0=50), including some 30 LIGs and two ULIGs already discovered (IRAS 07479+7832a and IRAS 10252+7013).

The objects have mostly emission-line spectra with strong Balmer lines (mainly Hα and Hβ in the observed range), [O III] 4959/5007, [N II] 6548/6584, and [S II] 6717/6731 lines. [O I] 6300/6363, [O II] 7320/7330, [Fe VII] 6087, He I 5876 emission lines and Na I 5890 absorption line are often present. The spectra were classified and activity types of the galaxies were determined on the basis of the emission line ratios, using well-known diagnostic diagrams (Veilleux & Osterbrock 1987). The types are as follows: Sy2 - 21 objects, LINER - 3, composite spectrum objects - 7, HII - 143, normal (absorption-line) galaxies - 15, unknown (mainly because of low S/N) - 24. There are also 6 AGN contained in interacting/merging systems, interesting for further studies. The normal absorption-line galaxies must be investigated also to understand the source of their excess IR radiation. Objects having a composite spectrum (two distinct emission nebulae, for instance an HII region and a Seyfert, being superimposed on the slit, Véron et al. 1997) should be observed with higher resolution (< 3 Å) to study the profile of the various emission lines and to identify the nature of the emission objects.

Morphology and the spectra obtained allow separating different interesting subsamples for further studies. They are important for understanding of certain processes taking place in galaxies. Evidence of activity, starburst and/or interactions in the same object allows the study of connections between these phenomena toward understanding what triggers intense starburst processes.

References

Afanasiev, V.L. et al. 1995, at http://www.sao.ru/bta/bta6m.html#instr

Balayan, S.K., Hakopian, S.A., Mickaelian, A.M., & Burenkov, A.N. 2001, AstL, 27, 284

IRAS 1988, IRAS Catalogs and Atlases, 2. The Point Source Catalog, Joint IRAS Science Working Group, NASA, Washington, DC: US GPO

Markarian, B.E., Lipovetski, V.A., Stepanian, J.A., Erastova, L.K., Shapovalova, A.I. 1989, Commun. Special Astrophys. Obs., 62, 5

Mickaelian, A.M. 1995, Ap, 38, 349

Mickaelian, A.M. 2001, Ap, 44, 185

Mickaelian, A.M., Hakopian, S.A., Balayan, S.K., Burenkov, A.N., 1998, AstL, 24, 736

Mickaelian, A.M., Hakopian, S.A., Balayan, S.K. 1999, Proceedings of the IAU Symposium No. 194., ASP, 156

Mickaelian, A.M., Véron-Cetty, M.-P., Véron, P. 2001, ASP Conference Series No. 232, 278

Veilleux, S., Osterbrock, D.E., 1987, ApJS 63, 295

Véron, P., Gonçalves, A.C., Véron-Cetty, M.-P., 1997, A&A, 319, 52

Search for Obscured *IRAS* Galaxies

A.M. Mickaelian, S.A. Hakopian and S.K. Balayan

Byurakan Astrophysical Observatory (BAO), Byurakan 378433, Armenia. E-mail: aregmick@bao.sci.am

Abstract. We have discovered a number of sources still remaining in the IRAS catalogs with no convincing optical identification. Their IR colors and high galactic latitudes indicate faint galaxies (fainter the DSS limit). These empty fields have been observed with the Byurakan Obs. 2.6m telescope in VRI. 22^m-23^m faint formations have been revealed.

1. Obscured *IRAS* galaxies

One of the major objectives of infrared, submillimeter and X-ray astronomy is to determine if there exists an extragalactic population so obscured as to be spectroscopically unidentifiable from optical observations. Such objects have been discovered recently by SCUBA, ISO far-infrared, and Chandra X-ray surveys providing evidence that an obscured population exists (e.g. Barger et al. 2000). Such objects may be revealed from IRAS sources, too. Half of them remain without any identification yet and their physical nature is unknown. The Byurakan-IRAS Galaxies (BIG) sample has been constructed on the basis of optical identifications of IRAS PSC sources (IRAS 1988) in the region $+61° < \delta < +90°$ ($|b| > 15°$) of the First Byurakan Survey (FBS). A brief description of the sample is given in these Proceedings (Mickaelian et al. 2002). The sample contains objects in the range 12^m-21^m: most of them are emission-line galaxies, some having AGN properties (Sy2, LINERs, composite spectrum objects). A number of observed galaxies revealed high IR luminosity ($L_{fir} > 10^{11}$ $L_\odot$), so that they are LIGs and ULIGs (Sanders & Mirabel 1996).

2. High-Luminosity *IRAS* Galaxies in the *BIG* Sample

The ultraluminous infrared galaxies (ULIGs) are defined as having $L_{fir} > 10^{12}$ $L_\odot$ (for H_0=50 and q_0=0.5). QSOs with M_b=-24.0 have this luminosity between 1 micron and 1 keV.

It is interesting that the IRAS PSC sources with colors typical of galaxies (except for the brightest objects) have approximately the same IR fluxes at $60\mu m$ and $100\mu m$ (0.4-1.5Jy and 1-3Jy, respectively), while their optical magnitudes fall in a large range, as mentioned above. Moreover, we can assume that 21^m is not the limit for such objects. Fainter galaxies in our BIG sample should appear to be LIGs and ULIGs. They are much fainter in the optical range and much farther on the average, hence their IR luminosities will be much larger.

The far-infrared sky is characterized by extended, filamentary structure, particularly at 100μm. For sources only detected at 100 μm, the sum of the two cirrus flags CIRR1+CIRR2 has to be less than 10 in order to discriminate against knots of the IR cirrus (Klaas & Elsässer 1993); 436 sources in our region have been eliminated for this reason (to be checked later?). However, we are left with 30 good sources which still have no optical counterparts, and which may be very distant hyperluminous IR galaxies (HyLIGs).

The BIG Redshift Survey (Mickaelian et al. 2002) revealed 30 new LIGs and two ULIGs (IRAS 07479+7832a and IRAS 10252+7013) even at 17^m–18^m. The total number of LIGs and ULIGs in the BIG sample is estimated to be about 300 (15%).

3. Search for Obscured *IRAS* Galaxies

25% of our sources are associated with galaxies fainter than 20^m. 30 IRAS sources (having colors typical of galaxies) seeming to be real extragalactic objects (not cirrus) have no optical counterparts at the IR coordinate positions. The corresponding optical objects must be beyond the DSS limit.

These empty fields are to be studied by deep imaging to reveal the faint objects responsible for the IR. Taking into account their IR colors typical of galaxies, they must be very faint galaxies in optical wavelengths. Taking into account that 12 of them also have radio counterparts (association with NVSS sources), it may be believed that there exist real objects at these positions. The radio objects may be both AGNs and ULIGs, a very important combination for study of the AGN/starburst phenomena.

The Byurakan Astrophysical Observatory (BAO) 2.6m telescope with the ByuFOSC focal reducer and the TM 1060×514 CCD has been used for deep imaging of the candidate "obscured" IRAS galaxies from our lists. Five objects (empty fields): IRAS 08596+6741, 09246+6956, 09247+6541, 09509+7641, and 09531+6955 were observed in November, 2000, in VRI colors (5 min exposures). A few faint formations have been revealed at the positions of the IRAS sources for all of them, 3-4 objects for each field, most probably distant galaxies. Their magnitude estimates give 22^m–23^m in V.

This program will reveal objects which must be studied in detail with larger telescopes and SIRTF in 2002 (in collaboration with J.Houck and B.Brandl, Cornell Univ., USA). We plan to take IRS spectra for these fields to confirm the IRAS sources, measure accurate IR fluxes, and measure their redshifts.

References

Barger, A.J., Cowie, L.L., & Richards, E.A. 2000, AJ, 119, 2092

IRAS 1988, IRAS Catalogs and Atlases, 2. The Point Source Catalog. Joint IRAS Science Working Group, NASA, Washington, DC: US GPO

Klaas, U., Elsässer, H. 1993, A&AS, 99, 71

Mickaelian, A.M., Balayan, S.K., & Hakopian, S.K. 2002, IAU Col. 184, ASP

Sanders, D.B., & Mirabel, I.F. 1996, ARA&A, 34, 749

Part 3
X-ray Surveys for AGN

Wolfgang Voges and Friends

The 2.2m Telescope Dome

The Obscured AGN Population Probed by X-ray Observations

K. Iwasawa

Institute of Astronomy, Madingley Road, Cambridge CB3 0HA, United Kingdom

Abstract. X-ray properties of obscured AGN are reviewd and recent results obtained from ASCA, BeppoSAX, and Chandra are presented. There is a population of AGN that can only be revealed by hard X-ray observations. Studies of the X-ray background suggest that obscured AGN should significantly outnumber their unobscured counterparts.

1. X-ray Spectra of Obscured AGN

1.1. Effects of X-ray Absorption

The energy distribution of AGN can be approximated by a power-law and is roughly flat over the X-ray band. However, we now know at least in well-studied bright Seyfert galaxies that there are a significant excess of soft X-ray (below half keV) emission and a roll-over of the hard X-ray continuum at a few hundred keV. The soft X-ray excess is believed to be related to the accretion disk while the high energy roll-over is probably due to a cut-off in energy of thermal electrons in a corona which Comptonizes the disk emission into the power-law continuum. There are other subtle spectral distortions due to reflection off the accretion disk surface. These phenomena are likely happening in a region within 100 gravitational radii of a central black hole.

When cold matter lies in the line of sight at radii much further out, a series of photoelectric edges of various elements as well as Hydrogen are imposed on the X-ray continuum and make a low energy cut-off. The energy of the absorption cut-off increases as the column density of the absorbing matter increases. Therefore, going up across the X-ray energy band has the same effect as increasing wavelength in the optical/near-infrared for obscured objects. Assuming the Galactic dust to gas ratio, the optical depth which a near-infrared observation (say, in the K band) can probe is equivalent to that at 2–3 keV. For example, ROSAT, whose bandpass was up to 2 keV, was sensitive to relatively unobscured AGN but no longer sensitive when an absorption column density exceeds $10^{22} cm^{-2}$ (of course, this restriction is relaxed for high redshift objects). X-ray observatories such as ASCA, BeppoSAX, Chandra and XMM-Newton have sensitivity above 2 keV, which enables these satellites to search for highly obscured AGN. It should, however, be noted that simply shifting the energy window to higher energies does not necessarily guarantee detection of more heavily obscured objects, because of a) increasing importance of Compton down-scattering, as explained below; b) approaching the intrinsic spectral roll-over at high energies.

1.2. What Are Compton-Thick Sources?

When a column density of an absorber exceeds $N_H = 1.5 \times 10^{24} \text{cm}^{-2}$, which corresponds to unity in Thomson depth, even high energy photons that have survived photoelectric absorption get scattered by cold electrons more than once in the absorber and lose their energy via Compton down-scattering. Some of these photons enter the photoelectric absorption regime after losing energy and will be absorbed eventually. This is an important effect (see e.g., Wilman & Fabian 1999) and the transmitted continuum is significantly suppressed by that. The suppression by the Compton down-scattering has a geometry dependence, i.e, the smaller the covering fraction of the absorber, the stronger the suppression, and this dependence is as large as a factor of 3 when N_H is $\sim 3 \times 10^{24} \text{cm}^{-2}$, for instance (see Matt, Pompilio & La Franca 1999 for details). This introduces an uncertainty when estimating the unabsorbed luminosity of a strongly absorbed source. In summary, when an X-ray absorber becomes optically thick to Compton scattering, the transmitted light is observed to be very faint.

At the same time, the energy of the absorption cut-off moves out above 10 keV. Naturally, detection of transmitted X-ray radiation requires high energy X-ray detectors, as opposed to the modern imaging X-ray telescopes of which the highest energy bandpass terminates around 10 keV. The first such object to be found was the nearby infrared galaxy NGC 4945 with Ginga (Iwasawa et al 1993). Subsequently, many more Compton-thick sources have been found with BeppoSAX (e.g, Risaliti, Maiolino & Salvati 1999; Matt et al 2000). In an extreme case, the absorption column is so large that any transmitted component becomes invisible. A classical example is the prototype Seyfert 2 nucleus of NGC1068 (Fig. 1). No absorbed continuum is detected up to 200 keV (Matt et al 1997).

So, how can Compton-thick sources be probed by Chandra or XMM-Newton? As found in optical light, reflected light from a hidden central source could be present below 10 keV, and its X-ray spectrum would exhibit some distinctive signatures such as a very strong iron Kα line. Possible problems are the flux level of this reflected light can be very low and can also be confused with emission not related to the active nucleus, e.g., a circumnuclear starburst (Fig. 2).

The major X-ray reflecting agent is, of course, electrons. However, there are more possibilities to reflect the radiation of a hidden source into the line of sight in X-ray than in optical light. The inner surface of the obscuring matter itself, which is optically thick, can be an X-ray reflector. X-ray reflection from a cold, optically thick medium has been studied well in terms of reflection from an accretion disk (e.g., George & Fabian 1991), and various studies of simulated reflection spectra have been performed in the context of the unified scheme (e.g., Ghisellini et al ; Krolik, Madau & Zycki 1994). The reflection from cold matter is characterized by a very hard continuum and a strong iron K line at 6.4 keV (with EW of ~ 1 keV for solar abundance). Highly (photo-)ionized gas with a much smaller optical depth, sometimes identified with the warm absorbers found in Seyfert 1 galaxies (Krolik & Kriss 2001), can also be an important reflector. In this case, the reflected continuum has a softer spectrum because

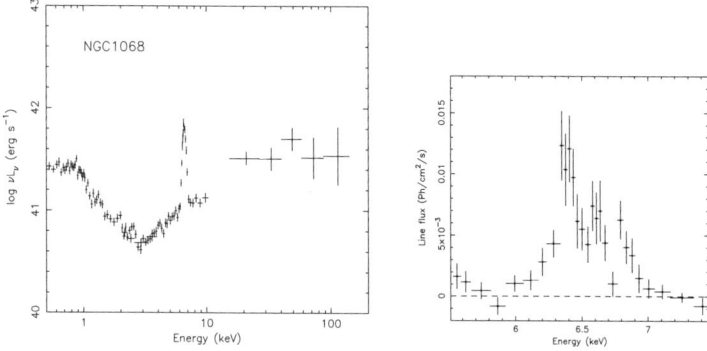

Figure 1. Right panel: The BeppoSAX spectrum of NGC1068. No direct emission from a hidden nucleus is visible up to 200 keV. The spectrum is dominated by reflected light, particularly in the hard X-ray band, reflection from optically thick cold matter produces a very hard spectrum (Matt et al 1997). Right panel: The iron K line complex in NGC1068 observed with the ASCA SIS (Iwasawa, Fabian & Matt 1997).

photoelectric absorption within the reflecting medium is less important at lower energies, and emission lines from ionized atoms are observed. Thus X-rays from a hidden nucleus may contain reflection from matter with a range of ionization. The iron K line complex is the most prominent spectral feature, and usually the 6.4 keV line is the major component of the line complex. This indicates that reflection from cold matter is important and a major component at least at energies around 6 keV. Although the reflected X-ray light may be relatively easy to identify from the spectral signatures, the X-ray flux we can observe is only a small fraction (1 per cent or likely to be much smaller) of a primary source. Combined with the reduced transmitted component due to the high absorption column density, the faint nature of Compton-thick sources makes it difficult to detect them at moderate to high redshift even with the advanced X-ray observatories.

2. What Fraction of AGN is Obscured?

Fig. 3 shows X-ray spectra of the three nearest AGN[1] obtained with BeppoSAX. They are all located within 4 Mpc, all strongly absorbed ($N_{\rm H} \sim 2\times 10^{23}$cm^{-2} for

---

[1] NGC 4395 is, for example, another AGN at a distance of 2.6 Mpc, but has a much lower luminosity (3×10^{39}erg s^{-1})

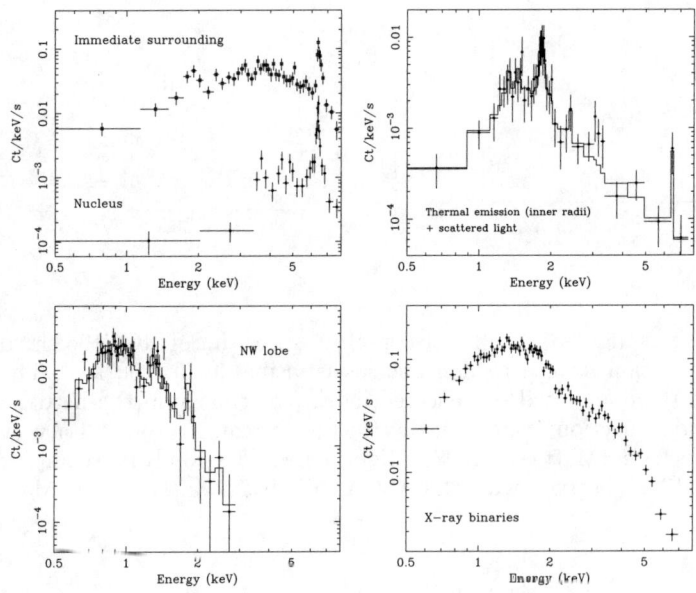

Figure 2. Chandra ACIS-S spectra taken from different areas of the X-ray emitting region of the nearby starburst/Seyfert 2 galaxy, NGC4945. The total X-ray emission from an obscured AGN like this galaxy can consist of the nuclear emission as well as many other components unrelated to the active nucleus. In this example, reflected AGN light from the hidden active nucleus, thermal emission induced by starburst winds extending towards the NW, which shows significant spatial variations, and X-ray binaries scattered across the galaxy are shown to demonstrate the possible complexity in an X-ray spectrum of an obscured AGN. It should be noted that these components cannot be resolved in distant objects.

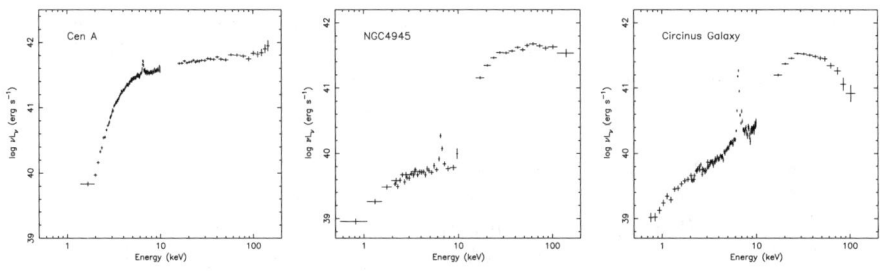

Figure 3. The BeppoSAX spectra of the nearest three AGN, Cen A, NGC 4945 and the Circinus Galaxy. All three X-ray nuclei are strongly absorbed.

Cen A, 3×10^{24}cm^{-2} for the Circinus Galaxy and 5×10^{24}cm^{-2} for NGC 4945) and their absorption-corrected luminosities exceed 5×10^{41}erg s^{-1}. Suppose our local Universe is representative of the whole Universe. Integrating the 0.5–2 keV ROSAT AGN luminosity function by Miyaji, Hasinger & Schmidt (2000) down to 5×10^{41}erg s^{-1} gives a density of 2×10^{-4} Mpc^{-3}. As mentioned above, the ROSAT AGN population are predominantly unobscured AGN. With a correction for local density in a sphere of radius corresponding to ~ 500 km s^{-1} relative to the mean density of the Universe (Schlegel et al 1994), the number of AGN with $L_\mathrm{x} \geq 5 \times 10^{41}$erg s^{-1} within 4 Mpc expected from the soft X-ray AGN luminosity function is then 0.05, 60 times smaller than observed (see Matt et al 2000). This estimate may be too simple but demonstrates that obscured AGN should outnumber their unobscured counterparts significantly.

This argument is supported by studies of the X-ray background (XRB). We now believe that the XRB originates from the integrated X-ray emission of AGN. Modeling of its flat spectrum implies considerable absorption in most AGN (Setti & Woltjer 1989; Madau, Ghisellini & Fabian 1994; Comastri et al 1995; Celotti et al 1995). Suppose the 30 keV peak of the XRB energy distribution is little affected by photoelectric absorption (Compton down-scattering means Compton-thick sources would not have significant contribution at this energy), we estimate 80–90 per cent of the total accretion power in the Universe is absorbed (Fabian & Iwasawa 1999). Correcting for this absorption, using Sołtan's cosmology-free argument, and assuming an accretion efficiency of 0.1 and a mean AGN redshift of 2, the XRB intensity is translated into a mean local density of black holes. The value we obtained is 6×10^5 $M_\odot$ Mpc^{-3}, much larger than estimates based on optical quasar counts, but in agreement with recent estimates based on X-ray counts (Salucci et al 1999) and Maggorian et al (1998) results (Haehnelt, Natarajan & Rees 1999). The absorbed accretion power probably emerges in the infrared bands, and also contributes some fraction of the infrared background light (Fixen et al 1998). Obscured AGN found in infrared luminous galaxies are discussed below.

3. X-ray Properties of Luminous Infrared Galaxies

Luminous Infrared Galaxies are found to be X-ray faint both at low and high redshift. Generally, starburst galaxies generate X-ray emission of luminosity 3–4 orders of magnitude below that of infrared emission. Whether the X-ray faintness of these objects is due to lack of active nuclei or heavy obscuration is a subject of debate.

Local Ultra-Luminous Infrared Galaxies (ULIGs) found by the IRAS survey (Soifer et al 1987) are divided into warm and cold objects based on their infrared colours. It has been known through optical spectroscopic surveys that warm IRAS objects tend to host Seyfert nuclei (e.g., De Grijp et al 1985) while cold IRAS objects are usually star-forming galaxies. X-ray observations are in agreement with this tendency: an absorbed X-ray source is often found in warm IRAS objects. However, there are exceptions unique to the X-ray observations. NGC 4945, as mentioned above, hosts a heavily obscured X-ray nucleus, yet this galaxy shows a cold IRAS color and no evidence for an active nucleus in the optical to near-infrared band (e.g., Marconi et al 2000), and the mid-infrared spectrum observed with ISO is typical of a starburst (Genzel et al 1998). A higher luminosity example is NGC 6240 (references for the X-ray observations are Iwasawa & Comastri 1998; Vignati et al 1999; Ikebe et al 2000). These two galaxies exhibit almost identical shapes of spectral energy distributions, albeit nearly two decades apart in luminosity (Fig. 4). These examples demonstrate that X-ray observations sometimes tell a different story from conventional emission-line spectroscopic classification.

Arp 220, the nearest of the ULIGs, however, does not show any AGN signature even in the X-ray band. A BeppoSAX observation failed to detect hard X-ray emission above 10 keV which imposes a rather strict constraint on the presence of an AGN: if this merger system were to host an energetically significant AGN, the absorption column density hiding a central source must exceed 10^{25}cm^{-2} (Iwasawa et al 2001).

Faint SCUBA sources are probably high-redshift analogs to ULIGs. The weak correlation of detection between SCUBA and Chandra (e.g., Fabian et al 2000; Hornschemeier et al 2000; Bautz et al 2000) suggests that the SCUBA sources are similar to Arp 220 in the X-ray band also (see Fig 2 in Fabian et al 2000). Clearly, it is important to identify what powers Arp 220 type objects (estimates of the AGN contribution to the FIR background can be found in Almaini et al 1999 and Risaliti et al 2002).

4. Obscured AGN at High Redshift

As in the case for low redshift objects, galaxies exhibiting evidence for hot dust, which is likely to be heated by AGN radiation, are among the first to be found to harbor an obscured AGN emitting at QSO luminosity. BeppoSAX detected a hard X-ray excess (Franceschini et al 2000) above the previously known cluster emission around IRAS 09104+4109, one of the hyperluminous infrared galaxies at $z = 0.442$ (Kleinmann et al 1988). A follow-up Chandra observation clearly detected a reflection-dominated X-ray nucleus, spatially separated from the surrounding cluster emission (Iwasawa, Fabian & Ettori 2001). This object shows

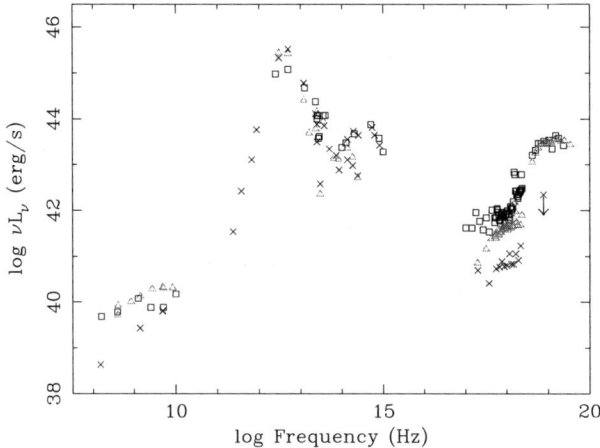

Figure 4. Spectral energy distributions of three infrared galaxies, Arp 220 (crosses), NGC 6240 (squares) and NGC 4945 (triangles). The SED of NGC 4945 has been mutiplied by 80 to match the far-infrared peak of Arp 220. NGC 4945 and NGC 6240 harbor heavily obscured active nuclei, which can only revealed in the hard X-ray band, and show almost identical SEDs. Note that Arp 220 shows considerably fainter X-ray emission compared with the other two.

a very warm IRAS color and is not detected with SCUBA at 850 μm (Deane & Trentham 2002), which indicates the lack of cool ($T \sim 40$ K) dust as seen in local ULIGs. Similarly, the two Chandra sources detected serendipitously in the field of the cluster Abell 2390 (with the aid of the lensing magnification of the cluster gravitational potential) have no SCUBA counterparts but are identified with ISOCAM 6.7 and 15 μm sources (Lémonon et al 1998; Altieri et al 1999). A broad-band study suggests that these X-ray sources are obscured AGN with hot dust at redshifts of 0.85 and 2.8 (Wilman, Fabian & Gandhi 2001). Objects similar to these at high redshift appear to make a significant contribution to the cosmic background radiation at 60 μm, while dusty star-forming galaxies as found with SCUBA could account for all the background radiation at longer wavelengths (Blain & Phillips 2002).

Recent deep Chandra surveys have resolved most of the XRB (Mushotzky et al 2000; Brandt et al 2001; Giacconi et al 2001). The log N–log S plot obtained from these surveys shows flattening of the slope for 2–10 keV sources below a flux level of 10^{-14}erg s^{-1}cm^{-2}. This means that X-ray sources at that flux level are the majority of those composing the XRB. A serendipitous source survey was conducted using various Chandra observations (mainly of clusters), aiming at detection of these X-ray sources, because the flux range can be reached with a moderate amount of Chandra exposure (30-40 ks). This survey has found several more obscured AGN with QSO luminosity (Crawford et al 2001; Gandhi et al 2002). An important remark about these sources is that some of them show no optical signature of an AGN or even no emission lines at all, particularly in heavily obscured objects (e.g., Mushotzky et al 2000; Gandhi et al 2002).

A Compton-thick Seyfert 2 galaxy at $z = 3.7$ has been found in the Chandra Deep Field South (Norman et al 2002). Such reflection-dominated sources appear to be rare even in the deep surveys and their contribution to the XRB may be small. A strongly absorbed ($N_{\rm H} \sim 10^{23} {\rm cm}^{-2}$), powerful X-ray source $L_{\rm X} \sim 10^{45}$ erg s^{-1}) has been found in a radio galaxy B2 0902+343 at $z = 3.4$ (Fabian Crawford & Iwasawa 2002). Thus the list of highly obscured AGN at high redshift is becoming longer as Chandra and XMM-Newton observations increase.

References

Altieri B., et al., 1999, A&A, 343, L65
Almaini O., Lawrence A., Boyle B.J., 1999, MNRAS, 305, L59
Bautz M.W., et al, 2000, ApJ, 543 L119
Blain A.W., Phillips T.G., 2002, MNRAS, 333, 222
Brandt W.N., et al, 2001, AJ, 122, 1
Celotti A., et al, 1995, MNRAS, 277, 1169
Comastri A., et al, 1995, A&A, 296, 1
Crawford C.S., et al, 2001, MNRAS, 324, 427
Deane J.R., Trentham N., 2001, MNRAS, 326, 1467
De Grijp M.H.K., et al, 1985, Nature, 314, 240
Fabian A.C., Iwasawa K., 1999, MNRAS, 303, L34
Fabian A.C., et al., 2000, MNRAS, 315, L8
Fabian A.C., Crawford C.S., Iwasawa K., 2002, MNRAS, 331, L57
Fixen D.J., et al, 1998, ApJ, 508, 123
Franceschini A., 2000, A&A, 353, 910
Gandhi P., et al, 2002, MNRAS, submmitted
Genzel R., et al, 1998, ApJ, 498, 579
George I.M., Fabian A.C., 1991, MNRAS, 249, 352
Ghisellini G., Haardt F., Matt G., 1994, MNRAS, 267, 743
Giacconi R., et al, 2001, ApJ, 551, 624
Haehnelt M.G., Natarajan P., Rees M.J., 1998, MNRAS, 300, 817
Hornschemeier A., et al, 2000, ApJ, 541, 49
Ikebe Y., et al, 2000, MNRAS, 316, 433
Iwasawa K., et al, 1993, ApJ, 409, 155
Iwasawa K., Comastri A., 1998, MNRAS, 297, 1219

Iwasawa K., et al, 2001, MNRAS, 326, 894
Iwasawa K. Fabian A.C., Ettori S., 2001, MNRAS, 321, L15
Kleinmann S.G., et al, 1988, ApJ, 328, 161
Krolik J.H., Kriss G.A., 2001, ApJ, 561, 684
Krolik J.H., Madau P., Zycki P.T., 1994, ApJ, 420, L57
Lémonon L., et al, 1998, A&A, 334, L21
Madau P., Ghisellini G., Fabian A.C., 1994, MNRAS, 270, L17
Magorrian J., et al, 1998, AJ, 115, 2285
Marconi A., et al, 2000, A&A, 357, 22
Matt G., et al, 1997, A&A, 325, L13
Matt G., Pompilio F., La Franca F., 1999, New Astronomy, 4, 191
Matt G., et al, 2000, MNRAS, 318, 173
Miyaji T., Hasinger G., Schmidt M., 2000, A&A, 353, 25
Mushotzky R.F., et al, 2000, Nature, 404, 459
Norman C., et al, 2002, ApJ, 571, 218
Risaliti G., Maiolino R., Salvati M., 1999, ApJ, 522, 157
Risaliti G., Elvis M., Gilli R., 2002, ApJ, in press
Salucci P., et al, 1999, MNRAS, 307, 637
Setti G., Waltjer L., 1989, A&A, 224, L21
Schlegel D., et al, 1994, ApJ, 427, 527
Soifer B.T., et al, 1987, ApJ, 320, 238
Sołtan A., 1982, MNRAS, 2000, 115
Vignati P., et al, 1999, A&A, 349, L57
Wilman R.J., Fabian A.C., 1999, MNRAS, 309, 862
Wilman R.J., Fabian A.C., Gandhi P., 2001, MNRAS, 318, L11

The AGN Content of Hard X-ray Surveys

Andrea Comastri

Osservatorio Astronomico di Bologna, via Ranzani 1, I–40127 Bologna, Italy

Cristian Vignali

Dept. of Astronomy and Astrophysics, Penn State University, 525 Davey Lab – University Park, PA 16802, USA

Marcella Brusa

Dipartimento di Astronomia, Universita' di Bologna, via Ranzani 1, I–40127 Bologna, Italy

on behalf of the HELLAS and HELLAS2XMM consortia[1]

Abstract.
Multiwavelength observations of the hard X–ray selected sources discovered by *BeppoSAX*, *Chandra* and XMM–*Newton* surveys have significantly improved our knowledge of the AGN population. The increasing number of X–ray obscured AGN so far discovered confirms the prediction of those AGN synthesis models for the X–ray background based on the Unified scheme. However, follow–up optical observations of hard X–ray selected sources indicate that their optical properties are quite varied and the simple relations between optical and X–ray absorption are by no means without exception. Moreover there is evidence of a substantial number of luminous X–ray sources hosted by apparently normal galaxies. In this paper the results obtained from multiwavelength observations of hard X–ray selected sources discovered by *BeppoSAX* and XMM–*Newton* are presented and briefly discussed.

1. Introduction

A large fraction of the energy density contained in the cosmic X–ray background spectrum (XRB) is accounted for by the summed contribution of Active Galactic Nuclei (AGN) if most of their high-energy radiation integrated over cosmic time is obscured by gas and dust. Several independent models based on the AGN unification scheme (Setti & Woltjer 1989, Madau et al. 1994, Comastri et

---

[1]HELLAS: F. Fiore, P. Giommi, G. Matt, F. La Franca, G.C. Perola, S. Molendi, R. Maiolino, A. Antonelli; HELLAS2XMM: F. Fiore, A. Baldi, S. Molendi, M. Mignoli, P. Ciliegi, F. La Franca, G. Matt, G.C. Perola, P. Severgnini, R. Maiolino

1995, Gilli et al. 2001) and energetic arguments (Fabian & Iwasawa 1999) lead to the conclusion that a fraction as high as 80–90% of the luminosity produced by accretion-powered sources is obscured at almost all wavelengths, emerging only in the hard X–ray band above a few keV. Hard X–ray surveys represent thus the most efficient method to search for and to trace the cosmological evolution of accretion-powered sources. A still unknown fraction of obscured radiation is reprocessed and re-emitted in the far–infrared band. Multiwavelength follow–up observations of hard X–ray selected sources would allow probing where the bulk of obscured accretion power is re-emitted and estimating the AGN contribution to the extragalactic background light in the infrared band.

Thanks to their revolutionary capabilities (arcsec imaging and high-energy throughput) *Chandra* and XMM–*Newton* have opened up a new era in the study of the hard X–ray sky. Deep *Chandra* surveys (Brandt et al. 2001, Rosati et al. 2001) have reached extremely faint fluxes in the 0.5–2 keV and 2–7 keV bands virtually resolving the entire XRB flux at these energies; while relatively deep XMM–*Newton* exposures (Hasinger et al. 2001, Baldi et al. 2001) have extended by a factor of 50 the sensitivity in the 2–10 and 5–10 keV bands with respect to previous *ASCA* and *BeppoSAX* observations. The X–ray source counts and their average spectral properties, which are now probed over several decades of fluxes and energy ranges, appear to be consistent with AGN synthesis model predictions. Although remarkable results have been obtained so far, deep *Chandra* and XMM–*Newton* surveys are limited by small area coverage (less than a quarter of a square degree) and by the extremely faint magnitudes of the optical counterparts, which make the identification of X–ray sources very difficult, if not impossible, even at 8m class telescopes (Giacconi et al. 2001, Tozzi et al. 2001, Norman et al. 2001). In order to fully characterize the nature and evolution of the X–ray source population it is customary to complement deep surveys with shallower observations carried out in larger areas (see for example the Einstein Medium Sensitivity Survey: EMSS; Gioia et al. 1990). This approach allows us to minimize the effects of field-to-field fluctuations (the cosmic variance) and makes the optical identification follow-up observations much easier, given the average brighter magnitude of the counterparts.

In this review the results obtained by two large area surveys carried out with *BeppoSAX* (HELLAS) and XMM–*Newton* (HELLAS2XMM) are summarized. The main scientific drive of this project is to probe the obscured accretion history of the X–ray Universe. The adopted strategy is a trade–off between the hardest X–ray band and the largest area which can be covered with *BeppoSAX* and XMM–*Newton*.

2. The HELLAS View of the Hard X–ray Source Population

The High Energy Large Area Survey (HELLAS; Fiore et al. 1999, 2001a) has provided, for the first time before the advent of the new X–ray observatories *Chandra* and XMM–*Newton*, a well-defined and large-area sample of hard (5–10 keV) X–ray selected sources obtained with an imaging instrument (the MECS instrument onboard *BeppoSAX*). At the flux limit of about 5×10^{-14} erg cm^{-2} s^{-1}, the integrated flux of the HELLAS sources accounts for some 20–30 % of the hard 5–10 keV XRB flux (Comastri et al. 2001).

The optical identification follow–up has been carried out on a subsample of 118 HELLAS sources covering $\sim$ 55 deg^2. The details of sample selection and optical identification breakdown are reported in Fiore et al. (2001a, 2001b) and La Franca et al. (2002). About 60% of the objects have been optically identified either by cross–correlation with public catalogs (25 objects) or at 4m class telescopes (49 objects). For 13 sources there are no clear counterparts down to R$\simeq$21. Even if the optical identifications are dominated by type 1 AGN, the relative fraction of type 2 objects (including Seyfert 1.8, 1.9 and emission–line galaxies) vs. type 1 is about 0.40, considerably higher than the value of 0.25 found in the *ROSAT* Ultradeep Survey (Lehmann et al. 2001a) and that of 0.20 in the *ASCA* Large Area Survey (LSS; Akiyama et al. 2000). If one assumes that the 13 "empty" fields host Type 2 AGNs, then the Type 2/Type 1 ratio becomes about 0.7. This shows the efficiency of revealing obscured X–ray sources in the 5–10 keV band with *BeppoSAX*.

Although for the majority of the HELLAS sources a "standard" spectral analysis is not possible due to the low photon statistics, the analysis of X–ray colors (hardness ratios) allows us to unambiguously reveal a substantial number of objects with flat and/or absorbed X–ray spectra. The most intriguing result is the presence of hard, presumably absorbed X–ray spectra in objects optically classified as type 1 AGN with a blue optical/UV continuum and broad lines. The number of X–ray absorbed type 1 quasars appears to increase with redshift and/or luminosity (Fig. 1).

In fact, while at redshifts lower than 0.3–0.4 X–ray obscured HELLAS sources are mainly associated with Type 2 AGNs and narrow emission-line galaxies, at higher redshifts the presence of X–ray absorption in a sizeable fraction of Type 1 AGNs represents a new, intriguing issue to be subjected to further *Chandra* and *XMM-Newton* observations, clearly suggesting that AGNs properties cannot be exhaustively interpreted without a multiwavelength approach. The decoupling between the optical and X–ray absorption properties of $z >$0.3–0.4 broad-line AGNs can be explained assuming a dust-to-gas ratio significantly lower than the Galactic value. From a physical perspective, this could imply the presence of a population of dust grains with different properties (e.g., large size) than previously known or thought (Maiolino et al. 2001). Strong X–ray absorption ($N_H > 10^{23}$ cm^{-2}) in high-redshift broad-line AGNs has also been suggested to explain the X–ray properties of four objects detected in the course of the *ASCA* LSS (Akiyama et al. 2000), and recent results from *Chandra* observations seem to confirm this result at lower X–ray fluxes (Fiore et al. 2000, Akiyama et al. 2001). The discovery of X–ray absorption in optically "normal" broad-line AGNs has important consequences for the AGN synthesis models for the XRB. According to these models, the sources responsible for the XRB must be characterized by a spectral energy density spanning a wide range of luminosities and absorption column densities, in order to reproduce both the XRB spectrum and the source counts in different energy ranges. In particular, the energetically dominant contribution comes from sources around the knee of the X–ray luminosity function ($L_X \sim$ a few 10^{44} erg s^{-1} at z=1) and with absorbing column densities of the order of 10^{23} cm^{-2} (Comastri 2000). If one relies on the AGN unified model, these objects, the so-called "QSO2", are expected to be the luminous counterparts of the local Seyfert 2 galaxies with narrow optical emission lines. Despite extensive searches only a handful of candidates have

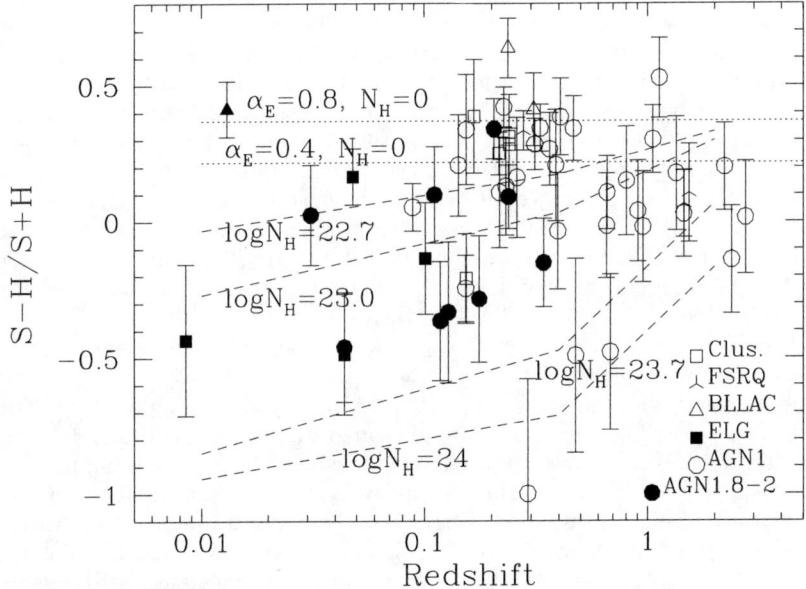

Figure 1. The softness ratio (S–H/S+H) as a function of redshift for the identified sources, where S and H are the counts in the soft 1.3–4.5 and hard 4.5–10 keV bands, respectively. The dotted lines show the expected hardness ratio for power law models, while dashed lines the expectations for absorbed power law models with $\alpha=0.8$ and different values of column density at the source redshift.

been found, the most remarkable example being the $z=3.7$ quasar discovered in the *Chandra* deep field south (Norman et al. 2001). The results obtained by HELLAS suggest that if the statement: obscured Type 2 ≡ narrow optical lines is not always true, the moderate/high-redshift absorbed Type 1 objects could have the same role of the so far elusive class of QSO2 in contributing to the XRB (Comastri et al. 2001).

Absorption column densities in excess of the Galactic one, often larger than 10^{22}–10^{23} cm^{-2}, are also derived combining archival *ROSAT* data with the *BeppoSAX* ones (Vignali et al. 2001). Interesting enough, this broad-band X–ray analysis has revealed that a fraction of sources which are thought to be absorbed in the hard X–rays are characterized by hard colors also in soft X–rays. This means that, although with lower efficiency, it is possible to pick up absorbed objects in the soft X–rays. It must be noted that some evidence of absorbed objects has also been obtained from the *ROSAT* Deep and Ultradeep Surveys in the Lockman Hole (Hasinger et al. 1998; Lehmann et al. 2000, 2001a). *ROSAT* observations of the HELLAS sources suggest that their average spectral properties are not well accounted for by a simple absorbed power-law model. More complex spectra are required in a high number of cases, and this is particularly true for very absorbed objects. Additional soft components possibly due to scattered nuclear flux and/or starburst components are required, in most cases involving about 10–50 % of the primary radiation (Vignali et al. 2001). Similar results have been obtained by *ASCA* (Della Ceca et al. 1999).

Remarkable results have been obtained also by the photometric follow-up observations of a subsample of HELLAS sources. Interestingly enough, the optical and the near-infrared properties of a fraction of intermediate (1.8–1.9) and Type 2 objects (Maiolino et al. 2000), as well as red quasars (Vignali et al. 2000), are dominated by the stellar component of the galaxies hosting the (obscured) X–ray active nucleus. This result has straightforward consequences, since it implies that a fraction of sources responsible for the hard XRB may be hosted by normal, passively-evolving galaxies, possibly being missed by previous optical surveys based on color-selection criteria.

3. From BeppoSAX to XMM–Newton: The HELLAS2XMM Survey

The HELLAS2XMM survey aims to cover a portion of the redshift–luminosity plane which cannot be probed by deep pencil-beam surveys. The main purpose of this complementary approach is to study the X–ray source populations at fluxes where a large fraction of the hard X-ray cosmic background (HXRB) is resolved ($\approx 50\%$ at $F_{2-10} > 10^{-14}$ erg cm^{-2} s^{-1}, see e.g. Comastri 2000), but where a) the area covered is as large as possible, to be able to find sizeable samples of "rare" objects; b) the X-ray flux is high enough to provide at least rough X-ray spectral information; and c) the magnitude of the optical counterparts is bright enough to allow, at least in the majority of the cases, relatively high-quality optical spectroscopy, useful to investigate the physics of the sources. Our goal is to evaluate for the first time the luminosity function of hard X-ray selected sources in wide luminosity and redshift ranges. By integrating this luminosity function we will compute the hard X-ray luminosity density per unit volume due to accretion as a function of the redshift. A comparison with the history of the UV luminosity density (proportional to the history of the star-formation) may give us a clue about the correlations between formation and evolution of AGN and supermassive black holes and formation and evolution of galaxies.

At present, the HELLAS2XMM sample, performed using 15 XMM–*Newton* public observations, consist of 1022, 495, and 100 sources detected down to minimum fluxes of about 6×10^{-16}, 3×10^{-15}, 6×10^{-15} erg cm^{-2} s^{-1} in the 0.5-2, 2-10 and 4.5-10 keV bands, respectively, over an area of about 3 deg^2 (Baldi et al. 2001). The source counts in these bands are in good agreement with previous determination by other satellites and XMM–*Newton* itself. In the hard 2–10 and 5–10 keV bands our survey samples a flux range neither accessible by shallower *ASCA* (Ueda et al. 1999, Della Ceca et al. 1999) and *BeppoSAX* surveys (Fiore et al. 2001a), which were limited to relatively bright fluxes, nor by deep *Chandra* (Brandt et al. 2001, Rosati et al. 2001) and XMM–*Newton* (Hasinger et al. 2001) surveys which are limited by the small area.

Four of the HELLAS2XMM fields were selected for follow–up observations in the optical band using the ESO 3.6m and the TNG 3.5m telescopes. The selected subsample contains 115 sources with 2–10 keV fluxes between $\sim 8 \times 10^{-15}$ and 10^{-13} erg cm^{-2} s^{-1}. Medium–deep R band images are available for all the sources in the sample. Optical counterparts brighter than R≃24 within 5 arcsec from the X–ray position (actually within 3 arcsec in most of the cases; see Fiore et al. 2001b) are present for about 80% of the sample. At the time of writing, optical spectra have been obtained for 46 out of the 115 sources in our sample. The number of faint hard X–ray selected sources with optical identification is comparable to that obtained in deep *Chandra* observations (Barger et al. 2001). The HELLAS2XMM sources populate a region of the luminosity–redshift diagram which is barely covered by deep surveys (Fig. 2). A uniform sampling over a large region of the $L_X - z$ parameter space is a key issue to compute the luminosity density and evolution of the X–ray sources.

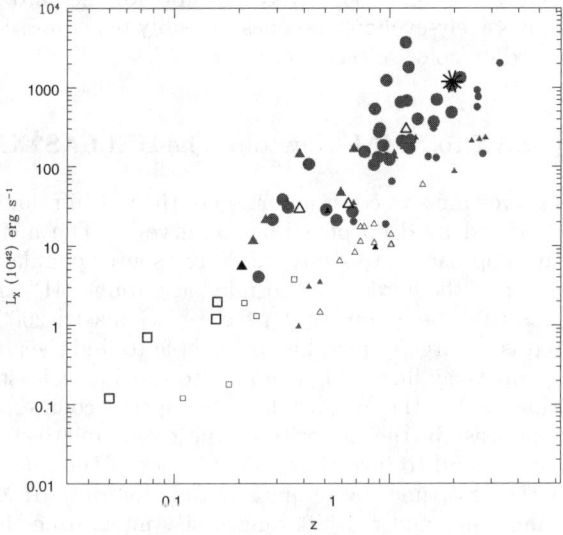

Figure 2. The luminosity-redshift diagram for the HELLAS2XMM sources (big symbols) and the *Chandra* SSA13, HDFN and A370 deep surveys (smaller symbols, data from Barger et al. 2001). Different symbols identify different source classes: filled circles = broad-line quasars and Sy1; filled triangles = narrow-line AGN; open squares = optically 'normal' galaxies; open triangles = emission-line galaxies; big star = the candidate type 2 QSO at z=1.955.

3.1. Optical Identification Breakdown

The most interesting results emerging from the optical identification program are the discovery of one X–ray luminous ($L_X \simeq 10^{45}$ erg s^{-1}) quasar at $z = 1.955$ with narrow optical emission lines (FWHM $\lesssim$ 1000 km s^{-1}) and of some X–ray luminous but apparently normal galaxies at low redshift. Although type 2 quasars were predicted by AGN synthesis models for the XRB, their space density and optical appearance is still a matter of debate (see the previous paragraph). The present finding suggests that hard X–ray selection coupled with large area provides an efficient method to uncover new objects of this class. More surprising is the presence of X–ray bright sources in the nuclei of otherwise passive "normal" galaxies. Their 2–10 keV luminosities, in the range 10^{42-43} erg s^{-1}, are more than one order of magnitude higher than that predicted on the basis of the L_X–L_{opt} relation of normal galaxies (Fabbiano et al. 1992) and more typical of Seyfert galaxies. Their count ratios indicate a hard X–ray spectrum. A detailed broad-band study has been recently performed on one of these objects (Comastri et al. 2002), suggesting that a heavily obscured, Compton-thick ($N_H > 10^{24}$ cm^{-2}) AGN may be responsible for the observed properties (Fig. 3).

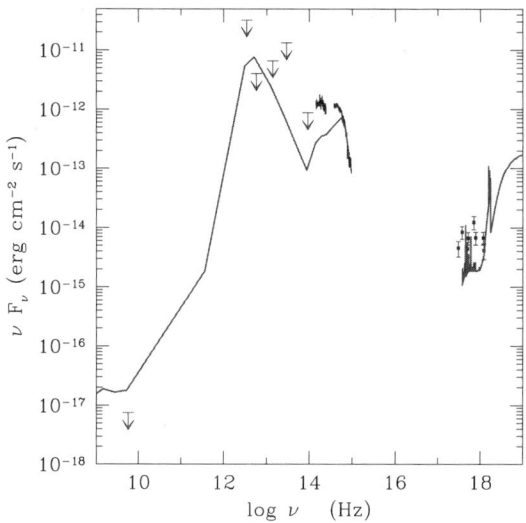

Figure 3. The spectral energy distribution (SED) of the X–ray bright optically quiet galaxy Fiore P3 (see Comastri et al. 2002 for details) is compared with that of the highly obscured Seyfert 2 galaxy NGC 6240 (solid line). The latter SED was normalized to the observed optical spectrum.

3.2. X–ray to Optical Properties

The results of photometric observations indicate that ~ 23 % of the X–ray sources have counterparts with $22 \lesssim R \lesssim 24$ and ~ 17 % with R > 24. The identification of the optically faintest counterparts will not probably be feasible even with 8m class telescopes, requiring deep multifilter observation to get reliable photometric redshifts. An alternative solution would be to search for redshifted iron Kα lines. Although such an approach may be feasible for the brightest X–ray sources, for the large majority of the objects in our sample the X–ray spectrum cannot be measured due to the low counting statistics. Useful constraints on the nature of the faint X–ray source population might be obtained from the analysis of the already available optical and X–ray fluxes and from an estimate of their average X–ray spectral properties inferred from the hardness ratio analysis. The R band magnitudes plotted versus the 2–10 keV flux are reported in Fig. 4.

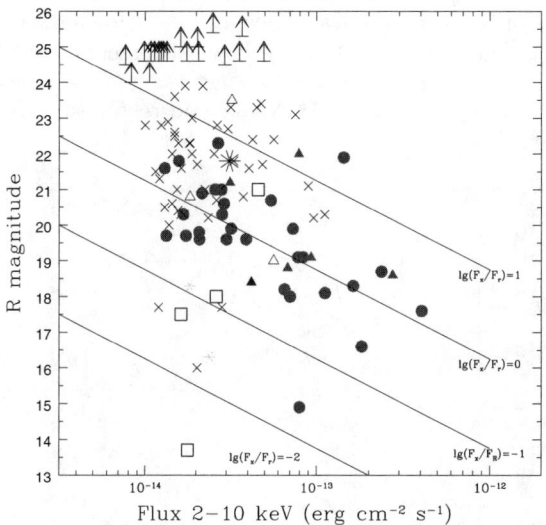

Figure 4. The 2–10 keV X–ray flux plotted against the R magnitude. The symbols are the same as in Fig. 2, while the crosses represent unidentified sources. Two Galactic stars are also plotted as asterisks. The loci of constant F_X/F_{opt} are reported with the values as labeled.

The AGN identified so far show a relatively well-defined correlation with optical magnitude around $F_X/F_{opt} \simeq 1$. This correlation is similar to that found by *ROSAT* for soft X–ray selected quasars (Hasinger et al. 1998) and confirmed by *Chandra* and XMM–*Newton* observations (Lehmann et al. 2001b) also for hard X–ray selected sources. On the other hand, the X–ray to optical flux ratio of unidentified sources is characterized by a larger scatter and skewed towards higher F_X/F_{opt} values. At faint fluxes there are several objects with $F_X/F_{opt} > 10$ suggesting the presence of highly obscured AGN. In order to quantify this possibility, we have computed the average hardness ratio as a function of the X–ray to optical flux ratio and the X–ray flux. The entire sample has been divided into bright and faint sources according to the median flux of the survey in the hard X–ray band: 2.5×10^{-14} erg cm^{-2} s^{-1}.

The results reported in Fig. 5 indicate a hardening of the X–ray spectrum for those sources with the highest values of F_X/F_{opt}, this effect being more pronounced in the faint sample.

4. Conclusions

The relatively high number of obscured AGN discovered by *BeppoSAX* and XMM–*Newton* makes high-energy large area surveys extremely well suited to

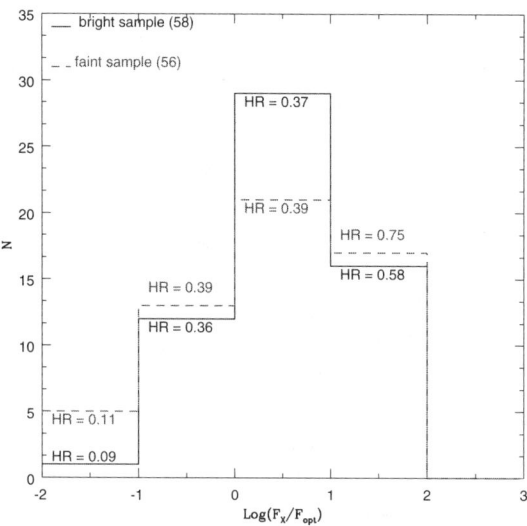

Figure 5. The distribution of X–ray to optical flux ratios for the bright and faint samples. The labeled values correspond to the average hardness ratio in the bin defined as H–S/H+S where H and S are the 2–10 keV and 0.5–2 keV fluxes respectively. Sources with $\log F_X/F_{opt} > 1$ are grouped in a single bin.

study the physics and the evolution of the sources responsible for the hard X–ray background. The hard X–ray sky is populated by AGN with extremely varied broad-band properties. The most important result concerns the optical appearance of X–ray obscured AGN as a function of redshift. In the local Universe a fraction of absorbed objects are associated with apparently normal, early-type galaxies. The lack of any AGN feature in their optical–infrared spectra suggests the presence of buried, probably Compton-thick nuclei. At higher redshift the presence of hard X–ray sources with broad optical lines indicates that the absorbing gas is dust–free. Finally a sizeable fraction of hard X–ray selected sources lacks an optical counterpart at the limit of 4m class telescopes. Multiwavelength observations of hard X–ray selected sources allow us to uncover AGN activity in a number of objects which would have not been classified as such on the basis of observations at other wavelengths. Larger samples of hard X–ray selected sources will provide new insights into the physics and the cosmic history of accretion.

Acknowledgments. This research has been partially supported by ASI contracts ARS–99–75 and I/R/107/00, and by the MURST grant Cofin-00–02–36. CV also aknowledges the financial support of NASA LTSA grant NAG5-8107.

References

Akiyama, M. et al. 2000, ApJ, 532, 700
Akiyama, M., Ohta, K., Ueda, Y. 2001, ApJ, in press (astro-ph/0111037)
Baldi, A., Molendi, S, Comastri, A., Fiore, F., Matt, G., Vignali, C. 2001, ApJ, in press (astro-ph/0108514)
Barger, A. et al. 2001, AJ, 122, 2177
Brandt, W. N. et al. 2001, AJ, in press (astro-ph/0108404)
Comastri, A., Setti, G., Zamorani, G., Hasinger, G. 1995, A&A, 296, 1
Comastri, A. 2000, in Proc. Conf. "X–ray Astronomy '999: Stellar Endpoints, AGN, and the Diffuse Background (astro-ph/0003437)
Comastri, A. et al. 2001, MNRAS, 327, 781
Comastri, A. et al. 2002, ApJL, submitted (see also astro–ph/0109117)
Della Ceca, R. et al. 1999, ApJ, 524, 674
Fabbiano G., Kim D. W., Trinchieri G., 1992, ApJS, 80, 531
Fabian, A. C., Iwasawa, K. 1999, MNRAS, 303, L34
Fiore, F. et al. 1999, MNRAS, 306, L55
Fiore, F. et al. 2000, New Astronomy, 5, 143
Fiore, F. et al. 2001a, MNRAS, 327, 771
Fiore, F. et al. 2001b, Proc. of the ESO/ECF/STSCI Workshop on "Deep Fields", Garching October 2000 (astro-ph/0102041)
Giacconi, R. et al. 2001, ApJ, 551, 624
Gilli, R., Salvati, M., Hasinger, G. 2001, A&A, 366, 407
Gioia, I. M. et al. 1990, ApJS, 72, 567
Hasinger, G. et al. 1998, A&A, 329, 482
Hasinger, G., Altieri, B., Arnaud, M. et al. 2001, A&A, 365, L45
La Franca, F. et al. 2002, ApJ, submitted
Lehmann, I. et al. 2000, A&A, 354, 35
Lehmann, I. et al. 2001a, A&A, 371, 833
Lehmann, I., Hasinger, G., Murray, S. S., Schmidt, M. 2001b, Proc. of the Symposium "X–rays at Sharp Focus", in press (astro-ph/0109172)
Madau, P., Ghisellini, G., Fabian, A. C. 1994, MNRAS, 270, L17
Maiolino, R. et al. 2000, A&A, 355, L47
Maiolino, R. et al. 2001, A&A, 365, 2
Norman, C., 2001, ApJ, submitted (astro-ph/0103198)
Rosati, P. et al. 2001, ApJ, in press (astro-ph/0110452)
Setti, G., Woltjer L. 1989, A&A, 224, L21
Tozzi, P. et al. 2001, ApJ, in press (astro-ph/0103014)
Ueda, Y. et al. 1999, ApJ, 518, 656
Vignali, C., Mignoli, M., Comastri, A., Maiolino, R., Fiore, F. 2000, MNRAS, 314, L11
Vignali, C., Comastri, A., Fiore, F., La Franca, F. 2001, A&A, 370, 900

AGN Populations from Optical Identification of *ASCA* Surveys

Masayuki Akiyama

Subaru Telescope, National Astronomical Observatory of Japan, Hilo, HI, 96720

Yoshihiro Ueda

ISAS, Sagamihara, Kanagawa, 229-8510, Japan

Kouji Ohta

Department of Astronomy, Kyoto University, Kyoto, 606-8502, Japan

Abstract. To understand luminous AGNs in the $z < 1$ universe, the *ASCA* AGN samples are the best at present. Combining the identified sample of AGNs from the *ASCA* Large Sky Survey and Medium Sensitivity Survey, the sample of hard X-ray selected AGNs has been expanded up to 108 AGNs above a flux limit of 10^{-13} erg s^{-1} cm^{-2} in the 2–10 keV hard X-ray band. We discuss the fraction of absorbed AGNs in the hard X-ray selected AGN sample, and the nature of absorbed luminous AGNs.

1. Introduction : Importance of a Bright Hard X-ray AGN sample

The fraction of absorbed AGNs, especially luminous absorbed AGNs, is a big issue in understanding the true number density of active nuclei in the universe. Recently many candidates for absorbed luminous AGNs have been found in AGN surveys in radio, X-ray, and near-infrared wavelengths (e.g., Webster et al. 1995). The discoveries imply that we have been missing a significant fraction of nuclei with high activity in traditional optical/UV-selection of AGNs due to absorption toward the nucleus. However, the fraction of absorbed AGN in the entire AGN population is not clear. Radio-selected samples are affected by red AGNs with red synchrotron components (Francis et al. 2001), soft X-ray selection is biased against heavily absorbed AGNs (Kim & Elvis 1999), and 2MASS-selected red AGNs are limited to the low-redshift universe (Cutri et al. in this volume).

In order to construct a complete sample of AGNs less biased against absorption toward the nucleus, we conducted optical follow-up observations for the *ASCA* Large Sky Survey (hereafter ALSS; Ueda et al. 1999) and the *ASCA* Medium Sensitivity Survey (hereafter AMSS; Ueda et al. 2001) in the hard X-ray band. Hard X-ray emission can penetrate the obscuring matter of absorbed AGNs and is very suitable to search for absorbed AGNs. Using 2–10 keV hard X-ray emission, we can detect AGNs with X-ray absorption up to hydrogen col-

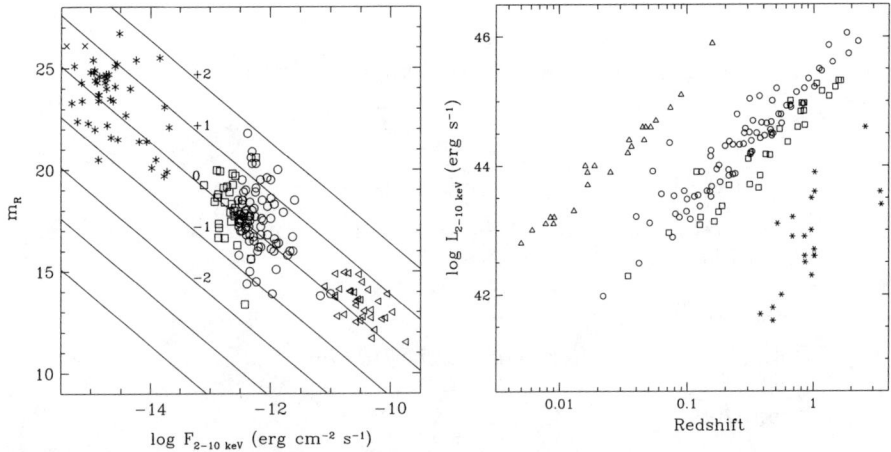

Figure 1. Left) R-band magnitudes of optical counterparts of ALSS (square) and AMSSn (circle) AGNs are plotted as a function of 2–10 keV hard X-ray flux. Dashed lines represent the X-ray to optical flux ratio of $\log f_X/f_V = +2, +1, 0, -1,$ and -2 from top to bottom. Triangles and asterisks indicate samples from *HEAO1* A2 (Piccinotti et al. 1982) and the *Chandra* survey in the HDF-N (Hornschemeier et al. 2001) Right) Hard X-ray luminosities of hard X-ray selected AGNs plotted as a function of redshift. Same symbols as in the left panel.

umn density of $10^{22\sim23}$ cm^{-2}, which corresponds to A_V of $20 \sim 50$ with the Galactic conversion factor, without bias. The ALSS is a survey in a continuous field of 5.4 square degrees near the North Galactic Pole. We selected 34 X-ray sources detected with SIS 2–7 keV significance greater than 3.5σ. The sources are identified with 30 AGNs, 2 clusters of galaxies and 1 galactic star (Akiyama et al. 2000). One X-ray source with a hard spectrum is still unidentified, and a *Chandra* follow-up observation is planned in Cycle 3. The AMSS is a serendipitous source survey based on *ASCA* pointed observations conducted in high galactic latitude regions ($|b| > 20°$). We conducted optical follow-up observations for 86 X-ray sources detected with GIS 2–10 keV significance greater than 5.6σ in the northern sky (declination above 20°; we call it the AMSSn sample). All of the X-ray sources are identified with 78 AGNs, 7 clusters of galaxies, and 1 galactic star (Akiyama et al. in preparation). In total, we constructed a sample of 108 hard X-ray selected AGNs with a flux limit of *ASCA*, $\sim 10^{-13}$ erg s^{-1} cm^{-2} in the 2–10 keV band. In Figure 1, we plotted the hard X-ray flux vs. optical magnitude (left) and the redshift vs. luminosity distribution (right) diagrams of ALSS and AMSSn AGNs. The *ASCA* samples are two orders of magnitude brighter and more luminous than the sample of deep *Chandra* and *XMM-Newton* surveys, and consists of luminous AGNs, i.e., QSOs, in the universe below redshift 1. The high completeness of the *ASCA* samples makes it possible to discuss the fraction of absorbed AGNs definitively.

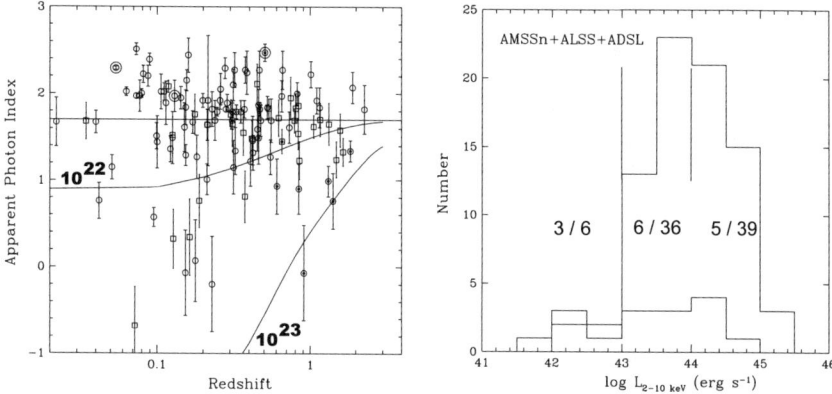

Figure 2. (Left) Apparent photon indices of ALSS (squares) and AMSSn (circles) AGNs in the 0.7–10 keV hard X-ray band are plotted as a function of redshift. BL Lac objects are marked with large circles. The solid lines show the apparent photon index of a power-law continuum with intrinsic photon index of 1.7 absorbed by hydrogen column density of 10^{22} cm^{-2} (top) and 10^{23} cm^{-2} (bottom) at each redshift. AGNs with a faint optical counterpart (log f_X/f_V larger than 1) are marked with dots. (Right) Luminosity distributions of all (thin) and significantly absorbed (thick) AGNs below a redshift of 0.6 from a combination of the AGN samples of the ALSS, AMSSn, and ADSL.

2. Fraction of Heavily Absorbed AGNs

Using the hardness of the X-ray spectrum of each source, we can estimate the X-ray absorption to the nucleus in each object. The 0.7–10 keV apparent photon index distributions of ALSS and AMSSn AGNs are plotted as a function of redshift in the left panel of Figure 2. The upper and lower solid lines in the figure correspond to the apparent photon index of an object with intrinsic photon index of 1.7 and X-ray absorption with hydrogen column density of $\log N_{\rm H} = 22 ({\rm cm}^{-2})$ and $\log N_{\rm H} = 23 ({\rm cm}^{-2})$ at each redshift, respectively. The X-ray sources with apparent photon index smaller than 1 can be regarded as significantly harder than canonical power-law spectra of broad-line AGNs (with photon index of 1.7). They correspond to intermediate redshift AGNs with X-ray absorption of $\log N_{\rm H} = 22 - 23 ({\rm cm}^{-2})$ and high-redshift AGNs with absorption of $\log N_{\rm H} > 23 ({\rm cm}^{-2})$. At high-redshift ($z \sim 1$), the apparent photon indices of highly absorbed objects become close to that of an object without absorption, because we observe very high energy photons from the source-frame, which are less affected by absorption than low-energy photons.

Based on the estimated amount of absorption to the nucleus, we examine the fraction of absorbed AGNs in the hard X-ray selected AGNs. For simplicity, we limit the sample to redshifts less than 0.6, and regard AGNs with $\log N_{\rm H} > 22 ({\rm cm}^{-2})$ as significantly absorbed AGNs. It should be noted that at high redshifts ($z > 0.6$), AGNs with hydrogen column densities of $\log N_{\rm H} = 10^{22-23}$ (cm^{-2}) cannot be regarded as significantly absorbed in the current sample. In the right panel of Figure 2, the luminosity distributions of all AGNs and significantly absorbed AGNs from the combination of the ALSS, AMSSn, and *ASCA* Deep Survey in the Lockman Hole (ADSL; Ishisaki et al. 2001) are plotted. The fraction of absorbed AGNs is higher in the lowest luminosity range, but there is no clear deficiency of absorbed AGN above 10^{44} erg s^{-1}, which is observed in the ALSS sample (Akiyama et al. 2000). The fraction of absorbed AGN is 6/36 and 5/39 in the luminosity range between 10^{43} erg s^{-1} and 10^{44} erg s^{-1} and in the luminosity range above 10^{44} erg s^{-1}, respectively. The fraction of absorbed AGNs is higher in the luminosity range below 10^{43} erg s^{-1} (3/6) than in the luminosity range above, but the number of AGNs in the low luminosity range is fairly limited. The fraction of luminous ($L_X > 10^{44}$ erg s^{-1}) AGNs with $\log N_H > 22 (\mathrm{cm}^{-2})$ in the sample of AGNs without bias up to $\log N_H = 23(\mathrm{cm}^{-2})$ (~15%) is clearly smaller than that expected from models of the cosmic X-ray background (45%; Comastri et al. 1995) or that observed in local low-luminosity Seyfert galaxies (40%; Risaliti et al. 1999).

3. Case Studies of Absorbed QSOs

The fraction of absorbed QSOs is not as large as expected, but we detected several candidates for absorbed QSOs in the *ASCA* surveys. Their counterparts are relatively faint and have larger X-ray to optical flux ratios than normal AGNs (see dotted objects in the left panel of Figure 2). Most of the high-redshift AGNs with hard X-ray spectra have large X-ray to optical flux ratios. The X-ray to optical flux ratio is similar to those of the optically-faint hard X-ray source population found in deep *Chandra* surveys (see left panel of Figure 1; e.g., Alexander et al. 2001), and the *ASCA* optically-faint AGNs can be low-redshift and/or high-luminosity cousins of the *Chandra* population.

Although the measured amount of X-ray absorption is large, most of the luminous absorbed QSOs show broad MgII 2800Å or Hα 6563Å emission lines. The origin of the discrepancy can be 1) broad MgII 2800Å from scattered nuclear light or 2) a discrepancy between the amount of X-ray photoelectric absorption and optical dust reddening. We show two examples of absorbed QSOs that fall in each category.

3.1. An absorbed QSO at $z = 0.65$ with a strong broad MgII 2800Å emission line

AX J131831+3341 is an absorbed radio-quiet QSO at a redshift of 0.65 found in the ALSS (Akiyama et al. 2000). Its X-ray luminosity is estimated to be $\sim 10^{45}$ erg s^{-1}, which corresponds to the luminosities of QSOs. The observed X-ray spectrum of the object in the 0.7–10 keV band is described by intrinsic absorption with a hydrogen column density of $N_H = 6.0^{+4.4}_{-4.2} \times 10^{21}$ cm^{-2} and an intrinsic photon index of 1.7. The hydrogen column density corresponds to the lower edge of the column density distribution of Seyfert 1.8-1.9 galaxies.

The optical spectrum of the object shows strong emission lines, such as broad MgII 2800Å, narrow [O II] 3727Å, and narrow [O III] 5007Å, but no broad Hβ emission line (see right panel of Figure 3). Its small Hβ-to-[O III] 5007Å equivalent width ratio is comparable to those of Seyfert 1.8-2 galaxies. Optical and near-infrared images show nuclear and extended components (see

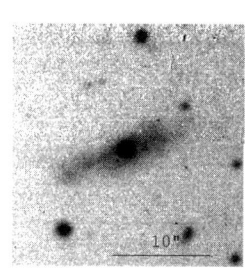 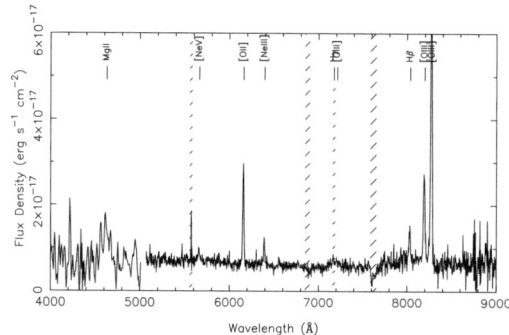

Figure 3. (Left) Optical R-band image of AX J131831+3341. (Right) Optical spectrum of AX J131831+3341. An 1800s FOCAS/Subaru spectrum above 5000Å with a 3600s MOSCA/Calar Alto 3.5m spectrum around MgII 2800Å emission.

left panel of Figure 3). Because the nuclear component has a very red $I - K$ color but blue $V - R$ and $R - I$ colors, the nucleus is thought to be absorbed with $A_V \sim 3$ and to emerge only in the K-band (Akiyama and Ohta 2001). The amount of absorption is consistent with the amount of X-ray absorption. The optical blue continuum and broad MgII 2800Å emission line can originate from scattered nuclear light (Akiyama et al. 2001).

3.2. A candidate for a type-2 QSO with large X-ray absorption and a strong broad-Hα emission line

AX J08494+4454 is a candidate for a type-2 QSO at $z = 0.9$ found in the course of optical identification of the *ASCA* deep survey in the Lynx field (Ohta et al. 1996). Recently, a deep *Chandra* hard X-ray spectrum and an IRCS/Subaru J-band spectrum of the object have been obtained (Akiyama et al. 2002). The 0.5–10 keV 150ks *Chandra* spectrum of AX J08494+4454 is hard, and is explained well with a power-law continuum absorbed by a hydrogen column density of $(2.3 \pm 1.1) \times 10^{23}$ cm^{-2}. The 2–10 keV luminosity of the object is estimated to be $7.2^{+3.6}_{-2.0} \times 10^{44}$ erg s^{-1}, after correcting for absorption, and reaches the range of hard X-ray luminosities of QSOs. The large X-ray absorption and the large intrinsic luminosity support the original identification of AX J08494+4454 as a type-2 radio-quiet QSO. Nevertheless, deep Subaru/IRCS J-band spectroscopic observation suggests the presence of a strong broad Hα emission line from AX J08494+4454 (left panel of Figure 4). The broad Hα emission line has a velocity width of 9400 ± 1000 km s^{-1}, which corresponds to a typical broad Balmer line velocity width of a luminous QSO. The existence of the strong broad Hα line means that the object is not a type-2 QSO, but a luminous cousin of a Seyfert 1.9 galaxy in the source-frame optical spectrum. The Balmer decrement of broad lines, the broad Hα emission to the hard X-ray luminosity ratio, and optical SED (right panel of Figure 4) suggest that the nucleus is affected by dust extinction with A_V of $1 - 3$ mag in optical wavelengths. The estimated amount of dust extinction is much smaller than that expected from the X-ray column density ($A_V = 130 \pm 60$ mag). The discrepancy can be explained with a smaller

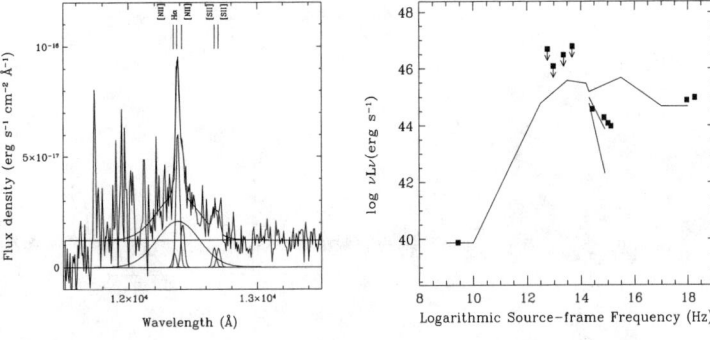

Figure 4. (Left) J-band spectrum of AX J08494+4454. The best fit models are also plotted with solid lines. (Right) Spectral energy distribution (SED) of AX J08494+4454. The solid line indicates the SED of an average radio-quiet QSO (Elvis et al. 1994) and is normalized at the data point observed at 1.4 GHz. Dashed lines represent optical SEDs with dust extinction with A_V of 1 mag (upper) and 2 mag (lower).

dust to gas mass ratio which may due to dust sublimation in the X-ray absorbing matter, the size difference between optical and X-ray emitting regions, or a different dust size distribution in AGNs (e.g., Maiolino et al. 2001).

Acknowledgments. The authors thank ALSS and AMSS members.

References

Akiyama, M., et al. 2000, ApJ, 532, 700

Akiyama, M., and Ohta, K. 2001, PASJ, 53, 63

Akiyama, M., et al. 2001, PASJ, 52, 577

Akiyama, M., Ueda, Y., and Ohta, K. 2002, ApJ, in press (astro-ph/0111037)

Alexander, D.M., et al. 2001, ApJ, in press (astro-ph/0107450)

Comastri, A., Setti, G., Zamorani, G., and Hasinger, G. 1995, A&A, 296, 1

Elvis, M., et al. 1994, ApJS, 95, 1

Francis P.J., et al. 2001, PASA, in press (astro-ph/0107235)

Hornschemeier, A., et al. 2001, ApJ, 554, 742

Ishisaki, Y., et al. 2001, PASJ, 53, 445

Kim, D., and Elvis, M. 1999, ApJ, 516, 9

Maiolino, R., Marconi, A., and Oliva, E. 2001, A&A, 365, 37

Ohta, K., et al. 1996, ApJ, 458, 57

Piccinotti, G., et al. 1982, ApJ, 253, 485

Risaliti, G., Maiolino, R., Salvati, M. 1999, ApJ, 522, 157

Ueda, Y., et al. 1999, ApJ, 518, 656

Ueda, Y., et al. 2001, ApJS, 133, 1

Webster, R., et al. 1995, Nature, 375, 469

The X-ray Variability of High-Redshift QSOs

J. Manners, O. Almaini, & A. Lawrence

Institute for Astronomy, University of Edinburgh, Royal Observatory, Blackford Hill, Edinburgh EH9 3HJ

Abstract. We present an analysis of X-ray variability in a sample of 156 radio quiet quasars taken from the ROSAT archive, covering a redshift range $0.1 < z < 4.1$. Through combining light curves in ensembles we are able to identify trends in variability amplitude with luminosity and with redshift. The decline in variability amplitude with luminosity identified in local AGN ($z < 0.1$) is confirmed out to $z = 2$. There is tentative evidence for an increase in QSO X-ray variability amplitude towards high redshifts ($z > 2$) in the sense that QSOs of the same X-ray luminosity are more variable at $z > 2$. The simplest explanation for this effect may be that high-redshift QSOs are accreting at a higher efficiency than local AGN.

1. Introduction

Rapid X-ray variability appears to be very common in AGN. Temporal power spectra show 'red noise' (i.e. more power at lower frequencies), with the form $P(f) \propto f^{-\alpha}$ where $\alpha \approx 1.5$ (Lawrence & Papadakis 1993, Green et al 1993). Departures from a featureless power spectrum are rare. Some evidence for quasi-periodic oscillations has been observed in NGC 5548, NGC 4051 (Papadakis & Lawrence 1993, 1995) and IRAS 18325-5926 (Iwasawa et al 1998), and in a handful of AGN a turnover has been seen at low frequencies (e.g. Edelson & Nandra 1999). A high frequency cut-off would indicate the size of the emission region, although this cannot yet be distinguished from the noise for even the most well studied AGN.

Variability studies of local AGN ($z < 0.1$), over a fixed timescale, indicate that more luminous sources vary with a lower amplitude. This may be explained if more luminous sources are physically larger in size and are actually varying more slowly. Alternatively, they may contain more independently flaring regions and so have a genuinely lower amplitude. The slope of this correlation has been calculated in a number of papers using overlapping samples of local AGN. Lawrence & Papadakis (1993) and Green et al (1993) analyzed samples of light curves from the EXOSAT database. The variability amplitude was found to vary with luminosity as $\sigma \propto L_X^{-\beta}$ with $\beta \approx 0.3$. The most comprehensive analysis of the variability-luminosity relation was carried out by Nandra et al, 1997 (hereafter N97) for 18 local Seyferts observed with the ASCA satellite. They find $\beta = 0.355 \pm 0.015$. Whether this well-defined correlation applies to high-redshift QSOs is not so clear. Observations of distant QSOs are generally of low signal-to-noise and measurements of variability in individual objects are poorly defined. Almaini et al (2000) developed a technique to measure the amplitude of variability for low signal-to-noise sources and thus high-redshift AGN. By combining light curves from a number of AGN they were able to measure the amplitude of variability over ranges in luminosity and redshift. They studied a sample of 86 QSOs from the Deep ROSAT Survey of Shanks et al (1991) spanning a wide range in redshift ($0.1 < z < 3.2$). The behaviour of variability amplitude with luminosity was found to be in rough agreement with the anti-correlation seen in local AGN but showing a possible upturn for the most luminous sources. Tentative evidence suggested this was due to increased variability at high redshifts, although a definite trend in the redshift behaviour could not be clearly confirmed.

In this paper we use the techniques of Almaini et al (2000) to determine the amplitude of variability in an expanded sample of QSOs taken from the ROSAT archive. QSOs at $z > 1$ are preferentially selected in order to constrain the redshift behaviour of X-ray variability. A cosmology with $q_0 = 0.5$, $H_0 = 50$ km s^{-1} Mpc^{-1} is used throughout.

2. QSO Sample & Data Reduction

The sample consists of 156 QSOs between $0.08 < z < 4.11$ taken from the ROSAT PSPC archive. It is made up of QSOs taken from a number of sources that all adhere to the following selection criteria: radio quiet quasar IDs, X-ray exposures $> 10,000$ seconds, flux signal-to-noise > 5, within 20 arcmin of the ROSAT pointing, and at least 3 time bins in the light curve.

The redshift distribution of the entire sample is displayed in Fig. 1. The data were obtained from the LEDAS online database facility at Leicester. Each source was extracted using a circular mask of radius chosen to include 90 per cent of the PSF. The data were then filtered to remove periods of high particle background and restrict the energy range to 0.1 – 2.4 keV. After careful background subtraction, an algorithm was used to construct light curves and bin up the data to allow meaningful Gaussian statistics.

The method used for measuring the amplitude of intrinsic variability is fully described in Almaini et al. (2000). To compare objects of different flux, the light curves are divided by the mean, so that measurements of *fractional* variability are made. A maximum likelihood technique is used to separate the intrinsic variations in the light curve from those due to noise. Individual measurements of variability were of low signal-to-noise, therefore in order to extract meaningful information from the sample it was necessary to combine the light curves into ensembles over given ranges of luminosity or redshift. Variability amplitudes are normalized to a timescale of 1 week, assuming the power spectrum slope seen in local AGN ($\alpha = 1.5$). Corrections are also made for time dilation and the effects of irregular binning.

3. The Results

Maximum likelihood estimates of the corrected intrinsic variability amplitudes were calculated for all 156 quasars. Nearly half the sample (69 QSOs) give a non-zero estimate of variability amplitude. The remaining QSOs can only provide

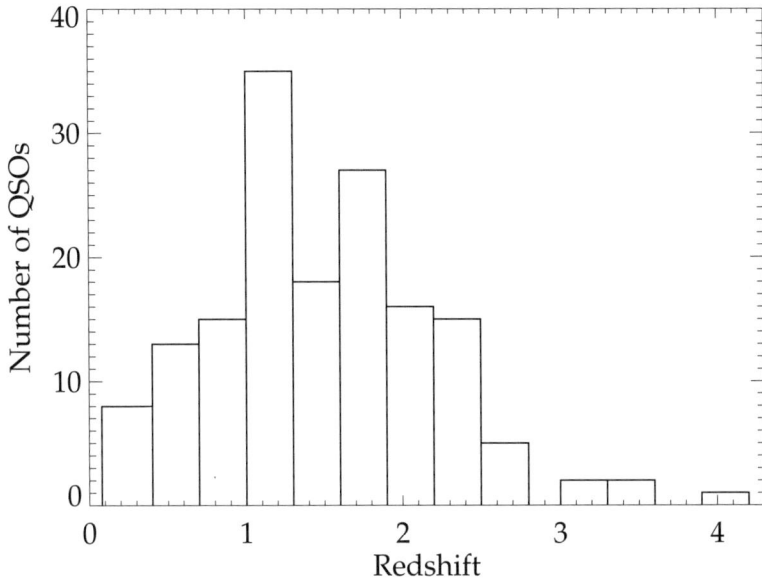

Figure 1. Redshift distribution of our sample of 156 RQQs.

upper limits, mainly due to low signal-to-noise. Treating the entire sample as one ensemble, the corrected mean ensemble variability is: $\sigma = 0.15 \pm 0.01$ ($\sigma = 0.16 \pm 0.01$ uncorrected) on a time-scale of 1 week.

The data were split into 6 luminosity bins each of ~ 0.5 dex, and 11 redshift bins each of $\delta z = 0.3$. For each ensemble, maximum likelihood estimates were calculated for the intrinsic variability amplitude in the combined light curve. Fig. 2 displays the ensemble results. Immediately apparent in Fig. 2(a) is an anti-correlation between quasar luminosity and variability amplitude. A power law fit to the individual quasars gives the best fit relation: $\sigma \propto L_X^{-\beta}$ where $\beta = 0.18 \pm 0.05$ (plotted as a dot-dash line). This gives a fairly poor fit to the ensemble points, mainly due to some highly variable, high luminosity QSOs. However, a power law fit to QSOs with redshift less than 2 (solid line, $\beta = 0.27 \pm 0.05$) passes through the majority of the ensemble points and is very close to the average relation found for local AGN (plotted as a dashed line). The errors quoted here are 68 per cent confidence limits for the slope on allowing the normalization to float to its optimum value.

For the redshift dependence (Fig. 2b), it is reasonable to expect a certain amount of degeneracy with luminosity. This effect appears to dominate at low and medium redshifts. At redshifts beyond ~ 2, an upturn in variability is observed which cannot be explained as a consequence of the known trend with luminosity.

In order to decouple the effects of luminosity and redshift, the variability amplitude was plotted as a function of luminosity for 3 redshift intervals (Fig. 3).

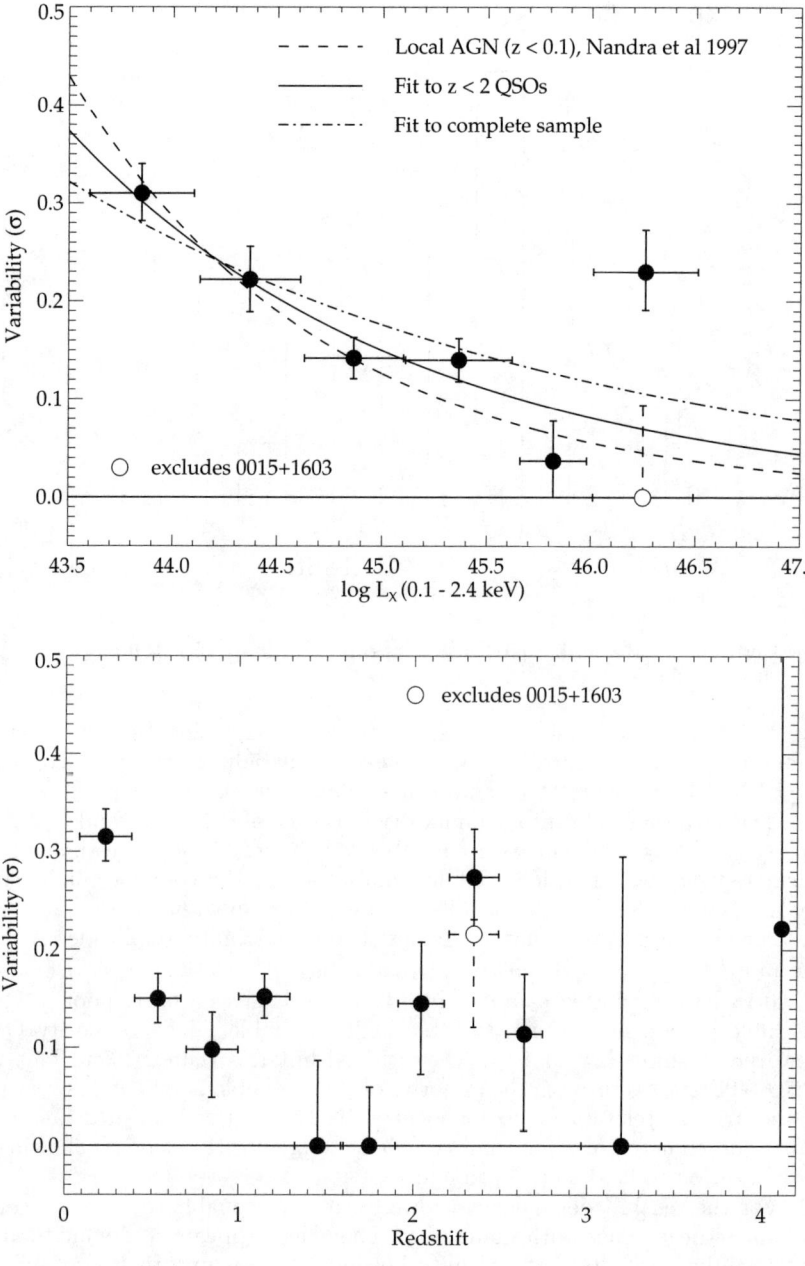

Figure 2. Variability amplitude as a function of luminosity (a) and redshift (b) for the 156 QSOs, displayed in ensemble form. The exclusion of QSO 0015+1603 is explained in Section 4.

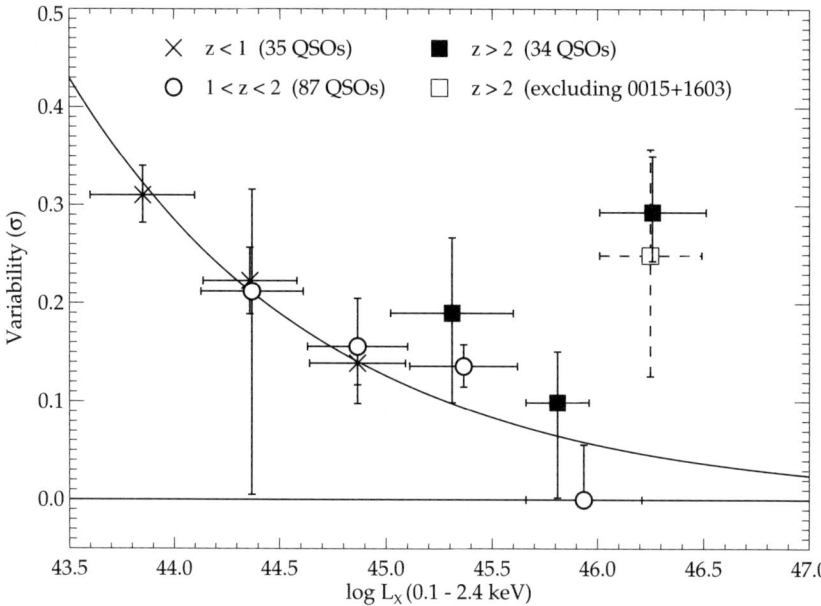

Figure 3. Variability amplitude as a function of luminosity over 3 redshift intervals. The line plotted is the best fit relation to local ($z < 0.1$) AGN found by Nandra et al (1997).

The luminosity bins for each redshift interval overlap, providing a measure of the change in variability amplitude as a function of redshift. Comparing the first 2 redshift intervals we find no significant difference in the $\sigma - L_X$ relation. The anti-correlation between variability and luminosity appears to be unchanged out to a redshift of 2. To compare this with the relationship for local AGN, the power-law slope found by N97 is plotted on Fig. 3. For the redshift interval $z > 2$ all 3 points show an 'excess' variability indicating that high-redshift QSOs may not be well characterized by the variability-luminosity correlation of local AGN.

4. Properties of High-z Variable QSOs

The apparent increase in variability seen in high redshift QSOs may be due to the inclusion of a new population of objects rather than differences in the 'typical' population. Narrow-line Seyfert 1s are known to exhibit enhanced variability (Boller et al 1996, Leighly 1999) and could be responsible for this upturn if their high-z equivalents were more prevalent than they are today. In order to test this hypothesis, the identifications for the 12 QSOs that exhibit detected variability at redshifts greater than 1.9 (i.e. the last 5 redshift intervals in Fig. 2b), were studied in more detail. All the QSOs with adequate optical spectra were found to contain broad permitted lines (> 3000 km s^{-1} FWHM). In one object, 0015+1603, the optical grism spectrum was not of sufficient quality to determine linewidths. This object also displays a steep X-ray spectrum, so is a (potential) NLS1 candidate. 0015+1603 was the highest luminosity object in the sample with a high significance detection of variability. To determine the effect of this steep-spectrum QSO on the characteristics of the sample, the variability analysis was repeated with the object removed. The affected bins are plotted as unfilled points in Figs. 2 & 3. Removing this highly variable object decreases the significance of the upturn in variability for the high-z sample, but does not affect the direction of the trend.

5. Discussion

There is still no definitive explanation for the anti-correlation between X-ray variability and luminosity. In particular, it is unclear whether variability is intrinsically linked with luminosity, or whether the link is to a third parameter that happens to scale with luminosity. If the upturn in variability observed at high redshifts is real, then an extra parameter must exist. X-ray variability timescale may directly scale with black hole mass if, for example, the emission occurs at a fixed number of Schwarzschild radii. This hypothesis has been supported by recent studies of Ptak et al (1998) and Iwasawa et al (2000). Low-luminosity AGN (LLAGN) that are expected to harbour relatively massive black holes display very little X-ray variability, while the dwarf Seyfert NGC 4395, which is thought to contain a small black hole, displays a large variability amplitude. If the amplitude of X-ray variability is linked to black hole mass, the upturn in variability seen at high redshifts would indicate that early quasars are undergoing far more efficient accretion than local AGN.

References

Almaini, O., Lawrence, A., Shanks, T., et al. 2000, MNRAS, 315, 325
Boller, T., Brandt, W.N., Fink, H. 1996, A&A, 305, 53
Edelson, R., Nandra, K., 1999, ApJ, 514, 682
Green, A.R., McHardy, I.M., & Lehto, H.J. 1993, MNRAS265, 664
Iwasawa, K., Fabian, A.C., Brandt, W.N., et al. 1998, MNRAS, 295, L20
Iwasawa, K., Fabian, A.C., Almaini, O., et al. 2000, MNRAS, 318, 879
Lawrence, A. & Papadakis, I. 1993, A&A, 414, L85
Leighly, K.M. 1999, ApJS, 125, 297
Nandra, K., George, I.M., Mushotzky, R.F., et al. 1997, ApJ, 476, 70
Papadakis, I.E. & Lawrence, A. 1993, Nat, 361, 250
Papadakis, I.E. & Lawrence, A. 1995, MNRAS, 272, 161
Ptak, A., Yaqoob, T., Mushotzky, R., et al. 1998, ApJ, 501, L37
Shanks, T., Georgantopoulos, I., Stewart, G.C., et al. 1991, Nat, 353, 315

New Results from the REX Survey

A. Caccianiga, M.J.M. Marchã

Observatório Astronómico de Lisboa, Tapada da Ajuda, 1349-018, Portugal

T. Maccacaro, A. Wolter, R. Della Ceca

Osservatorio Astronomico di Brera, via Brera 28, I-20121, Milan, Italy

I.M. Gioia

Istituto di Radio Astronomia del CNR, via Gobetti 101, Bologna, Italy

Abstract. The REX survey is an ongoing project aimed at the selection of new samples of QSOs and BL Lac objects. Spectroscopic identification of the ~1600 sources in the survey is in progress reaching, so far, about 40% of complete identification level. In this paper, the most recent results derived from the REX survey are briefly summarized.

1. The REX Survey: Sampling the BL Lac Population

The main scientific goal of the REX survey is the selection of new samples of AGNs, like radio-loud QSOs and BL Lac objects. One of the most interesting characteristics of the REX survey is that it has been designed to sample the BL Lac population more homogeneously than previous radio or X-ray surveys. The EMSS survey (Morris et al. 1991), for instance, has selected objects with a flat X-ray-to-radio spectral index (α_{RX} <0.7, defined as High Energy Peaked BL Lacs, HBL) while the radio selected 1 Jy sample (Stickel et al. 1991) has found sources with α_{RX} >0.7 (Low Energy peaked BL Lacs, LBLs). In Figure 1 the α_{RX} distribution of the BL Lacs discovered so far in the REX survey is compared to that of the BL Lacs in the EMSS and the 1 Jy samples. The REX survey is clearly finding many objects with intermediate properties between HBLs and LBLs, thus suggesting that the presence of a dichotomy in the BL Lac class, often claimed in the literature, could be artificially created by selection effects.

2. The XB-REX Sample: Studying the Cosmological Evolution

A first completely identified sub-sample of 237 REXs has already been selected by imposing an X-ray flux limit greater than 4×10^{-13} erg s^{-1} cm^{-2} and a magnitude (B) brighter than 20.5. The resulting sample (the X-ray Bright REX sample, XB-REX) has been used to estimate the cosmological evolution of emission

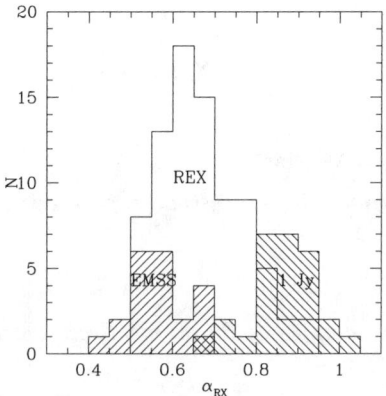

Figure 1. The α_{RX} distribution of the BL Lacs discovered so far in the REX, in the EMSS and in the 1Jy samples

line AGNs and BL Lac objects through a V_e/V_a analysis. A detailed description of the results is presented in Caccianiga et al. (2001).

• *BL Lacs.* The XB-REX sample contains 55 BL Lac objects and it is one of the largest complete samples of BL Lacs available for statistical studies. The V_e/V_a test indicates that the population is not affected by a strong cosmological evolution. Moreover, the V_e/V_a analysis applied to the subsample of HBLs does not reveal any significant departure from a uniform distribution. This result confirms that the cosmological properties of BL Lac objects are significantly different from those of emission line AGNs (see below).

• *AGNs.* The XB-REX sample contains 95 EL AGN. The V_e/V_a test is indicative of a strong positive evolution. We have then estimated the evolution by assuming a pure luminosity evolution of the form: $L(z) = L(0)(1+z)^k$ under the hypothesis that the values of k in the three bands (radio, optical and X-ray) are the same. The best fit value is $k = 3.0$ with a 1σ interval of (2.5, 3.3). The value of k is intermediate (but still consistent at 1σ) between that found for the X-ray selected (k=2.92, Della Ceca et al. 1994) and the optically selected (e.g. k=3.49, Padovani 1993) radio-loud AGNs.

Acknowledgments. This work has received partial financial support from the Portuguese FCT and from the Italian Space Agency (ASI).

References

Caccianiga, A., et al. 2001, ApJ, in press
Caccianiga, A., et al. 1999, ApJ, 513, 51
Della Ceca, et al. 1994, ApJ, 430, 533
Morris, S.M., et al. 1991, ApJ, 380, 49
Padovani, P. 1993, MNRAS, 263, 461
Stickel, M., et al. 1991, ApJ, 374, 431

Optical Identification of X-ray Sources in a High X-ray Flux Sensitivity Area from the RASS

Raúl Mújica

INAOE-Tonantzintla, Apdos. Postales 51 y 216 C.P. 72000 Puebla, Pue., México

Franz-Josef Zickgraf

Hamburger Sternwarte, Gojenbergsweg 112, 21029 Hamburg, Germany

V. Chavushyan, Y. Juárez, A. Serrano

INAOE-Tonantzintla, Apdos. Postales 51 y 216 C.P. 72000 Puebla, Pue., México

I. Appenzeller, J. Krautter

Landessternwarte-Königstuhl, D-69117 Heidelberg, Germany

Abstract.
We have optically identified a complete sub-sample of ROSAT All-Sky Survey (RASS) X-ray sources contained in 6 study areas. Of the original 12° × 12° area near the NEP, only one-quarter was observed until now in order to keep the number of sources to a manageable size. This area is of particular interest because the RASS integration time is about a factor of ten longer than in the other areas and therefore a factor of 3-4 deeper in flux. We have started to observe the RASS sources in the remaining 3/4 of this area. First results are presented.

1. Introduction

ROSAT detected about 60,000 sources during the All-Sky Survey, i.e., about 1.5 sources per square degree. Complete, spectroscopically identified subsamples of the RASS are extremely important because, first, it is impossible to do spectroscopy on all sources; and second, it is the only way to achieve an unbiased sample of optically identified X-ray sources, in order to carry out statistical studies and to obtain identification criteria to extrapolate towards all X-ray sources.

We have optically identified a complete sub-sample of RASS X-ray sources (Appenzeller et al. 1998, Paper III) contained in 6 study areas. Originally each area measured 12° × 12° in size. Due to the large number of X-ray sources in two of the six areas, one close to the North Galactic Pole (NGP, area IV) and one near the North Ecliptic Pole (NEP area V), the original sample was reduced in size by observing only part of the latter two areas (Zickgraf et al. 1997, Paper

II) in order to get a manageable number of sources. The statistical analysis was done by Krautter et al. (1999, Paper IV).

2. The Area

The sample was redefined in terms of the minimum fluxes or countrates: 0.01cts/s in area V and 0.03 cts/sec in the other areas. The samples in the NEP and NGP were also reduced in size to smaller sub-areas. The resulting areas were called Va (western quarter of field V) and IVac (western half of area IV). Area Va was later extended to the east to obtain a field of 6.2° × 6°. Area V contained 671 sources; 183 were included in area Va. The goal of the present project is to identify the x-ray sources in the remaining 3/4 of the original area, that contains ∼ 300 sources with countrate ≥ 0.01cts/sec, following the same ID criteria as in the first part of the project (see Paper II).

3. Observations

The observations were carried out at the 2.1m telescope of the Guillermo Haro Observatory, operated by INAOE. The LFOSC focal reducer was used; it allows direct CCD imaging, filter photometry, and multi-object spectroscopy. A setup covering the spectral range 4200-9000Å with a dispersion of 8.2Å/pix was adopted. Usually, all possible counterparts within a circle of about 50" to 60" radius around the X-ray position were observed.

4. Preliminary Results

We started to observe all the sources inside the error circle of the RASS positions in the remaining 3/4 of area V with the same limiting count-rate (0.01 cts/sec). Since May 1999 we have observed more than 100 RASS positions, and we obtained in this way more than 300 spectra of candidate optical counterparts. We are following the method for optical identification as explained in detail in Paper II. The summary of new identifications is as follows: 38 stellar counterparts, 28 AGN, 1 Cluster and 2 Multiple. We have another 50 positions already observed, but not analysed or requiring complementary observations.

References

Appenzeller et al. 1998, ApJS, 117, 319, Paper III
Condon et al. 1998, AJ, 115, 1693
Krautter et al. 1999, A&A, 350, 743, Paper IV
Stocke et al. 1991, ApJ, 348, 141
Zickgraf et al. 1997, A&AS, 123, 103, Paper II

AGN Search From Multicolor Photometric Observations of Faint ROSAT X-ray Sources in a One Square Degree Field

Suijian Xue, Xu Zhou and Haotong Zhang

National Astronomical Observatories, Chinese Academy of Sciences. A20 Datun Rd., Chaoyang District, Beijing 100012

Abstract. We report that a volume-limited, X-ray selected AGN sample, including 49 QSO like and starburst galaxies (SBs) at the 0.1-2.4 keV flux level of $\gtrsim 5.3 \times 10^{-14}$ ergs cm^{-2}s^{-1} was formed with photometric redshifts.

1. Introduction

75 cataloged X-ray sources from a medium deep ROSAT survey (Molthagen et al. 1997) were found in a 1 sq.deg. field, centered at 09:56:24.46, +47:35:08.4 (J2000). The field has been heavily CCD imaged with 15 intermediate-band filters covering the wavelength range 3360-9745 Å. Full identification of the sample sources will not only extend the results of the ROSAT All-Sky Survey to a fainter limit, but also test recent results of Chandra Deep Surveys in a different larger sky field.

2. Results

In total we have made 67 optical identifications from 75 X-ray sources with several extragalactic populations. The result is summarized in Table 1., and the procedure of identification illustrated in Figure 1.

Table 1. Identification Summary

Population	Numbers	Fraction (%)
All	75	100.0
AGNs	25	33.3
SBs	24	32.0
Galaxies	9	12.0
Stars	9	12.0
Unknown	8	10.7

References

Molthagen, K., Wendker H.J., & Briel U.G. 1997, A&AS, 126, 509

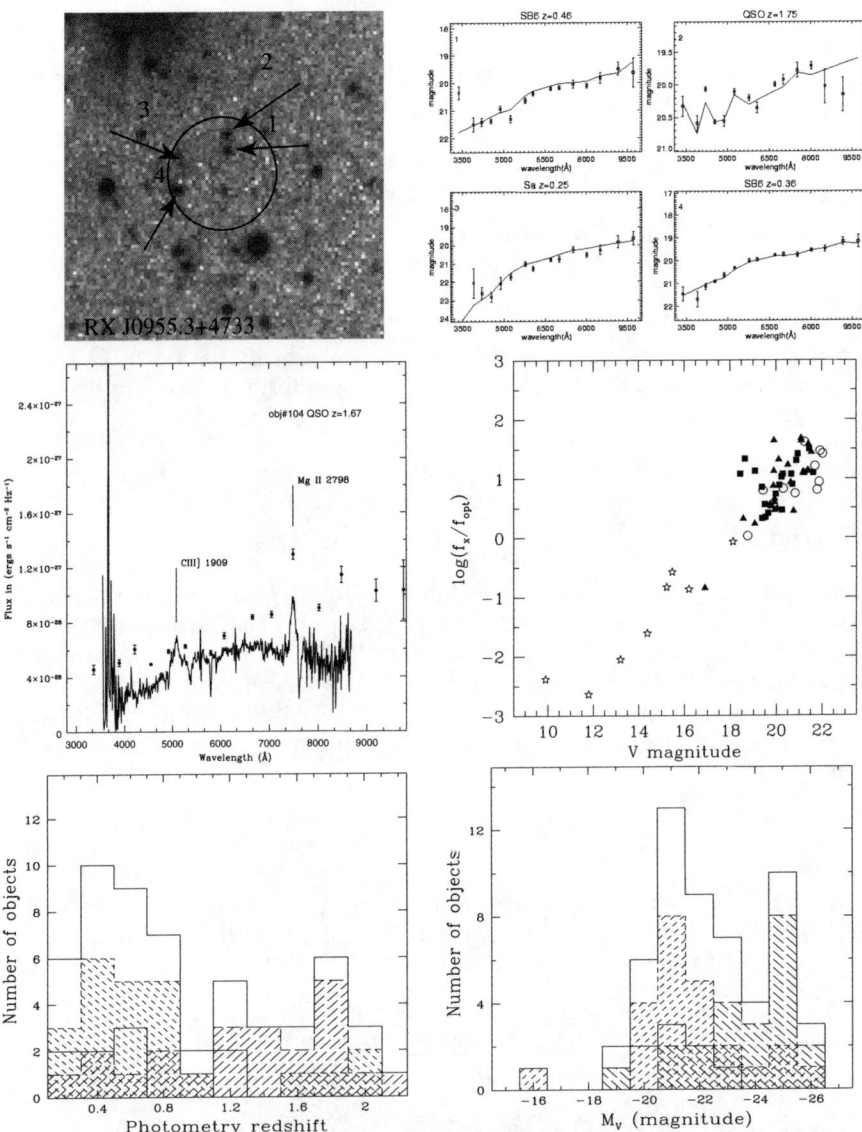

Figure 1. The procedure of source identification shown as an example in *top left*: 4 optical associations were found within the error circle of RX J0955.3+4733 and their SEDs were measured in 15 colors; *top right*: 4 sources were classified based on SED fitting with AGN, normal galaxy and star SED templates; *middle left*: Object 2 was identified with RX J0955.3+4733, and the follow-up spectrum confirms it is a new QSO at $z = 1.67$. *Middle right:* X-ray to optical flux ratio vs. visual magnitude of the sample. AGNs: solid squares; SBs: solid triangles; normal galaxies: circles; stars: stars. *Bottom left & right*: the redshift and visual luminosity distributions of identified extragalactic populations. Thick Solid: the total; long dashed: QSOs; short dashed: SBs; thin solid: normal galaxies.

Part 4
Radio Surveys for AGN

Yoshi Taniguchi outside the 2.2m dome

The Cognac Factory: Carlos De Breuck, Brigitte Rocca-Volmerange

Surveys of Parsec-scale Radio Structures in AGN

L.I. Gurvits

Joint Institute for VLBI in Europe, P.O. Box 2, 7990 AA Dwingeloo, The Netherlands

Abstract. Several recent global and Space VLBI surveys of quasars and other types of AGN provide a wealth of material on milli- and sub-milliarcsecond radio structures in hundreds of sources. Results of these projects are presented with an emphasis on the statistics of redshift- and angular-scale-dependent properties of the milli- and sub-milliarcsecond radio structures. These studies make possible disentanglement of intrinsic (possibly, evolutionary) phenomena of parsec-scale radio structures and the imprints of the cosmological model.

1. Introduction

Over the last decades it has become commonplace to call AGN the most luminous powerhouses of the Universe. Indeed, their radiation, with as high an intensity as 10^{48} erg/s, originates in regions on the scale of parsecs and smaller. Not surprisingly, AGN are the most popular targets of high angular resolution studies in all domains of the electromagnetic spectrum. Of the presently available astronomical techniques, Very Long Baseline Interferometry (VLBI) offers an unrivaled angular resolution reaching milliarcsecond (mas) and sub-mas scales (see review by Kellermann & Moran 2001). As any other astronomical technique, VLBI began by observing just a few "famous" sources. However, its rapid progress since the first observations in the late 1960s has resulted in VLBI surveys of thousands of extragalactic sources, mostly AGN. Table 1 presents the major imaging VLBI surveys published to date.

Present-day VLBI systems operate at meter to millimeter wavelengths (frequencies from ~ 0.3 to ~ 100 GHz, respectively) at baselines comparable to the diameter of the Earth. The resolution of a VLBI system is defined by the difraction limit, λ/B, where λ is the wavelength and B is the projection of the baseline on the picture plane. For a typycal wavelength of $\lambda = 6$ cm and a practical maximum baseline length between two Earth-based radio telescopes of $B \approx 10000$ km, the achievable angular resolution is $\theta \approx 1.5$ mas. Since 1997, the first dedicated Space VLBI mission VSOP (Hirabayashi et al. 1998) makes it possible to observe with baselines up to 30000 km, which gives about a three times "sharper" image than the longest global Earth-based interferometers.

There is good reason for radio-loud AGN to be primary targets of VLBI studies: in order to be detected on baselines of $\sim 10^4$ km with "modest" 25-m class radio telescopes operating with a bandwidth of tens of MHz, a radio

Table 1. On-line VLBI survey data bases

Survey	Freq [GHz]	URL
PR and CJ[1]	5 & 1.6	www.astro.caltech.edu/ tjp/cj/
USNO RRFID[2]	2.3, 8.4 & 15	rorf.usno.navy.mil/rrfid.html
VLBA Calibrators[3]	2.3 & 8.4	magnolia.nrao.edu/vlba_calib/index.html
VLBApls[4]	5	www.jive.nl/jive/jive/svlbi/vlbapls/
VSOP Survey[5]	5	oj287.vsop.isas.ac.jp/survey/
VLBA 2 cm Survey[6]	15	www.cv.nrao.edu/2cmsurvey/

[1] Pearson–Readhead (Pearson & Readhead 1988) and Caltech–Jodrell Bank (Taylor et al. 1996 and references therein) surveys;
[2] USNO Radio Reference Image Database (Fey & Charlot, 2000, and references therein)
[3] VLBA Calibrator Survey, Peck & Beasley (1998)
[4] VSOP/VLBA Prelaunch Survey of Extragalactic Radio Sources, Fomalont et al. 2000
[5] VSOP Survey Program, Hirabayashi et al. 2000a
[6] 2 cm Survey of Extragalactic Radio Sources, Kellermann et al. 1998; Gurvits et al. 2001

source must have a brightness temperature[1] $T_B \geq 10^7$ K (see Thompson, Moran & Swenson 2001 for deep insight into the technique of radio interferometry in general and VLBI in particular). Kellermann and Pauliny-Toth (1969) have shown that, for a stationary source, there is an inverse Compton cooling limit on brightness temperature at $T_B \sim 10^{12}$ K. As became clear in early VLBI observations, some bright sources associated with AGN (mostly quasars) indeed have components as bright as 10^{12} K. As an example of a strange cosmic coincidence, the size of our planet permits intereferometers which can just barely resolve sources of that brightness. I leave out of this short review the question of whether or not this mysterious coincidence indicates a very special place occupied by our civilization in the Universe...

As pointed out by Zensus (1997), the de-facto paradigm of the AGN phenomenon, the relativistic jet model (Blandford and Königl, 1979), is based on three pillars: (i) accretion onto the massive central object, (ii) relativistic ejection from nuclei, and (iii) relativistic beaming. Accretion theory descends from the work done in the 1950s-1970s and has received its present form in direct application for AGN by Begelman et al. (1984; also references therein). Relativistic ejection from galactic nuclei has been suggested by Rees (1966) as a mechanism of powering extended (kpc-scale) radio structures in extragalactic sources. Earlier, the idea of relativistic ejection from the core of M87 (Vir A) was mentioned by Ambartsumian (1958) as a possible mechanism for fueling its synchrotron emission (similar to the Crab Nebula). Finally, relativistic beam-

---

[1] It is a historical tradition in radio astronomy to present the source's brightness in terms of its brightness temperature, which is equal to the physical temperature of a black body of the same brightness at the particular wavelength. This "thermal" flavor of the definition is in sharp discord with the non-thermal synchrotron mechansm which dominates the radio emission of AGN.

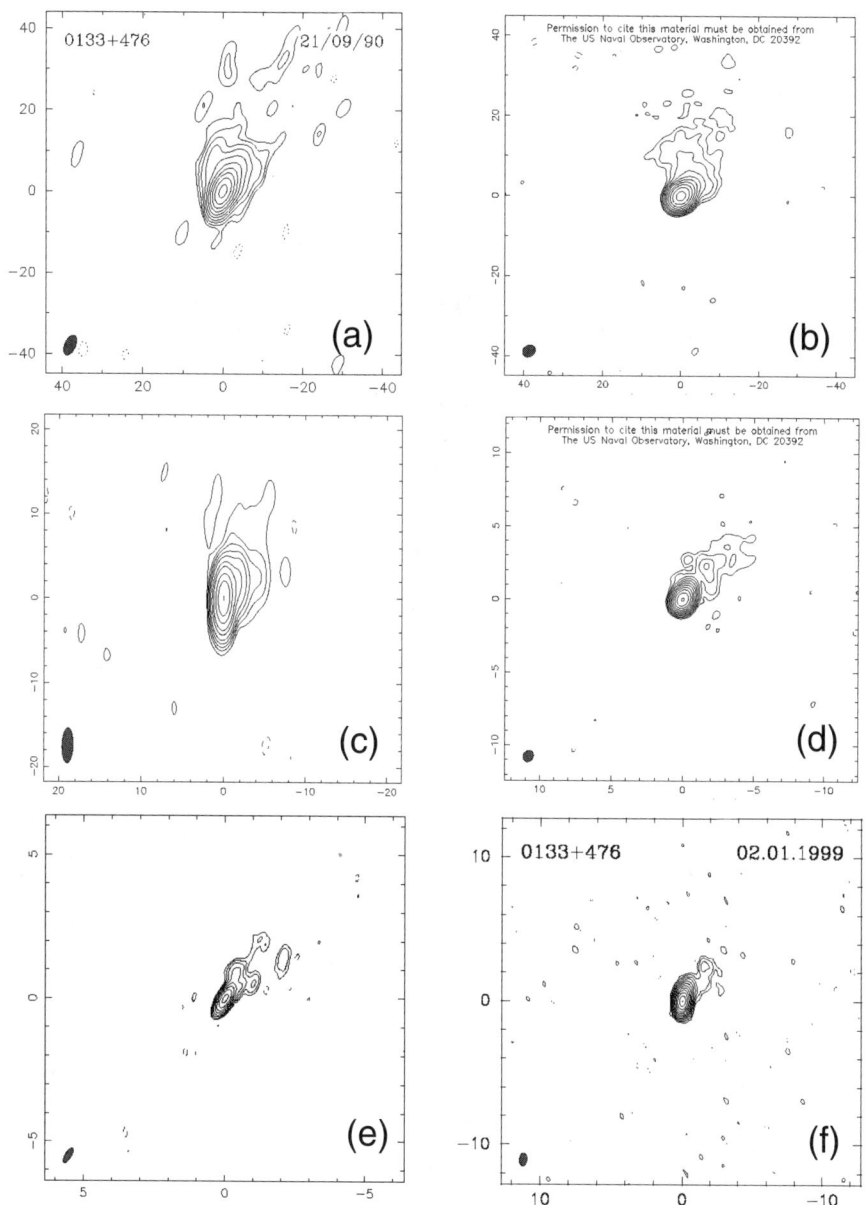

Figure 1. Images of the quasar J0136+4751 (0133+476) from VLBI surveys. See Table 2 for details.

ing was proposed by Shklovskii (1963) as an explanation of the apparent high brightness of some extragalactic radio sources.

Table 2. Parameters of the VLBI images shown in Fig. 1

	Freq GHz	Survey	Epoch	B_S Jy/bm	b_{LC} mJy/bm	Beam FWHM [mas] P.A.	Ref.
a	1.6	CJ1	1990.72	1.0	1.5	5.2×2.7, $-24°$	[1]
b	2.3	RRFID	1998.61	1.25	1.8	3.5×2.8, $-59°$	[2]
c	4.9	VLBApls	1996.43	2.29	2.3	4.2×1.4, $-2°$	[3]
d	8.6	RRFID	1998.61	2.11	2.0	0.8×0.7, $-35°$	[2]
e	4.8	VSOP	1999.62	1.21	7.8	0.5×0.2, $-31°$	[4]
f	15.3		1999.01	2.65	1.6	0.9×0.5, $-7°$	[5]

[1] Polatidis et al. 1995
[2] USNO Radio Reference Frame Image Database (RRFID)
[3] Fomalont et al. 2000
[4] Hirabayashi et al. 2000a
[5] Gurvits et al. 2001

2. Toward Understanding the AGN Power Plant Design

Since the formulation of the relativistic jet model more than twenty years ago, considerable progress has been achieved in its fine tuning, not the least owing to the data supplied by VLBI studies of particular AGN or their limited samples. An example of the "deliverable product" of VLBI studies of AGNs is presented in Fig. 1. It shows mas- and sub-mas-scale radio images at different wavelengths of the quasar J0136+4751 ($z = 0.859$, Véron-Cetty & Véron, 1998). Charcteristics of these images are listed in Table 1. This example represents the so called "core-jet" morphology typical for an overwhelming majority of mas-scale radio emitting areas in AGN. In almost all such known structures, there is a component characterized by a flat or inverted radio spectrum which dominates the radio emission at frequencies above several GHz. This component, dubbed a "core" is believed to lie at the base of a jet. In the example shown in Fig. 1 the core is clearly identifiable as the brightest spot in the maps. Understanding of the physics of the central engine of AGN requires, in particular, a better knowledge of the "design" of its emitting region with adequate linear resolution. The need for further improvement in resolution at all wavelengths has been reviewed recently by Rees (2001).

One of the opportunities to address this question is being offered by imaging VLBI surveys. They allow us to determine the characteristic scales of the central engine, or at least of its appearance in radio waves. Fig. 2 represents the mean value of the correlated flux density versus projected baseline for a subsample of about 100 radio-loud AGN. The value of correlated flux density can be considered a measure of radio emission coming from an area defined by the synthesized beam of the interferometer. The longer the baseline, the smaller the beam, and for a resolved source, the smaller the correlated flux density. An unresolved source has a constant flux density at all baselines.

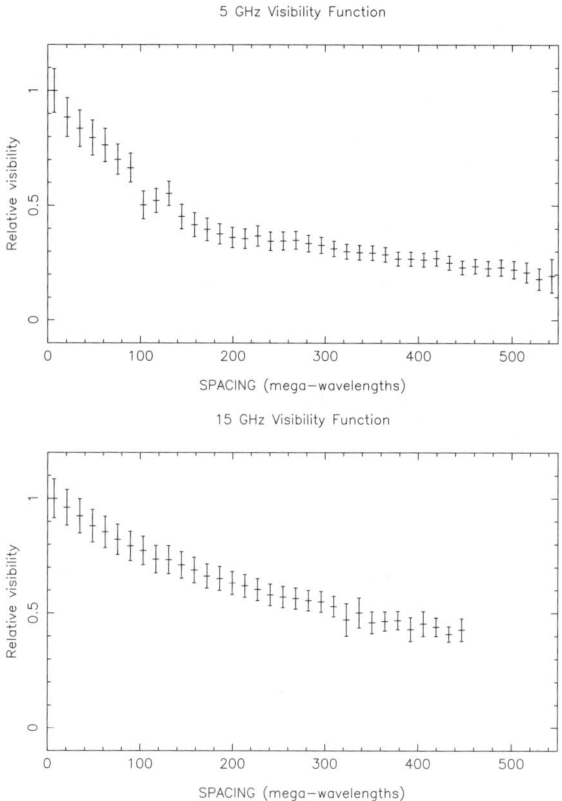

Figure 2. The mean value of correlated flux density versus projected baseline for a combined sample of VLBApls and VSOP Survey sources observed at (*top*) 5 GHz and (*bottom*) 15 GHz. The 5 GHz data used are from Hirabayashi et al. (2000) and Fomalont et al. (2000), 15 GHz data are from Kellermann et al. (1998) and Gurvits et al. (2001). Courtesy E.B.Fomalont.

Both plots in Fig. 2 indicate that on average the sources are resolved significantly on baselines up to $\sim$ 500 Mλ at both frequencies. A small "kink" visible at about $\sim$ 120 Mλ at 5 GHz and $\sim$ 340 Mλ at 15 GHz most likely is caused by a sampling effect of the Very Long Baseline Array used in both cases: the array seems to have a peculiar geometrical configuration on the scale of about 6600 km corresponding to $\sim$ 120 Mλ and $\sim$ 340 Mλ at 5 and 15 GHz, respectively.

More significant for the physics of the sources is a clear flattening (change of the slope) of the dependence at around $\sim$ 200 Mλ at 5 GHz (Fig. 2, top) which corresponds to an angular scale of about 1 mas. Indeed, as known from hundreds of available VLBI images at 5 GHz, AGN radio structures on the scale of several milliarcseconds (baselines shorter than $\sim$ 200 Mλ) typically represent a combination of core and extended jet (e.g. Gurvits, Kellermann & Frey 1999,

Fomalont et al. 2000). Not surprisingly, at these baselines the slope is steep. The flatter slope at baselines ≥ 200 Mλ seems to indicate that on sub-mas scale the core starts to be resolved, and its inner structure is not self-similar to the large-scale "core-jet" morphology.

The change of slope described above is almost absent at 15 GHz (Fig. 2, bottom). This comes as no surprise if the change of morphology from "core-jet" to "core only" indeed takes place around the angular scale of ~ 1 mas. As the extended mas-scale jets have significantly steeper radio spectra than the cores, the contribution of jets to the correlated flux density at 15 GHz is smaller than that at 5 GHz. Thus, the "disappearance" of jets on sub-mas scales at 15 GHz is less prominent than at 5 GHz.

I note that the result at 5 GHz shown in Fig. 2 (top) is a combination of ground-based VLBA observations (up to a baseline length of ~ 150 Mλ, Fomalont et al. 2000) and the VSOP Survey (from ~ 50 Mλ to ~ 550 Mλ, Hirabayashi et al. 2000). The significance of the VSOP results is obvious, because the change in structural regime takes place at baselines exceeding those available for ground-based VLBI systems at 5 GHz.

It has to be noted also that the change of structural regime described here is purely qualitative and is based on a sample of sources at different redshifts (thus, observed at different rest frame frequecies). Further detailed investigation must account for redshift-dependent effects.

3. AGN Pc-Scale Radio Structures and Cosmology

VLBI data on radio-loud AGN provide measurable parameters which could be looked at in a cosmological perspective. These parameters are the source's characteristic angular size, apparent proper motion of its structural components and the source count statistics.

The radio emission from the AGN cores on the milliarcsecond scale ought to be controlled by a limited number of parameters (such as, e.g., the mass of the central black hole, the ambient and intrinsic magnetic fields in the base of jet, the accretion rate and the angular momentum of the black hole). Kellermann (1993b) argued that the milliarcsecond-scale radio structures in AGN are much less dependent on source evolution and properties of the interstellar/intergalactic medium compared to the arcsecond (kiloparsec) scale structures. Thus, mas-scale radio "cores" could be considered as non-ideal cosmological standard rods. The dependence of their apparent angular sizes on redshift (the so- called "$\theta - z$" test, Fig. 3) contains an imprint of the cosmological model. Attempts by Gurvits (1993, 1994) and Kellermann (1993a, 1993b) to extract the cosmological information from independent ad hoc VLBI samples favored a value of the cosmological deceleration parameter $q_0 \leq 0.5$ (under assumption that the cosmological term $\Lambda = 0$). Further work on the milliarcsecond scale "$\theta - z$" test by several authors (reviewed by Gurvits, Kellermann & Frey 1999) addressed various "pros" and "cons" of the approach. In particular, Dabrowski et al. (1995) pointed out the difficulty in obtaining a meaningful constraint on the average density Ω due to the effects of relativistic beaming in limited (let alone ad hoc) source samples. They estimated that cosmologically conclusive results could be obtained with VLBI samples containing several thousand sources. Recently Lima & Alcaniz

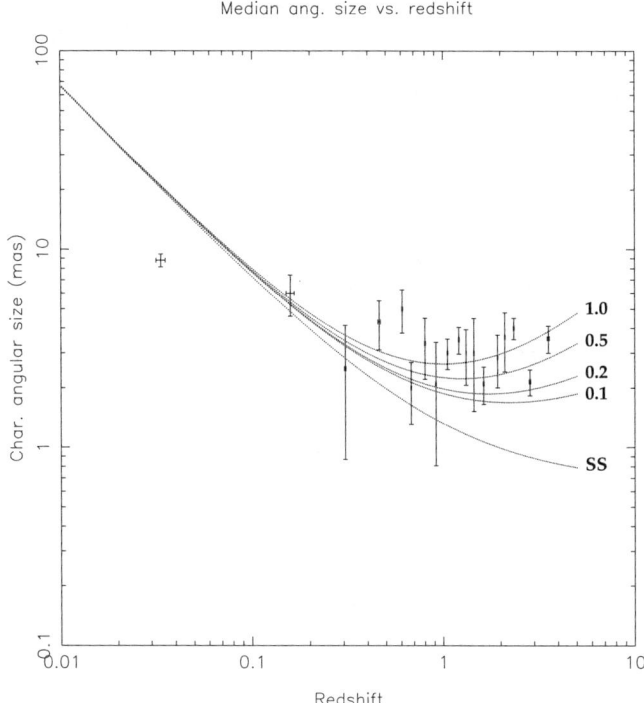

Figure 3. An example of the "$\theta - z$" test: median angular size of the sample of 330 AGN versus redshift (Gurvits, Kellermann & Frey 1999). The full length of the error bars corresponds to 1σ. The solid lines correspond to the assumed linear size of 9.6 pc, the Steady-state model (SS) and models of a homogeneous, isotropic Universe with $\Lambda = 0$ and values of $q_\circ = 1.0, 0.5, 0.2, 0.1$ (as marked on the plot). Data are binned into 18 bins nearly equally populated (18–19 sources per bin). The curves are shown as examples only. none of them represents the best fit.

(2000, 2001) applied the "$\theta - z$" data from Gurvits, Kellermann & Frey (1999) to constrain the parameters of the cosmic equation of state. Their result favors the conventional flat ΛCDM model ($\omega = -1$) with $\Omega_m = 0.2$. Wiik & Valtaoja (2001) modified the "$\theta - z$" test by assuming shocks in AGN jets as standard objects. Applying this approach to 14 AGNs with both total flux density monitoring and VLBI data available, they were able to demonstrate the potential of the method for estimating the dynamical parameters of the cosmological model.

The statistics of apparent velocity of VLBI components in AGN jets as a function of redshift (the "$\mu - z$" test) has been analysed by Vermeulen & Cohen (1994) and refined later by Vermeulen (1995) as a promising tool for studying the beaming model under various unification scenarios as well as a means of measurement of the Hubble constant H_0 and the deceleration parameter q_0. This method is based on the assumption of a "standard velocity" of moving features in milliarcsecond scale jets translated into observed apparent, sometimes superluminal, velocities (Pelletier & Roland 1989). An effort to exploit the "$\mu - z$" test with multi-epoch VLBI data for the CJF sample of AGN (several hundred sources) is underway (S.Britzen, 2001, private communication).

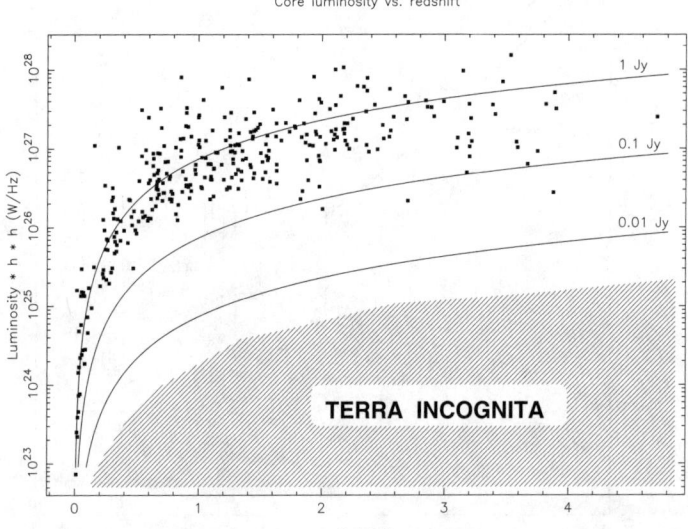

Figure 4. Luminosity (Lh^2) as a function of redshift for a sample of 330 sources (almost exclusively AGN of various types) at 5 GHz, adopted from Gurvits, Kellermann & Frey (1999). The solid lines show luminosities of sources with flux densities of 1, 0.1 and 0.01 Jy, calculated assuming spectral index $\alpha = 0$. The plot indicates a strong luminosity selection effect in VLBI samples presently available. In spite of the ad hoc nature of the sample based on various publications, the sources' luminosity distribution closely follows the lines which correspond to flux densities within the range from 300 mJy to several Jy.

Finally, direct number counts of correlated flux densities of radio loud AGNs detected at various VLBI spacings (i.e. projected baseline lengths) analysed in much the same way as "traditional" source counts offer a cosmological hint which might become significant if conducted on a large enough sample of sources (Gurvits 2001).

4. Future Outlook

The experience of VLBI AGN surveys reviewed above has shown convincingly that their astrophysical and cosmological applications are far from being exhausted. Further progress will require improvement of angular resolution and sensitivity. The former is needed to confirm and exploit in full the trend hinted by the dependences shown in Fig. 2. The significance of the latter is illustrated by Fig. 4. The data points in this plot show the dependence of luminosity on redshift for a typical flux density limited sample (the data taken from Gurvits,

Kellermann & Frey 1999). It is clear that a large volume of the "luminosity – redshift" space (dashed in Fig. 4) requires inclusion in all-sky VLBI imaging surveys objects with flux densities in the range of 100 – 1 mJy (and perhaps less in specially selected deep fields), which correspond to luminosities $\leq 10^{25}$ W/Hz. This will increase the size of VLBI imaging samples by 2–3 orders of magnitude. Several ongoing instrumental developments will bring about higher angular resolution and higher sensitivity of VLBI imaging.

Higher resolution will become possible with the next generation Space VLBI missions, such as VSOP-2 (Hirabayashi et al. 2000b) or ARISE (Ulvestad 2000). Compared to the VSOP mission, their higher angular resolution will be achieved owing to a slightly higher orbit (by a factor of ~ 2) and higher observing frequencies (by a factor of $\sim 8-16$). Millimeter VLBI is also promising to achieve higher sensitivity thus allowing en masse imaging of AGN with sub-mas angular resolution. The improvement of sensitivity will materialize with the advent of the Square Kilometer Array – a new radio telescope with collecting area of ~ 1 km^2 distributed over baselines up to thousands of kilometers (Taylor 1999). Both these developmets will bring new excitement in surveying compact radio structures in the powerhouses of the Universe.

Acknowledgments. I thank the organizers of the Colloquium for all their efforts and hospitality. I am grateful to E.Ye. Khachikian and Ian Avruch for very useful comments, and Alan Fey for the data from the USNO RRFID database. This review has made use of the USNO Radio Reference Frame Image Database (RRFID), the data obtained at (*i*) the National Radio Astronomy Observatory, operated by Associated Universities, Inc. under a Cooperative Agreement with the NSF; (*ii*) the European VLBI Network, a joint facility of European and Chinese radio astronomy institutes funded by their national research councils; (*iii*) the VSOP Project, led by the Institute of Space and Astronautical Science (Japan) in cooperation with many organizations around the world. Several results reported here were obtained in collaboration with Ken Kellermann, Sandor Frey, Ed Fomalont, Matt Lister and Zhang Hai-Yan.

References

Ambartsumian, V.A. 1958, Izv. AN ArmSSR, ser. fiz.-mat. nauk II, No. 5, 9
Begelman, M.C., Blandford, R.D. & Rees, M.J. 1984, Rev. Mod. Phys. 56, 255
Blandford, R. & Königl, A. 1979, ApJ, 232, 34
Dabrowski, Y., Lasenby, A. & Saunders, R. 1995, MNRAS, 277, 753
Fey, A. L. & Charlot, P. 2000, ApJS, 128, 17
Fomalont, E.B., Frey, S., Paragi, Z., Gurvits, L.I., Scott, W.K., Taylor, A.R., Edwards, P.G., & Hirabayashi, H. 2000, ApJS, 131, 95
Gurvits, L.I. 1993, in Sub-arcsecond Radio Astronomy, eds. R.J. Davis & R.S. Booth, (Cambridge: Cambridge Univ. Press), 380
Gurvits, L.I. 1994, ApJ, 425, 442
Gurvits, L.I. 2001, in Galaxies and their constituents at the highest angular resolution, eds. R.T.Schilizzi, S.N.Vogel, F.Paresce & M.S.Elvis (San Francisco: ASP), 146

Gurvits, L.I., Kellermann, K.I. & Frey, S. 1999, A&A, 342, 378
Gurvits, L.I., Kellermann, K.I., Fomalont, E.B., Zhang, H.Y. 2001, in prep.
Hirabayashi, H., Hirosawa, H., Kobayashi, H. et al. 1998, Science, 281, 1825
Hirabayashi, H., Fomalont, E.B., Horiyuchi, S. et al. 2000a, Publ. Astron. Soc. Japan, 52, 997
Hirabayashi, H., Murphy, D.W., Murata, Y., Edwards, P.G., Avruch, I.M., Kobayashi, H & Inoue, M. 2000b, in Astrophysical Phenomena Revealed by Space VLBI, eds. H. Hirabayashi, P.G. Edwards & D.W. Murphy, (Sagamihara: ISAS), 277
Kellermann, K.I. 1993a, in Sub-arcsecond Radio Astronomy, eds. R.J.Davis & R.S.Booth, (Cambridge: Cambridge Univ. Press), 386
Kellermann, K.I. 1993b, Nature, 361, 134
Kellermann, K.I., Cohen, M.H., Zensus, J.A. & Vermeulen, R.C. 1998, AJ, 115, 1295
Kellermann, K.I., & Moran, J.M. 2001, ARA&A, in press
Kellermann, K.I., & Pauliny-Toth, I.I.K. 1969, ApJ, 155, L71
Lima, J.A.S. & Alcaniz, J.S. 2000, A&A, 357, 393
Lima, J.A.S. & Alcaniz, J.S. 2001, ApJ, in press
Pearson, T.J. & Readhead, A.C.S. 1988, ApJ, 328, 114
Peck, A.B. & Beasley, A.J. 1998, in Radio Emission from Galactic and Extragalactic Radio Sources, eds. J.A.c Zensus, G.B.Taylor & J.M.Wrobel, (San Francisco: ASP), 155
Pelletier, G. & Roland, J. 1989, A&A, 224, 24
Polatidis, A.G., Wilkinson, P.N., Xu, W., Readhead, A.C.S., Pearson, T.J., Taylor, G.B., Vermeulen, R.C. 1995, ApJS, 98, 1
Rees, M.J. 1966, Nature, 211, 468
Rees, M.J. 2001, in Galaxies and Their Constituents at the Highest Angular Resolution, eds. R.T.Schilizzi, S.N.Vogel, F.Paresce & M.S.Elvis (San Francisco: ASP), 2
Shklovsky, I.S. 1963, Sov. Astron., 6, 465
Taylor, G.B., Vermeulen, R.C., Readhead, A.C.S. et al. 1996, ApJS, 107, 37
Taylor, A.R. 1999, in Perspectives on Radio Astronomy: Science with Large Antenna Arrays. ed. M.P. van Haarlem, (Dwingeloo: ASTRON), 2
Thompson, A.R., Moran, J.M., & Swenson, G.W. 2001, Interferometry and Synthesis in Radio Astronomy, (New York: Wiley–Intersci.), 2nd ed.
Ulvestad, J.S. 2000, Adv. Sp. Res., 26, 735
Vermeulen, R.C. 1995, Proc. Natioanl Academy of Sci., 92, 11385
Vermeulen, R.C. & Cohen, M.C. 1994, ApJ, 430, 467
Véron-Cetty, M.-P. & Véron P. 1998, A Catalogue of Quasars abd Active Nuclei (Garching: ESO), 8th ed.
Wiik, K. & Valtaoja, E. 2001, A&A, 366, 1061
Zensus, J.A. 1997, ARA&A, 35, 807

Radio AGN Surveys

Carlos De Breuck

Institut d'Astrophysique de Paris, France

Wil van Breugel

IGPP/LLNL, Livermore, CA, USA

Huub Röttgering

Sterrewacht Leiden, The Netherlands

Chris Carilli

NRAO, Socorro, NM, USA

Abstract. We present a short overview of radio surveys for AGN, including the 'complete' flux limited surveys and 'filtered' surveys. We also describe our ultra-steep spectrum search for the highest redshift radio galaxies, and our follow-up VLA and ATCA observations of the most distant ($z = 5.19$) and the most luminous $z < 2$ radio galaxy known.

1. Flux-limited and Filtered Radio Surveys

Radio surveys play a crucial role in the study of AGN. Searches for AGN at radio wavelengths have several major advantages over other wavelength regimes, including: (i) the radio emission is often extremely powerful, and can be detected out to the highest redshifts (e.g. SDSSp J083643.85+005453.3 at $z = 5.82$, Fan et al. 2001), (ii) the most powerful radio sources at the highest redshifts pinpoint the most massive and luminous galaxies at such redshifts, allowing a detailed study, (iii) radio emission is not affected by dust emission, which often affects optical selection criteria.

Several major observational efforts have therefore been performed during the last 4 decades to identify the host galaxies of large samples of radio sources (Table 1). These samples can be divided between 'complete' flux-limited surveys, and 'filtered' surveys, designed to select the highest redshift objects. The advantage of the flux-limited surveys is that, given sufficient spectroscopic redshift information, they can be used to derive the radio luminosity function (RLF; e.g., Dunlop & Peacock 1990, Willott et al. 2001). This RLF shows a strong increase between $z = 0$ and $z \sim 2$, and a possible decrease at larger redshifts is the subject of considerable debate (e.g. Jarvis et al. 2001b, Waddington et al. 2001). The main problem for establishing the high-z RLF is the small number of $z > 2$ radio galaxies and quasars in the flux-limited samples.

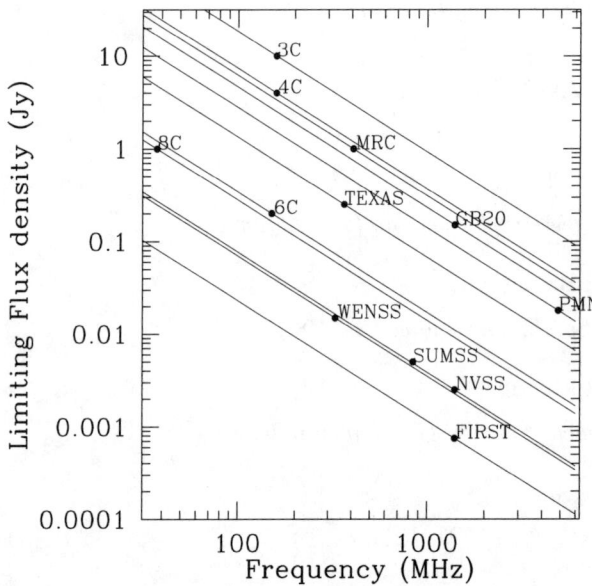

Figure 1. Limiting flux density of all major radio surveys. Lines are of constant spectral indices of −1.3. Note that WENSS and SUMSS are ideally matched to NVSS to construct samples of USS sources.

To find more high redshift radio galaxies (HzRGs; $z > 2$), additional selection criteria have been used to construct 'filtered' surveys. The most successful of these is the selection of sources with ultra-steep radio spectra (USS), although others such as angular size upper limits have also been used. While the main explanation for the success of the USS criterion is a simple k−correction of the generally concave radio spectrum, other effects could strengthen the $\alpha - z$ correlation, including (i) the steepening of the rest-frame radio spectral index with radio power, and (ii) more important inverse Compton losses at high redshift.

Table 1 shows that these filtered surveys reach fainter flux densities, and find objects with a mean redshift $z \sim 2$, while the flux-limited surveys target brighter sources at $z \sim 1$.

2. Ultra-Steep Spectrum Search for High Redshift Radio Galaxies

With the advent of several major radio surveys during the last decade (Table 2), it is now possible to construct large, uniform samples of radio sources. We have started a program to find significant numbers of $z > 3$ radio galaxies. Our 3 samples consist of 669 USS sources, covering the entire sky outside the Galactic plane (see De Breuck et al. 2000). To derive spectral indices, we combined the WENSS and TEXAS survey with the NVSS. Figure 1 and Table 2 show that WENSS is ideally matched in both resolution and sensitivity for such a USS sample. In the regions not covered by the WENSS survey ($\delta < +28°$), we have used the shallower TEXAS survey.

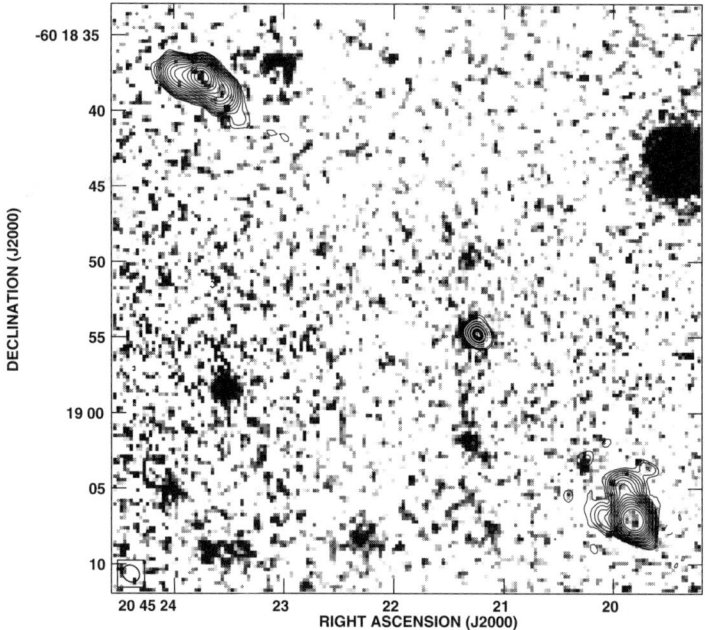

Figure 2. MP J2045−6018 ($z = 1.464$): *Greyscales:* CTIO K−band image; *Contours:* deep ATCA 8.64 GHz image. This is the most powerful radio source known at $z < 2$. Note the asymmetric structures in the radio lobes, and the weak radio core. The separation between the radio lobes is $32\rlap{.}''6$.

More recently, we have also constructed a deeper sample in the $-25° < \delta < -8°$ region using the southern extension of the WENSS, the Westerbork in the Southern Hemisphere (WISH, De Breuck et al., in prep.). At $-40° < \delta < -30°$, we shall combine the SUMSS with the NVSS to construct a first sensitive USS sample in the deep southern hemisphere.

At $\delta < -40°$, the extragalactic radio sky is even less explored, with several extremely luminous radio sources (comparable to northern 3C sources) remaining to be identified. Burgess & Hunstead (1994) describe the only observational effort to date to identify these intriguing objects. We have therefore also constructed the first USS sample at $\delta < -40°$, using the MRC and PMN surveys. From this sample, we have already discovered one of the most powerful radio sources known: MP J2045−6018 at $z = 1.464$ (Fig. 2).

To identify the radio sources in our USS samples, we first obtained high resolution radio images from the VLA and ATCA. After an initial campaign of optical identifications, we switched to K−band imaging using the Keck and CTIO telescopes (De Breuck et al. 2002). Even with moderately deep imaging ($K < 20$ to $K < 22$), we find an identification rate of >95% for the 86 sources observed in K−band. Subsequent optical spectroscopy of 46 sources with 3-10m class telescopes has yielded redshifts for 72% of these, with a mean redshift of

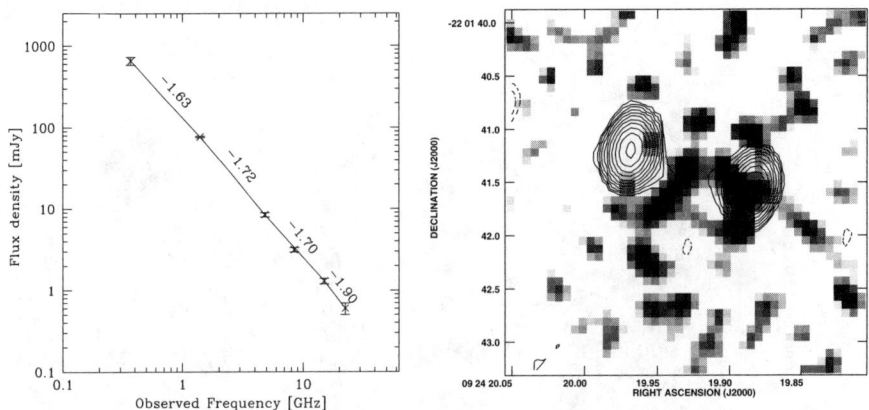

Figure 3. TN J0924−2201 ($z = 5.19$): *Left:* radio spectrum, based on VLA observations. Note the steeper spectral indices at higher frequencies. *Right:* VLA 8.6 GHz map overlaid on a Keck/NIRC K−band image. The relative astrometric uncertainty is $< 0\rlap{.}''2$.

$z \sim 2.5$ (De Breuck et al. 2001). From this sample, we have discovered the most distant radio galaxy known: TN J0924−2201 at $z = 5.19$ (van Breugel et al. 1999). Figure 3 shows the radio spectrum determined from multi-frequency VLA imaging. The spectrum curves at high frequencies, consistent with what is seen in most powerful radio galaxies. Our deep VLA images (Fig. 3) show a $1\rlap{.}''25$ double radio source, but no radio core.

Interestingly, despite integration times of 1h or more with the Keck telescope, 5 sources show only a faint continuum emission, but no identifiable emission lines between 4000 Å and 9000 Å. These sources are possibly obscured AGN, as suggested by the detection of several of these at (sub)mm wavelengths (Reuland et al., in preparation).

Acknowledgments. This work was supported by a Marie Curie Fellowship of the European Community programme 'Improving Human Research Potential and the Socio-Economic Knowledge Base' under contract number HPMF-CT-2000-00721, and by the Research and Training Network 'The Physics of the IGM' set up by the Human Potential Programme of the European Commission. The work by WvB was performed under the auspices of the U.S. Department of Energy, National Nuclear Security Administration by the University of California, Lawrence Livermore National Laboratory under contract No. W-7405-Eng-48.

References

Becker, R., White, R., & Helfand, D. 1995, ApJ, 450, 559
Bock, D., Large, M, & Sadler, E. 1999, AJ, 117, 1578
Burgess, A. & Hunstead, R. 1994, ASP Conference Series, Vol. 54, p.359
Condon, J., et al. 1998, AJ, 115, 1693

De Breuck, C., van Breugel, W., Röttgering, H., & Miley, G. 2000, A&ASup, 143, 303
De Breuck, C., et al. 2001, AJ, 121, 1241
De Breuck, C., et al. 2002, AJ, in press, astro-ph/0109540
Douglas, J., et al. 1996, AJ, 111, 1945
Dunlop, J. & Peacock, J. 1990, MNRAS, 247, 19
Fan, X., et al. 2001, AJ, in press, astro-ph/0108063
Griffith, M. & Wright, A. E. 1993, AJ, 105, 1666
Jarvis, M., et al. 2001a, MNRAS, 326, 1563
Jarvis, M., et al. 2001b, MNRAS, 327, 907
Lacy, M., et al. 1999, MNRAS, 308, 1096
McCarthy, P., et al. 1996, ApJSup, 107, 19
Rawlings, S., Eales, S., & Lacy, M. 2001, MNRAS, 322, 523
Rengelink, R., et al. 1997, A&A, 124, 259
Röttgering, H., et al. 1994, A&ASup, 108, 79
van Breugel, W., et al. 1999, ApJ, 518, L61
Waddington, I., et al. 2001, MNRAS, 328, 882
Willott, C., et al. 2001, MNRAS, 322, 536

Table 1: Recent Radio Redshift Surveys

Survey	Selection in S [mJy]	Selection in α	Median S_{1400} [mJy]	Mean S_{1400} [mJy]	% z_{spec}	Mean z	# sources	Reference
MRC	$S_{408} > 950$	All	463	641	67.0	0.74	558	1
6CE	$2000 < S_{151} < 3930$	All	391	427	94.9	1.14	59	2
MP	$S_{408} > 700; S_{4850} > 35$	$\alpha_{408}^{4850} < -1.20$	311	395	15.5	0.88	58	3
USS	varies[a]	$\alpha < -1.0$[a]	281	318	2.3	1.27	1165	4
7C-III	$S_{151} > 500$	All	184	211	81.5	1.19	54	5
6C*	$960 < S_{151} < 2000$	$\alpha_{151}^{4850} < -0.98$	131	127	100.0	1.96	29	6
TN	$S_{365} > 150; S_{1400} > 10$	$\alpha_{365}^{1400} < -1.30$	74	118	7.1	2.10	268	3
WN	$S_{325} > 18; S_{1400} > 10$	$\alpha_{325}^{1400} < -1.30$	18	41	4.4	1.87	343	3

[a] The USS sample of Röttgering et al. (1994) consists of 9 sub-samples, each with different flux density and spectral index limits. See Table 4 of Röttgering et al. (1994) for details.

REFERENCES: (1) McCarthy et al. (1996); (2) Rawlings et al. (2001); (3) De Breuck et al. (2000a); (4) Röttgering et al. (1994); (5) Lacy et al. (1999); (6) Jarvis et al. (2001a)

Table 2: Major Radio Surveys

	WENSS	TEXAS	MRC	SUMSS	NVSS	FIRST	PMN
Frequency (MHz)	325	365	408	843	1400	1400	4850
Sky region (J2000)	$\delta > +29°$	$-35°.5 < \delta < +71°.5$	$-85° < \delta < +18°.5$	$\delta < -30°$	$\delta > -40°$	$b > 35°$	$-87°.5 < \delta < +10°$
# of sources	229,576	67,551	12,141	in progress	1,814,748	771,000	50,814
Resolution	$54'' \times 54''\,\mathrm{cosec}\delta$	$10''$	$\sim 2''.7$	$43'' \times 43''\,\mathrm{cosec}\delta$	$45'' \times 45''$	$5'' \times 5''$	$4'.2$
Position uncertainty (strong sources)	$1''.5$	$0''.5 - 1''$	$3'' - 10''$	$1''$	$1''$	$0''.1$	$6''$
RMS noise	~ 4 mJy	20 mJy	700 mJy	1 mJy	0.5 mJy	0.15 mJy	~ 8 mJy
Flux density limit	18 mJy	150 mJy	670 mJy	5 mJy	2.5 mJy	1 mJy	20 mJy
Reference	1	2	3	4	5	6	7

REFERENCES: (1) Rengelink et al. (1997); (2) Douglas et al. (1996); (3) Large et al. (1981); (4) Bock et al. (1999); (5) Condon et al. (1998); (6) Becker et al. (1995); (7) Griffith & Wright (1993)

AGN Selection by Size of Dominant Emission

Pedro Augusto

Universidade da Madeira, Centro de Ciências Matemáticas, Caminho da Penteada, 9000-390 Funchal, Portugal

J. Ignacio Gonzalez-Serrano

Instituto de Física de Cantabria (CSIC-Universidad de Cantabria), Facultad de Ciencias, 39005 Santander, Spain

Abstract. The potential of AGN surveys extends to structure-related objectives: e.g. to study the size and properties of the NLR in AGN. From a complete sample of 1665 radio sources, we selected fifty-five with features on $\sim$0.2–2 kpc scales (core/jet flux ratio <7:1) for further study. Here, we summarize the radio interferometric selection technique used and speculate on the prospects for optical surveys (using a size criterion).

1. Introduction

Typically, brightness related criteria (only) are used to select sub-samples from AGN surveys. Why is a physical size selection criterion (almost) never adopted? Such selection is particularly useful when used in statistical/systematic studies of some type of AGN (or regions within) with a given size range; also, to determine the type of AGN associated with some physical phenomenon (dependent on size). A careful size selection has relevant scientific output (see Section 3).

The standard model of AGN, so far consistent with observations, besides two powerful opposing radio emitting jets, contains a (likely) spherically symmetric NLR, even though we normally only see the excited NLR cones (e.g. Mrk 3 in Cappetti et al. 1996). The radius of the NLR is in the range 0.01–1 kpc, although it has been well studied only in nearby Seyfert galaxies ($z \lesssim 0.2$), namely by using the HST. Correlations of the NLR with the radio jets have been identified (e.g. Augusto et al. 2001); simulations of jets going through physical interfaces such as the NLR/rest of the ISM show that it is possible to associate certain radio features with the physical size of the NLR. However, since only nearby Seyferts had NLR studies done, the sample of Section 3 is an example of how to approach the answer to questions such as: How is the NLR in the rest of the Universe? is it omnipresent in all AGN types? is its size always the same? are its physics the same? is there any evolution, perhaps correlated with the general AGN evolution? How does the standard model change when we know more about the NLR evolution?

2. The ~Kpc-Scale Phenomena in AGN

2.1. CSO/MSOs

Compact (medium) symmetric objects (CSO/MSOs) show flat or inverted spectra ($\alpha < 0.5$; $S_\nu \propto \nu^{-\alpha}$) and have symmetric components (lobes) on each side of a putative core. CSOs are smaller than 1 kpc, while MSOs have sizes of 1–20 kpc. There are many cases of ultra-small pc-scale CSOs (VLBI scale; e.g. Taylor et al. 1996) and a few candidates for large sub-kpc CSOs (Augusto et al. 1999). A few CSOs had kinematics studied (e.g. Owsianik, Conway, & Polatidis 1998) showing them to have typical ages of $\sim 10^3$ yrs. If they are the early stages of FRII radio galaxies, Augusto, Wilkinson & Browne (1998) have shown them to evolve self-similarly, i.e., their lobes expanding as they grow and most[1] becoming Compact Steep Spectrum Sources (CSSs) when reaching sub-kpc sizes.

2.2. NLR Probes

CSO/MSOs are excellent probes of the NLR/ISM interface, since they likely lie in the plane of the sky (unbeamed, symmetrical structure). Finding their redshifts and studying them better could mean understanding the NLR over the billions of years of evolution of AGN.

As regards other types of objects, e.g. core-plus-one-sided-jets (CJs), orientation issues are of more concern, although these can be solved with enough data (Fig. 1). Their potential as NLR probes, however, is improved when they present not only features (knots and strong blobs) on the right scales for a NLR/ISM interface (e.g. Fig. 1 in Augusto et al. 2001) but also heavily bent jets ($> 90°$): hint of strong shocks which should be confirmed from radio polarization and optical spectroscopy observations.

By probing the NLR with radio jets at different redshifts we can learn about: i) the size of the NLR; ii) the actual sub-structure of the NLR (cloud formations, etc.) — from radio shocks; iii) the (mean) density of the gas clouds; iv) bolometric emission characteristics; v) importance of magnetic fields — from radio polarization maps; vi) all of i)-v) in redshift space, tracing the evolution of the NLR in several parameters.

3. Sampling by Size: an Example on ~ 0.2–2 Kpc Scale

3.1. The Sample Selection

Starting from a parent sample of all radio sources in the northern sky with $S_{8.4\text{GHz}} > 100$ mJy, $\alpha_{1.4}^{4.85} < 0.5$ and $\left|b^{II}\right| > 10°$ (totaling 1665 objects), Augusto et al. (1998) have selected a sub-sample of 55 radio sources with structure on 90–300 milliarcseconds (mas) scale, which translates into projected linear structures (strong multiple radio features with <7:1 flux density ratio) of ~0.2–2 kpc at $z > 0.2$.

---

[1] As for MSOs, they could be the descendents of the few CSOs that reach maturity without getting through the CSS stage.

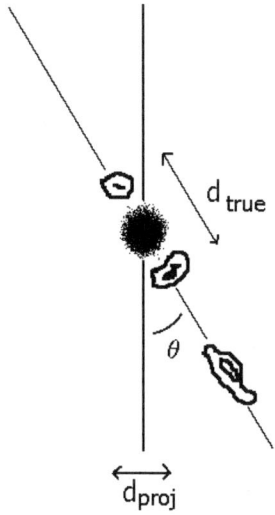

Figure 1. Knowing the orientation of a radio source is essencial for translating projected sizes of the NLR into actual sizes (dproj ≤ dtrue). For core-jets we can (e.g.): a) determine the Lorentz factor, if a counter-jet is detected, and beaming exists (likely); b) use limb brightening of the jets/blobs; c) use polarization effects.

As described in Augusto et al. (1998), the size selection was *not* made by looking at the radio maps of the 1665 radio sources but rather by looking at their radio *visibilities*. Any radio interferometer can actually 'see' below the formal beam (resolution). At 8.4 GHz, the VLA-A (which was used in this case) has a resolution of $\sim$ 200 mas and yet Augusto (1996) has shown that a compact 50 mas equal double can actually be detected via visibility, although it is impossible to see more than an unresolved source in the map. However, this detection depends on the observing conditions (namely elevation of source — Augusto et al. 1998) and it may not work all the time. This is why, with 100% confidence, only 90 mas equal doubles are detectable from VLA-A visibilities (Augusto et al. 1998; Fig. 2).

The techique used for detecting 50 mas doubles (or larger) from the visibilities was as shown in Fig. 3. In the case of a double radio compact source with components S_1 and S_2 where $S_1 \leq 7S_2$ we have (interferometry theory) $S_{peak} = S_1 + S_2 \leq 8S_2$ and $S_{min} = S_1 - S_2 \geq 6S_2$. Hence:

$$\frac{S_{peak} - S_{min}}{S_{peak}} = 1 - \frac{S_{peak}}{S_{min}} \geq 1 - \frac{6}{8} = 0.25$$

So, all we did was to look for sources that showed a *greater than* 25% decrease in correlated flux from the shorter to the longer baselines.

3.2. Observational Status

An extensive follow up has been conducted in the optical (imaging — Gonzalez-Serrano et al., in prep.) and radio (high resolution dual frequency maps — Augusto et al., in prep.) for most of the 55 objects. We also have 22 GHz data

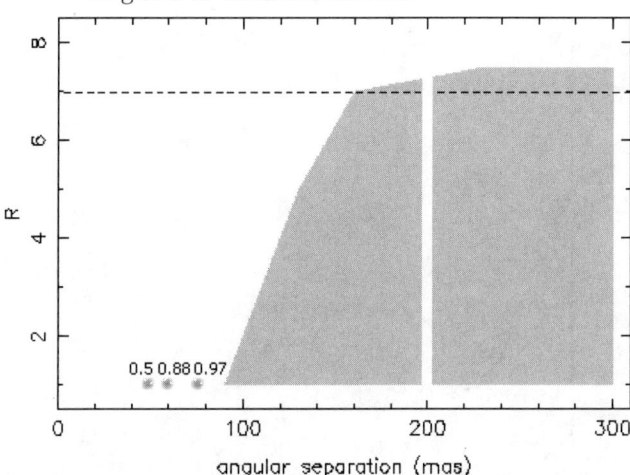

Figure 2. The size selection of compact doubles with flux density ratio R as function of separation. The grey area means 100% reliability, while the fraccional reliabilities (labels) refer to equal doubles. The VLA-A 8.4 GHz formal beam is marked (white line at 200 mas). Note that larger-than-the-beam compact doubles with $\lesssim$ 7.5:1 flux density ratio are also detected (adapted from Augusto et al. 1998).

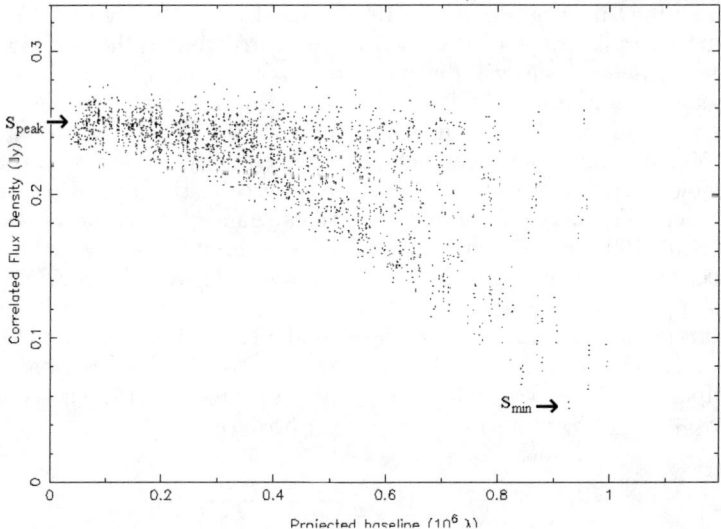

Figure 3. The technique for selecting radio sources by size on 90–300 mas (with 100% confidence — see Fig. 2). This was looking for sources with a $\geq$25% decrease in correlated flux from the shorter to the longer baselines: $(S_{peak} - S_{min})/S_{peak} \geq 0.25$. Note that S_{min} may not necessarily correspond to the longest baselines (adapted from Augusto 1996).

(VLA, MERLIN) on some objects. Reduction of this extensive amount of data is underway. Preliminary optical results indicate that, of the 55 objects, 3 are fainter than 23 mag. Their morphology is not clear, since most appear too small for ground-based telescope observations.

A spectroscopic programme, vital for redshift determination and shock confirmations, has yet to be started. Nevertheless, we can make a preliminary analysis of the (biased) literature information existing for 20 of the 55 sources. At the moment we get $<z>= 0.66\pm0.64$ suggesting a medium-to-high redshift sample of objects with large spread. In fact, the redshift range is $0.015 \leq z \leq 2.249$ with a Sy2 galaxy and a QSO at the closer and farther extremes, respectively. The most distant confirmed radio galaxy in the sample is at $z = 1.152$. It is then probable that the overall average redshift of the sample of 55 sources will be larger than 1. More important is the fact that a large spread exists and thus we might be able to probe the evolution of the NLR in AGN (and testing AGN models in general) in redshift space, as intended.

4. Size Selection

4.1. Its Potential

We have just shown relevant scientific output from a size selection. Although this paper (like the Colloquium) is on Extragalactic Astronomy (AGN), we should stress that any size selection from surveys can have the specific aim of Galactic objects and phenomena. In Section 3 we have virtually excluded 'contamination' by Galactic objects ($\left|b^{II}\right| > 10°$ was one of the criteria imposed to get the 1665-source sample from the original surveys). However, we might be interested, for example, in finding very young Supernova Remnants in our galaxy (flat spectrum and small). In Table 1 we show the range of astronomy exploration that can be made, even unintentionally, with a size selection.

Table 1. The Potential of Size Selection for Different Areas of Astronomy.

Linear projected size	Distance (example)	Type of study (example)
$\sim$0.2–2 kpc[a]	$z \sim 0.5$ (AGN)	NLR probes
$\sim$4–24 pc	~ 20 Mpc (Virgo Cluster)	Supernova Remnants (SNR)
$\sim$0.2–1.2 pc	~ 1 Mpc (Andromeda galaxy)	very young SNR
$\sim$40–240 AU	~ 1 kpc (Galaxy)	Stellar wind, planets, IPM/ISM

0.″5 – 0.″3 angular size selection

[a]Depends on cosmological model. All currently accepted models are included in this order-of-magnitude range.

4.2. At Different Wavelengths

Radio There are many advantages in selecting radio sources by size using 'raw' radio interferometric data (calibrated only), before any CLEANing is done (the

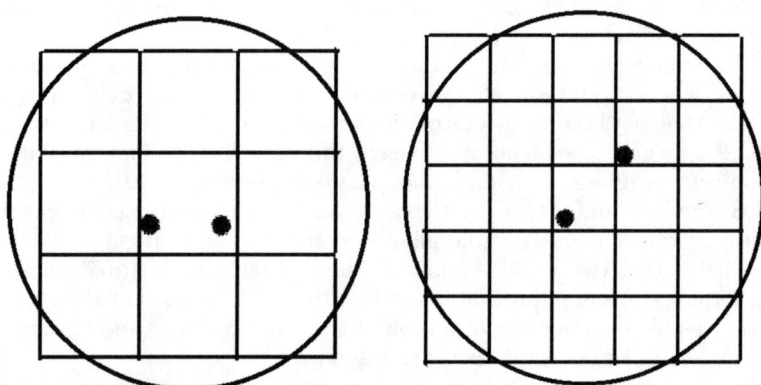

Figure 4. (**left**) Using pixels with a size of 60 mas, a 50 mas compact double may fit inside one pixel (the VLA-A 8.4 GHz formal beam is marked with a circle 200 mas in diameter). (**right**) This time we use 35 mas pixels making sure that a 50 mas compact double will *never* have both components in the same pixel.

most popular algorithm for producing radio maps). For example, although the VLA-A 8.4 GHz can 'see' down to a 50 mas size (separation) — Section 3.1, this structure is only found on the visibility data, either visually (e.g. Fig. 3) or by automatically applying models.

An example of how information still can be preserved has to do with the so-called superresolution: typically, precisely because a sensitive radio interferometer like the VLA can 'see' smaller than the formal beam, we can get a feeling of smaller-than-the-beam radio structure by convolving the final CLEANed data with a beam smaller than the natural beam (e.g. for VLA-A 8.4 GHz maybe a 70 mas beam). Although risking misinterpretation of radio features if taken too seriously, superresolution allows us to know how really unresolved a given radio source (component) is, without having to look at the radio visibility. CLEANing usually starts off with deciding on the pixel size for the algorithm. This, typically, is about one-third of the formal beam, i.e., $\sim$ 60 mas for the VLA-A 8.4 GHz. However, using this size of pixels will elude 'detection' with superresolution of, e.g., any $\sim$ 50 mas compact equal doubles, even though they are clearly seen in visibility (Augusto et al. 1998) — Fig. 4. On the other hand, using smaller pixels ($\leq$ 35 mas) may increase the computational time for reducing the data but will allow detection of $\sim$ 50 mas compact doubles using superresolution (Fig. 4), since each component now always ocupies different pixels. To prevent aliasing by inefficient sampling, the pixels cannot be too small, however.

Optical (philosophy) It is obvious that interferometry should work pretty much the same as radio interferometry and so, detection of structure smaller than the formal resolution will also be possible in sensitive instruments. As regards single-telescope use (with CCD), we must split into; 1) ground-based and 2) space-based:

1. although the pixels are, e.g., $0''\!.1$ on a side, the seeing limits the resolution to, say, $0''\!.6$; worse than that, there is a complex rapid-time-varying 'beam' resulting from the convolution of the PSF of a point source and the seeing effect. With the use of 'smart' adaptive optics, we may be able not only to compensate the seeing (as it is currently done successfully) but also to go below the formal resolution of the instrument (using smaller CCD pixels and a way to get extra-information on the source from the complex 'beam'); at the moment we certainly can use wide field, large pixel (e.g. $0''\!.6$) CCDs to go around the seeing problem and separate structure (compact doubles: binary stars, for example). For extended sources, we may play around with isophotes, modeling them and 'guessing' structure smaller than $0''\!.6$, even without the use of adaptive optics;

2. HST is already using the dithering technique which allows a slight improvement in resolution; when we are working with single pixels, however, the issue of cosmic-ray contamination becomes relevant and we must exercise care for detecting $\sim 0''\!.05$–$0''\!.1$ structure, distinguishing it from cosmic rays — either using several frames (as it is done currently in most observations) or by simulating the likelihood of neighboring events.

Acknowledgments. We acknowledge the grant from the Fundação para a Ciência e a Tecnologia (ESO Programme): PESO/P/PRO/15133/1999. The European Commission, under contract ERBFMGECT 950012, gave us partial support for this research.

References

Augusto, P. 1996, Ph.D. thesis, Univ. Manchester, UK
Augusto, P., Wilkinson, P. N., & Browne, I. W. A. 1998, MNRAS, 299, 1159
Augusto, P., Gonzalez-Serrano, J. I., Edge, A. C., Gizani, N. A. B., Wilkinson, P. N., & Perez-Fournon, I. 1999, New Astron. Rev., 43, 663
Augusto, P., Gonzalez-Serrano, J. I., Gizani, N. A. B., Perez-Fournon, I., & Edge, A. C. 2001, Astron. & Astroph. Trans., in press
Cappetti, A., Axon, D. J., Macchetto, F., Sparks, W. B., & Boksenberg, A. 1996, ApJ, 469, 554
Owsianik, I., Conway, J. E., & Polatidis, A.G. 1998, A&A, 336, L37
Taylor, G. B., Vermeulen, R. C., Readhead, A. C. S., Pearson, T. J., Henstock, D. R., & Wilkinson, P. N. 1996, in: Snellen, I., Schilizzi, R. T., Rottgering, H. J. A., & Bremer, M. N. (eds), 2nd Workshop on Gigahertz Peaked Spectrum and Compact Steep Spectrum Radio Sources, Univ. Leiden

The orientation of the Seyfert nucleus in Mrk 348

Sónia Antón
Observatório Astronómico de Lisboa, Tapada da Ajuda 1349-018, Portugal

Andrew Thean, Ian Browne, Alan Pedlar
Jodrell Bank Observatory, Cheshire SK11 9DL, U.K.

Abstract. We present new data on Mrk348: 5 GHz data with MERLIN and infrared data with ISO. The radio properties of Mrk 348 are unusual among Seyfert galaxies, and we discuss the orientation of the AGN axis with respect to the line of sight.

1. Introduction

Mrk 348 is a nearby object, hosted by a giant spiral galaxy (Antón 2000 and references therein). The optical total intensity spectrum shows narrow emission lines, while spectropolarimetric observations reveal a hidden broad line region (Tran 1995). Mrk 348 has a linear triple radio structure 0.2″ in size at 5 GHz, and the core is variable on scales of months (Neff & de Bruyn 1983). Two VLBA epochs show that the components are expanding at sub-relativistic speeds (Ulvestad et al. 1999). Recently, H_2O megamaser emission was detected (Falcke et al. 2000). In order to further discuss the properties of Mrk 348, we gathered new data to (1) analyse the spectral energy sistribution (SED) of the nuclear emission and (2) search for polarised radio emission. Here we present ISOPHOT photometry at 170, 90, 60, 25 μm, radio 5 GHz maps with MERLIN, and photometry at 1350 μm from the JCMT-SCUBA archive.

2. Results

The SED reveals that Mrk 348 has a flat spectrum up to the millimetre band ($S \sim \nu^{0.09}$) – see Figure 1 Left – suggesting that the emission up to that band is synchrotron in origin. A 5 GHz MERLIN polarisation map reveals that only the southern component is polarised – see Figure 1 Right – with P=5%. The polarisation seen in the southern component is unusual as most Seyferts show little polarisation.

3. Discussion

The peculiarity of Mrk 348 comes from the SED and the polarisation seen in the southern component: both are unusual among radio-quiet objects, but compa-

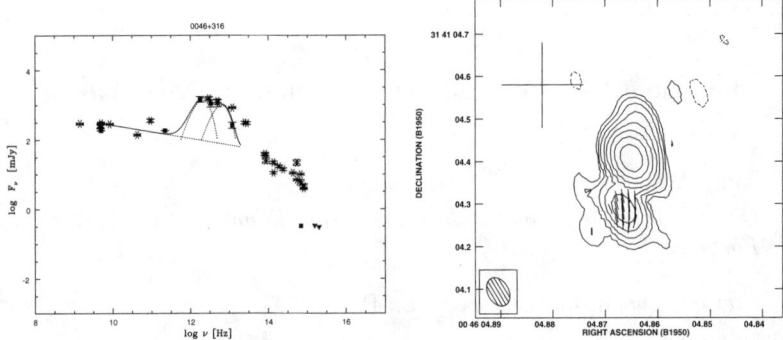

Figure 1. Left – SED. Flux densities from the literature are shown as star symbols. Our data are represented by square symbols. Inverted triangles represent upper limits. The solid line represents the sum of a power law spectrum with two greybody spectra of temperatures ~ 20 K and ~ 62 K (each component is represented by dashed lines). Right – Naturally–weighted 5 GHz map. The root-mean-square noise level (σ) is 0.158 mJy/beam. The contour levels are chosen as $6\sigma \times (-2,-1,1,2,4,8,16,32,64,128)$. Polarisation vectors with a scale of 1 mJy/beam to 0.015 arcsec are overlayed.

rable to properties of low-luminosity radio-loud objects. The flatness of the SED is very similar to that of objects in which the radio emission is related to the presence of a relativistic jet. A key problem is the orientation of the system with respect to the line of sight. The evidence points in different directions: some observations (spectropolarimetric observations, megamaser studies, sub-luminal motions of the radio components) are consistent with the axis of the system being at a large ($\geq 45°$) angle to the l.o.s, whereas other observations (flatness of the SED, core-dominated radio structure, radio variability, polarisation asymmetry; see Antón 2000) are consistent if the radio jet axis makes a small angle ($\leq 10°$) to the l.o.s. If the radio emission in Mrk348 is beamed, then Mrk 348 is intrinsically a radio-quiet object. But if the angle to the line of sight is large and the Doppler beaming is not a dominant factor, Mrk 348 is close to, but slightly above, the radio-loud/radio-quiet boundary. In this case, the peculiarity of Mrk 348 comes from the combination of a radio-loud object and a spiral host galaxy.

References

Antón, S. 2000, PhD thesis, University of Manchester.
Falcke, H., Henkel, C., Peck, A. B., Hagiwara, Y., Almudena Prieto, M. & Gallimore, J. F. 2000, A&A, 358, L17.
Neff, S. G. & de Bruyn, A. G. 1983, A&A, 128, 318.
Ulvestad, J. S., Wrobel, J. M., Roy, A. L., Wilson, A. S., Falcke, H. & Krichbaum, T. P. 1999, ApJL, 517, L81.
Tran, H.D. 1995, ApJ, 440, 578.

Optical Identification of Weak and Compact Radio Sources

V. S. Artyukh

PRAO, FIAN, 142290, Pushchino, Moscow Reg., Russia

M. A. Hovhannisyan, A. P. Mahtesyan, and V. H. Movsesyan

Byurakan Astrophysical Observatory, 378433, Byurakan, Aragatzotn Province, Armenia, e-mail: martin@bao.sci.am

A sample of 289 compact radio sources selected from the 7C Catalogue, covering an area of 0.097 steradian, was surveyed at 102 Mhz by the scintillation method. The observations show that the sizes of these sources are less than $0.1''$ and their flux densities do not exceed 2 Yn. These sources are identified with objects from the FIRST catalogue of radio sources. 99 of these objects have scintillations and have no neighbor in the surrounding $5'$. 34 of those 99 are identified with objects on the POSS within $10'' \times 10''$ areas. 17 of the last 34 are closer to identified POSS objects (within $2'' \times 2''$ areas). So we suppose that these 17 radio sources are clearly identified with optical sources. Most of them are probably bright quasars. The other 17 will not be discussed here, as we didn't succeed in their optical identification. There are no POSS objects brighter than 21^m in the close surroundings of the remaining 65 radio sources. We have started optical observations of areas close to those 65 objects with the 2.6-m telescope of the Byurakan observatory. Up to now 15 areas were observed. In 5 cases optical objects were found fainter than $22^m - 24^m$ within $3''$ of the compact radio sources (one example is shown in Fig. 1). We are sure that there are no optical objects brihgter than 25^m that may be identified with the remaining 10 objects. Those five identifications give us an opportunity to suppose that they are remote quasars. More powerful telescopes than the 2.6-m are needed to perform photometric and spectral observations of similar objects in order to prove our suggestion.

References

Artyukh V.S., Tyulbashev S.A. 1996, Russian Astron J., 73, 661
Artyukh V.S., Tyulbashev S.A. 1998, Russian Astron J., 75, 655
Artyukh V.S., Tyulbashev S.A., Isaev E.A. 1998, Russian Astron J., 75, 323
McGilchhrist, M. M., Baldwin, J. E., & Riley J. M., 1990, MNRAS, 246, 110

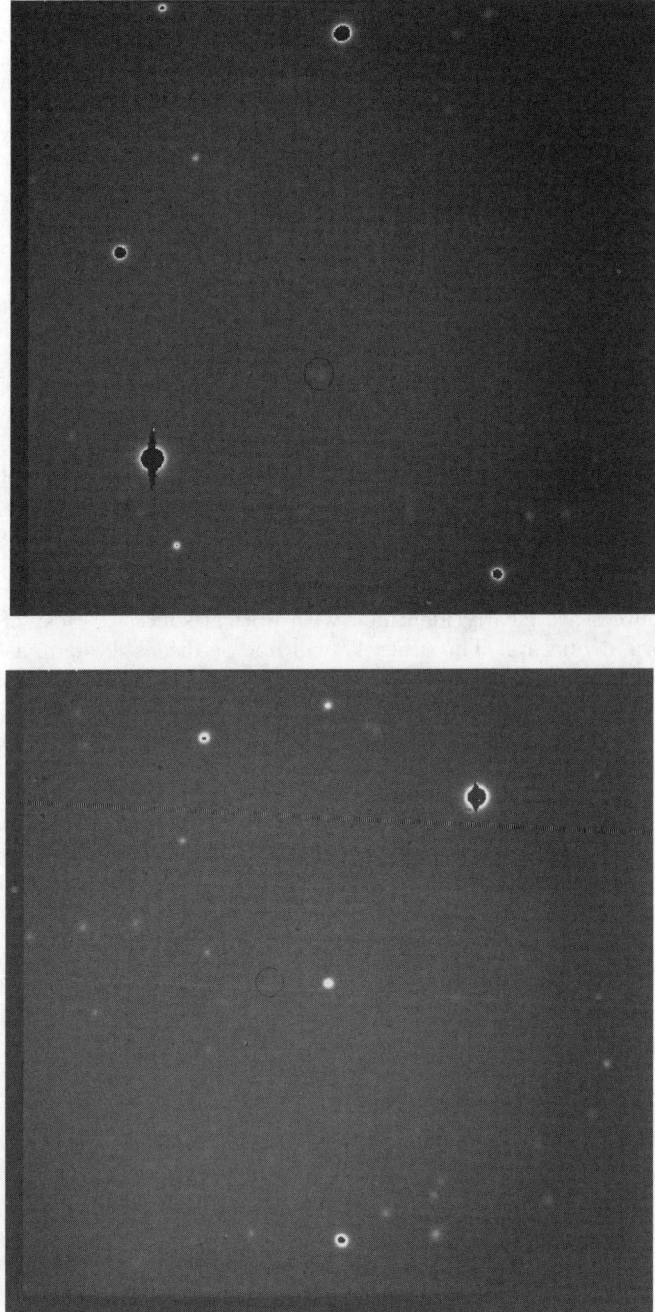

Figure 1. There are optical objects brihgter than 25^m within $3''$ of the radio sources (top) and there are no such objects in the lower field.

The FIRST-APM QSO Survey (FAQS) in the SBS Region. Current Status

V. Chavushyan, R. Mújica, J.R. Valdés, L. Carrasco

INAOE, A.P. 51 y 216. C.P. 72000. Puebla, Pue., México.

J. Stepanian

IA-UNAM, A.P. 70-264, México D.F. 04510, México.

O. Verkhodanov

SAO RAS. Karachai-Cherkesia, 357147. Russia.

Abstract. As the first step of the Multiwavelength AGN Survey (MWAS), we have started the FIRST-APM QSO Survey (FAQS). The main goal of FAQS is to compile the most complete sample of bright QSOs, located in the area of the sky covered by the Second Byurakan Survey (SBS). Here we report the current status of an ongoing study based on the cross-identification of the FIRST radio catalog and the APM optical catalog. The overlapping sky area between FIRST and SBS is about 700 deg^2. The compiled list of sources for this overlapping region contains $\sim$ 400 quasar candidates brighter than $B = 18^{m}_{.}5$. About 90 objects were already spectroscopically classified. During 1999-2000, we observed spectroscopically more than 150 FAQS objects with the 2.1m telescope of the Guillermo Haro Astropysical Observatory (GHAO).We have found 51 new QSOs (4 BAL QSOs), 13 Seyfert Galaxies (5 NLSy1's), 23 emission line galaxies, 3 BL Lac objects and 57 stars.

1. Introduction

Every survey technique has redshift and luminosity-dependent selection biases (Wampler & Ponz 1985). Taking into account the selection effects inherent to every technique adopted to search for QSOs, it is obvious that only a combination of different search techniques in different spectral ranges will yield a complete sample of quasars with an adequate representation of all the properties inherent to them (Hartwick & Shade 1990).In order to circumvent the problem of bias, we have started a multi-wavelength search for QSOs in the well investigated Second Byurakan Survey sky area (Stepanian et al. 1999, 2001).

As the first step of the Multiwavelength AGN Survey (see Stepanian & Chavushyan, contribution to this colloquium), we cross-identified the FIRST (Becker et al. 1995) radio and APM optical catalogs. The FIRST radio survey provides a new resource for constructing a large quasar sample (White et al. 2000). However, the drawback of radio selection is that most of the radio-quiet quasars will not be included in the sample. Therefore, the sample will not be a good representation of the quasar population. An example is that only 25% of SBS QSOs were detected in the FIRST survey.

2. Sample, Observations, and Preliminary Results

We named this sample FAQS – FIRST-APM Quasar Survey (Chavushyan et al. 2001). The overlapping area between the FIRST and SBS surveys is quite significant, covering about 700 sq. degrees, within the limits defined by $07^h40^m < \alpha < 17^h15^m$, and $+48°52' < \delta < +57°36'$. The cross-identification between the FIRST and APM catalogues was done adopting the following selection criteria: a) position coincidence between FIRST and APM objects within $3''$ radius; b) Stellar-like classification on APM; c) B-magnitude between 14.5 and 18.5 on APM. The resulting list of sources, in the overlapping region, contains 412 objects, 90 of which were previously known AGNs, mainly discovered by the SBS.

The spectroscopy was carried out with the 2.1m telescope of the Guillermo Haro Astropysical Observatory (INAOE-Mexico) and the LFOSC focal reducer (Zickgraf et al. 1997). A set-up covering the spectral range 4200-9000 A, with a dispersion of 8.2 Å/pixel was adopted. The effective instrumental spectral resolution was about 16 Å. During 1999-2001, we have carried out spectroscopic observations for ~ 150 FAQS objects. The BV photometry for the objects classified as AGN ($B < 17.5$) was carried out with the 1.5m telescope of San Pedro Martir Observatory (UNAM-Mexico), and reduction is in process.

In the subsample studied until now, we have found 51 new QSOs, 13 Seyfert Galaxies (5 NLSy1's), 23 emission line galaxies (ELG), 3 BL Lac objects and 57 high galactic latitude stars. Amongst the 51 QSOs, we have found 4 broad absorption line (BAL) QSOs.

We expect to complete spectroscopy and photometry of the total sample by the end of 2001.

Acknowledgments. This work was supported by CONACyT research grants 28499-E, J32178-E and, G28586-E.

References

Becker, R.H., White, R.L., & Helfand, D.J., 1995, ApJ, 450, 559
Chavushyan, V.H. Mujica, R., Carrasco, L., Valdes, J.R., Verkhodanov, O., & Stepanian, J., 2001, ASP Conf. Ser., 232, 102.
Hartwick, F.D.A., & Shade, D., 1990, ARAA, 28, 437
Stepanian, J.A., Chavushyan, V.H., Carrasco, L., Tovmassian, H.M, & Erastova, L.K., 1999, PASP, 111, 1099
Stepanian, J.A., Green, R.F., Foltz, C.B., Chaffee, F., Chavushyan, V.H., Lipovetsky, V.A., & Erastova, L.K., 2001, AJ (in press).
Wampler, E.J. & Ponz, D., 1985, ApJ, 298, 448
White, R.L., Becker, R.H., Gregg, M.D, et al., 2000, ApJS, 126, 133
Zickgraf, F.J., Thiering, I., Krautter, J. et al., 1997, A&AS, 123, 103

Energy Density and Radiation Losses in Giant Radio Galaxies

Marek Jamrozy & Jerzy Machalski

Astronomical Observatory of the Jagiellonian University, ul. Orla 171, 30-244 Krakow, Poland

Abstract. The volumes, equipartition energy density, and radiation losses due to the synchrotron and inverse Compton interactions as a function of redshift and projected linear size of Giant radio galaxies are discussed. The new results are based on data from three samples: 1) Ishwara-Chandra & Saikia (1999), 2) Schoenmakers (Ph.D. Thesis, 1999), and 3) Machalski, Jamrozy & Zola (2001).

1. Introduction

Giant radio sources with linear sizes greater than 1 Mpc form an extreme class of extragalactic radio sources. The majority of them are radio galaxies at $z < 0.3 \sim 0.4$. They are of special interest for studying a number of astrophysical problems. One of them is under what circumstances some sources evolve to "giant" sizes. These Giants must be extremely old and/or located in a very underdense environment, and/or have a stronger central source of energy as compared to smaller radio sources. It is very important to compare various physical parameters of normal-size and giant sources. Here we present a contribution to the following problems: 1) how does the energy density in the lobes of FRII-type radio sources evolve with redshift, and how does it relate to the evolution of energy and pressure of the intergalactic medium (IGM), and 2) how do radiative losses differentiate the lobes of giant sources from normal ones.

2. Results

Correlation between the Energy Density and Redshift

Using the Spearman partial rank correlation coefficients between the equipartition energy density in lobes u_{me} and redshift and/or power $P_{1.4}$ for 49 Giants with $D > 1$ Mpc, we deduced that the apparent correlation $u_{me} - z$ arises from two others: $u_{me} - P$ and $P - z$. We also checked the correlation between u_{me} and size D. The relevant partial correlations for 32 Giants with $10^{24.2} \leq P_{1.4}[\mathrm{W\,Hz^{-1}sr^{-1}}] < 10^{25.0}$ show that there is a significant correlation $u_{me} - D$ which implies the same between the energy density and volume. Therefore, there is no evidence that the undetected dependence between the energy density and redshift does not exist; it can be much weaker than the detected $u_{me} - P$ and $u_{me} - D$ correlations. The apparent distribution of the energy density vs. redshift is shown in Fig. 1, left panel.

Correlation between the Pressure Ratio and the Lobe Size

A non-relativistic, diffuse and uniform IGM in thermal equilibrium has an electron pressure, p_{IGM}, which should increase with redshift as $p_{\mathrm{IGM}}(z) = p_o(1+z)^5$

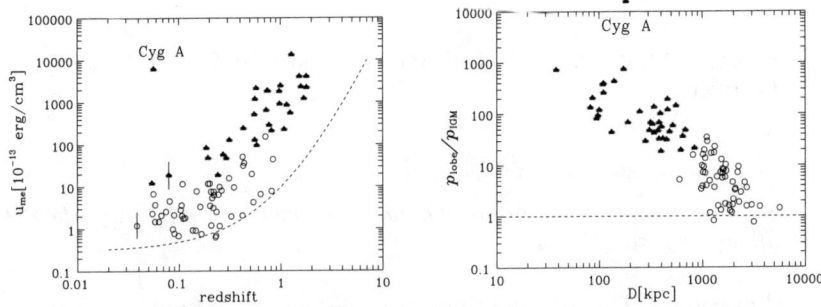

Figure 1. **left panel:** Apparent distribution of $u_{me} - z$ in the sample of giants (open circles) and 3C sources (filled triangles) (cf. Alexander & Leahy 1987, Leahy et al. 1989, and Liu et al. 1992). Vertical bars mark the typical error of u_{me}. The dashed line indicates the minimum energy density of the intergalactic medium (cf. Subrahmanyan & Saripalli 1993). **right panel:** Distribution of p_{lobe}/p_{IGM} vs. D. Symbols as in the left panel.

(Subrahmanyan & Saripalli 1993). On the other hand, pressure in the lobes must be higher and can be determined from the equipartition energy density as $p_{lobe} = u_{me}/3$. We found that the correlation between p_{lobe}/p_{IGM} and D is very strong and significant. The newly discovered Giants confirm that the pressure in the lobes of these sources is close to that of the intergalactic (even intercluster) medium (see Fig. 1, right panel.

Correlation between the Rate of Radiation Losses and Size

Following Ishwara-Chandra & Saikia, the dependence of the radiation losses due to synchrotron and inverse Compton interactions on the linear size of Giant and 3C sources has been analysed. The new data confirm that (i) there is a distinct upper limit for these losses which can be described by $B_{me}^2 + B_{IGM}^2 \propto D^{1.4}$ and (ii) the synchrotron losses diminish with increasing size of sources and tapers to a minimum value for Giants. This minimum is determined by the equivalent energy density of the microwave background radiation.

Acknowledgments. This contribution is supported by the State Committee for Scientific Research (KBN) under the contract 2PO3D 015 17 and 2PO3D 008 20.

References

Alexander P., Leahy J.P., 1987, MNRAS 225, 1
Ishwara-Chandra C.H., Saikia D.J., 1999, MNRAS 309, 100
Leahy J.P., Muxlow T.W.B., Stephens P.W., 1989, MNRAS, 239, 401
Liu R., Pooley G., Riley J., 1992, MNRAS 257, 545
Machalski J., Jamrozy M., Zola S., 2001, A&A, 371, 445
Schoenmakers A.P., 1999, Ph.D. Thesis, Univ. of Utrecht
Subrahmanyan R., Saripalli L., 1993, MNRAS 260, 908

AGN SURVEYS
ASP Conference Series, Vol. 284, 2002
R.F. Green, E.Ye. Khachikian, D.B. Sanders

Compact Jets in 100 AGNs with the Strongest Broad-Band Variability of 1–22 GHz Spectra in 1997–2001

Yu. A. Kovalev[1], Y. Y. Kovalev[1], N. A. Nizhelsky[2], A. V. Bogdantsov[2]

[1] *Astro Space Center of the Lebedev Physical Institute RAS, Profsoyuznaya 84/32, 117997 Moscow, Russia*
[2] *Special Astrophysical Observatory RAS, Nizhnij Arkhyz, 369167 Russia*

Abstract. Results of monitoring observations at the radio telescope RATAN–600 and a model interpretation of instantaneous 1–22 GHz spectra at six frequencies for 100 selected AGNs are presented. The index of variability at these frequencies is shown for 550 sources monitored in 1997–2001 at 11 epochs. The spectra of the selected sources exhibit flux density variations of about 50% and up. The type of spectral evolution is similar for all the selected objects, favoring the same basic physical model. Model analysis shows that the nature of the radio sources and the observed variability behavior of the spectra can be explained by a model with a relativistic jet of parsec scale in a longitudinal magnetic field.

1. Observations and Results of Analysis

Since 1997, we have monitored instantaneous 6-frequency 1–22 GHz spectra of 550 compact extragalactic radio sources with milliarcsecond components. Monitoring sets take place several times per year at the RATAN–600 ring radio telescope. The sources were selected from the Preston et al. (1985) VLBI survey. Every 1–22 GHz spectrum was measured at six wavelengths of 1.4, 2.7, 3.9, 7.7, 13, and 31 cm over a period of a few minutes (see details in Kovalev et al. 1999).

From the distributions of 550 sources over the $V_\nu = \frac{S_{\nu,\max} - S_{\nu,\min}}{S_{\nu,\max} + S_{\nu,\min}}$ variability indexes at 22–2.3 GHz over the four-year period (1997, March, — 2001, May; 11 epochs; Fig. 1) it follows that: (i) the distributions have one peak at indexes of about 0.17 at 22 GHz and 0.10–0.05 at 11–2.3 GHz; (ii) the mean index of variability for the full sample decreases as 0.24, 0.19, 0.19, 0.15, 0.15 with decreasing frequency; (iii) more than 20% of the sources exhibited very strong variability at the higher frequencies in this period: with 1.5–15 times greater changes in the flux (V_ν is more than 0.25).

The 100 sources with the strongest variability ($V_\nu > 0.25$, see Fig. 1) ever observed have been selected from the 550 objects monitored. The behavior of the spectra is similar for the 100 AGNs selected: a wave of a perturbation moves along the spectrum from higher to lower frequencies. As a result, a spectrum can be observed as a variable one in the entire 1–22 GHz band or only in the higher frequency part of the band (see examples in Fig. 1).

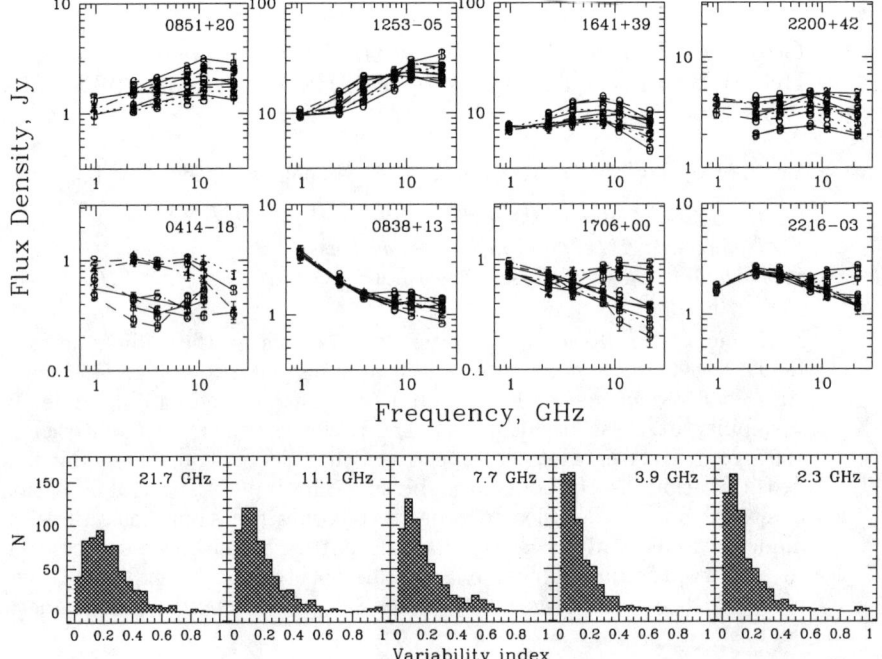

Figure 1. On the top: typical examples of multi-epoch instantaneous spectra for 8 of 100 AGNs with the strongest variability in 1997–2001. Epochs: 1997, March, June, September, December; 1998, March; 1999, April, September; 2000, April, August; 2001, March, May (11 epochs, labeled 0–9, a). On the bottom: histogram of indexes of the long-term variability at 22–2.3 GHz for 550 VLBI compact extragalactic radio sources monitored during the same 11 epochs.

Such behavior of the variable spectra and the variability index–frequency statistical dependence can be explained by a variability of the flow of the emitting particles at the beginning of a jet in a relativistic jet model with a quasi-radial magnetic field (for details of the model interpretation, see Kovalev et al. 2000).

Acknowledgments. This work has been partly supported by the Russian State Program "Astronomy" (project 1.2.5.1), the NASA JURRISS Program (project W-19611) and the Russian Foundation for Basic Research (projects 99-02-17799, 01-02-16812, 01-02-06084).

References

Kovalev, Y.Y., Nizhelsky, N.A., Kovalev, Yu.A., et al. 1999, A&AS, 139, 545
Kovalev, Yu.A., Kovalev, Y.Y., and Nizhelsky, N.A. 2000, PASJ, 52, 1027
Preston, R.A., Morabito, D.D., Williams, J.G., et al. 1985, AJ, 90, 1599

Survey and Analysis of 1–22 GHz Spectra for the Full Sample of 660 AGNs North of Declination −30°

Y. Y. Kovalev[1], N. A. Nizhelsky[2], Yu. A. Kovalev[1], G. V. Zhekanis[2], A. V. Bogdantsov[2]

[1] *Astro Space Center of the Lebedev Physical Institute RAS, Profsoyuznaya 84/32, 117997 Moscow, Russia*
[2] *Special Astrophysical Observatory RAS, Nizhnij Arkhyz, 369167 Russia*

Abstract. Measurements of broad-band 6-frequency 1–22 GHz spectra of 660 compact extragalactic radio sources were performed in 1997 (for declinations $-30° < \delta < +43°$) and 1998 (north of +49°) at RATAN–600. "Average rest-frame" statistical spectral shapes for different subsamples of sources are analyzed. These shapes and each spectrum observed can be represented as the sum of a spectrum of an extended optically thin component (magnetized envelope/lobe), constant or slowly variable and dominating at lower frequencies, and a spectrum of a compact component (relativistic jet), dominating at higher frequencies, with any type of variability. We have revealed specific radio features of the EGRET subsample; this favors the models suggesting a relation between the emission mechanisms in radio and gamma-ray ranges. Sources are sampled for which the most compact VLBI structure is expected.

1. The Sample, Observations and Discussion of Results

About 700 sources north of declination −30° were selected from the Preston et al. (1985) VLBI survey with correlated flux density more than 0.1 Jy at 13 cm. Observations reported here were done in December, 1997, at the northern sector of RATAN–600 for the declination range from −30° to +43° (Kovalev et al. 1999), and in July/September, 1998, at the southern sector for sources north of declination +49°, at six wavelengths of 1.4, 2.7, 3.9, 7.7, 13, and 31 cm. Observations of the sources, located in the declination gap from +43° to +50°, are planned to the end of 2001 at an upgraded sector of the telescope.

The analysis of one-epoch spectra allows one to identify the broad-band emission of a compact continuous jet in the longitudinal magnetic field for almost all the objects of the sample. In addition, a second component at lower frequencies with a steep spectrum is visible in many sources. The second component can be explained by the emission of an extended magnetized envelope/lobe which accumulates the relativistic particles from the jet in peripheral regions of an object. The distributions of spectral parameters obtained testify in favor of the common physical nature of quasars and BL Lacs. The peak histogram value of the turnover frequency is about 10–20 GHz in the rest frame for the total sample investigated.

In Figure 1 the average rest-frame statistical spectral shapes for several subsamples of objects are presented. These spectra are constructed by averaging the spectral indices. The frequency range for each two-point spectral index was corrected by a factor of $(1+z)$ before calculations, where z is the redshift. The most flat cm spectral shape is observed for γ-ray-bright and highly polarized objects, supporting the suggestion that the majority of them are the most active amongst all of the compact extragalactic sources. The high frequency portion of the spectra is the most flat for γ-ray bright objects. This fact gives an advantage to models implying a physical relation between γ-ray and radio emission (by Compton or synchrotron mechanisms) and indicates the peculiarity of the EGRET sample in comparison with the parent sample of flat-spectrum radio sources. The mean spectral shapes obtained are interpreted by the combined emission from an envelope/lobe and a continuous jet with quasi-stationary ejection, each of them contributing differently to the total emission.

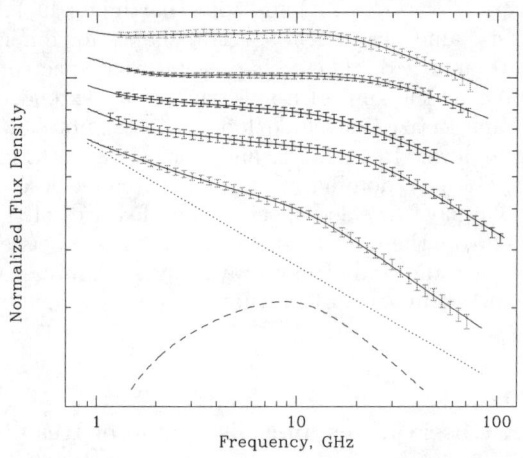

Figure 1. The average rest-frame statistical spectral shapes for several subsamples of sources with error bars. Spectra are vertically shifted for the sake of convenience. The total sample of about 660 objects is involved. From top to bottom are spectra for highly polarized objects, EGRET detected objects, BL Lacertae objects, quasars, and galaxies. Fitted total model spectra are shown by solid curves. For galaxies model spectra are also plotted for the jet (dashes) and extended component (short dashes), the sum of which gives the resulting model spectrum.

Acknowledgments. This work has been partly supported by the NASA JURRISS program (project W–19611), the RFBR (project 01–02–16812), and the Russian State Program "Astronomy" (project 1.2.5.1).

References

Kovalev, Y. Y., Nizhelsky, N. A., Kovalev, Yu. A., et al. 1999, A&AS, 139, 545

Preston, R. A., Morabito, D. D., Williams, J. G., et al. 1985, AJ, 90, 1599

Different Types of Radio Sources and Possible Evolution of Radio Galaxies

Gabriel A. Ohanian

Byurakan Astrophysical Observatory, 378433, Byurakan, Aragatzotn province, Armenia

Extragalactic radio sources have been studied for many years, but it is still unclear how they are formed and evolve. The sizes of the most powerful radio emitters in the Universe vary from less than one parsec to more than 1 Mpc. This large range of sizes has been interpreted as evidence for the evolution of the linear sizes of radio structure (e.g., O'Dea and Baum, 1997). A crucial element in the study of their evolution is the identification of the young compact counterparts of "old" FRI/FRII extended objects. Good candidates for young radio sources are those with peaked spectra (Gigahertz Peaked Spectrum - GPS and Compact Steep Spectrum - CSS, e.g., O'Dea 1998). Radio sources are presumably born in the very compact GPS phase, then they expand beyond 1 kpc into the CSS regime and finally, they reach a size of 20 kpc, and afterwards evolve into large-scale radio sources (young scenario, e.g., O'Dea 1998). Alternatively, GPS sources may be compact because a particularly dense environment prevents them from growing larger (old scenario, e.g., O'Dea 1998). In either scenario, the radio source host galaxy determines the time evolution of the radio structure. By studying the optical environments and host galaxies we hope to obtain clues to the evolution of the radio sources. Similarities or differences in host galaxy properties over a range of radio source types and sizes enable us to investigate possible differences or similarities of the radio size class as a whole.

It is hence of crucial importance to enlarge a multispectral sample of different classes of radio sources. Proceeding from this position we have formed GPS, CSS and VSS (Very Steep Spectrum) radio source samples selected from the Ooty lunar occultation survey at 327 MHz, which have flux limits about 30 times fainter than that of the 3C (Singal 1987, and references therein). About 300 radio sources from the Ooty lists have been observed by us at 102 MHz on the Large Phased Array (Pushchino, Russia) and on the RATAN-600 radiotelescope (Zelenchukskaya, Russia) at 968, 2300, 3950 and 7700 MHz. Details of these observations, the spectra and the list of radio sources are presented by Ohanian, 1998. Of these observed sources, 232 have spectra determined by 4 or more flux density points, at different frequencies. From those 232 classified spectra we select the above mentioned samples of radio sources.

In this poster we present a comparison of the 3 samples properties and a preliminary qualitative discussion. We have compared the properties of the VSS, CSS and GPS classes of radio sources in Table 1.

If the steepness of the optically-thin part of the radio spectrum in CSS and GPS sources is caused by radiative and adiabatic losses, then the break in the spectrum will move to lower frequencies as time increases (Kardashev 1962). The steepness of the spectra in GPS sources may be evidence for inverse Compton losses, which are significant in the early epoch of evolution of radio

Table 1. Some parameters of different samples of radio sources

Samples (number of r.s.)	Mean spec. index.	Relative number of gal.	Relative number of EF r.s.	Relative number of QSO
GPS(26)	1.25	20%	48%	32%
CSS(47)	1.06	32%	51%	17%
VSS(48)	1.11	20%	67%	13%

sources (Kardashev 1962). As we see from Table 1 the mean spectral indices of the VSS and CSS sources are similar. This corresponds to Kardashev's assumption, that in time the steepness of the spectrum stays constant or rises. That is, energy losses are offset by relativistic electron injection. The injection of relativistic electrons takes place by continuous and/or by recurrent bursts. From the point of view of V. A. Ambartsumian (1968), bursts which are accompanied by high energy release take place in galaxies where the core luminosity is not high (in this case radio luminosity, i.e. radio-quiet objects), and more quiet forms of activity are associated with bright cores (in this case - radio loud objects). Further, in the last decade several studies have shown that variability is anti-correlated with luminosity, in the sense that luminous QSOs have smaller amplitude variations than low luminosity ones (e.g. Hook et al., 1994). For three elliptical galaxies - NGC 1052, NGC 5077 (core is GPS type), and IC 1459 - that have been observed for at least 20 yr., it is noticed that the radio emission varies monotonically on time-scales of decades, though bursts in NGC 1052 and 5077 clearly occur on much shorter timescales (Slee et al., 1994). However, in radio-quiet objects monotonic variations are not typical. It is possible that in the radio-quiet objects, re-supply of relativistic electrons takes place only by separate bursts, i.e., the continuum injection component is negligible. Hence, in these objects we do not observe large-scale radio structure. In radio-loud objects, the presence of large-scale radio structures implies that a continuous resupply of electrons must take place. Superposed on them could be a burst component. That is, if a radio-loud object exists in a dense interstellar medium (like Seyfert type galaxies), then the medium cannot impede the outward propagation of the radio source, due to the continuous resupply of sufficient energy to the radio source structure.

If it's true that the GPS and CSS sources are young versions of the large radio sources, it means that large radio sources in host galaxies would pass through the GPS and CSS phase. The maximum lifetime of radio sources is $10^{7.5}$ years (Alexander and Leahy, 1987). In that time the host galaxies of large scale radio sources must live through a GPS-CSS-VSS evolutionary sequence. As we see from Table 1, the relative number of QSOs continually decreases when passing through the GPS-CSS-VSS sequence, and consequently, at that time the relative number of galaxies plus EFs (EFs are extremely faint - m>21 galaxies) are increasing at the same rate. It means that QSOs are changed to the galaxies. Thus, if it's true that the radio source evolutionary sequence is a transition from the GPS phase to the CSS phase and then to the VSS phase, then Table 1 is the possible evolutionary scheme of radio galaxies.

References

Alexander P. & Leahy J. P. 1987, MNRAS, 225, 1

Ambartsumian, V. A. 1968, IAU Symp. 29, 11

de Vries W. H., O'Dea C.P., Baum S. A., Perlman E., Lehnert M. D. & Barthel P. D. 1998, STSI prep. No 1226, 1

Hook J. M., McMahon R. G., Boyle B. J. & Irwin M. J. 1994 MNRAS, 268, 305

O'Dea, C. P. and Baum, S. A. 1997, AJ, 113, 148

O'Dea, C. P. 1998, STSI prep. No 1216

Ohanian G. A. 1997, PhD Thesis, Byurakan

Kardashev, N. S. 1962, Soviet Ast., 6, 317

Singal, A. K. 1987, A&AS, 69, 91

Slee, O.B., Sadler Elaine M., Reynolds J. E. & Ekers R. D. 1994, MNRAS, 269, 928

5-GHz VLBI Imaging Observations of 7 Equatorial AGNs

Z.-Q. Shen

ISAS, Yoshinodai 3-1-1, Sagamihara, Kanagawa 229-8510, Japan

D. R. Jiang, Y. J. Chen, T.-S. Wan

Shanghai Observatory, Shanghai 200030, P. R. China

Abstract. Since 1992 we have been conducting a 5-GHz VLBI imaging survey of southern and equatorial radio sources. So far, we have published the results of two observing sessions with 26 southern radio sources imaged in total (Shen et al. 1997; 1998). In this paper, we present the preliminary results of the third session of observations of 7 equatorial sources in the sample.

1. Introduction

We have been conducting a 5-GHz VLBI imaging survey of southern and equatorial radio sources to improve the understanding of the collective properties of compact radio sources in the southern hemisphere. The major goals are: 1) to fill the gap in southern hemisphere VLBI imaging observations, 2) to provide ground monitoring of the potential targets for space VLBI (e.g. VSOP) and 3) to search for southern superluminal candidates.

We selected the sources from the one-(intercontinental)-baseline surveys at 2.3 and 8.4 GHz (Morabito et al. 1986) with the following criteria: 1) declination (B1950.0) $-45°<\delta<+10°$, 2) correlated flux densities $S^c_{2.3GHz}>0.6$ Jy and $S^c_{8.4GHz}>0.6$ Jy, and 3) total flux density $S^t_{5GHz}>1.0$ Jy. There are 36 sources that met these criteria (Shen et al. 1997).

In November 1992 and May 1993, we performed two observing sessions using the Southern Hemisphere VLBI Experiment (SHEVE) network consisting of antennas in Australia, South Africa and the Shanghai station of China, which provided better (u,v) coverage for the southern sources. As a result, milliarcsecond (mas) structures were obtained for 26 sources, of which 11 had no previously published VLBI images (Shen et al. 1997; 1998). This has added appreciably to the number of sources showing the bending of jet-like structure between arcsecond (kpc) and mas (pc) scales. The misalignment seems to be a common feature in AGNs.

Of the 10 sources left in our 36-source sample, only seven sources (0736+017, 1055+018, 1354−152, 1548+056, 2121+053, 2131−021 and 2318+049) were selected for the 3rd observing session because two of the others (0727−115 and 1749+096) have been imaged with the VLBA and one (3C 279) has been extensively studied with VLBI. Since, except for 1354−152, the other six sources are

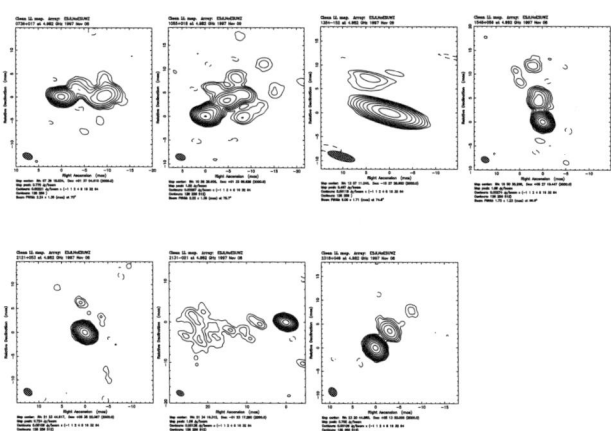

Figure 1. 5-GHz VLBI images of 7 quasars observed in November 1997

located around the equatorial plane, the European VLBI Network (EVN) plus Hartebeesthoek in South Africa is the most suitable for the observations.

2. Observations and Results

The 5.0 GHz EVN observations were carried out on 1997 November 8-9. Sources were observed in a snapshot mode, i.e. 5-12 thirteen-minute scans for each source. The left-hand circular polarized radio signals were recorded in Mark III mode C with a total bandwidth of 14 MHz at each station. The data correlation was done on the MPIfR MK III Correlator in Bonn, Germany with an output averaging time of 4 s. Standard post-correlation data reduction, including fringe fitting, imaging and model fitting, was made using the NRAO AIPS and Caltech Difmap analysis packages. We present the preliminary imaging in Figure 1.

Of the seven equatorial quasars imaged, 5 (0736+017, 1055+018, 1548+056, 2131−021 and 2318+049) showed a core-jet structure while the remaining 2 (1354−152 and 2121+053) were dominated by single compact core emission.

The tentative detection of component superluminal motions in 1055+018 (Attridge, Roberts, & Wardle 1999) were confirmed by our new data. By comparing our observations with the 5 GHz VSOP prelaunch VLBA survey (Fomalont et al. 2000), superluminal motion was inferred for the first time for three other sources: 0736+017 (5.1-10.9c), 1548+056 (3.3c) and 2318+049 (7.5c) (assuming $H_0 = 65$ km s^{-1} Mpc^{-1} and $q_0 = 0.5$).

References

Attridge, J. M., Roberts, D. H., & Wardle, J. F. C. 1999, ApJ, 518, L87
Fomalont, E. B. et al. 2000, ApJS, 131, 95
Morabito, D. D. et al. 1986, AJ, 91, 1038
Shen, Z.-Q. et al. 1997, AJ, 114, 1999
Shen, Z.-Q. et al. 1998, AJ, 115, 1357

Decametric AGNs: FIRST and NVSS Maps and Radio Spectra

O. V. Verkhodanov, N. V. Verkhodanova

Special Astrophysical Observatory, Nizhnij Arkhyz, Russia, 369167

H. Andernach

Depto. de Astronomía, Apdo. Postal 144, Univ. Guanajuato, Mexico

Abstract. Radio sources from the decametric UTR–2 catalog were cross-identified with other radio catalogs at higher frequencies. We used the CATS database to extract all sources within the UTR beam size ($\sim 40'$) to find candidate radio identifications. Using the least squares method, we fitted the spectrum of each source with one of a set of curves. We extracted NVSS and FIRST radio images for the radio-identified sources, and looked for a possible relation between size and spectral index.

A radio survey obtained with the UTR telescope (Kharkov, Ukraine) at frequencies 10–25 MHz has resulted in a catalog of 1822 sources (Braude et al. 1978–1994; www.ira.kharkov.ua/UTR2). Covering about 30% of the sky north of $-13°$ declination, this survey is presently the lowest-frequency source catalog of its size, and thus provides an ideal basis to study the little known optical identification content of sources selected at decametric frequencies. The rather large uncertainties of UTR positions ($\sim 0.7°$) require an iterative process for finding radio counterparts at successively higher frequencies (and thus higher positional accuracy). In this we were aided by selecting previously cataloged sources from the CATS database (Verkhodanov et al. 1997) in a box of RA×DEC = $40' \times 40'$ centred on the nominal UTR position. The "raw" spectra given by these fluxes were refined using computer charts of source locations around UTR positions. All counterparts from TXS, GB6 and PMN within circles of $1'$ radius were considered one source. Groups of sources lying further apart were assigned separate spectra, each with the UTR flux as an upper limit.

We were able to fit spectra for all but 7 of the 2314 radio counterparts to UTR sources. Fits were either straight (S), convex (C^-), or concave (C^+) curves in the $\lg \nu$–$\lg S$ plot. The distribution of radio source spectra among the various spectral types is given in Table 1. The resulting catalog (Verkhodanov et al. 2000) includes information from a large number of electronically available catalogs of radio, infrared, optical and X–ray sources.

The majority of UTR sources (97%) have an identification with NVSS objects (Condon et al., 1998) (a total of 2253 IDs), and all UTR objects with $\delta > 30°$ (1143 objects) have IDs in FIRST (White et al., 1997). 552 sources were resolved into components either in the FIRST ($5''$ beam) or NVSS ($15''$ beam) catalogs within a circle of $60''$.

Table 1. Distribution of radio continuum spectral types of 2307 radio counterparts to UTR sources, where $X=\log_{10}(\text{frequency/MHz})$, and $Y=\log_{10}(\text{flux density in Jy})$

Spectral class	Fitting function	N	%
Straight (S)	$Y = +A + B*X$	894	39
Convex (C^+)	$Y = +A \pm B*X - C*X^2$	184	8
Concave (C^-)	$Y = +A - B*X + C*X^2$	1150	50
or	$Y = \pm A \pm B*X + C*EXP(-X)$	79	3

Table 2. Flux and size distribution of UTR counterparts in FIRST and NVSS: median value for various ranges of spectral index. N is the number of objects.

Sp. index range	N	median FIRST flux (mJy)	size ('')	N	median NVSS flux (mJy)	size ('')
$-0.9 \div -1.0$	88	514	26	183	574	24
$-1.0 \div -1.1$	93	363	19	168	365	22
< -1.1	75	193	12	185	212	17

We extracted NVSS and VLA maps of sources with steep and straight power law spectra, and grouped them according to their spectral slopes: (1) $-0.9 > \alpha > -1.0$, (2) $-1.0 > \alpha > -1.1$ and (3) $\alpha < -1.1$ ($S \propto \nu^\alpha$). NVSS objects were taken outside of the Galactic plane ($|b| > 15°$), while all the FIRST sources have $|b| > 20°$. 536 NVSS objects and 256 FIRST objects were selected. Parameters of source samples are provided in Table 2. Surprisingly, the median source size decreases with steepening radio spectrum. However, the dispersion in size is so large that there is no significant correlation between spectral index and source size.

References

Braude S.Ya. et al. 1978,1979,1981,1985,1994, Ap&SS, **54**,37; **64**,73; **74**,409; **111**,1; **213**,1

Condon J.J., Cotton W.D., Greisen E.W., Yin Q.F., et al. 1998, AJ 115, 1693

Verkhodanov O.V., Trushkin S.A., Andernach H., & Chernenkov V.N. 1997, in "Astronomical Data Analysis Software and Systems VI", eds. G.Hunt & H.E.Payne, ASP Conf. Ser. **125**, 322 (astro-ph/9610262)

Verkhodanov O.V., Andernach H., Verkhodanova N.V. 2000, Bull. Spec. Astroph. Obs. **49**, 53 (astro-ph/0008431)

White R.L., Becker R.H., Helfand D.J., & Gregg M.D. 1997, ApJ 475, 479

AGN SURVEYS
ASP Conference Series, Vol. 284, 2002
R.F. Green, E.Ye. Khachikian, D.B. Sanders

IRAS F02044+0957: A Radio Source in an Interacting System of Galaxies

O. V. Verkhodanov

Special Astrophysical Observatory, Nizhnij Arkhyz, Russia

V. H. Chavushyan, R. Mújica and J. R. Valdés

Instituto Nacional de Astrofísica Óptica y Electrónica, Puebla, México

S. A. Trushkin

Special Astrophysical Observatory RAS, Nizhnij Arkhyz, Russia

Abstract. The steep spectrum of IRAS F02044+0957 was obtained with the RATAN-600 radio telescope at four frequencies. Optical spectroscopy of the system components was carried out with the 2.1m telescope of the Guillermo Haro Observatory. The observational data allow us to conclude that this object is a pair of interacting galaxies, a LINER and an H II galaxy, at $z = 0.093$.

Trushkin & Verkhodanov (1995) compiled a list of about 750 objects by using the CATS, as a result of cross-identifications of infrared IRAS catalogues and the source catalogue of the Texas survey at 365 MHz. From the sample of steep spectrum sources we selected those without classification in public databases (CATS, NED, ADS and LEDA). One of these objects is IRAS F02044+0957, identified with the NVSS radio source J020706+101147.

Radio observations were carried out on April 23–25, 1999, with the North Sector of the RATAN–600 telescope. The wide-band radiometer complex was used at four frequencies: 2.3, 3.9, 7.7 and 11.2 GHz. The nearby non-variable radio source PKS 1345+12 was used as the flux density calibration source. A linear fitting to the radio spectrum was used to estimate a spectral index by the least squares method. Each point of the spectrum was weighted proportionally to the value of $1/(\Delta S/S_\nu)^2$; where $\Delta S/S_\nu$ is the relative error of the flux density. The radio spectrum is described by a power-law $S_\nu[Jy] = 85\,\nu^{-0.94\pm0.02}$ MHz. The radio luminosity in the frequency interval from 365 to 10^4 MHz is $L = 9.381 \times 10^{34}$ erg sec^{-1} cm^{-2} (assuming H_0=64 km Mpc^{-1} sec^{-1} and q_0=0.8).

Optical spectroscopy of the objects marked by A, B, C, D, and E in Fig. 1 was obtained in August and November 1999. We used the 2.1m telescope of the Guillermo Haro Observatory (GHO) in Cananea, Sonora, Mexico, operated by the National Institute of Astrophysics, Optics and Electronics (INAOE). The Faint Object Spectrograph and Camera (LFOSC) (Zickgraf et al. 1997) was used. A setup, covering the spectral range 4200 – 9000Å with a dispersion of 8.2 Å/pix was adopted. The effective instrumental spectral resolution is about 15 Å.

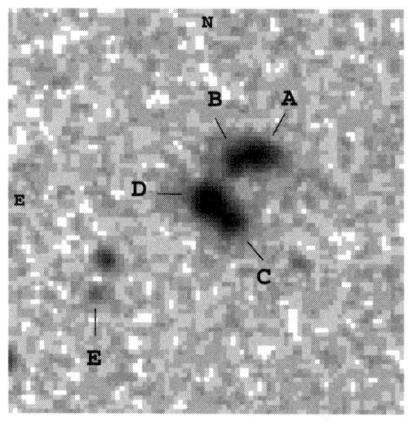 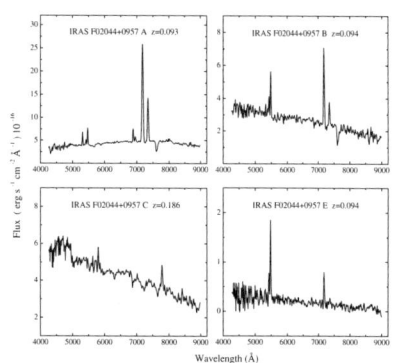

Figure 1. DSS image of IRAS F02044+0957 and corresponding spectra

The data reduction was done using the IRAF packages and included bias and flat field corrections, cosmic rays cleaning, wavelength linearization and flux transformation.

The detailed spectroscopy showed that objects A and B are apparently a pair of interacting emission-line galaxies at z=0.093. The second pair (objects C and D) is a geometrical projection of a star (D) and an emission line galaxy (C) at z=0.186. It is not possible that the G type Main Sequence star, is the source of infrared and radio emission.

IRAS F02044+0957 is located at 16′ from the center of the galaxy cluster ZwCL 0203.6+1008 (Zwicky 619). The radius of the cluster is 12′.6. This means that IRAS F02044+0957 must be out the cluster boundaries.

In order to investigate the nature of the interacting system AB, we used the diagnostic diagrams $\text{Log}([\text{N\,II}]\lambda 6583/\text{H}\alpha)$ vs $\text{Log}([\text{O\,III}]\ \lambda 5007/\text{H}\beta)$ and $\text{Log}([\text{S\,II}]\lambda\lambda 6717+6731/\text{H}\alpha)$ vs $\text{Log}([\text{O\,III}]\lambda 5007/\text{H}\beta)$.

The values for component A, $\text{Log}([\text{N\,II}]/\text{H}\alpha) = -0.15$ and $\text{Log}([\text{S\,II}]/\text{H}\alpha) = -0.23$, put the object on both diagrams in the AGN region, very close to the boundary with the H II region-like galaxies. According to the criteria for spectral classification proposed by Ho, Filippenko & Sargent (1997) (HFS97), the component A is a LINER because the value of $[\text{O\,I}]\lambda 6300/\text{H}\alpha = 0.26$. For component B, we obtained the values: $\text{Log}([\text{N\,II}]/\text{H}\alpha) = -0.76$ and $\text{Log}([\text{S\,II}]/\text{H}\alpha) = -0.55$. Therefore, according to HFS97, the component B is a H II galaxy. Summarizing, the interacting system AB is composed of a LINER and an H II galaxy.

Acknowledgments. This work was partially supported by CONACyT grants 28499-E, J32178-E, and 32106-E.

References

Heckman, T. M. 1980, A&A, 87, 152
Ho, L. C., Filippenko, A. V., Sargent, W. L. W. 1997, ApJS, 112, 315
Sanders, D. B., Soifer, B. T., Elias, J. H., et al., 1988, ApJ, 325, 74
Trushkin, S. A., Verkhodanov, O. V. 1995, *Bulletin of SAO*, 39, 150.
Zickgraf, F. J., Thiering, I., Krautter, J., et al., 1997, A&AS, 123, 103.

Photometric Study of Radio Galaxies in the RATAN–600 "Cold" Survey

O. V. Verkhodanov, Yu. N. Parijskij, N. S. Soboleva, A. I. Kopylov, A. V. Temirova, O. P. Zhelenkova

Special Astrophysical Observatory, Nizhnij Arkhyz, Russia

W. M. Goss

National Radio Astronomy Observatory, Socorro, NM, USA

Abstract. About 100 steep-spectrum radio sources from the RATAN–600 RC catalog were mapped by the VLA and identified with optical objects down to 24^m–25^m in the R band using the 6-m telescope. An updated list of calibrators with known redshifts of the same class of RGs was compiled to evaluate the accuracy of photometric redshift estimates. BVRI photometry for 60 RC objects was performed with the 6-m telescope, and by standard model fitting we have estimated color redshifts and ages of the stellar populations of the host gE galaxies. The mean redshift of FR II RGs from the RC list turned out to be ≈1. Several objects were found in which active star formation began in the first billion years after the Big Bang.

A catalogue of 1145 radio sources has been obtained in the RATAN–600 "Cold" (RC) survey carried out in 1980–1988 with the radio telescope RATAN–600 in a strip of sky, 24^h long and $0°.6$ wide ($\delta \approx +5°$), at 6 frequencies from 1 to 22 GHz with a sensitivity of about 3 mJy at 3.9 GHz.

To select candidates for distant radio galaxies (RGs) in the "Big Trio" project (RATAN–VLA–SAO 6-m) we used the most distant population of radio sources in the 10–50 mJy range, where the $log(N)-log(S)$ curve has a maximum slope, and selected about 100 steep spectrum (SS) radio sources ($\alpha \geq 0.9$).

In the following stages we did the selection of the FR II type RGs with VLA maps and optical identification and photometry from the 6-m telescope. 65 objects look like FR II and about 20 of them belong presumably to the most distant generation of RGs. 19 objects were classified as quasars by their stellar appearance on CCD images. Apart from the normal FR II RG class, some objects with SS happened to be of quite a different nature. 16 were not resolved even with the VLA in "A" configuration at 3.7 cm, being less than $0''.2$ –$1''.0$ (CSS class). 16 "subgalactic" doubles, with sizes less than $4''$, are of separate interest. A few of them are very complex and cannot be considered as young progenitors of FR II objects.

The technique of multicolor photometry has became in the past few years the main method in selecting candidates for distant galaxies, and the only approach at very high redshifts. Determination of the age of high redshift stellar systems may be the only way of estimating the formation redshift of the first galaxies if star formation begins at redshifts larger than the z of secondary ionization. Direct observation of the protogalaxies predicted by some recent computer simulations is not possible.

The method of multicolor photometry was checked on a sample of 45 radio galaxies published in the literature and carefully selected as steep spectrum high z objects of the FR II type (Verkhodanov et al., 1999). It was shown that $BVRI$ colors are sufficient for accurate estimates of z and age in the redshift range 0.5–3.5.

We have implemented this approach for 60 FR II and CSS radio galaxies of the RC catalogue that had been observed in B, V, R_c, I_c bands in 1994–1998 with the CCD camera on the 6-m telescope. (Our sample of SS radio galaxies is now the largest one with four-band optical photometry.) The data were used to estimate color photometric redshifts and ages of host galaxies by comparison with the PEGASE (Fioc and Rocca-Volmerange, 1997) and GISSEL'98 (Bruzual and Charlot, 1993) models of evolution of the spectral energy distribution (SED). For a few typical cases, the reality of the color z determinations was confirmed spectroscopically (Dodonov et al., 1999).

- It was shown that photometric redshifts give tolerable (10–20% err) agreement with spectral ones for powerful RGs.
- A redshift–magnitude diagram shows much larger dispersion at $m_R \geq 22^m$.

- The distribution in z of our objects has a maximum at $z \sim 1$, i.e., near the maximum activity stage in the Universe, and $\sim$10 objects at $z>2.5$.
- A galaxy age is model-dependent and detected uncertainly (within the limits of 0.5–10 Gyr). One can set a lower limit to the galaxy age and, hence, z of its formation. This age is always larger than the standard estimate of radio source lifetime.

The mean multicolor ($BVRI$) age of the stellar population of the RC radio source host galaxies is of order 1 Gyr, and at least in some objects, active star formation had begun in the first Gyr after the Big Bang. Such distant objects must be of high density contrast, and modern cosmology has to explain this very early appearance of dense and massive (Tera–solar masses) protogalaxies, with quickly formed massive black holes inside them to produce FR II structures. From our 100 square degree area sample we can estimate that more than 10000 very early objects, born before the QSO epoch, are available on the sky and accessible for present-day optical and radio facilities. They can help us to penetrate into the "Dark Age" of the Universe, between the recombination epoch and the epoch of appearance of QSOs.

Future activity connected with the "Big Trio" project will be concentrated on direct spectroscopy of the most probable candidates for the first galaxy generation selected from our RC list.

References

Bruzual G. & Charlot S. 1993, ApJ, **405**, 538
Dodonov S.N., Parijskij Yu.N., Goss W.M., Kopylov A.I., Soboleva N.S., Temirova A.V., Verkhodanov O.V. & Zhelenkova O.P., 1999, AZh, **76**, 323.
Fioc M. & Rocca-Volmerange B, 1997, A&A, **326**, 950
Verkhodanov O.V., Kopylov A.I., Parijskij Yu.N., Soboleva N.S. & Temirova A.V. 1999.

System to Estimate Ages and Redshifts for Radio Galaxies

O. V. Verkhodanov, A. I. Kopylov, N. V. Verkhodanova, O. P. Zhelenkova, V. N. Chernenkov, Yu. N. Parijskij, N. S. Soboleva, A. V. Temirova

Special Astrophysical Observatory, Nizhnij Arkhyz, Russia

Abstract. The system allowing a user to operate at the server **sed.sao.ru** with simulated curves of spectral energy distributions (SED) and to estimate ages and redshifts from photometric data is described.

The existence of a huge volume of observational data in the optical and infrared wavelength range increases significantly our knowledge about the distant Universe. However, information about the space distribution of extragalactic objects is not yet accessible because of the limits of observational possibilities: the direct measurement of redshifts is possible with spectroscopic methods having a sensitivity 2 magnitudes lower than photometric ones. Special interest lies in the study of distant objects allowing astronomers to investigate both large-scale structures andthe evolution of active galactic nuclei (AGNs) (which are connected with black holes). Spectroscopy of such objects is rather difficult. However, using photometric data one can essentially simplify this problem since it allows an astronomer to make the initial selection.

To accelerate a procedure of age (and photometric redshift) estimation we have begun a project "Evolution of radio galaxies", which has to allow a user to obtain age and photometric redshift estimates. The main tasks of the system are: 1) estimate of ages with fixed redshift z; 2) estimate of both ages and z; 3) archiving of optical observations of RC radio galaxies (in FITS, JPEG, PS formats with text comments); 5) archiving of the main publications by the current topic; 6) development of HTPP and e-mail access; 7) local SED operation to simulate an observational process.

To estimate ages and redshifts with photometric data we operate with simulated curves of spectral energy distributions (SED) for different types of galaxies of two models PEGASE (Fioc, Rocca-Volmerange, 1997, 1999), GISSEL96 (Bruzual, Charlot, 1993; Bolzonella et al., 2000).

Before the estimate of values we smooth a SED with a filter transmission curve to simulate observational data using a "compressing" filter with the growth of redshift: $S_{ik} = \frac{\sum_{j=0}^{n} s_{i-n/2+j} f_{jk}(z)}{\sum_{j=0} f_{jk}(z)}$. Here s_i is the initial synthetic SED, S_{ik} is the smoothed SED in the k-th filter, $f_k(z)$ is the transmission in the k-th filter, "compressed" by $(1+z)$ times "moving" along the SED, $j = 1, n$ is the pixel index in the curve of filter transmission.

The estimate of ages and redshifts is performed by selection of the optimum location on the SED curves of the measured photometric points obtained when observing radio galaxies in different filters. We use the already computed tabulated SED curves for different ages. Using discrepancies we construct a probability function in the form of $p = \frac{1}{max} exp(\chi^2)$, where max is the maximum value of the calculated function. χ is the discrepancy calculated by sliding the photometric points along the SED curve:

$$\chi^2 = \sum_{k=1}^{Nfilters} \left(\frac{F_{obs,k} - pSED_k(z)}{\sigma_k} \right)$$

Here $F_{obs,k}$ is the observational magnitude in the k-th filter, $\text{SED}_k(z)$ is the simulated magnitude for the given SED in the k-th filter at the given redshift z, p is the free coefficient, σ_k is the error of the observed magnitude.

In order to take account of the absorption, we apply the maps (as FITS-files from the paper "Maps of Dust IR Emission for Use in Estimation of Reddening and CMBR Foregrounds" (Schlegel et al., 1998).

The system is situated on the special Web-server **http://sed.sao.ru** operating in the Linux Red Hat (6.2) system, unifying various type resources and accessed by FTP, HTTP and e-mail. Typical e-mail form for a request looks like this:

```
seds start
object 3C65; model PEGASE, type=E
z_limits: 0 6, age_limits: 200 16000
B=23.73+0.21 V=23.57+0.2 R=22.36 I=20.81
extinction off
seds end
```

Here `seds start` and `seds end` are opening and closing keywods of the form. Keywords `object`, `model` and `type` determine an object name, a type of a model and a type of a galaxy, respectively. Available galaxy types are E, S0, Sa, Sb, Sc, Sd. `z_limits` and `age_limits` determine the limits of search for a redshift and an age (in Myr). The observed magnitudes age given with B, V, R, I, J, H, K, g, r, i, etc. keywods corresponding to the filter names. The error of the magnitude detection is given via plus '+' after a value of the magnitude. Extinction in this example is not calculated.

Other supported possibilities are a) sorted bibliographical collection of papers for different stages of radio galaxy evolution; b) archive of radio galaxy data in various wavelength ranges (both observed in the Special Astrophysical Observatory and taken from the Internet), containing information on the objects and figures in FITS, JPEG and PostScript formats.

The project was supported by the Russian Foundation for Basic Research (Grant No 99-07-90334).

References

Bolzonella M., Miralles J.-M., Pelló R. 2000. Astron. Astroph., **363**, 476

Bruzual G., Charlot S. 1993, ApJ, **405**, 538

Fioc M., Rocca-Volmerange B., 1997, Astron. Astroph., **326**, 950, 1999: astro-ph/9912179

Schlegel, D., Finkbeiner, D., Davis, M., 1998, ApJ, **500**, 525

Part 5
AGN Phenomena

Sailing on Lake Sevan:

Joe Mazzarella, Herve Aussel, Carlos De Breuck, Wolfgang Duschl, Yoshi Taniguchi, Yuko Kakazu, John Hucthings, Luis Ho

The Conference Banquet: Vahe Petrosian, Richard Green

Surveys of High-Redshift QSO Hosts

J.B. Hutchings

Herzberg Institute of Astrophysics, 5071 West Saanich Rd, Victoria, B.C., Canada

Abstract.
Recent investigations of high-redshift QSO morphology are reviewed and compared. The PSF-removal is difficult and results are given in several different forms. The reliability of results from HST and ground-based AO are compared, and several caveats to conventional wisdom are given.

1. Introduction

The detection and measurement of QSO host galaxies at redshifts in the range about 1 to 2.5 have been accomplished by a number of investigations in the past decade. I have elsewhere given a review of most of these (see Hutchings 2001 and references therein), and attempted to intercompare the results. My own experience with this kind of work and the very different ways in which the data processing has been undertaken by the various investigations show that a) we are far from having a statistical 'survey' of these objects, and b) we should take a careful look at some of the general claims made about the results.

Similar investigations of radio galaxies - which are claimed to be the same as radio-loud QSOs by many unified scenarios for active galaxies - are also restricted to small samples and diverse observational datasets, as well as significant selection effects by their radio properties. Thus, again, we need to be careful in attempting statistical comparisons of the different groups' samples in 'testing' purely orientation-dependent models.

From my review of the results (Hutchings 2001), I claim that we can believe the following broad properties of higher redshift QSO hosts.

- Host galaxies are resolved in many objects to redshifts 3 and higher. This is because of high resolution data and also the fact that the hosts are very luminous galaxies (see Fig 1).

- While luminous, many host galaxies are compact and blue compared with local universe galaxies. Many are also asymmetric. They appear to have young stellar populations, and are very like the non-AGN Lyman-break galaxies.

- The radio-loud hosts are brighter and have richer companion galaxy environments, and are like radio galaxies with similar radio properties.

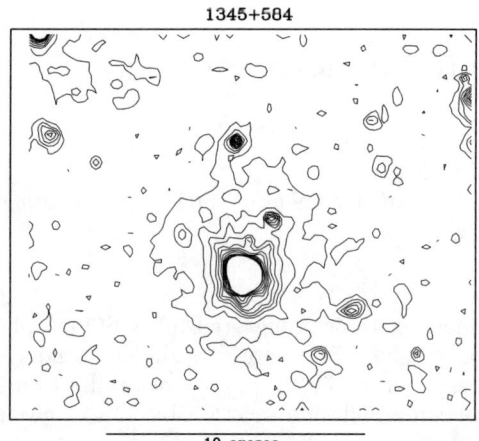

Figure 1. HST CCD (rest wavelength UV) image of z=2 QSO (Hutchings 1998) showing luminous host galaxy

- The QSOs are associated with compact groups of compact galaxies (see Fig 2). Such environments are not found in the local universe.

- Merging and interactions appear to be common in sufficiently deep images.

- Emission-line gas is associated spatially with radio structure. Emission line velocities are large in the few cases investigated.

2. Space Imaging or Ground-Based AO?

The study of high-redshift QSO hosts requires a combination of signal level and spatial resolution, and knowledge of the point spread function. This has led to use of HST and ground-based adaptive optics cameras as the approaches of choice. Since each has strengths and problems, I have attempted to make a comparison of the various critical aspects of the two datasets, as I see them. The table gives this in a cryptic form, and also shows my own scoring system for judging the effects. Overall, I think that we have learned different and complementary things from space and ground-based observing, and neither one is clearly superior. As NGST and large telescope AO systems are put into service, the comparison will need to be remade, but I doubt that a clear 'winner' will emerge.

3. Caveats to Host Galaxy Folklore

The above comparison and my look at the papers of recent years lead me to the following cautionary statements.

- 'Host galaxies are essentially all ellipticals.'

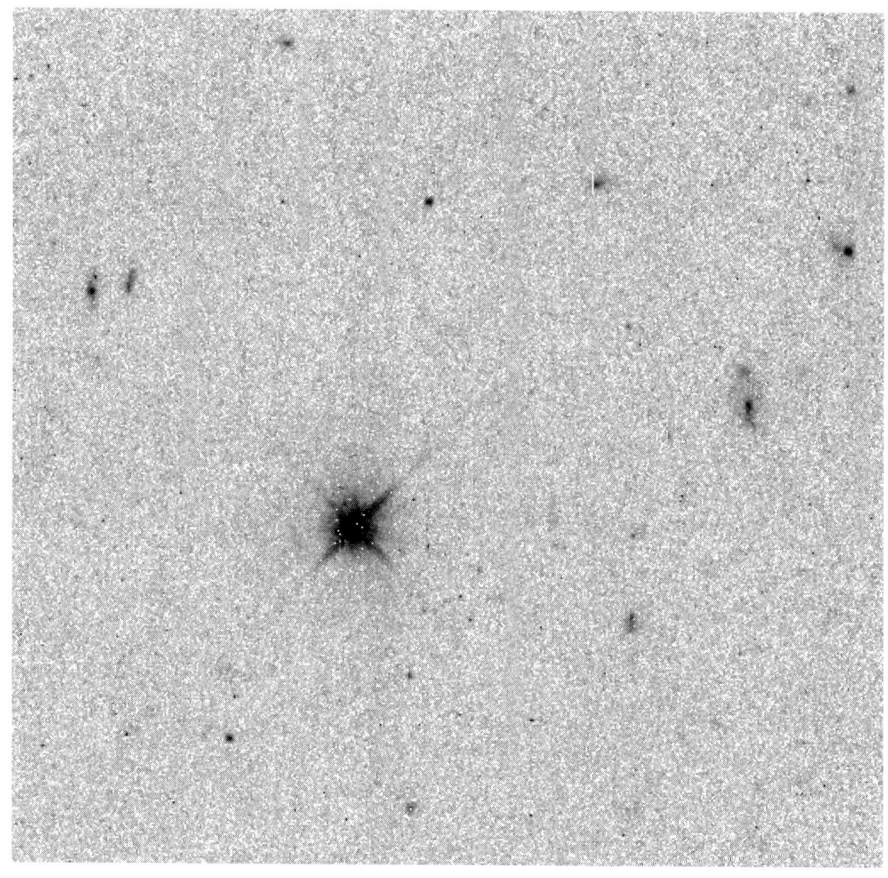

Figure 2. HST CCD image of z=2 QSO 0225-014 with compact companions. Field is 23 arcsec.

Table 1. Comparison of HST and AO data

	HST		Ground-based AO
+	High strehl over whole field	−	Good correction over small field
+	Field size 2-3 arcmin	0	Field size 20 arcsec
−	Undersampled	+	Well sampled
++	Whole sky available	−	Need bright guide star
+	Same strehl at all wavelengths	−−	Strehl increases with wavelength
0	PSF ∼const over field and time	−	PSF varies over field and time
−	PSF complex cannot be modelled	0	PSF simple may be modelled
−−	Optical artifacts mask host	−	Low strehl masks host
−−	Small effective area	+	Large effective area
0	Moderate spatial resolution	+	(Much) higher spatial resolution
−−	Noisy, ageing detectors	++	State of the art detectors
++	Dark sky in IR	−	Bright IR sky
−	Scattered light problems	+	Low scattered light
-2	Score	-1	

Does well

Sees 'spheroidal' hosts	Sees tails and disks
Good colour resolution	Sees faint companions
Good for low z	Good for high z

HST is good at seeing the bright bulges but not the faint disks and tidal tails. Thus, necessarily shallow HST imaging data will not detect or measure structures beyond the spheroidal population (see Fig 3). We also find (Hutchings et al 2001) from comparing intermediate-redshift galaxy clusters and field galaxies, that galaxies in an interaction-rich environment have brighter bulges and fainter disks as a result.

- 'Radio-loud QSOs have central BH masses of $10^9 M_\odot$.'

At high redshift, this is based on extrapolating the 'Magorrian relationship'. The growth of central black holes and growth and evolution of bulge populations are unlikely to be related by the same proportionality over the cosmic evolution of galaxies.

In addition, there are now claims that there is a continuum of radio power - not a dichotomy (e.g. Lacy et al 2001). Thus, the concept of a minimum central mass for radio-emitting jets is not well established.

- 'Many hosts do not look irregular or peculiar: interactions are rare.'

This statement is usually made from HST data. As noted above, HST 'loses' inner irregular structure because of its complex PSF, and does not reveal faint outer disks or tidal debris by not having enough sensitivity. Deep imaging by larger telescopes do reveal these structures in many QSOs.

- Any claims of a standard ('elliptical' or 'disk') host galaxy type are suspect,

  1. Because of imperfect PSF-removal, leading to incorrect apparent morphology (see Figs 3,4).
  2. Because of azimuthal averaging in models and profiles, which are then forced to fit a pure spheroid or disk model.
  3. Because at high z all galaxies are irregular and not standard Hubble types.

4. Current Survey Questions

In spite of my cautions, we have definitely made some exciting progress, as summarized in Section 1. Thus, we do have an exciting set of possibilities ahead of us in this field, and some important issues to address.

Because of the clear connection now established between central BH mass and the properties of the spheroidal stellar population in the low redshift universe, the study of host galaxies of QSOs (and AGN in general) are of interest in understanding the formation and evolution of galaxies, as well as the mechanisms of the central energy sources. Thus, I list the following questions as important ones for a statistical survey of higher redshift QSO hosts.

1. How and when do central massive black holes form?
2. How and when do QSO host galaxies form?
3. How are the two related a) at galaxy formation, and b) over a galaxy lifetime?

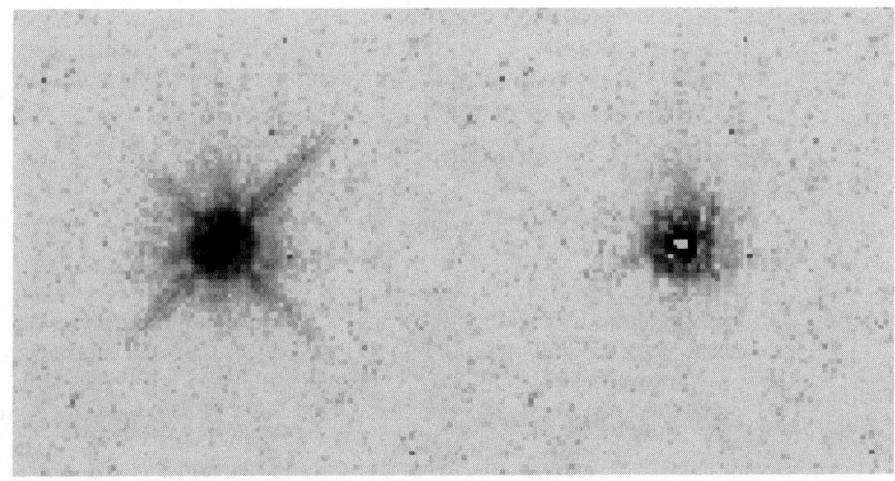

Figure 3. HST image of z=2.3 QSO 0820+296 (left), PSF-subtracted (right). Large telescope images are resolved to beyond the full 4.5 arcsec field shown here.

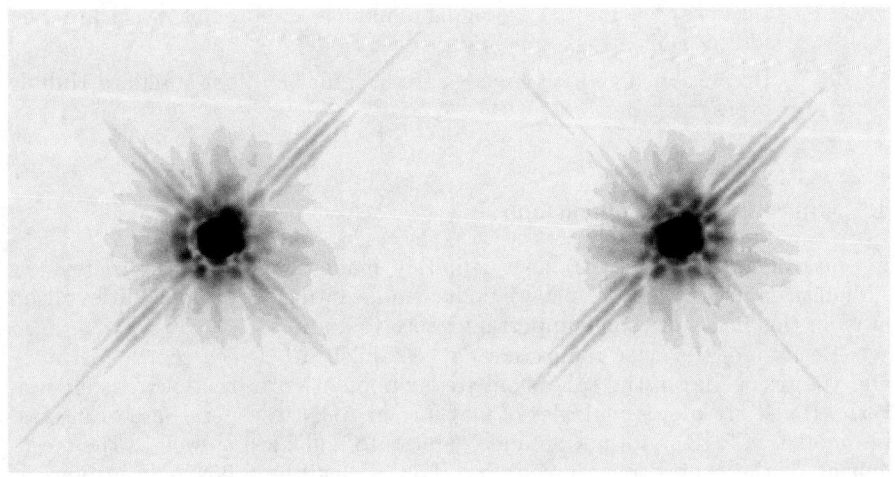

Figure 4. HST PSF models showing sensitivity to optics decentre of 2 mm. Inner PSF structure is about 0.7" across

4. What is the role of merging in galaxy evolution and BH growth?

5. Is there a radio-loud radio-quiet dichotomy or a continuous range of radio luminosity and structure?

6. Is there a general scenario for central mass flows and spectral absorbers? Does it change with cosmic evolution of galaxies?

I look forward to a real separation of the effects of source-evolution and line of sight orientation with the large and unbiased databases that are being established, and with the new generation of 8m-class telescopes both on the ground and in orbit.

The material for Figures 2, 3, and 4 is from work recently completed with Philip Dumont and Danielle Frenette, and will be published in full elsewhere.

References

Hutchings J.B. 1998, AJ, 116, 20

Hutchings J.B. 2001, IAA workshop on QSO Hosts (astro-ph/0107157)

Hutchings J.B., Saintonge A., Schade D., Frenette D. 2001, AJ (submitted)

Lacy M., Laurent-Muehliesen S.A., Ridgway S.E., Becker R.H., White R.L. 2001, ApJ, (astro-ph/0103087)

Are There AGNs in the Nearby Dwarf Galaxies?

I. D. Karachentsev

Special Astrophysical Observatory, Russian Academy of Sciences, N.Arkhyz, 369167, Russia

V. E. Karachentseva

Astronomical Observatory of Kiev University, 04053, Observatorna 3, Kiev, Ukraine

Abstract. The presence or absence of a nucleus is considered for a sample of galaxies limited by distance, not by flux. For the 365 nearest galaxies situated within a distance of ~ 7 Mpc, classification of their central regions is based on large-scale images obtained with the Hubble Space Telescope as well as with the 6-m and other large ground-based telescopes. Occurrence of nucleated galaxies is considered as a function of their luminosity, morphological type and other global properties.

1. Introduction

The problem of studying galactic nuclei during the time elapsed since the Byurakan symposium "Nonstationary Events in Galaxies" (1965) has come to be more sharply defined. The apearance of numerous new data are due to the improvement of observational facilities; and using these data, in turn, resulted in the development of theoretical insight into the processes that occur in the nuclei of galaxies. Note, however, that the subject of investigation itself is not infrequently defined either by mutual agreement as a certain central structure in a galaxy, a "nucleus", or — depending on the task stated, observational possibilities or the taste of authors — it breaks down into quite a few definitions. For illustration, let us present a few classification schemes with no pretence to their completeness.

Byurakan Classification (Kalloglian & Tovmassian 1964; Tovmassian 1965; Tovmassian 1966; Sahakian 1968, and papers cited therein.) The galaxies here are divided into five classes: "1" — no nucleus (i.e. in the central parts of the galaxy there are no signs of any condensation); "2" — a faint nucleus is supposed to exist against the background of the central part; "3" — the nucleus does not stand out against the bright central condensation; "4" — the nucleus is not completely star-like; "5" — the nucleus is practically star-like, the contrast to the central part is high.

Vorontsov-Velyaminov's Classification (Vorontsov-Velyaminov 1965). It divides the "nuclear regions of galaxies" into: disk "D" whose thickness is roughly equal to the thickness of the spiral arms. Found only in late spirals; lens "L" — more or less dense structure, with brightening towards the center. Encountered only in Sa and Sb galaxies; bulge "B" — somewhat different from the lens by greater sphericity and a more rapid drop of brightness from the center towards the edge. Characteristic of S0 and SB0 galaxies; nucleus "N" — structure resembling a bulge, but relatively more compact. Characteristic of Sb, Sbc, Sc; kernel "K" — star-like condensation of very small size. They are found in some E galaxies and are likely to be absent in diffuse dwarf Sculptor-type galaxies in the Local group.

From the papers of the 1960s the classification according to Deitch (1966) can be mentioned. Papers by Kalloglian & Tovmassian (1964); Tovmassian (1965); Tovmassian (1966); Sahakian (1968); Vorontsov-Velyaminov (1965); and Deitch (1966) were based on photographic data. In each series of classification, about 200–300 galaxies of different types were examined. The authors noted satisfactory agreement of the characteristics which describe the central regions of galaxies in different schemes. Note that these early papers did not seek to classify galaxies from a homogeneous (flux-limited or volume-limited) sample. Their object was the search for the regularities associated with the variation of the class/brightness of the nucleus depending on the morphological type/luminosity of the host galaxy.

Van den Bergh's Classification (van den Bergh 1995). Using the data of "The Atlas of Galaxies" (Sandage & Bedke 1988), 342 central regions of late-type galaxies in the Shapley-Ames Catalog with $m_{pg} \leq 12\overset{m}{.}0$ (flux-limited sample) are classified here. The classification scheme: NN – no nucleus, N — star-like nucleus, SSN — semistellar nucleus, CB — a small central bulge or a disk is visible in the center of the image, NB — bar-like structure in the center, Tr — transient objects, intermediate between spirals, having a central bulge, and objects with the central regions resolved into stars and knots.

In the 1980s-1990s many new papers appeared which were concerned with CCD photometry of early-type galaxies (Binggeli & Cameron 1991) in the Virgo and Fornax clusters, as well as individual galaxies (Kormendy 1985; Lauer et al. 1995). Without touching upon the interpretation of data, which is not infrequently fundamentally different, as in papers by Binggeli & Cameron (1991) and Caldwell & Bothun (1987), it will be noted that their authors point to the presence of either a rather dense and contrasty "bulge" in the central regions, or a diffuse "bulge", or a star-like "spike" superposed on the bulge (in the case of E galaxies), or on the body of the galaxy (in the case of dSph).

Detailed studies of 42 E galaxies with the HST (Lauer et al. 1995) allow their nuclear regions to be classified by the character of the surface brightness distribution as "core", "power law" and "nuclear star clusters". We will return later to the question of existence of nuclei in dwarf spheroidal galaxies.

2. The Local Volume Sample

Before passing to the description of our sample, two basic remarks will be made. 1. Any statistical study of galaxies (in our case it is classification and determination of the frequency of occurrence of nuclei in galaxies of different types) must be performed from a volume-limited sample as complete as possible. With a diameter-limited or a magnitude-limited sample allowance should be made for different selection effects. 2. Dwarf galaxies of low luminosity and surface brightness are practically not represented in all general whole-sky catalogs. This refers also to the catalog of 179 nearby galaxies of Kraan-Korteweg and Tammann (1979) with $D \lesssim 10$ Mpc, (KKT). The KKT sample has been complemented by Karachentsev (1994) from literature data and enhanced to N = 216 (galaxies of the Local Volume, LV, with $V_{LG} < 500$ km/s). On the basis of using the sky surveys POSS-II and ESO/SERC, the authors started quests for dwarf galaxy candidates with an angular diameter $a \gtrsim 0\rlap{.}'5$ in vicinities of the LV galaxies in order to make the sample still more complete (Karachentsev 1994). 260 galaxies (KK-galaxies) predominantly of low surface brightness have been found; half of them have been cataloged for the first time. H I observations of KK-objects with the 100-m radio telescope in Effelsberg (Huchtmeier et al. 1997; Huchtmeier et al. 2000) confirmed many of them as very nearby galaxies. By the present time, the whole-sky searches for dwarf candidates have been practically accomplished (Karachentseva et al. 1999) - KKR-objects; (Karachentseva & Karachentsev 2000) — southern KKs-objects; (Karachentsev et al. 2000) — KKSG-objects; (Karachentsev et al. 2001) — KKH-objects). A total of over 600 candidates for the LV members were detected, about 300 of them for the first time.

Further observations added about 100 new dwarf galaxies to the LV. Together with the objects found by other researchers, the number of Local Volume galaxies, as compared to the KKT sample, has increased by more than a factor of two, and to-date N = 365. When preparing this new sample, we united and arranged in an ordered fashion our results obtained in papers (Karachentseva & Karachentsev 1998), [18–21] (Karachentseva et al. 1999; Karachentseva & Karachentsev 2000; Karachentsev et al. 2000; Karachentsev et al. 2001) as well as new data from the literature. As a result, we now have the following characteristics of the LV galaxies: the galaxy name, the equatorial coordinates, apparent total magnitude, radial velocity reduced to the Local group according to (Karachentsev & Makarov 1996), morphological type of the galaxy in the digital code (so, T = 10 corresponds to irregular dwarfs, while N = −5 to the faintest dwarfs dSph), distance of the galaxy in Mpc, luminosity logarithm in units $L_\odot$. We will comment in detail on our classification of the nuclear region of the galaxy in the next section.

3. Observational Data and Classification Scheme

As our experience shows, the photographic sky surveys POSS-II, ESO/SERC and SERC EJ on the whole serve for the morphological classification of nearby galaxies. However, the classification of their nuclei requires other data because the galaxy central regions usually appear to be over-exposed and dusty.

We have collected and analyzed the images accesible to us of LV galaxies from different sources (we enumerate them in the order of preference): 2MASS Catalog (Jarrett et al. 2000), the archives of the HST data, CCD images obtained with the 6-m and Nordic telescopes, the Atlas of Galaxies (Sandage & Bedke 1988), POSS-II, ESO/SERC and SERC EJ sky surveys. We have also used photographs of LV galaxies from different publications. We succeeded in doing classification of many objects from our list independently for several sources.

For the problem stated it turned out to be sufficient to isolate three main types of nuclear regions:
- "**K**" (kernel) — star-like "nucleus"
- "**B**" (bulge) — rather dense, but extended "nucleus"
- "**A**" — no "nucleus".

The following types in other classification schemes correspond approximately to these main types (Table 1):

Table 1. A correspondence between the KK and other authors galaxy nuclei classification scheme

This work	Byurakan	Vorontsov-Velyaminov	van den Bergh
K	5	N, K	N
B	2:, 3, 4:	L, B, N:	CD, NB, SSN:
A	1, 2:	-	NN
coincidence,	7/11	6/9	31/35
%	63%	67%	89%

Thus, the concepts "star-like nucleus" or "no nucleus" in the schemes considered are described practically by one symbol. If a nuclear region is seen as extended, then several designations of other authors correspond to our symbol "B".

The results of the comparison are quite explicable: for classifying central regions of galaxies, large-scale images are most suitable. The inclusion of the 2MASS survey, where the influence of dust is practically removed, provides additional advantages for classification. Figure 1 gives some examples of classification of nuclear regions.

4. Results and Conclusions

Fig.2 presents the distribution of the number of galaxies in the LV with different types of central regions versus the distance of the galaxy, D, Mpc. As one can see, the median values of D for "**K**", "**B**" and "**A**" objects are about the same, i.e., no strong selection with distance. At median D = 5 Mpc $((m-M)_0 = 28\overset{m}{.}5)$ the detection limit of a nucleus $m_{lim} \simeq 18\overset{m}{.}5$ corresponds to $M \simeq -10^m$. The angular resolution of the 2MASS survey at the same distance is approximately equal to 40 pc, and $\simeq 2$ pc corresponds to the angular resolution of the HST, $\simeq 0\overset{''}{.}1$.

The occurrence of different nuclear types per galaxy in in the Local Volume is $f_K = 23/365 = 6\%$ for galaxies with kernels, and $f_B = 33/365 = 9\%$ for galaxies with bulges. The Local Volume density of nuclei is $V_K = 0.014/\text{Mpc}^3$, and $V_B = 0.023 /\text{Mpc}^3$ within $D \sim 7$ Mpc.

The distribution of the number of galaxies "**K**", "**B**" and "**A**" vs. host galaxy morphological type is displayed in Fig.3. Galaxies with bulges occur among all the types, kernels are observed in a few E galaxies and in late spirals. K-type nuclei are absent in irregular and spheroidal dwarfs.

The distribution of the number of galaxies with different type nuclei vs. galaxy luminosity is shown in Fig.4. Galaxies with "**K**" and "**B**" have approxi-

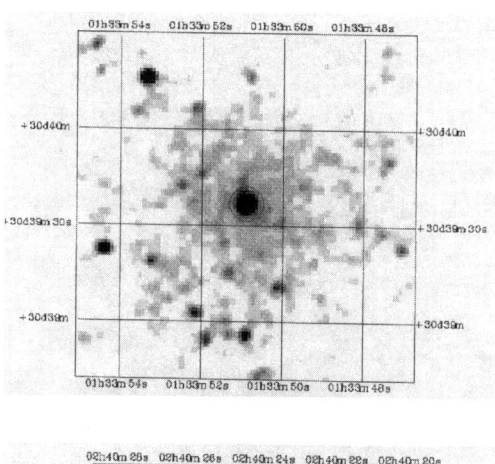

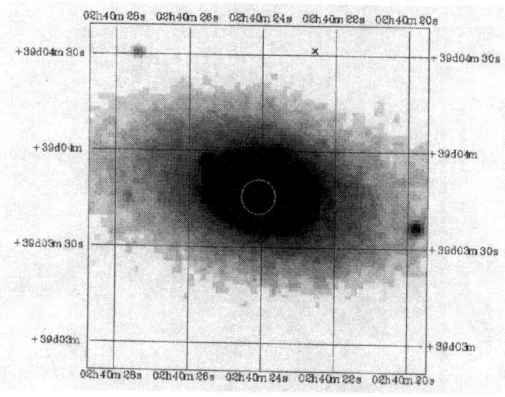

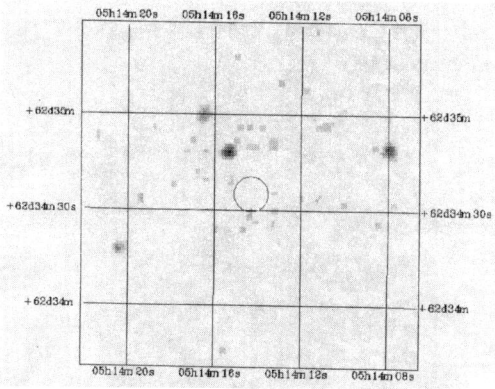

Figure 1. Examples of classification of nuclear regions (top: M33 - K, middle: NGC 1023 - B, bottom: UGCA 105 - A).

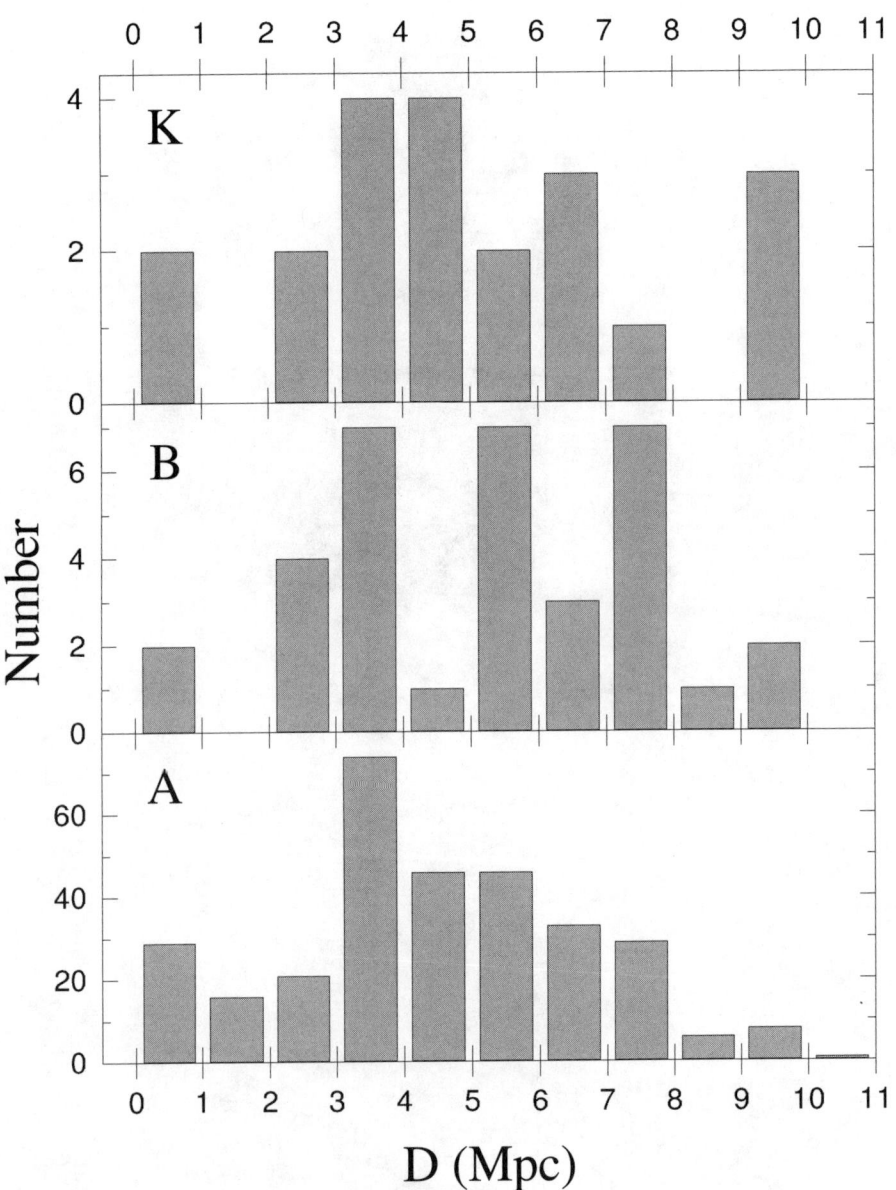

Figure 2. The distribution of the number of the LV galaxies with different nuclear region types versus the distance of the galaxy.

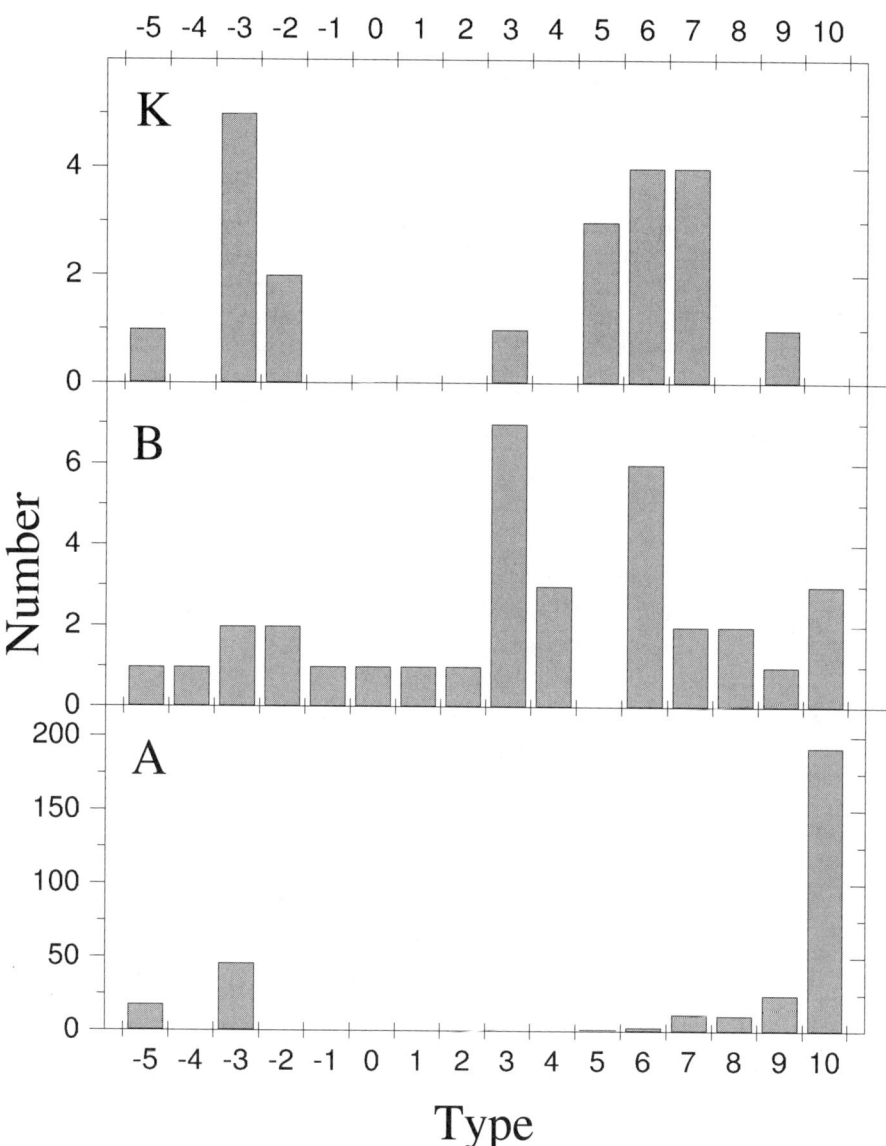

Figure 3. The distribution of the number of galaxies "K", "B", and "A" versus host galaxy morphological type.

mately the same distributions in luminosity shifted by 5^m with respect to "**A**" galaxies. There are no "**K**" and "**B**" among dwarf galaxies (with an absolute magnitude $M > -15\overset{m}{.}0$). A few exceptions are dwarfs brighter than -15^m, which are likely to belong to the Virgo cluster. Among giant galaxies with $lgL > 10.0$, or $M < -19\overset{m}{.}6$ the cases "**B**" (N =14) and "**K**" (N = 5) predominate. Only three giant galaxies: NGC 253, NGC 4945, and NGC 5236 are noted as "**A**", but they are heavily dusty and may have hidden nuclei.

Our data do not confirm the existence of "nucleated" dSph/dE dwarfs in the Local Volume, which are found in a great number in Virgo (Binggeli et al. 1985) and Fornax (Ferguson & Sandage 1988). Examples of dSph dwarfs in the M 81 group, which look "nucleated", are BK5N, K 64 — here a distant galaxy is projected onto their center, and K 61 — a globular cluster is situated near its center.

As is well known, dwarf spheroidal galaxies in the Local Group have no nuclei. The global properties of these objects in the LG, in the M 81 Group, and in the Virgo and Fornax Clusters are alike. For this reason, it seems to us that the existence of "nuclei" in dwarf spheroidal members of the Virgo and Fornax cluster is still open to question.

References

Binggeli, B., & Cameron, L. M. 1991, A&A, 252, 27
Binggeli, B., Sandage, A., & Tammann, G. A. 1985, AJ, 90, 1681
Caldwell, N. & Bothun, G. D. 1987, AJ, 94, 1126
Deitch, A. N. 1966, Publ. Main Astron.obs.(Pulkovo), N.179, 95
Ferguson, H. C., & Sandage, A. 1988, AJ, 96, 1520
Huchtmeier, W. K., Karachentsev, I. D. & Karachentseva, V. E. 1997, A&A, 332, 375
Huchtmeier, W. K., Karachentsev, I. D., Karachentseva, V. E. & Ehle, M. 2000, A&AS, 141, 469
Jarrett, T. H., Chester, T., Cutri, R., Schneider, S. Skrutskie, M., & Huchra, J. P., 2000, AJ, 119, 2498
Kalloglian, A. T. & Tovmassian, H. M., 1964, Communic. Byurakan. obs., 36, 31
Karachentsev, I. D., 1994, Astron. Astrophys. Trans., 6, 1
Karachentsev, I. D., Karachentseva, V. E., & Huchtmeier, W. K. 2001, A&A, 366, 428
Karachentsev, I. D., Karachentseva, V. E., Suchkov, A. A., & Grebel, E. K. 2000, A&AS, 145, 415
Karachentsev, I. D. & Makarov, D. I., 1996, AJ, 111, 794
Karachentseva, V. E., & Karachentsev, I. D. 1998, A&AS, 127, 409
Karachentseva, V. E., Karachentsev, I. D. & Richter, G. M. 1999, A&AS, 135, 221
Karachentseva, V. E., & Karachentsev, I. D. 2000, A&AS, 146, 359
Kormendy, J. 1985, ApJ, 295, 73

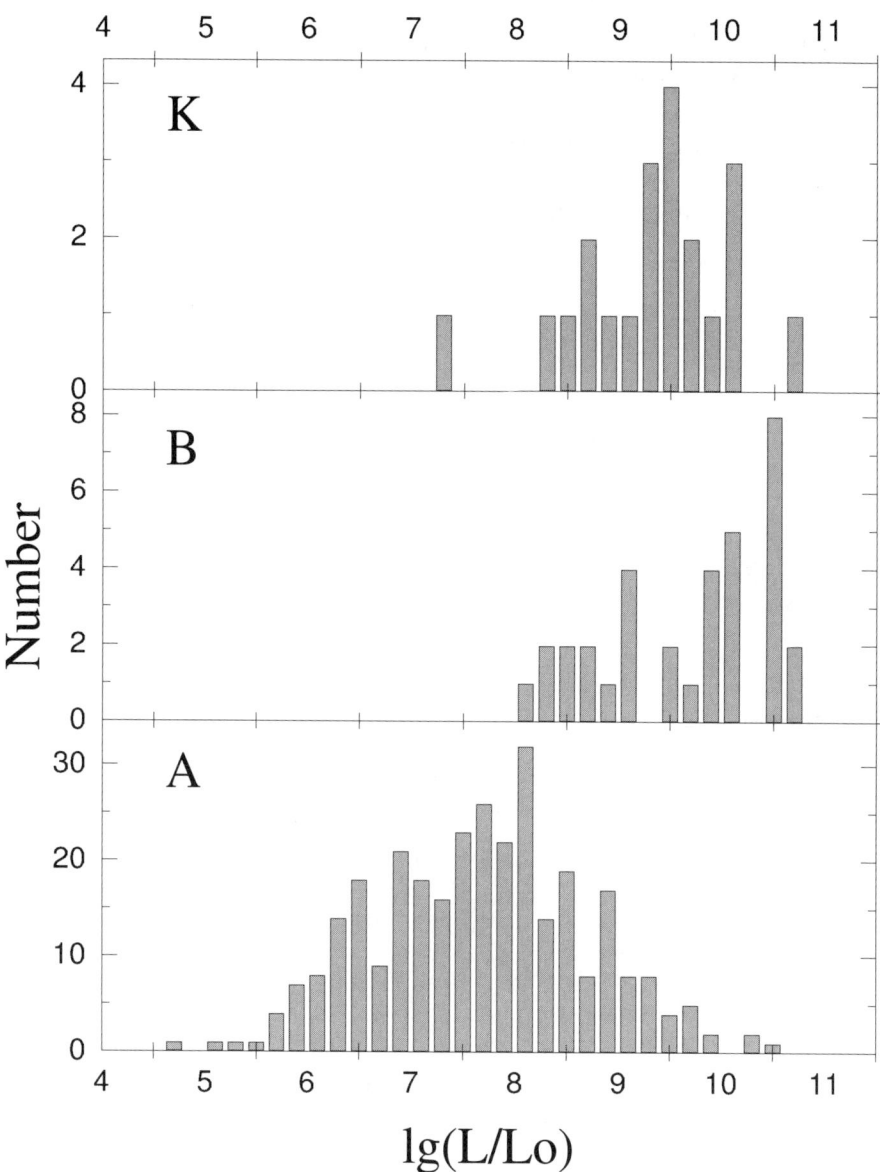

Figure 4. The distribution of the number of galaxies with different nuclear region types versus galaxy luminosity.

Kraan-Korteweg, R. C., & Tammann, G. A. 1979, AN, 300, 181
Lauer, T. R., Ajhar, E. A., Byun, Y. I. K., Dressler, A, Faber, S. M., Grilimair, C., Kormendy, J., Richstone, D., & Tremaine, S. 1995, AJ, 110, 2622
Sahakjan, K. A. 1968, Astrofizika, 4, 41
Sandage, A. R. & Bedke, J. 1988, Atlas of galaxies, Washington, DC
Tovmassian, H. M. 1965, Astrofizika, 1, 197
Tovmassian, H. M. 1966, Astrofizika, 2, 317
Vorontsov-Velyaminov, B. A. 1965, AZh, 42, 1168
van den Berg, S. 1995, AJ, 110, 613

AGN and the Demographics of Supermassive Black Holes

Richard F. Green[1,2]

Kitt Peak National Observatory, Tucson, AZ, USA; rgreen@noao.edu

[1] *On behalf of the STIS Instrument Definition Team Galaxy Nuclei Group: Gary Bower, John Hutchings, Charles Joseph, Mary Elizabeth Kaiser, John Kormendy, Charles Nelson, Douglas Richstone, Donna Weistrop, Bruce Woodgate.*

[2] *On behalf of the Nuker Team: Gary Bower, Alan Dressler, Sandra Faber, Alex Filippenko, Karl Gebhardt, Carl Grillmair, Luis Ho, John Kormendy, Tod Lauer, John Magorrian, Jason Pinkney, Douglas Richstone, Christos Siopis, Scott Tremaine.*

Abstract.

High angular resolution observations from WFPC and STIS now allow well-constrained dynamical measurement of the masses of supermassive black holes (SMBH) in nearby galaxies. An initial statistical analysis by Magorrian et al. showed that 97% of bulges host SMBH. Black hole mass is correlated moderately with bulge luminosity and strongly with the velocity dispersion of the whole bulge, suggesting that black hole formation may be an intrinsic aspect of bulge formation. Black hole masses for AGN determined from reverberation mapping fall on the same relationship with bulge velocity dispersion as those determined from stellar dynamical measurements. The prospect is therefore that the large-scale distribution of black hole masses in distant quasars may be determined through relatively straightforward measurement. Integral constraints show consistency between the total AGN luminosity density and the total volume density in SMBH contained in galaxy bulges. The strong peak of the high-luminosity quasar luminosity function at early cosmic time is consistent with the association of the build-up of SMBH through accretion and bulge formation. Alternate scenarios requiring substantial build-up of the most massive black holes at later cosmic times are more difficult to reconcile with the evolution of the LF.

1. Gas and Stellar Dynamical Evidence for Central Black Holes

There are two observational signatures and one requirement for the presence of a central SMBH in a galaxy nucleus. Spectra of adequate angular and velocity resolution will show rapid Keplerian rotation of gas and/or stars and a strong peak in the central velocity dispersion. The enclosed mass-to-light ratio within that region of enhanced velocities will be well in excess of that for a normal stellar population. The shape of the central gravitational potential can then be modeled from the motions of gas disks and/or the ensemble of stellar orbits.

For the M31 group, ground-based spatial resolution relative to the size of the central dynamically affected region actually exceeds that of HST for galaxies at the distance of the Virgo Cluster. Dressler and Richstone (1988) and Kormendy (1988) both found the strong rotational and dispersion peak signature for stellar motions in M31. They derived a best fitting central black hole mass of 7×10^7 $M_\odot$. Dressler also pointed out the utility of the near-infrared Ca II triplet absorption feature for obtaining a clean dynamical signature that was not strongly dependent on the mix of stellar populations producing the integrated spectrum.

As is now well known, our Instrument Definition Team designed the STIS spectrograph with a performance goal of measuring galaxy nuclear dynamics (Woodgate et al. 1998, Kimble et al. 1998). The long (50") slit and 0.05" pixels were chosen to address that problem, with a balance between angular resolution and S/N per pixel achievable on nearby galaxies.

In order to publicize the instrument after launch, NASA Headquarters requested the team to discover a new black hole. Gary Bower selected M 84, a Virgo cluster elliptical, for its suggestive kinematics in ground-based spectra, its central gas disk, and favorable radio properties. M 84 is an FR II source, with its two radio jets nearly equal in power and extent. Given the normal model of relativistic beaming, the two jets must be nearly in the plane of the sky. The axis of rotation of the inner accretion disk and probably the more extended gas disk must also lie nearly in the plane of the sky. Measured radial velocities will therefore yield nearly the full amplitude of rotation.

There is a weak AGN in the center of M 84, complicating the determination of velocity centroids. Bower et al. (1998) did derive a velocity curve for [N II], which showed both the slowly rotating outer disk and a rapidly rotating inner disk. A simple model of a cold gas disk in Keplerian rotation around a hidden dark mass, with velocities mixed through the HST+STIS PSF sampled by the 0.05" pixels, fits the data well. That model yielded a central massive dark object of $1.5 \times 10^9 M_\odot$. Neglect of line broadening (asymmetric drift, e.g., Barth et al. 2001) probably makes that value an underestimate for the mass of the SMBH. This evidence provides a direct connection between the presence of a supermassive black hole and the activity that powers quasars and radio galaxies.

Firm claim for a detection of an SMBH requires sufficient spatial resolution that other exotic high-density configurations can be ruled out. M31 is one such case; NGC 4258 provides another. It is a low-power AGN found to have strong water maser emission. Observations by Miyoshi et al. (1995) found a rapidly rotating disk of gas. The extremely high precision of radio VLBI shows a Keplerian velocity law to 1% from 0.13 – 0.26 pc. The central dark mass is $3.6 \times 10^7 M_\odot$. The central *density* is greater than $10^9 M_\odot pc^{-3}$, which rules out stable alternatives to a central black hole (Maoz 1998).

Well-ordered rotating gas disks are likely to be the exception for near-nuclear configurations. Sarzi et al. (2001) found such disks in only 15% of their sample of bulges, which were already known to contain emitting gas. Currently, stellar dynamical determinations of mass are likely to be more generally reliable. Smaller elliptical galaxies and spiral bulges are rotationally supported.

The trend of body rotation can be removed cleanly from the local rapid rotation near the nucleus. Given the run of rotation and velocity dispersion near the nucleus, stellar dynamical modeling is then required for a determination of the mass of the central Massive Dark Object. The solution of the collisionless Boltzmann equation can be derived with 2-integral or 3-integral constraints.

The most convincing case for a SMBH from stellar motions is the Galaxy itself, through proper motion measurements around Sgr A* (Genzel et al. 2000). A. Ghez et al. (1998) used the Adaptive Optics system on the Keck Telescope to obtain a time series of near-IR images. They show that the fastest transverse motion is 1350 ± 40 km/s; the ensemble yields a black hole mass for the Milky Way of $2.6 \pm 0.2 \times 10^6 M_\odot$. The central density is greater than $10^{12} M_\odot pc^{-3}$. Alternative clusters of compact objects would quickly evaporate, so the Massive Dark Object is demonstrated to be a Supermassive Black Hole. Emboldened by these nearby cases, we'll call other, more distant MDOs SMBH as well.

For other galaxies, the nuclear regions cannot be resolved into individual stars. The ensemble stellar motions producing the observed line profile shapes can be described by the Line of Sight Velocity Distribution (LOSVD), which can be fit parametrically with Gauss-Hermite polynomials to yield the moments of the distribution. The far-red Ca II triplet feature near 8600 Å provides a robust template that is not very sensitive to modest reddening, dominant spectral type, or metallicity. The LOSVD can be derived from template matching to the broadened galaxy profiles through either the Fourier Correlation Quotient method (Bender 1990) or the Maximum Penalized Likelihood method (Merritt).

2. The Black Hole Census and Correlation with Bulge Properties

John Magorrian and his "Nuker Team" collaborators (1998) analyzed 36 nearby bulges with surface brightness profiles from HST images and spectra from Kitt Peak and other ground-based telescopes. They applied an axisymmetric 2-integral model (i.e., constraints of only conservation of energy and angular momentum), with inclination angle, M/L, and central dark mass as the free parameters. The technique uses Maximum Likelihood to deproject the luminosity profile, calculate the gravitational potential, calculate second moments of the velocity distribution from the Jeans equation, then fit the absorption spectrum in each spatial bin. Fits can be projected as probability contours in the M/L black hole mass plane for a fixed inclination angle. Of the sample of 36 bulges, 30 contained a non-zero SMBH at high significance and 5 were consistent with a non-zero SMBH. There was only one object in which a central SMBH was ruled out, and that was distant, with a dusty core. Population statistics based on this sample showed that 97% of massive galaxies with bulges host central black holes.

They also identified a correlation, since called the "Magorrian relation", between central black hole mass and bulge luminosity. Conversion through M/L suggested that the black hole mass was a constant fraction of total bulge mass, however with substantial scatter. The Nuker Team had initiated their investigations with HST wondering which bulges would be most likely to contain a SMBH. Within a few HST proposal cycles, the paradigm had shifted to the view that a SMBH is an integral part of every massive galaxy with a bulge. The

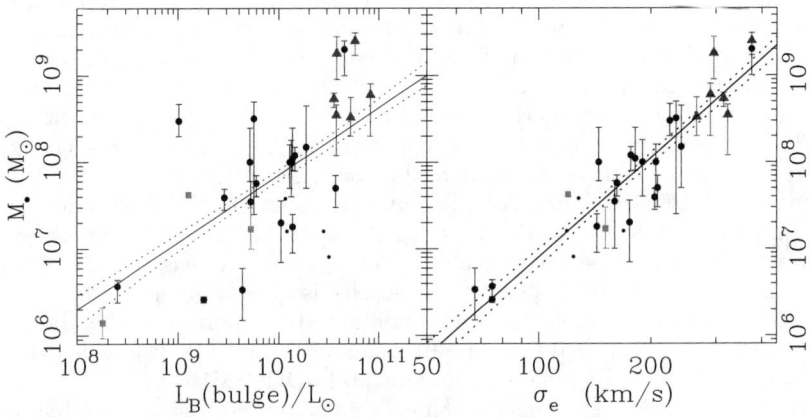

Figure 1. Left: Black hole mass vs. blue luminosity of the host bulge in solar units. Right: Black hole mass vs. luminosity-weighted velocity dispersion of the bulge.

upper limit of $\leq 10^4 M_\odot$ for M 33 (Gebhardt et al. 2001, Joseph et al. 2001) argues that black holes of that scale are not found in galaxies without bulges.

An anisotropic distribution of radial orbits near the nucleus can lead 2-integral models to overestimate the central mass. Richstone and Gebhardt (2002) devised a practical realization of the Schwarzschild method for finding an axisymmetric 3-integral model fit built up from pre-computed orbital families. There are now some 30 high-quality stellar kinematic determinations of central mass, a dozen good gas disks and 3 maser sources. An update to the Magorrian relation by Kormendy and Gebhardt in their thorough review (2001) gives $M_{dark} = 0.78 \times 10^8 M_\odot (L_{B,bulge}/10^{10} L_\odot)^{1.08}$. Since $M/L \propto L^{0.2}$, the dependence implies $M_{dark} \propto M_{bulge}^{0.9}$. The dispersion in the relation is large: the rms is a factor of 2.8, and the total range in M_{dark} is two orders of magnitude at a fixed M_{bulge}.

Even more significantly, Gebhardt et al. (2000) and Ferrarese & Merritt (2000) discovered that there is a tighter correlation between black hole mass and the luminosity-weighted velocity dispersion of the bulge inside its effective radius. Gebhardt et al. (2000) and Tremaine et al. (2002) found that the scatter in the relationship may be entirely attributable to measuring uncertainties when only the best measurements are considered. That correlation is therefore more fundamental than the $M_{dark} - L_{bulge}$ correlation, although both suggest a close connection between bulge and central black hole growth.

Exceptions to the Faber-Jackson relation (like NGC 4486B) still satisfy the $M_{dark} - \sigma$ relation. We could speculate that dynamically hotter galaxies have higher surface brightnesses and smaller effective radii. If their formation

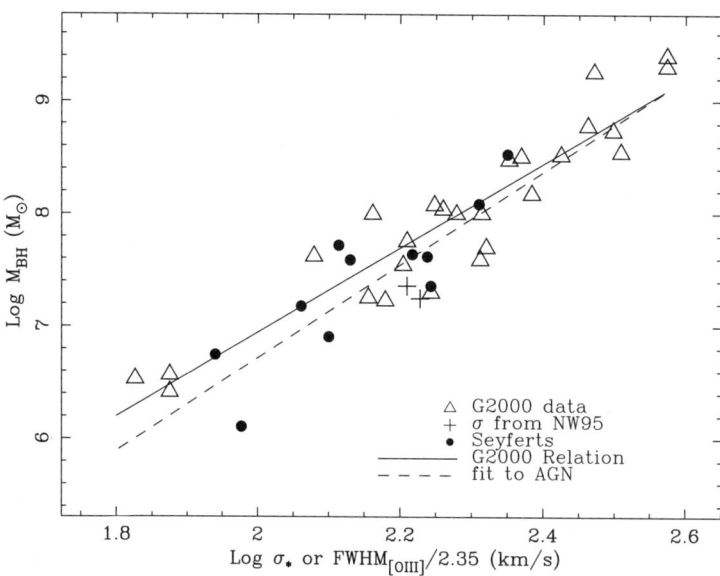

Figure 2. Black hole mass vs. host bulge velocity dispersion. Open triangles mark stellar dynamical determinations from Gebhardt et al. (2000) and the solid line marks the best fit relationship for those data. Filled circles show black hole masses derived from reverberation mapping, with the dashed line the best-fit relationship for AGN only. The agreement in zero point and slope is very close.

involved more dissipation inside their dark halos, the central black hole mass was increased proportionally. In general, the dissipative collapse and mergers that form bulges must also build the SMBH. Growing a black hole through accretion is accepted as the source of AGN luminosity. The epoch of luminous quasars is likely to be the epoch of major bulge and elliptical galaxy formation.

The current capabilities of HST and STIS allow well-constrained SMBH mass determinations out to about the distance of the Virgo Cluster for quiescent black holes. For AGN, the near-nuclear stellar dynamics cannot be measured, and the glare of the central source currently limits the ability to make direct measurements of stellar absorption in the host galaxies to all but the most modest redshifts. There is therefore active interest in finding proxies for both central dynamical measurements and probes of bulge potential. Nelson et al. (2002) and Ferrarese et al. (2001) both used the Kitt Peak 4-m to test one proxy, broad emission line width for central stellar dynamics. They used a sample of low-redshift AGN monitored for reverberation mapping of emission-line response to continuum variations. They then measured the bulge velocity dispersions for the host galaxies of those AGN. Nelson et al. found that the black hole masses from reverberation mapping and the newly measured host bulge dispersions fell along the Gebhardt et al. relation from stellar dynamics, with only a minor difference in slope.

The agreement between the relationship based on stellar dynamical determinations and that from reverberation mapping validates the strong assumptions underlying reverberation mapping techniques:

- Line cooling from a given species lies in a relatively restricted zone at a definable effective distance from the nucleus.

- The motions of the emission-line gas are dominated by response to the gravitational potential.

- The cross-correlation derivation of time lag of response of emission-line flux to continuum variation is unambiguous.

A good correlation with moderate scatter exists between reverberation masses and photoionization model masses (Wandel, Peterson, & Malkan 1999). The photionization derivation requires only observed spectrophotometry at one epoch. The potential therefore exists to gain a first impression of the black hole mass spectrum in AGN as a function of cosmic time, along with the typical accretion rates relative to the Eddington limit. Large-scale spectroscopic surveys such as 2DF and SDSS can in principle provide significant samples matched in luminosity for different slices of cosmic time with which the hypothesis of black hole growth with time can be tested.

3. Implications of Integral Constraints

Soltan (1982) and Chokshi & Turner (1992) pointed out that the Universe should be populated today by relic black holes that were active in the quasar era with an average mass density that matches or exceeds the mass-equivalent of the energy emitted by them. Iwasawa (these Proceedings) showed us the calculation of the black hole mass density inferred from the hard X-ray background, with a local co-moving density of $6 \times 10^5 M_\odot pc^{-3}$. If that value is distributed according to the black hole mass vs. bulge luminosity relation through the M/L of present-epoch bulges, it exceeds the local black hole mass density by a factor of only 1.4. That excellent agreement, given the uncertainty of assumptions such as bolometric corrections and accretion efficiency, gives confidence in the connection between AGN in the early Universe and today's quiescent central black holes.

It is also possible to test for general consistency that the luminous quasars of early cosmic time correspond to the most massive quiescent black holes in the current epoch. Richstone et al. (1998) presented the form of this argument, here updated with more current estimates. We assume that at turn-on quasars radiate at the Eddington limit, which specifies the black hole mass, given a measured luminosity. As summarized by Wisotzki in these Proceedings, the quasar optical luminosity function at the highest redshifts is sampled down to luminosities still higher than the characteristic luminosity, capturing only the most luminous, unobscured objects. For the faint limit of the quasar LF of $M_B < -26 mag$, the bolometric correction of Elvis et al. (1994) gives a black hole mass of $4 \times 10^8 M_\odot$. The black hole mass – bulge mass correlation then maps to $M_{B,bulge} < 21.2 mag$. The local galaxy luminosity function for bulges that luminous gives an integrated volume density of $10^{-5} Mpc^{-3}$.

Estimates of the high-redshift quasar LF from the first SDSS data (Fan et al. 2001) are in surprisingly good agreement with the earlier results from the transit grism survey of Schmidt, Schneider & Gunn (1991). When combined with lower redshift data for high luminosity objects, the quasar density as a function of time shows a strong peak at a fractional lookback time of ≈ 0.85. This "quasar epoch" had an extent in time of $\approx 10^9$ yrs. The peak co-moving volume density is $\approx 10^{-6} Mpc^{-3}$. The luminous lifetime per object is approximated by the density of quasars / density of host galaxies $\times$ the extent of the "quasar epoch", which is $\approx 10^8 yrs$. That timescale represents ≈ 3 e-folding timescales for an accreting black hole with 10% efficiency, according to Salpeter (1964). The luminous phase of the most luminous objects is then reasonably consistent with the time required to build a SMBH. For the black hole mass – bulge mass correlation to be a consistent assumption, large bulges had to be large established by $z = 2$. That assumption is also consistent with deep surveys, such as the HDF.

Yu and Tremaine (2002) have recently performed a detailed calculation of the integral constraints. They use the quasar LF from the 2DF (Boyle et al. 2000) to estimate the black hole mass density accreted during optically bright quasar phases. They then solve a continuity equation to relate the evolving AGN LF to the local mass density of black holes inferred from the local LF of bulges. They find that the local density of high mass black holes ($> 10^8 M_\odot$) is completely accounted for by the optically luminous quasars, and that such quasars must radiate with an efficiency of 0.2-0.3, acceptable in the model of thin-disk accretion onto a Kerr black hole. They find two options for lower mass black holes. If their growth also occurs mainly during an optically luminous stage, their accretion must occur with lower efficiency (< 0.1). The alternative is that they accrete with high efficiency, but that a significant fraction is obscured.

The apparent consistency of black hole mass density in luminous quasars at higher redshift and in luminous bulges locally challenges the claim that substantial growth of the most massive black holes is attributable to highly luminous obscured sources at intermediate redshifts (cf. Barger et al. 2001). The close implied connection of black hole and host bulge formation would also require substantial bulge formation or major growth in relatively recent cosmic times. Such a late build-up could require recent major mergers at a rate potentially at variance with the typical ages inferred for the majority stellar populations in luminous bulges. Alternative scenarios may mitigate the apparent conflict. Yu and Tremaine's models are consistent with lower mass, hence lower power objects accreting later or over longer time intervals. Perhaps the hard X-ray sources at intermediate redshifts are more ordinary AGNs with relatively low luminosities at OUV wavelengths. Optically faint hard X-ray sources at $z \approx 1$ observed at good ground-based resolution may still harbor optical AGNs of moderate power that are diluted by luminous host galaxies to the point of non-detection. The hard-X to optical luminosity ratio may in that case be slightly anomalous, but not necessarily require strongly Compton-thick absorption in the line of sight of intrinsically high luminosity sources.

Resolving the entangled issues of the evolution of low-mass black holes, of low- luminosity AGNs, and of the contribution of obscured sources to the total radiated accretion energy poses a challenge for the next generation of surveys. One critically important task will be to determine the optical quasar luminosity function on the faint side of the characteristic luminosity for $z > 1$. An important anticipated step will be the deep Chandra/XMM/SIRTF/optical surveys, which should expand the multi-wavelength census and bring us closer to an unbiased count of all the manifestations of activity around SMBH.

References

Barger, A. et al. 2001, AJ, 122, 2177.
Barth, A., et al. 2001, ApJ, 555, 685.
Bender, R., 1990, å, 229, 441.
Bower, G. A., Green, R. F. et al., 1998, ApJ, 492, L111.
Boyle, B. J. et al, 2000, MNRAS, 317, 1014.
Chokshi, A. & Turner, E. L., 1992, MNRAS, 259, 421.
Dressler, A. & Richstone, D.O. 1988, ApJ, 324, 701.
Genzel, R. et al., 2000, MNRAS, 317, 348.
Elvis, M. et al. 1994, ApJS, 95, 1.
Fan, X. et al., 2001, AJ, 121, 54.
Ferrarese, L. & Merritt, D., 2000, ApJ, 539, L9.
Ferrarese, L., et al. 2001, ApJ, 555, 79..
Gebhardt, K. et al., 2000, ApJ, 539, L13.
Gebhardt, K. et al., 2001, AJ, 122, 2469.
Ghez, A. et al., 1998, ApJ, 509, 678.
Joseph, C. et al., 2001, Science, 293, 1116.
Kimble, R. A. et al., 1998, ApJ, 492, L83.
Kormendy, J., 1988, ApJ, 325, 128.
Kormendy, J. & Gebhardt, K., 2001, in AIP Conf. Proc. 586, The 20th Texas Symposium on Relativistic Astrophysics, eds, J. C. Wheeler & H. Martel (Melville:AIP), 363.
Magorrian, J. et al., 1998, AJ, 115, 2285.
Maoz, E., 1998, ApJ, 491, L181.
Miyoshi, M. et al. 1995, Nature, 373, 127.
Nelson, C., Bower, G. & Green, R. F., 2002, in preparation.
Richstone, D. et al. 1998, Nature, 395, A14.
Richstone, D. & Gebhardt, K., 2002, in preparation.
Salpeter, E. E., 1964, ApJ, 140, 796.
Sarzi, et al., 2001, ApJ, 550, 65.
Schmidt, M., Schneider, D. P., & Gunn, J. E., 1991, AJ, 101, 2004.
Soltan, A., 1982, MNRAS, 200, 115.
Tremaine, S. et al., 2002, ApJ, 574, 740.
Wandel, A., Peterson, B. & Malkan, M., 1999, ApJ, 526, 579.
Woodgate, B. E. et al., 1998, PASP, 110, 1183.
Yu, Q. & Tremaine, S., 2002, MNRAS, in press.

The Formation and Feeding of Massive Black Holes in the Early Universe

Wolfgang J. Duschl

Institute for Theoretical Astrophysics, University of Heidelberg, Germany; wjd@ita.uni-heidelberg.de

Peter A. Strittmatter

Steward Observatory, The University of Arizona, Tucson, AZ, USA; pstrittmatter@as.arizona.edu

Abstract. It is still an open question whether the super-massive black holes thought to be present in quasars are of primordial nature, or whether there is a viable way of forming them in the very short time scale (less than a billion years) permitted by the observational data. In this contribution, we present a way in which a galaxy-galaxy merger can provide not only the "fuel" for quasar activity, but can also build a super-massive black hole, i.e., "the engine", in the first place.

1. Introduction

During the last two decades dynamical studies of the nuclei of galaxies in the local Universe have lead to the conclusion that super-massive black holes of some $10^{6...9}\,M_\odot$ are commonly present (Kormendy 2001, and references therein). Actually, no firm counterexample has been found yet. The growth of these black holes to their current mass may be understood as a consequence of (intermittent) accretion lasting over the entire lifetime of the galaxies (Duschl 1988a, 1988b) at average rates of $10^{-4...-1}\,M_\odot\,\mathrm{yr}^{-1}$.

Soon after its discovery, the quasar phenomenon was attributed to accretion onto super-massive black holes. Accretion rates of $10^{-1...+1}\,M_\odot\,\mathrm{a}^{-1}$ and black hole masses of typically $10^{8...11}\,M_\odot$ (Lynden-Bell 1969) were inferred. However, the predominant presence of quasars at red-shifts $z > 1$ and their paucity in the local, contemporary Universe, demand a much more rapid and closed ended growth process of quasar black holes.

In this contribution, we will argue that the quasar phenomenon is a direct consequence of a major merger. As a "major merger" we define the coalescence of two galaxies of about equal mass resulting in the deposition of large amounts of gas in a disk close to the ccenter of the merged galaxy (Barnes & Hernquist 1996). Other galaxies which did not undergo such major mergers may also exhibit phases of nuclear activity, for instance as Seyfert galaxies, but in general harbor central black holes of considerably smaller masses than galaxies which hosted quasars.

While the occurrence of mergers as an important process for feeding quasars has been under discussion now for quite some time (e.g., Stockton 1999), we will show that the merger is not only instrumental in "providing the fuel", but even more for "building the engine" to produce the quasar phenomenon and accounting for the absence of quasars at the current epoch.

Before proceeding, however, we will briefly describe recent developments in the theory of self-gravitating accretion disks, which suggest accretion rates which exceed by more than an order of magnitude previous estimates. Then, we will give estimates which show how—as a consequence of a major merger—a super-massive black hole may develop in a galactic center within approximately $5 \cdot 10^8$ years, and how the quasar activity may set in, reach its maximum, and cease after about 10^9 years. Finally, we will note some open questions and work in progress.

2. Massive gas/dust disks and accretion

Whenever in an astrophysical flow angular momentum is not negligible, accretion disks come into play. The qualitative concept of accretion disks dates back to Kant (1755) and Laplace (1796) and their models for the formation of our planetary system. The modern quantitative description was first developed by Lüst (1952). The major obstacle to progress has been the lack of detailed understanding of the physical processes giving rise to the radial transport of mass (towards the disk's center) and angular momentum (in the opposite direction). A very successful, albeit originally purely heuristic, parameterization of the viscosity coefficient ν is due to Shakura (1972) and Shakura & Sunyaev (1973). They assumed that the viscosity in the flow is due to turbulence, and that the turbulent velocity and length scales are limited by the sound velocity c_s and the disk's thickness h. Together with a constant $\alpha \leq 1$ this led to the now famous "α-viscosity":

$$\nu = \alpha h c_s. \qquad (1)$$

Based on this viscosity prescription, the outbursts of dwarf novae and related systems could be explained to a surprisingly detailed level (Warner 1995). It turned out that for most astrophysical applications where such models were applicable at all, α was generally of order or slightly less than unity.

Later, a physical basis for this parameterization was re-discovered for the case of magnetic disks ("Balbus-Hawley instability", Balbus & Hawley 1991; but see also Velikhov 1959, Chandrasekhar 1960), but not for purely hydrodynamic disks. The fact that a physical process was known only for the magnetic case, should not, however, have led to the conclusion that the interaction of a magnetic field and the disk material is a necessary condition for angular momentum and mass transfer to occur.

Again, based on α-viscosity it has been shown that accretion disk models lead to exceedingly long evolution time scales for disks in the centers of AGN. To remedy this, various—mostly non-axisymmetric—processes (bars, spiral waves, etc.) were investigated in order to obtain sufficiently short time scales (e.g., Shlosman, Frank, & Begelman 1989; Chakrabarti & Wiita 1993) to account for the observed phenomenon.

In the meantime, however, it has been noted that, independent of the origin of the viscosity, α-disk models lead to physically altogether inconsistent results as soon as the mass of the disk is no longer small compared to the central, accreting body's mass, i.e., when the disks are self-gravitating (Duschl, Strittmatter, & Biermann 2000, hereafter DSB). DSB have also pointed out that experimental data on rotating fluids suggest an alternative, hydrodynamic origin of turbulence and hence of the viscosity.

2.1. Massive vs. mass-less accretion disks

For an accretion disk, self-gravity becomes important globally as soon as, at some radius s from the disk's center, the disk mass $M_{\text{disk}}(s)$ enclosed within this radius becomes comparable to or larger than the accreting body's mass M_a at the disk center. Then in the radial range where this is fulfilled, the disk is *fully self-gravitating* (FSG). But already at lower disk masses, self-gravity becomes important locally (in vertical direction). This happens when $M_{\text{disk}}(s)$ becomes larger than $(h/s) \cdot M_a$. For disk masses where local, but not yet global self-gravity is important, the azimuthal velocity v_φ is still given by a Keplerian rotation law, while the vertical structure is already dominated by self-gravity. Such disks are of the *Keplerian self-gravitating* type (KSG). Only for even lower masses the disks do not experience any important influence due to the gravitational forces of their own mass distributions, i.e., they are non-self-gravitating (NSG).

2.2. Viscosity parameterization in self-gravating disks

In both, KSG and FSG disks, a naive extrapolation of the α-parameterization leads to physically inconsistent results in that the disks would be isothermal in the radial direction. Recently a generalization of this parameterization was proposed (DSB) which solves the problem in the KSG and FSG cases, but recovers the α-parameterization for NSG disks (where this ansatz is so successful). Based on laboratory experiments and on theoretical considerations (Wendt 1933, Taylor 1936a, 1936b; see also Richard & Zahn 1999), the unconstrained viscosity ν is written as

$$\nu = \frac{1}{\Re_{\text{crit}}} s v_\varphi = \beta s v_\varphi \quad (2)$$

with the critical Reynolds number $\Re_{\text{crit}} \sim 10^{2...3}$ and $\beta = 1/\Re_{\text{crit}}$. As an additional constraint, the corresponding turbulent velocity $v_{\text{turb}} \sim \sqrt{\beta} v_\varphi$ is always required to be less than or equivalent to the sound speed. In the limiting case of a NSG accretion disk this latter condition leads to the α-viscosity prescription but gives a different result in all other cases (i.e., for subsonic turbulence or shock limited selfgravitating disks).

The timescale τ_{visc} can then be estimated to be of the order of

$$\tau_{\text{visc}} = \frac{s^2}{\nu}. \quad (3)$$

3. How to build and run a quasar engine

In the following, we assume that due to a major merger tidal forces have driven a large amount ($10^{9...10}\,M_\odot$) of accretable matter into the central regions (within a few 10^2 pc from the center) of the newly formed merged galaxy (Barnes & Hearnquist 1996) , where—due to its angular momentum—the material takes up a disk equilibrium configuration. This is a fairly robust assumption, unless the rare case happens that the net angular momenta of the two merging galaxies point (almost) exactly in the same direction. We also assume that there is no pre-existing super-massive black hole at the center of the merged galaxies. This scenario provides the starting point for our model.

3.1. A brief description of the physical model

We envisage that this self-gravitating gas (and dust) mass will start to accrete towards the center, independent of whether a seed black hole (of comparatively small mass) is present or not. Initially, the accretion disk is capable of radiating all energy liberated through dissipation. The huge amounts of mass moving towards the disk's center (see next paragraph for details) will lead to *(a)* the formation of a seed black hole (if none was present before), and *(b)* an initial phase of accretion into it which is Eddington-limited. While the details of the formation process of a seed black hole remain to be investigated, the assumption that it will form and that it does so very quickly seems to us to be unavoidable.

The black hole accretes at its Eddington rate as long as the disk delivers enough mass to maintain this rate. It is still an open question what happens to the mass which the accretion process supplies to the black hole, but which cannot be swallowed by it due to the effect of radiation pressure. One may speculate that this is an ideal source of material and energy for a jet and a broad line region to form. This process, however, is beyond the scope of this paper.

Ongoing accretion will deplete the mass of the accretion disk and thus decrease the mass delivery rate towards the black hole, while—at the same time—the black hole is growing in mass due to the same accretion process. This will continue in the way described above until the mass flow rate from the disk to the black hole has become smaller than the Eddington accretion rate. From this point on, free accretion sets in, and all the incoming mass may be accreted by the black hole. At the beginning of this phase, the accretion disk is still able to radiate all its liberated energy.

In the course of this evolution, however, the accretion rate drops, both in absolute terms as well as in units of the corresponding Eddington accretion rate. When the actual accretion rate falls below roughly 0.3 % of the Eddington rate, the flow becomes advection dominated (Beckert & Duschl 2002), and the radiation efficiency of the accretion process falls very quickly by several orders of magnitude. The luminosity decreases correspondingly.

Altogether, this leads to a three stage evolution of a quasar:

- **Eddington-limited phase:** During the phase in which accretion is limited by the Eddington rate, a sizable fraction of the central black hole is built up. This phase lasts as long as disk accretion is efficient enough that it exceeds the Eddington rate.

- **Free accretion phase:** As soon as the Eddington rate is no longer a relevant limit, all mass supplied by the disk can be accreted into the black hole. All the dissipated accretion energy can now be radiated. During this phase, the black hole accretion process runs at its maximum efficiency, although the rate itself and hence the luminosity declines with time (see Sect. 3.2.).

- **Advection-dominated phase:** When advection takes over, the actual quasar phase comes to an end (unless the galaxy encounters another major merger), and does so fairly abruptly. Due to the lack of large masses of accretable material, the growth of the black hole also effectively ceases. Minor episodes of accretion events may lead occasionally to comparatively short phases of enhanced activity. Such phases can also be experienced by non-quasar host galaxies and would be descibed as normal AGN of, for instance, Seyfert type.

A quasar model of this kind requires a very efficient underlying accretion mechanism. The process must be efficient enough to deliver sufficient mass from the disk towards its center so that the central black hole can grow quickly enough. Most quasars must reach their peak activity at a red-shift $z \sim 2$ (Hasinger 1998) in order to be compatible with the observed distribution of luminous quasars. Furthermore, the process involved must happen much more often in the $z > 1$ Universe and must almost shutdown at later epochs. This is indeed the case for major mergers.

3.2. Order-of-magnitude estimates

In the following, we will discuss a somewhat extreme example in order to show that the process discussed above is capable of giving rise to very massive quasar black holes. We assume an accretion disk of outer radius $s_{\text{disk}} = 100\,\text{pc}$ and initial mass of the disk of $10^{10}\,M_\odot$. The choice of the seed black hole mass is not crucial as long as it is considerably smaller than the disk mass. We estimate the accretion rate $\dot{M}$ at time t as

$$\dot{M}(t) = \frac{M_{\text{disk}}(t)}{\tau_{\text{visc}}(t)}. \tag{4}$$

In the same spirit, we estimate v_φ at the disk's outer radius s_{disk} as

$$v_\varphi(t) = \sqrt{\frac{GM_{\text{disk}}(t)}{s_{\text{disk}}}}. \tag{5}$$

Equations 2 – 5 then lead to $\dot{M}(t) \propto M_{\text{disk}}^{3/2}$. For the present estimates, we assume that the turbulent velocity is always subsonic. This, of course, has to be justified in later numerical model calculations.

With the above initial conditions, the mass flow rate deduced from Eq. 4 surpasses the Eddington rate by far at the beginning of the evolution. Subsequently, two counteracting effects come into play: The accretion process increases the black hole's mass and decreases the disk's mass. As a consequence, the Eddington rate increases, while the mass flow rate from the disk decreases.

In Eddington terms, the mass flow rate according to Eq. 4 becomes smaller and approaches unity. For the assumed initial disk parameters this transition point between Eddington-limited and free accretion is reached after $\sim 7 \cdot 10^8$ years, at a time when the central mass has reached $\sim 7 \cdot 10^9 \, M_\odot$. This is also the instant of peak accretion rate, which in the present case amounts to $\sim 100 \, M_\odot \, \mathrm{yr}^{-1}$.

From this point onwards, the mass flow rate is smaller than the Eddington rate, and the black hole accretion process is no longer Eddington-limited. What follows is the free accretion phase during which practically all mass supplied by the disk is accreted into the black hole. With a growing central mass and a shrinking disk mass, the mass flow rate decreases both in absolute as well as in Eddington terms. Again with the chosen parameters, the free accretion phase comes to an end after another $\sim 3 \cdot 10^8$ years when the mass flow rate drops to $\sim 0.5 \, M_\odot \, \mathrm{yr}^{-1}$, i.e., below approximately 0.3 % of its Eddington value and the flow becomes advection dominated. During this phase the central mass grows to $\sim 9 \cdot 10^9 \, M_\odot$, and the disk radiates all the dissipated energy.

The onset of advection dominated flow leads to a sharp drop in the radiation efficiency of the disk. Already within the first $\sim 1.5 \cdot 10^8$ years of the advection dominated phase, the radiation efficiency drops by about an order of magnitude while at the same time, the accretion rates drops by an additional factor of the same order. The disk radiates almost two orders of magnitude less than at the end of the free accretion phase, bringing the high luminous quasar phase to an abrupt end. In the course of time, the effect of advection becomes even stronger.

If we define—somewhat arbitrarily—the observable quasar phase to be the period during which the mass accretion rate is above $0.1 \, M_\odot \, \mathrm{yr}^{-1}$, then for this quasar its lifetime would have been somewhat less than 1 billion years.

4. Conclusions and Outlook

The essence of this scenario is that, within approximately one billion years the outcome of the merger has not only succeeded in "providing the fuel" for quasar activity, but also in "building the engine". As a consequence one then has to conclude that quasars occur in those galaxies which encountered major mergers during which large amounts of gas and dust were driven into the inner galaxies on short tidal time scales. Normal galaxies, in contrast, are those which never experienced a major merger. As a consequence, in today's Universe, the galaxies which once harbored a quasar have considerably more massive black holes in their centers ($10^9 \, M_\odot$ and more) then other galaxies.

For quasars an efficient accretion process is required which delivers enough material to the galaxy's center so that a black hole can grow quickly. This accretion process also has to be efficient enough not only to grow the black hole, but also to accrete away most of the available gas and dust and thus lead to a rapid end of the quasar phenomenon due to a combination of a drop in the accretion rate as well as in the disk's radiation efficiency. For this all to happen, the disk's viscosity is *the* crucial quantity. With the newly proposed generalization (DSB), self-gravitating disks are capable of doing this.

Currently we are carrying out numerical model calculations for the evolution of such disks in order to investigate in more detail the scenario presented here.

The results must also be compared with both quasar and merger statistics in order to test the assumed underlying physical processes.

While the scenario described above does not exclude the existence of primordial super-massive black holes in the young Universe, it makes their existence unnecessary as far as the quasar phenomenon is concerned.

References

Balbus, S.A., Hawley, J.F., 1991, ApJ 376, 214
Barnes, J.E., Hernquist, L., 1996, ApJ 471, 115
Beckert, T., Duschl, W.J., 2002, A&A (in press)
Chakrabarti, S.K., Wiita, P.J., 1993, ApJ 411, 602
Chandrasekhar, S., 1960, Proc. Natl. Acad. Sci. USA 46, 253
Duschl, W.J., 1988a, A&A 194, 33
Duschl, W.J., 1988b, A&A 194, 43
Duschl, W.J., Strittmatter, P.A., Biermann, P.L., 2000, A&A 357, 1123 (= DSB)
Hasinger, G., 1998, Astron. Nachr. 319, 37
Kant, I., 1755, Allgemeine Naturgeschichte und Theorie des Himmels oder Versuch von der Verfassung und dem mechanischen Ursprung des ganzen Weltgebäudes, Nach Newtonschen Grundsätzen abgehandelt
Kormendy, J., 2001, ASP Conf. Ser. 230, 247
Laplace, P.-S. de , 1796, Exposition du système du Monde
Lüst, R., 1952, Naturforsch. 7a, 87
Lynden-Bell, D., 1969, Nature 223, 690
Richard, D., Zahn, J.-P., 1999, A&A 347, 734
Shakura, N.I., 1972, AZh 49, 921 (translated in 1973, Soviet Ast. 16, 756)
Shakura, N.I., Sunyaev, R.A., 1973, A&A 24, 337
Shlosman, I., Frank, J., Begelman, M.C., 1989, Nature 338, 45
Stockton, A., 1999, IAU-Symp. 186, 311
Taylor, G.I., 1936a, Proc. Roy. Soc. London 157, 546
Taylor, G.I., 1936b, Proc. Roy. Soc. London, 157 565
Velikhov, E.P., 1959, Sov. Phys. JETP 36, 995
Warner, B., 1995, Cataclysmic Variable Stars, Cambridge University Press, Cambridge, UK
Wendt, F., 1933, Ingenieur-Archiv IV, 577

Quasar Variability: New Surveys and New Models

M.R.S. Hawkins

University of Edinburgh, Royal Observatory, Blackford Hill, Edinburgh EH9 3HJ, Scotland, UK

Abstract. In this paper results from a monitoring programme of a large sample of quasars comprising regular yearly observations over a period of 23 years are presented. Structure functions of the light curves are calculated and compared with predictions for models of quasar variability of current interest. These include recently published models of variability from accretion disk instability, variability from starbursts or supernovae, and variations caused by the microlensing effect of compact bodies along the line of sight. The analysis favours the accretion disk model for low luminosity AGN, but suggests that the variations of more luminous quasars are dominated by microlensing.

1. Introduction

It has long been understood that one of the keys to understanding the structure of active galactic nuclei (AGN) lies in the nature of the observed variability of their flux. Short term variations were used shortly after the discovery of the first quasars to put limits on the size of the emitting region, but it seems fair to say that since then, optical variability has been hard to characterise and use to constrain AGN structure. In fact much effort has gone into quasar monitoring programmes (Cristiani et al. 1996; Hawkins 1996; Hook et al. 1994) but the results have been hard to interpret. There appear to be several reasons for this. Perhaps the most significant is that quasars vary with most power on a timescale of decades, and monitoring programmes have been in existence this long. A related problem concerns the need for evenly spaced homogeneous data. Few monitoring programmes have achieved this, presenting serious difficulties in analysis.

There are also other difficulties independant of the quality of the data available. To distinguish between various quasar models, predictions of the spectrum of fluctuations must be made which can be compared with the observations. Until recently, little was published in this area, and even general expectations seemed to be at odds with observed timescales and amplitudes of variation. However, a big step forward was made with the publication (Kawaguchi et al. 1998) of predicted structure functions for a model of accretion disc instability, together with similar predictions for a starburst model of variability. In this paper we test these models together with predictions for microlensing against the observed structure function for a large sample of quasars and Seyfert galaxies.

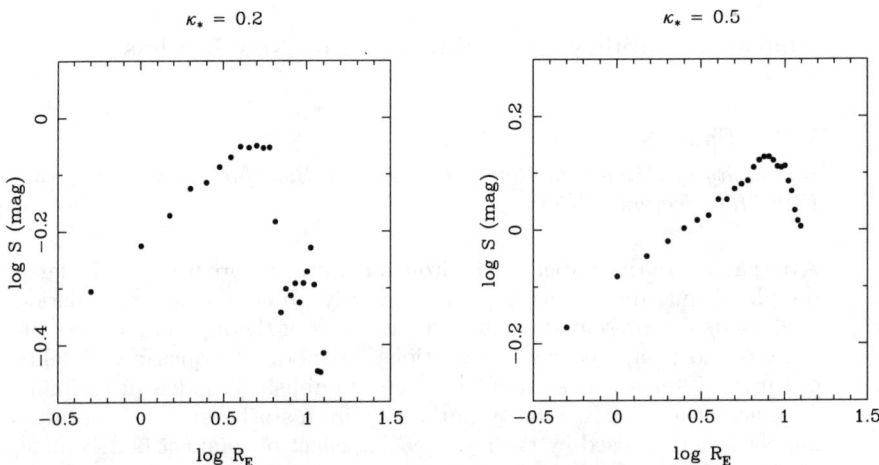

Figure 1. Structure functions for simulated microlensing light curves fo model b of Schneider & Weiss (1987) (left hand panel), and from Fig. 2(b) of Lewis et al. (1993) (right hand panel).

2. Model Predictions

In the standard model of AGN, the central power source is taken to be a massive black hole powering an accretion disc. The most obvious source of the observed flux variations is some form of instability or oscillation in the accretion disc, but so far this has proved very hard to model in such a way that the predicted variations are at all like the observed ones, even for basic parameters such as timescale and amplitude. Furthermore there has so far been little published in the form of quantitative statistical predictions which can be compared with observations.

A great step forward has recently been made by Kawaguchi et al, (1998) with their cellular-automaton model of accretion disk instability. They use this model to generate artificial light curves from which they calculate structure functions which can be compared with those of observed AGN light curves. The overall shape of the model structure functions indicates a power law relationship with inreasing power on long timescales, until an eventual turnover. Although the amplitude of the variations and the timescale of the break are dependent on the free parameters in their model, the logarithmic slope of the structure function is robust, with a value of 0.44 ± 0.03.

Kawaguchi et al. (1998) also carry out a similar analysis for the starburst model of AGN variability (Aretxaga & Terlevich 1994). In this model the nuclear light of AGN is the result of a sequence of supernova explosions, and the observed variations come from the stochastic nature of the burst timings. Kawaguchi et al. (1998) produce simulated light curves based on this model, and as for the accretion disc model, calculate structure functions. Again, the overall shape shows a power law relationship with a long timescale turnover, the position of which is dependent on model parameters. However, as with the accretion disc model, the logarithmic slope of the structure function is relatively stable with a value of 0.83 ± 0.08.

As well as models for the intrinsic variability of AGN, it has also been argued (Hawkins 1993) that the observed variations in quasars may be dominated by the effects of gravitational microlensing. In this picture it is postulated that a population of compact bodies criss-cross the line of sight to the quasar, causing the light to fluctuate from the effects of gravitational lensing. A number of groups have carried out simulations of the resulting light curves (Lewis et al. 1993; Schneider & Weiss 1987), and they have been published in such a form that it is possible to calculate the structure functions for comparison with the data.

Fig. 1 shows structure functions for long runs of simulated data from two different groups, digitised to enable comparison with observations. The data is presented as a function of R_E, the Einstein radius of the lenses, and to turn this into a timescale an average velocity across the line of sight must be applied. The relation has a power law form with a break towards long timescales. The power is dependent on the optical depth of the lenses but the logarithmic slope appears to be robust for different simulations with a value of 0.25 ± 0.03.

It will be seen that the three models we have considered in this section all predict structure functions with a power law form increasing in power towards an eventual turnover at long timescales. However, each model makes a different prediction for the logarithmic slope of the structure function, that is, 0.83 ± 0.08, 0.44 ± 0.03 and 0.25 ± 0.03 for the starburst, disc instability and microlensing models respectively. All three slopes are significantly separated from each other, and are well suited for comparison with observations to determine the most favoured model of variability.

3. Observed Structure Functions

Over the last 25 years or so, a large scale quasar minitoring programme has been undertaken using automated measures of UK 1.2m Schmidt plates (Hawkins 1996) taken on a variety of timescales from hours to years. Of particular interest for the present work is an unbroken series of yearly observations from 1977 to 2000 in the B_J passband (IIIa-J/GG395 emulsion/filter combination). The plates have a useful limiting magnitude of $B_J \sim 21.5$ and the monitoring area comprises the central 19 deg^2 of the Schmidt field. Some 180,000 objects were detected in this area of which around 1500 are likely to be quasars. So far 610 of these have been confirmed as quasars with measured redshifts. Each of these has a light curve covering 23 years and together they form the parent sample for the present investigation.

Initial examination of the light curves from the monitoring programme suggested that the low luminosity AGN with $M_B > -23$ (Seyfert galaxies) varied in a qualitatively different way to the more luminous objects with $M_B < -23$ (quasars). Accordingly, in the first instance the observed structure function was calculated for quasars alone. The result is plotted in Fig. 2, and it will be seen that it has an approximately power law form, with a logarithmic slope of 0.20 ± 0.01.

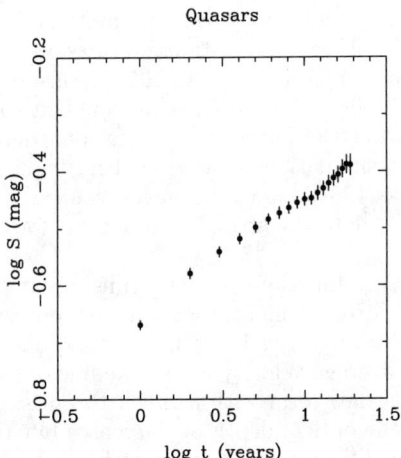

Figure 2. Structure functions for the light curves of a sample of 401 quasars from the survey of Hawkins (1996).

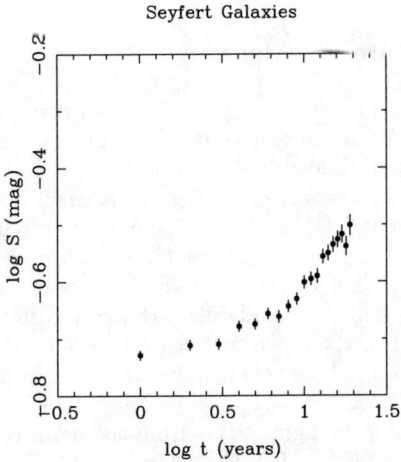

Figure 3. Structure functions for the light curves of a sample of 45 Seyfert galaxies from the survey of Hawkins (1996).

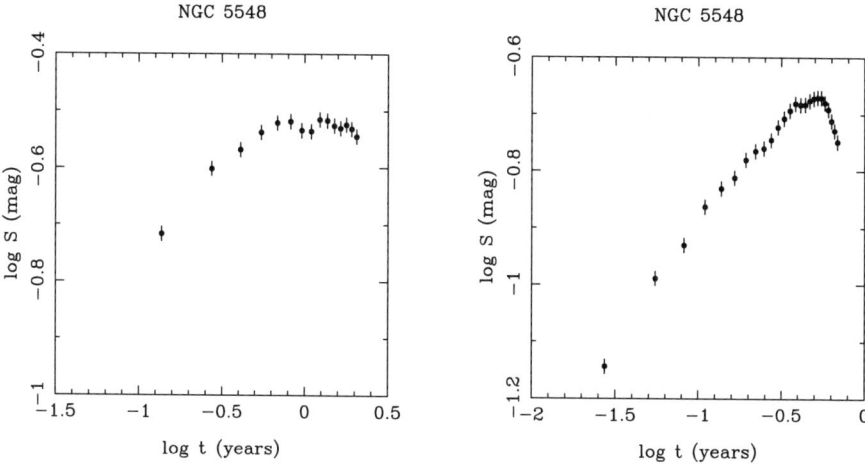

Figure 4. Structure functions for the light curve of NGC 5548 from Peterson et al. (1999) and references therein for 50 day intervals (left panel) and 10 day intervals (right panel).

If we now turn our attention to the Seyfert galaxies in the sample a rather different picture emerges. The structure function for these low luminosity objects is shown in Fig. 3 and it is immediately clear that there is much less power than for the quasars. Of more importance for the present investigation is the fact that the slope of the linear part is 0.36 ± 0.02, significantly steeper than for quasars. Towards short timescales the structure function becomes flat, which appears to be where the signal becomes dominated by noise.

This result can be tested by making use of results from the extensive monitoring programme of NGC 5548 by Peterson et al. (1999). This Seyfert galaxy has been monitored for many years on a timescale of a few days in optical wavelengths, and Fig. 4 shows the result of calculating structure functions from parts of the data sampled in a uniform manner. The left hand panel sampled at 50 day intervals shows a structure function with a long timescale turnover, which never really becomes truly linear at shorter timescales. In the right hand panel the data is sampled every 10 days, and here it is clear that there is an extended linear portion, with logarithmic slope of 0.38 ± 0.01. This is close to the value obtained for the Seyfert galaxies in the wide field monitoring programme.

4. Discussion

The results of the preceding two sections are summarised in Table 1. For Seyfert galaxies it is clear that the favoured model for the observed variations is accretion disc instability. The starburst model is ruled out at a high level of significance, and there appears to be no reason to attribute the variations to microlensing.

Table 1. Structure Function Slopes

Model Predictions

Model	Slope
Starburst	0.83 ± 0.08
Disc Instability	0.44 ± 0.03
Microlensing	0.25 ± 0.03

Observations

AGN Class	Slope
Seyfert Galaxies	0.36 ± 0.02
Quasars	0.20 ± 0.01

For quasars, the position is different. Here again the starburst model is ruled out, even more strongly, but the accretion disc instability model is also inconsistent with the observations. It is of course possible that the parameter space explored by Kawaguchi et al. (1998) is too small, or that other disc instability models might make different predictions for the structure function slope; future work may well clarify this. For the moment, the best match for the observed quasar structure function is from the predictions for microlensing.

The implication of Table 1 is that, while Seyfert galaxy variation is caused by accretion disc instability, quasar variability is dominated by the effects of microlensing. The most plausible explanation for this would appear to be that at low luminosity the observed variations in AGN are caused by accretion disc instability, but for more luminous objects intrinsic variations become smaller in amplitude. However, at the higher redshifts at which quasars are found, microlensing becomes dominant due to the greater optical depth of lenses.

References

Aretxaga, I. & Terlevich, R. 1994, MNRAS, 269, 462

Cristiani, S., Trentini, S., La Franca, F., Aretxaga, I., Andreani, P., Vio, R., Gemmo, A. 1996, A&A, 306, 395

Hawkins, M.R.S. 1993, Nature, 366, 242

Hawkins, M.R.S. 1996, MNRAS, 278, 787

Hook, I.M., McMahon, R.G., Boyle, B.G., Irwin, M.J. 1994, MNRAS, 268, 305

Kawaguchi, T., Mineshige, S., Umemura, M., Turner, E.L. 1998, ApJ, 504, 671

Lewis, G.F., Miralda-Escudé, J., Richardson, D.C., Wambsganss, J. 1993, MNRAS, 281, 647

Peterson, B.M. et al. 1999, ApJ, 510, 659

Schneider, P., Weiss, A. 1987, A&A, 171, 49

Rapid Variations in the Broad Hβ Profile of the Radio Galaxy 3C 390.3: Possible Evidence for Turbulence in the Accretion Disk.

N. S. Asatrian, E. Ye. Khachikian

Byurakan Astrophysical Observatory, 378433 Byurakan, Armenia

P. Notni

Astrophysikalisches Institut Potsdam, An der Sternwarte 16, D-14482 Potsdam, Germany

We report on implications for the geometrical and kinematic parameters of BLR gas on the basis of short timescale variability in the broad $H\beta$ profile.

Data on rapid variations have been obtained at the 6-m telescope of the SAO (Asatrian, Khachikian & Notni, 1999). To search for variations in the profile shape, difference spectra (first *minus* second epoch) were examined. We believe that the structure of the underlying stellar continuum and the atmospheric features do not affect the $H\beta$ difference profiles of 3C 390.3 significantly.

Variations occurred simultaneously on the blue and red sides of $H\beta$ on a timescale of $\sim$ 1.452 hours and take the form of three narrow, positive and negative small bumps drifting across the line profile in the difference spectrum. The positions of the bumps are -2300, +4700 (negative) and -3700 $km\,s^{-1}$ (positive).

These changes may indicate the response of circularly rotating emitting gas at two orbits to a light pulse from a central source. In this case the two bumps observed at -2300 and 4700 $km\,s^{-1}$ are formed in two opposite zones at the outer orbit close to the line of nodes. On the assumption that the inner and the outer orbits lie in the same plane around a central massive object, orbital parameters (radius, velocity and the inclination angle of the orbital plane) of the clouds and the central mass can be calculated. The shift of a bump is defined by the combination of the relativistic Doppler effect due to the Keplerian orbital motion and the gravitational redshift. The three observed radial velocities are determined by three parameters: the inner and outer orbital radii, R_{in}, R_{out}, (or velocities, V_{in}, V_{out}) and the inclination angle i of the rotation plane. Thus, the expressions for the radial velocities form a system of three nonlinear algebraic equations with three unknowns and can be solved numerically. Using the difference of the orbital radii in absolute units ($R_{out} - R_{in} = \Delta t\, C$, where $\Delta t \simeq 1.452$ hours and C is the speed of light) we can derive the central mass M. The results are: $R_{in} \simeq 63^{+18}_{-14} R_g$, $R_{out} \simeq 247^{+3}_{-4} R_g$,
$V_{in} \simeq 26600^{+3300}_{-3100} km\,s^{-1}$, $V_{out} \simeq 13500^{+110}_{-90} km\,s^{-1}$,
$i \simeq 15.01°\,^{+0.07}_{-0.04}$ and $M \simeq 2.9^{+0.3}_{-0.2} \times 10^6 M_\odot$.

However, the value of the mass obtained is smaller by two orders of magnitude than estimates for the dynamical mass of the nucleus of 3C 390.3 obtained in different ways (Wandel et al, 1999). Such a low mass is excluded, therefore.

The discrepancy in the mass may be due to possible macroturbulence in the BLR. The sizes of macroturbulence cells are assumed to be greater than or equal to the sizes of the bump emission regions. The gas associated with each bump is assumed to take part in two motions: the common rotation with the local Keplerian velocity and the macroturbulent motion. These two motions are either added or subtracted for each bump emission region. If we assume the presence of such a macroturbulence then the radial velocity V_r of a bump will be defined by

$$1 + \frac{V_r}{C} = \frac{1 - (V \pm V_t \cos\varphi)\cos\theta/C}{\sqrt{1 - (V \pm V_t \cos\varphi)^2/C^2}} \left(1 + \frac{R_g}{2R}\right),$$

where V and V_t are the orbital and turbulent velocities of the cloud, φ is the azimuthal angle of V_t about V, $|\varphi| < 90°$, θ is the angle between the direction of orbital movement and the line-of-sight at the nodes of orbits, $i = |90 - \theta|$, $R_g = 2GM/C^2$ is the gravitational radius, $R = GM/V^2$, and G is the gravitational constant. Adopting a value $M = 3.91^{+1.2}_{-1.5} \times 10^8 M_\odot$ (Wandel et al. 1999) for the mass, we can solve for the remaining parameters trying possible combinations of the directions of the velocities V and $V_t \cos\varphi$ at the two nodes. Only one out of 8 possibilities leads to real solutions. Error limits for the results are determined by the uncertainty of the input radial velocities, $300\,km\,s^{-1}$.

The overall solution is given in Table 1. At these high masses, it is nearly independent of the mass, and the inner and outer radii are barely different in R_g units. The mean radius of orbits is $R_{orb} \simeq 526^{+253}_{-136} R_g$ or 24^{+22}_{-13} light-days. The solution is independent of φ as long as the resulting V_t is less than the escape velocity of the turbulent cells, which we assume as the limiting allowable velocity. This is the case for $|\varphi| < 48°$. Outside of this limit, the ranges of solutions for R_{orb}, i and V_t become successively smaller. R_{orb} and i remain within the upper and lower limits given in Table 1, and the upper limit of V_t increases to $4450\,km\,s^{-1}$. The maximum φ at which we have found a solution is $66°$.

Table 1. Parameters of BLR gas

R_{in} in (R_g)	R_{out} in (R_g)	i in (°)	$V_t \cos\varphi$ in (km s^{-1})
526^{+253}_{-136}	525^{+253}_{-137}	$21.8^{+4.2}_{-2.6}$	2050^{+800}_{-700}

The results are consistent with the predictions for an accretion disk model for 3C 390.3 proposed by Eracleous & Halpern (1994). In addition, our estimates of R_{orb} and i are in agreement with the size of the maximum emission region (to order of magnitude) and its inclination derived from Hβ profile fitting (Shapovalova et al. 2001) and with the BLR size measured by the reverberation-rms method (Wandel et al. 1999). Thus, our results support the presence of a relativistic accretion disc with supersonic turbulence in the nucleus of 3C 390.3.

References

Asatrian, N. S., Khachikian, E. Ye., & Notni, P. 1999, in IAU Symp. 194, Activity in galaxies and related phenomena, ed. Y. Terzian, D. W. Weedman & E. Ye. Khachikian, (San Francisco: ASP), 406

Shapovalova, A. I. et al. 2001, A&A, 376, 775

Eracleous, M., & Halpern, J. P. 1994, ApJS, 90, 1

Wandel, A., Peterson, B. M., & Malkan, M. A. 1999, ApJ, 526, 579

Rapid Profile Variations in the Broad Hα Line of the Seyfert Galaxy Markarian 6: Possible Evidence for Turbulence in the Accretion Disk

N. S. Asatrian, E. Ye. Khachikian

Byurakan Astrophysical Observatory, 378433 Byurakan, Armenia

P. Notni

Astrophysikalisches Institut Potsdam, An der Sternwarte 16, D-14482 Potsdam, Germany

We report on implications for the geometrical and kinematic parameters of BLR gas on the basis of short timescale variability in the broad $H\alpha$ profile shape. Data on rapid variations have been obtained at the 2.6-m telescope of the BAO (Asatrian, Khachikian & Notni, 1999). To search for variations in the profile, difference spectra (second *minus* first epoch) were examined. We believe that the structure of the underlying stellar continuum and the atmospheric features do not affect significantly the $H\alpha$ difference profiles of Mark 6. Variations occurred simultaneously on the blue and red sides of $H\alpha$ on a time scale of $\simeq 50.7$ minutes and take the form of three narrow, positive small bumps on each side in the difference spectrum. The positions of the bumps are -4400, -3100, -1700 and +1900, +4200 and +6600 $km\,s^{-1}$. These changes may indicate the response of circularly rotating emitting gas at three orbits to a light pulse from a central source. In this case the pairs of blue and red bumps observed at -4400 and +6600, -3100 and +4200, and -1700 and 1900 $km\,s^{-1}$ are formed in two opposite zones of gas close to the line of nodes. On the assumption that these orbits lie around a central massive object, orbital parameters (radii, velocities and inclination angles of orbital planes) of the clouds and the central mass can be found. The shift of each bump is defined by the combination of the relativistic Doppler effect due to the Keplerian orbital motion and the gravitational redshift. The six observed radial velocities are determined by six parameters: the orbital radii, R_1, R_2, R_3 (or velocities, V_1, V_2, V_3) and the inclination angles i_1, i_2, i_3 of the rotation planes. Thus, the expressions for the radial velocities form a system of six algebraic equations with six unknowns and can be solved. Using the difference of the orbital radii in absolute units ($R_3 - R_1 = \Delta t\, C$, where $\Delta t \simeq 50.7$ minutes and C is the speed of light) we can derive the central mass M. Analytical solution gives: $R_1 \simeq 200^{+80}_{-40} R_g$, $R_2 \simeq 410^{+500}_{-130} R_g$, $R_3 \simeq 2250^{+7750}_{-1350} R_g$, $V_1 \simeq 14800^{+1900}_{-2200} km\,s^{-1}$, $V_2 \simeq 10500^{+2100}_{-3400} km\,s^{-1}$, $V_3 \simeq 4500^{+2500}_{-2400} km\,s^{-1}$, $i_1 \simeq 22°^{+5}_{-4}$, $i_2 \simeq 20°^{+13}_{-4}$, $i_3 \simeq 24°^{+38}_{-10}$ and $M \simeq 1.5^{+3.3}_{-1.2} \times 10^5 M_\odot$. Error limits for the results are determined by the uncertainty of the input radial velocities, $300\,km\,s^{-1}$. However, the value of the mass obtained is smaller by two orders of magnitude than the estimate for the dynamical mass of the nucleus of Mark 6 ($M = 1 \times 10^7 M_\odot$, Dibai 1984). Such a low mass is excluded, therefore.

The discrepancy in the mass may be due to possible macroturbulence in the BLR. The sizes of macroturbulence cells are assumed to be greater than or equal to the sizes of the bump emission regions. The gas associated with each bump is assumed to take part in two motions: the common rotation with the local Keplerian velocity and the macroturbulent motion. These two motions are either added or subtracted for each bump emission region. If we assume the presence of such a macroturbulence, then the radial velocity V_r of a bump will be given by

$$1 + \frac{V_r}{C} = \frac{1 - (V \pm V_t \cos\varphi)\cos\theta/C}{\sqrt{1 - (V \pm V_t \cos\varphi)^2/C^2}} \left(1 + \frac{R_g}{2R}\right),$$

where V and V_t are the orbital and turbulent velocities of the cloud, φ is the azimuthal angle of V_t about V, $|\varphi| < 90°$, θ is the angle between the direction of orbital movement and the line-of-sight at the nodes of orbits, $i = |90 - \theta|$, $R_g = 2GM/C^2$ is the gravitational radius, $R = GM/V^2$, and G is the gravitational constant. Adopting the value $M = 1 \times 10^7 M_\odot$ for the mass, we can solve for the remaining parameters trying various combinations of the velocities V and $V_t \cos\varphi$ at the two nodes. Only one out of 64 possibilities leads to real solutions. The overall solution is given in Table 1. The solution is independent of φ as long as the resulting V_t is less than the escape velocity of the turbulent cells, which we assume as the limiting allowable velocity. This is the case for $|\varphi| < 5°$. At greater $|\varphi|$, the ranges of solutions for R, i and V_t remain within the upper and lower limits given in Table 1. The maximum φ at which we have found a solution is 48°.

Table 1. Parameters of BLR Gas

R_1 (R_g)	R_2 (R_g)	R_3 (R_g)	i_1 (°)	i_2 (°)	i_3 (°)	$V_t \cos\varphi$ (km s^{-1})
269^{+21}_{-40}	297^{+13}_{-43}	300^{+21}_{-40}	19^{+2}_{-3}	25^{+2}_{-3}	9^{+1}_{-1}	4000^{+1400}_{-700}

Note that if we assume the same inclination angle for the three different orbital planes, we can form a system of equations with unknowns R_1, R_2, R_3 (or V_1, V_2, V_3), i, $V_t cos(\varphi)$, M, which do not contain any free parameters and include the mass. However, the numerical solution of this system gives an extremely low value for the mass ($\sim 10 M_\odot$) and must be rejected, therefore. We are forced to interpret the different values of the inclination angles as resulting from inhomogeneities in the accretion zone. Further high sampling rate spectroscopic monitoring at higher quality is evidently required for a more reliable study of the structure and kinematics of the BLR. The parameters of the BLR gas found here for Mark 6 are very similar to those found for 3C 390.3 (see our other poster in this volume) and favor models of a relativistic accretion disk with supersonic turbulence in the BLR.

References

Asatrian N. S., Khachikian E. Ye., & Notni P. 1999, in IAU Symp. 194, Activity in galaxies and related phenomena, ed. Y. Terzian, D. W. Weedman & E. Ye. Khachikian, (San Francisco: ASP), 409

Dibai, E. A. 1984, AZh, 61, 417

Spectral Variability of Some Seyfert Galaxies

E.K. Denissyuk, V.N. Gaisina, R.R. Valiullin

Fesenkov Astrophysical Institute, Kamenskoje Plato, Almaty, 480020, Kazakhstan. Electronic mail: (denis,gaisina,rashit)@afi.south-capital.kz

Abstract. The very first spectrograms of 40 Seyfert galaxies are discussed. All these objects were taken from Markarian's lists of galaxies. They were first investigated and classified at Fesenkov Astrophysical Institute during 1973-1988. Spectral study was carried out with a slit spectrograph, attached to the 70-cm telescope and equipped with a three cascades image-tube. Repeated observations of 22 galaxies were obtained in 1985-2000. An analysis of possible spectral variability of the studied galaxies over 10-26 years was made. A list of objects which are the most interesting for further observations is presented.

Spectral Observations and Results

The very first spectrograms of 40 Seyfert galaxies from Markarian's lists of galaxies were studied and classified at Fesenkov Astrophysical Institute during 1973-1988. Spectral observations were carried out with a slit spectrograph, attached to the 70-cm telescope and equipped with a three cascades image-tube. Spectrograms cover the wavelength region 6200-6800ÅÅ.

The list of galaxies: numbers from Markarian's lists, dates of the first observations, values of V magnitudes and Z are given in the Table together with the main characteristics of the spectra, such as availability of emission lines and widths of the broad emission profiles.

During the last ten years, spectra of more than half of this group were obtained repeatedly, using the same instrument and optical systems. A comparison of the early and recent observational data reveals any variations in the spectra during 20-30 years. Thus we can identify the objects that are rather promising for follow-up detailed investigations on the basis of their variability.

For example, changes in the FWHM of broad Hα, as high as 30 or more percent were registered in the spectra of Mrk 595, 705, 841 and 1040. These objects are considered to be candidates for further detailed investigations.

One more object –the galaxy Mrk 926– looks very promising for the searching of variations. It is rather bright and has a very broad emission Hα.

References

Lipovetsky, V.A. et al. 1987, Communications of SAO, No55

Table 1. Seyfert Galaxies Studied in the Fesenkov Astroph.Inst.

Mrk Number	V (mag)	$Z(FAI)/Z^a$	Date of the First Observ.	FWHM $H\alpha$ $km\,s^{-1}$	b	c	d
463	13.81	0.0506/0.0506	08.02.73	2650	-	B	+
464	16.12	0.0510/0.0510	03.02.73	6000	-	B	
474	15.25	0.0360/0.0396	01.03.73	4500	+	B	
486	14.78	0.0397/0.0397	18.10.73	2500	+	B	+?
504	15.78	0.0373/0.0373	20.10.73	4380	+	B	
595	14.42	0.0275/0.0275	08.02.73	4360	+	M	
609	14.4	0.0345/0.0345	18.01.74	4300	+	M	+?
618	13.56	0.0357/0.0357	26.01.73	2900	+	B	
620	11.92	0.0068/0.0069	24.01.73	2500	+	M	+
646	15.28	0.0537/0.0537	21.01.74	1100	-	M	+?
668	14.98	0.0797/0.0797	15.02.74	2100	-	M	
699	15.11	0.0348/0.0348	21.02.74	4000	+	B	
704	13.51	0.0290/0.0294	23.02.76	5500	+	B	+?
705	14.52	0.0282/0.0288	23.02.76	3200	+	M	+
707	14.88	0.0492/0.0505	26.11.76	2200	-	M	
715	15.45	0.0841/0.0841	27.02.76	2900	-	F	+?
716	16.5	0.0580/0.0574	27.02.76	3700	+	F	
720	15.1	0.0450/0.0451	27.02.76	2000	+	F	
734	14.81	0.0492/0.0497	23.02.76	2900	+	F	+
739	13.81	0.0300/0.0296	23.02.76	1800	-	M	+?
744	13.85	0.0097/0.0092	02.03.76	3300	+	F	+?
745	14.56	0.0101/0.0101	27.11.76	1500	-	F	+
766	12.64	0.0128/0.0128	23.02.76	3000	+	M	+?
771	14.93	0.0630/0.0632	02.03.76	2300	-	M	
817	13.62	0.0321/0.0321	23.02.76	3000	+	M	
841	14.48	0.0365/0.0365	23.01.76	4300	+	M	
871	14.22	0.0337/0.0337	25.02.76	2700	+	F	
896	14.61	0.0268/0.0262	20.11.76	2300	-	M	+?
926	13.82	0.0478/0.0475	21.11.76	10400	+	M	
928	14.14	0.0249/0.0249	26.11.76	2400	-	B	+
937	13.8	0.0301/0.0301	23.11.76	1750	-	M	+
955	14.01	0.0352/0.0351	24.11.76	3300	+	F	+?
957	15.14	0.0740/0.0711	20.11.76	1900	-	F	+?
975	14.19	0.0498/0.0498	20.11.76	5300	+	F	+?
993	13.39	0.0169/0.0169	24.11.76	2700	+	F	+?
1040	13.26	0.0173/0.0164	20.11.76	3800	+	M	+
1044	13.67	0.0164/0.0164	19.11.76	2400	+	B	+?
1048	13.38	0.0424/0.0424	23.11.76	2000	+	M	+?
1058	14.35	0.0174/0.0174	21.11.76	1400	-	M	+?
1095	12.87	0.0330/0.0327	21.11.76	4500	+	B	

[a] Lipovetsky et al. (1987).
[b] Column b shows presence of our repeated observations.
[c] Column c determines contrast of maximal intensity of $H\alpha$ relatively to continuum: B-"bright", M-"middle", F-"faint".
[d] Column d shows a presence of other emission lines besides $H\alpha$.

Analysis of Color Variability of BL Lac during the 1997 and 1999 Outbursts

V. A. Hagen-Thorn, V. M. Larionov, A. V. Hagen-Thorn, S. G. Jorstad & G. O. Temnov

Astronomical Institute, St. Petersburg University, St. Petersburg, Petrodvorets, 2, Bibliotechnaya pl. 198504, Russia

Abstract. The analysis of multicolor observations of BL Lac obtained in the 1997 and 1999 outbursts shows that in both cases the spectral energy distribution of the variable source was unchanged in the course of outburst. The sources have power-law spectra with slightly different spectral indices (the spectrum was flatter in the more powerful outburst of 1997). The variable source is most probably of synchrotron nature.

For the solution of the problem of nuclear activity in galaxies, it is very important to know the spectral energy distribution of the central point source responsible for the activity. Unfortunately, this source is not observed in a pure state because of the contribution of the underlying galaxy to the total observed flux. But in some cases, the distribution may be found on the basis of multicolor photometric observations of variability without knowledge of the contribution of the variable source to the total flux.

As we pointed out many times (for details see Hagen-Thorn & Marchenko 1999), the location of the points corresponding to the fluxes observed in different spectral bands in the "flux-flux diagrams" on straight lines allow the conclusion that the spectral energy distribution of the variable source is constant. The slopes of the straight lines give its relative spectral energy distribution.

The 1997 outburst of BL Lac was the most powerful during the last 20 years. Its multicolor photometric observations were made by several teams. Here we analyze the data published in Webb et al. (1998) and Sobrito et al. (1999). In 1999, another outburst of BL Lac occurred. In this case, we analyze the results of our own observations. To find fluxes from magnitudes, the absolute calibration from Mead et al. (1990) was used.

Figs. 1a,b give "flux-flux diagrams" for both outbursts. It is evident that in both cases the observed points lie on straight lines quite well. Thus we conclude that in both outbursts the spectral energy distribution of the variable source was constant. The slopes of the lines give the observed ratios $(F_i/F_R)^{obs}$, which were then corrected for interstellar reddening using the normal extinction curve.

The logarithms of the ratios $(F_i/F_R)^{corr}$ are compared in Fig. 1c with $\log \nu$ giving the relative spectral energy distributions of the variable sources. Because of the logarithmic scale, we may give an arbitrary vertical shift to one distribution relative to another. The distribution for 1997 is shifted upwards in accordance with the difference in maximum flux in the R band.

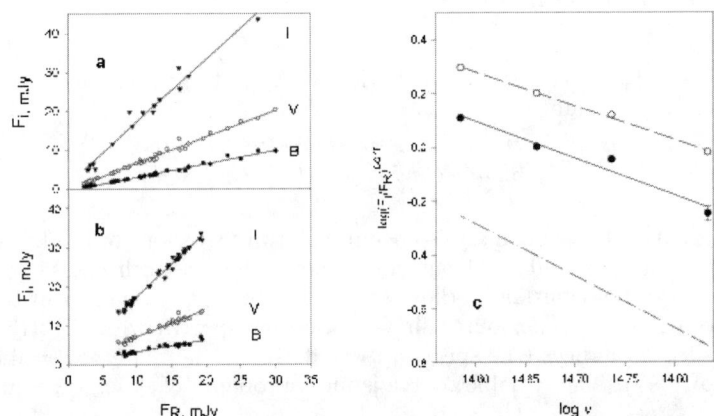

Figure 1. "Flux-flux diagrams" for 1997 (a) and 1999 (b). Spectra of the variable sources (c): filled circles – 1999, open circles – 1997, dashed line – 1983-84.

In both cases the points lie on straight lines showing power-law spectra. Bearing in mind the high observed polarization of BL Lac, one can conclude that the variable sources are of synchrotron nature.

We find the slopes of the lines (spectral indices): $\alpha = -1.29 \pm 0.06$ for 1997 and $\alpha = -1.41 \pm 0.11$ for 1999. The spectrum appears to be more flat in more powerful outbursts. The tendency is confirmed by the data published in Hagen-Thorn, Marchenko, & Mikolaichuk (1990). They have found $\alpha = -1.94 \pm 0.07$ for the small outburst of 1983-84. This spectrum is plotted in Fig. 1c by a dashed line. It is shifted downward relative to the 1999 spectrum also in accordance with the flux level in the R band at maximum brightness. For a synchrotron source, such a dependence means that in more powerful outbursts the spectrum of relativistic electrons in the source is harder. Earlier, a similar result was obtained for the blazar OJ 287 (Hagen-Thorn et al. 1998).

Acknowledgments. This work was supported by the Russian Federal Program "Integration" via grants K0232 & A0007.

References

Hagen-Thorn, V.A., Marchenko, S.G. 1999, Baltic Astronomy, 8, 575
Webb, J.R., Freedman, I., Howard, E. et al. 1998, AJ, 115, 2244
Sobrito, G., Villata, M., Raiteri, C.M. et al. 1999, Blazar Data, 1, No 5
Mead, A.R.G., Ballard, K.L., Brand, P.W.J.L. et al. 1990, A&AS, 83, 183
Hagen-Thorn, V.A., Marchenko, S.G., & Mikolaichuk, O.V. 1990, Astrophisics, 32, 244
Hagen-Thorn, V.A., Marchenko, S.G., Takalo, L.O. et al. 1998, A&AS, 133, 353

AGN From the Perspective of New Approaches in Astrophysics

S.G. Iskudarian

378433, Aragatsotn, Byurakan Observatory, Byurakan, Armenia.

Abstract. In all probability, AGN are one of the most convincing observable gleams of the Superforce in the Universe.

What are the new approaches in astrophysics? How do we look at AGN from the perspective of these new approaches (Iskudarian 2000a)? This postertext is connected with the sixth point of Iskudarian (2000a) - with the idea about similar behavior of galaxies and elementary particles (Iskudarian 2000b). Some physicists think that the Universe is ruled by a Superforce (Davies 1985). We don't see the last one, but if we want to see it, we must be ready to resist the great disaster. Such a huge disaster was the Big-Bang in the Universe.

It is not news if I say that there is something that is impossible on Earth, but in the Universe it is common. For example, there are the decay of the proton and the existence of the magnetic monopole. They are the very mechanism which I offer as the explanation of the huge energies in AGN. The image of NGC5128 (Sandage 1964) is very much like the decay of a proton, only on macro scales. And the fact that the first AGN were Seyferts in giant spirals (Suchkov 1988) allows us to think that here will be the participation - the function of the monopole also. And so, from the perspective of new approaches in astrophysics, we look at the AGN not only as agents which carried the huge energies of the Big-Bang to the corners of the observable Universe (Iskudarian 1998a), but also as large-scale elementary particles - conglomerates of protogalactic matter with the "genes" of elementary particles (Iskudarian 1998a). The activity of the nuclei of galaxies according to Ambartsumian's theory (1964, 1966) is just the activity of protogalactic matter with the combination of the processes of decay, explosion and ejection of some portions of protogalactic matter and gas. These processes are accompanied by the eruption of huge energies. According to theoretical physicists, only such energies can provide the decay of the proton or monopole (Davies 1985, Okun 1984). They have elaborated this part of their theory in such detail and so deeply, that one can say with great confidence that AGN are of largescale merge display of the aforementioned two particles. Such a supposition we made on the basis of the following facts:1.The first galaxies with active nuclei were Seyfert galaxies and as a rule, they were giant spiral galaxies (Suchkov 1988), 2.By the Byurakan ideas (Ambartsumian 1958, Iskudarian 1975), there is an intimate connection between the mechanism of formation of galaxies and the method of origin of their individual structural peculiarities 3.Two worlds of the Universe submit to the one general regularity, which is the ejection of the first type stellar population from the entrails of the second type one(in protogalactic stage, of course (Iskudarian 1996,1997)),4.The separate physical groups of galaxies with AGN, in all probability, are different evolutionary stages of the same objects (B.V.Komberg,preprint),5.M87 itself with its wide environment(look at Fig.1 from Iskudarian (1993), where the distribution of the brightest and faintest members of the N94 and N106 rich groups of galaxies (Huchra & Geller 1987) is given by their l and b coordinates) is a huge monopole with its fields of forces.These fields of forces are:1.The jet of M87,with blue condensations in it,2.The net of the rich globular cluster population (Iskudarian 1993),3.Elliptical galaxies around M87 (Fig.Ic of Iskudarian 1993), 4.Two symmetric details(in Fig.1c they are in contour line). It is interesting that these details have the same content of galaxies, only with contrary arrangement.

The whole picture with its fields of forces is only reminiscent of a huge monopole, but not in outward appearance. Theoretical physicists suppose that a monopole looks like an onion. The distribution of the faintest members of N106 looks more like an onion (look at Fig.1b of Iskudarian 1993). The last one is one more field of forces.

In fact, on outward appearance, in the center of the "monopole" (M87 with its extended environment) there is a proton (the nucleus of M87), which suffered decay, maybe not just once. As the groups N94 and N106 are connected physicaly, one may think that the huge monopole around M87 is physically connected with a closed looplike superstring (Iskudarian 1998b) to the nucleus of M87. The author admits that such a supposition is very bold. She is not a physicist, but she asks theoretical physicists to pay attention to this problem, especially to the part where they say that the monopole plays the role of catalyst in the process of the decay of the proton.

References

Ambartsumian, V.A., 1958, XI Solvay Conference, Brussels.
Ambartsumian, V.A., 1964, XIII Solvay Conference, Brussels.
Ambartsumian, V.A., 1966, IAU Symposium N29 Byurakan.
Davies, P., 1985, Superforce, New York.
Huchra, J.P. & Geller, M.J.,1987, ApJS, 52, 61.
Iskudarian, S.G., 1975, Byur.obs.Contr.,46, 73.
Iskudarian, S.G., 1993, International Workshop on"Galaxy Clusters and Large Scale Structures in the Universe", Sesto Pusteria (Bolzano,Italy).
Iskudarian, S.G., 1996, International workshop on "Hubble deep fields", Baltimore, USA.
Iskudarian, S.G., 1997, Meeting Astro-4, Moscow.
Iskudarian, S.G., 1998a, IAU Symposium N194, Byurakan.
Iskudarian, S.G., 1998b, Euroconference,"The Evolution of Galaxies on Cosmological Timescales", Tenerife, Spain.
Iskudarian, S.G., 2000a, JENAM-2000, Moscow.
Iskudarian, S.G., 2000b, Internatl. workshop on "Hubble deep fields", Muenchen.
Okun, L.B., 1984, Physics of Elementary Particles, Science, Moscow.
Sandage, A.R., 1961, The Hubble Atlas of Galaxies, Washington.
Suchkov, A.A., 1988, Science, Moscow.

Helical Structures in Seyfert Galaxies

A. V. Moiseev, V. L. Afanasiev, S. N. Dodonov

Special Astrophysical Observatory, Nizhnij Arkhyz, Karachaevo-Cherkesia, 369167, Russia

V. V. Mustsevoi and S. S. Khrapov

Volgograd State University, Volgograd, 400062, Russia

Abstract. Seyfert galaxies with Z-shaped emission filaments in the Narrow Line Region (NLR) are considered. We assume that observable Z-shaped structures and the velocity pattern of the NLR may be explained as tri-dimensional helical waves in the ionization cone.

1. Introduction

Numerous emission-line images of Seyfert galaxies show the existence of cone-like NLRs with broad opening angles and spatial sizes from 10 pc to 18 kpc (Wilson & Tsvetanov 1994). These galaxies also have highly collimated, elongated radio structures (radio-jets) coinciding with the cone axis (Wilson & Tsvetanov 1994; Nagar et al. 1999). An ordered Z-shaped emission pattern is a frequent feature in the NLR. We found more then 20 such objects on published emission-line images of nearby Seyferts. There is no common point of view on the origin of such regular structures in the NLR. Different models were proposed for individual objects: a bent bipolar outflow (Mulchaey et al. 1992), a strongly collimated precessing twin-jet (Veilleux, Tully, & Bland-Hawtorn 1993), a system of inclined gaseous disks (Morse et al. 1998), etc.

2. Kinematic Features of the Z-Shaped NLR

A sample of galaxies with Z-shaped NLRs were observed at the 6 m telescope. The scanning Fabry-Perot Interferometer and integral field spectrograph MPFS were used for study of the 2D kinematics of stars and ionized gas. Some systems of gas clouds with velocity differences more than $100 \div 200\,\mathrm{km\,s^{-1}}$ are present on the line of sight in the central regions of the cones, but the outer emission filaments have only one component of the emission lines. The gas velocity fields are strongly non-circular in comparison with the stellar fields. A large gradient in the line-of-sight gas velocities appears as a Z-shaped spiral. We note that similar features are also observed in the velocity fields of other Sy galaxies (see references in Moiseev et al. 2000) and these could be evidence for a common Z-shaped pattern.

368   Moiseev et al.

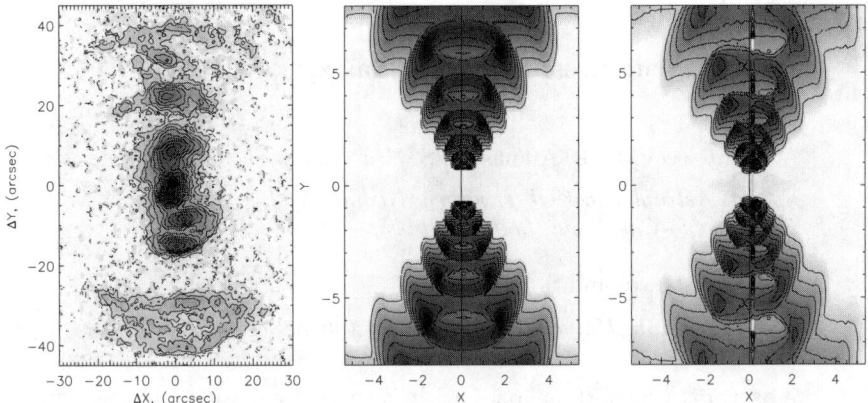

Figure 1. [OIII] image of NGC 5252 obtained at the 6 m telescope (*left*) and contours of the model luminosity for pinch (*middle*) and for helical (*right*) shock wave modes.

3. Non-Linear Simulations

We suggest that the Z-shaped spiral filaments in NLRs have a common wave origin and are generated by a hydrodynamic instability due to the velocity break between the galactic interstellar medium and outflowing gas from the central AGN. A collimated radio jet corresponds to the direction of outflow and matches with the cone's axis. A linear analysis shows that jets are unstable relative to waveguide-resonance development of pinch and helical internal gravity waves. Development of the instabilities leads to creation of a set of shock waves in the ambient medium. Our 2D and 3D non-linear hydrodynamic simulations show that the shock waves penetrate outward from the jet boundaries to the ambient medium. The resulting shock-wave structure covers a broad cone with opening angle $30° - 80°$ and appears in the sky-plane as a NLR with a bright emission pattern (Fig. 1). Pinch (axisymmetric) and helical modes of the shock waves develop in the NLR. Helical wave modes provide the Z-shaped emission structures which are observed in the ionized cones.

Acknowledgments. This work was supported by federal program "Astronomy" (Project 1.2.3.1)

References

Moiseev A.V., Afanasiev V.L., Dodonov, S.N., et al. 2000, astro-ph/0006323
Morse, J. A., Cecil, G., Wilson, A.S., Tsvetanov, Z.I. 1998, ApJ, 505,1 59
Mulchaey, J.S., Tsvetanov, Z., Wilson, A.S., et al., ApJ, 394, 91
Nagar, N.N., Wilson, A.S., Mulchaey, J.S., Gallimore, J.F. 1999, ApJS, 120, 209
Veilleux, S., Tully, R.B., Bland-Hawthorn, J. 1993, AJ, 105, 1318
Wilson, A.S. & Tsvetanov, Z.I. 1994, AJ, 107, 1227

Correlation of Optical and X-ray Radiation of NGC 4151 and 3C 390.3. Preliminary Results

Olga A. Novikova (1), Nikolai G. Bochkarev (2,3)

(1) Department of Astrophysics, Physical Faculty, Lomonosov Moscow State University (MSU), Russia

(2) Sternberg Astronomical Institute (SAI), Moscow, Russia

(3) Euro-Asian Astronomical Society (EAAS), Moscow, Russia

Abstract. The correlation of optical and X-ray radiation of some AGN is studied. We used low signal-to-noise data from the All-Sky Monitor of the Rossi X-ray Timing Explorer (2-10 KeV) and broad-band photometry of the optical continuum obtained mainly during the last several years through the "AGN Watch" program.

1. Introduction

Presently achievable angular resolution is not enough to get an image of the AGN "central engine". So, indirect methods are the only way to study the region of AGN energy output. The reverberation mapping method (RMM) is the most developed one. For its realization, it is necessary to use long, well sampled data sets, which can be provided only by an international monitoring program.

The theoretical basis of the RMM suggests analysis of the optical line profile response following variations of the X-ray continuum. Actually, the optical continuum is used instead of the X-ray one. Therefore it is very important to know the relationship between variations of the X-ray and optical continua.

2. Observational Data

We used X-ray data (2-10 keV) from the All-Sky Monitor (ASM) of the Rossi X-ray Timing Explorer for 1997-2000 for NGC 4151 and 3C 390.3. Note particularly that we used the ASM measurements in contrast to other investigations. The reason is that it is the only long-term (from 1986 up to now) homogeneous set of X-ray data, but the signal is rather small relative to the ASM background.

Optical photometric observations were made mainly in 3 observatories: the Crimean Laboratory of the Sternberg Astronomical Institute, the Special Astrophysical Observatory (North Caucasus) and Mt.Maidanak Observatory (Uzbekistan). V.M. Lyuty reduced the full list of data to one photometric system for an aperture of 27.5".

3. Results

A preliminary estimate of the correlation coefficient of the X-ray and optical radiation (U-band for NGC 4151 and B-band for 3C 390.3) was made. On account of the rather low S/N ratio for the X-ray data, we carried out a several-day average in order to increase the contribution of the X-ray signal relative to the noise.

From the optical light curves we removed the long-term ($\sim$ years) component, which has no X-ray analog. The relationship of variations in the optical and X-ray continua was considered after that.

For NGC 4151, the maximum estimate of the correlation coefficient corresponds to 5-day averaged X-ray data and is equal to 0.53. Cross-correlation analysis shows that the time lag of the optical continuum with respect to the X-ray flux is $< 1^d$. Therefore, in this particular case the existing photo-ionizing source seems to have a dimension less then one light-day and the reverberation mapping method may at least be used to search for structures significantly larger than one light-day.

For 3C 390.3 we did not find a significant correlation of the optical light curve and the X-ray flux. Possible reasons may be: 3C 390.3 has a low mean signal – about 0.2 count/sec, as opposed to 0.6 for NGC 4151. Perhaps the absence of correlation is connected with insufficiently accurate registration of the background or its variability. The signal distribution shows that within 2σ limits, the statistics are approximately normal, with a standard deviation about 0.3 c/s (for one-day average data), but for large signal absolute differences ($> 2\sigma$) the negative wing of the distribution is stronger than the positive one.

It could be evidence of inadequate calculation of background variations produced by solar activity events. The other possible reason for a negative result is some peculiarity of source structure (scattering in the disk corona, binary black hole, etc.).

4. Conclusion

Preliminary estimates of the correlation coefficient of the optical and X-ray continua for 3C 390.3 (in this case, we did not find a significant correlation) and for NGC 4151 (the estimate is about 0.53, sigma is about 0.22) were made. No time delay greater then one day was found and the upper limit is $\sim$ 1 day.

Acknowledgments. The work was partially supported with grants of RFBR-GFEN 99-02-39120 and RFBR 00-02-16272.

References

Blandford, R.D., McKee, C.F. 1982, ApJ, 255, 419

Bochkarev, N.G., Antokhin, I.I. 1982, Astron. Tsirk., 1238

Dietrich, M., et al. 1998, ApJSS, 115, 185

O'Brien, P., et al. 1998, ApJ, 509, 163

Shapovalova, A.I., Burenkov, A.N., Bochkarev, N.G. 1996, Bull. Spec. Astroph. Obs., 41, 28

Wozniak, P.R., et al. 1998, MNRAS, 299, 449

AGN SURVEYS
ASP Conference Series, Vol. 284, 2002
R.F. Green, E.Ye. Khachikian, D.B. Sanders

Energy Releases from Accreting Superdense Compact Objects in AGNs

Avetis Abel Sadoyan

Yerevan State University, Alex Manoogian 1, Yerevan 375025, Armenia,
Email:asadoyan@www.physdep.r.am

Abstract. We consider an accreting central symmetric object that is near to the Eddington limit, i.e., the luminosity pressure is balancing the gravity. In special cases, a layer of accreting matter will be formed near the surface of a centrally symmetric object. We discuss the conditions of formation of the layer, its inner structure and stability. If the luminosity pressure will be reduced for some reason, then the layer will collapse immediately, releasing an enormous quantity (up to 10^{42} erg/sec) of energy. We discuss some possible scenarios for energy releases.

1. Introduction

We are considering spherically symmetric accretion on superdense configurations that are near the Eddington Limit. Let's take Shchwarzshild coordinates with
$$g_{00} = 1 - \frac{r_g}{r} \; ; \; g_{11} = -\frac{1}{1-\frac{r_g}{r}} \; ; \; g_{22} = -r^2; \; g_{33} = -r^2 sin^2\theta;$$
where $r_g = \frac{2GM}{c^2}$, M is the mass of the central body, and c is the speed of light. The pressure of radiation has no evident impact on the parameters of the superdense configuration, except the thin layer of the Ae-shell near the surface. One can easily derive the equation for stationary accreting matter with radiation pressure as
$$\frac{dP}{dr} = -\frac{GM\rho\delta}{r^2\left(1-\frac{r_g}{r}\right)}$$
where ρ is the density of infalling matter, $\delta = 1 - \frac{L}{L_e}$, and L is the flow of radiated energy at a distance r from the center in units of time.

The definition for the Eddington Limit L_e in the fully relativistic case is
$$L_e(r) = \frac{4\pi cGM}{\chi\sqrt{1-\frac{r_g}{r}}}$$
where χ is the optical opacity of infalling matter, while in Newtonian Gravitation L_e is constant and equals $L_{EN} = \frac{4\pi GMc}{\chi}$;

As one can see in the region where δ is positive we have accretion; where negative, the resulting forces break down the accretion and could stop the infalling matter and even throw it back. The evolution can go by different scenarios depending on the differential accretion, the kind of the infalling matter, and the initial velocity at infinity.

2. Possible Scenarios for Bursts

There are several scenarios for the evolution of a central configuration that is near the Eddington Limit; let's discuss four interesting cases:

Scenario 1 ("Greenhouse Effect")

Energy release is conditioned by the thermal state of the central configuration, that is an enormous container of thermal energy. The accreting matter stops, forming a "hanging layer" that closes the way out for radiation. The hadron nuclei heat up, a burst takes place, nuclei lose their thermal energy through the work carried out during the outburst of the mass and a significantly small part through emission. The burst stops and the accretion to the surface starts just at the very moment when the luminosity becomes smaller than the Eddington Limit. After a while (quiet period of the burster life) the flow of energy to the hadron nuclei decreases and when the luminosity reaches the Eddington Limit a new burst occurs.

This scenario could be repeated several times and does not need a permanent source of accreting matter, i.e., a companion to feed the acretion, because falling matter plays the role of simple regulator for the radiation.

Scenario 2 ("Falling Matter")

Energy release is conditioned by the temp of accreting matter. The accretion rate decreases due to the gap, the luminosity becomes less than the Eddington limit and the hanging matter collapses onto the surface of the central engine, releasing all the gravitational energy for the accumulated mass. The luminosity increases and stops the further infall of matter.

Scenario 3 (X-Ray Bursters)

This a hybrid of the first and second scenarios. When the accumulated accreting matter falls down it heats up the central configuration, which outbursts, throwing back also the thin layer of Ae-shell and opening the hot core (G. Sahakyan, G. Alojants, A. Sarkissian 1991). We record an X-ray burst.

Scenario 4 (Soft Gamma Repeaters)

If in Scenario 3 the energy of the outburst is high enough to accelerate the burst matter to velocites such that a shock wave will occur, then we shall have a soft gamma-ray burst. The shock wave will pass through the accreting matter and will produce non-thermal gamma emission (Sadoyan & Alodjanc 2000).

As the opacity for electrons is much higher than that for protons, this is a good solution for the "Baryon Load Problem" in GRBs. Computer simulations of the model make it possible to understand the origin of intense trains of recurrent bursts, when the intervals between the bursts decrease at times to such an extent as to become comparable to the duration of the bursts themselves.

References

Sahakyan, G., Alojants, G., Sarkissian,A., 1991, Astrofizika, v.34,21-40
Sadoyan, A.A., Alodjanc G.A., 2000, in Texas 20 Conf. Proc.,112

Intermediate Resolution Hβ Spectroscopy and Photometric Monitoring of 3C 390.3. I. Further Evidence of a Nuclear Accretion Disk

A.I. Shapovalova, A.N. Burenkov, O.I. Spiridonova, V.V. Vlasuyk, V.P. Mikhailov

Special Astrophysical Observatory of the Russian AS, Nizhnij Arkhyz, Karachaevo-Cherkesia, 369167, Russia

L. Carrasco, V.H. Chavushyan, J.R. Valdes, F. Legrand

Instituto Nacional de Astrofisica, Optica y Electronica, Apartado Postal 51. C.P. 72000, Puebla, Pue., Mexico

V.T. Doroshenko, V.M. Lyuty, N.G. Bochkarev

Sternberg Astronomical Institute, University of Moscow, Universitetskij Prospect 13, Moscow 119899, Russia

A-M. Dumont, S. Collin

DAEC, Observatorie de Paris, section de Meudon, Place Jansen, F-92195 Meudon, France

O. Kurtanidze, M.G. Nikolashvili

Abastumani Observatory, Republic of Georgia

Abstract. The variations of the Hβ broad emission line and continuum flux in 3C 390.3 are investigated. Our results favor the formation of the broad Hβ line of 3C 390.3 in an accretion disk.

1. Observations

Spectroscopic monitoring of 3C 390.3 was carried out at the 6-m and 1-m telescopes of SAO RAS (Russia) and the 2.1-m GHO telescope at Cananea (México), in the range 4000-8000 Å, with a spectral resolution of 5-15 Å, and a S/N ratio of ≥ 50 in the Hβ region. Broad-band BVRI CCD photometric monitoring was carried out at the 1 m and 60 cm telescopes of SAO RAS, at the 60 cm telescope of the Crimean Laboratory of SAI (Russia) and at the 70cm meniscus telescope of Abastumani Observatory (Georgia).

2. Results

i. The continuum flux and the Hβ broad component varied by a factor of about three during the period 1995-1999 (Shapovalova et al., 2001).

ii. From a cross-correlation analysis, we found two values for the time lag in the emission line response relative to the continuum variations. In addition to the delay of ~100 days, obtained at any given time, we found a second value of ~35 days in periods of time when the sampling was better.

iii. The observed flux of the Hβ wings varied quasi-simultaneously, with the same lag relative to the continuum variations. This behavior excludes models of broad line formation in biconical gas streams or jets.

iv. The shift of the blue peak in integral Hβ profiles nearly follows the trend found by Eracleous et al. (1997). Their fitted model for a binary black hole with a very large mass ($\geq 10^{11} M_\odot$) seems to fit our data as well. Hence, our results provide further support for the dismissal of the binary black hole hypothesis for 3C 390.3, on the basis of the masses required.

v. The observed Hβ profiles are best reproduced by an inclined accretion disk (i=25°) whose region of maximum emission is located roughly at 200 R_g.

vi. From the difference profile of the Hβ line we found that the velocity variations of the blue and red bumps and their differences are anticorrelated with the Hβ flux variations and with the lag in the continuum (Shapovalova et all, 2001). Thus, the maximum of the line emission energy moves across the disk and corresponds to smaller radii when the continuum flux decreases (bump velocities increase), and to larger radii when the continuum flux increases (bump velocities decrease). These transient phenomena are expected to result from the variable accretion rate close to the black hole.

3. Conclusion

Our results do not support either the models of outflowing biconical gas streams or those of a supermassive binary black hole. Instead, our results are consistent with an accretion disk model.

Acknowledgments. This paper has had financial support from INTAS (grant N96-0328), RFBR (grants N97-02-17625, and N00-02-16272a), RFBR + CHINE (grant 99-02-39120), and CONACYT research grants G28586-E, 28499-E and 32106-E (México).

References

Eracleous, M., et al. 1997, ApJ, v. 490, p. 216
Shapovalova, A.I., et al. 2001, A&A, v. 376, p. 775

Rapid Variations in the Seyfert 1 Galaxy Mrk 474

R. R. Valiullin

Fesenkov Astrophysical Institute, Kamenskoye Plato, Almaty, 480020, Kazakhstan. Electronic mail: rashit@afi.south-capital.kz

Abstract. Spectra of Mrk 474 have been obtained in order to search for short-term (less than one day) variability in the broad Hβ and Hα profiles. Observations were carried out using a slit spectrograph and a three-cascade image tube attached to the 70-cm reflector. The processing of the series of spectrograms, obtained on 1999 April 25, showed a rapid (less than 20 minutes) and powerful increase of the flux in the region of [O III]λ5007. The increase of the flux near maximum (5060$\mathring{A}$ ± 10$\mathring{A}$) appeared to be equal approximately to 300%. The value similar to the FWHM for the long-wavelength excess of the flux was equal to 170$\mathring{A}$ ± 20$\mathring{A}$. The dying-out of the long-wavelength excess lasted approximately two hours.

1. Analysis and Results

1. The third spectrum of the series if compared with the first and the second ones has a significant excess of the flux above the 'average' spectrum in the region of the [O III]λ5007 emission line. The maximum of a long-wave flux excess corresponds to the 5060$\mathring{A}$ ± 10$\mathring{A}$. The increase of the flux near the maximum of the long-wavelength excess (5060$\mathring{A}$) appeared to be approximately equal to 300%. The value similar to the FWHM for the long-wavelength excess appeared to be equal to 170$\mathring{A}$ ± 20$\mathring{A}$. The dying-out of the long-wavelength flux excess lasted approximately two hours with a variable velocity: firstly, during a transition from the third spectrum of the series to the fourth one, it goes quickly, and then—slower. The position of the center of symmetry of the long-wavelength flux excess during the transition from the third spectrum of the series to the seventh one gradually shifted towards the center of the Hβ emission line.

2. During the transition from the fifth spectrum of the series to the sixth, a noticeable excess of the flux appeared in two regions. The center of symmetry of the first short-wavelength flux excess corresponds to 4750$\mathring{A}$ ± 10$\mathring{A}$, and the increase of the flux in the first short-wavelength excess near 4750$\mathring{A}$ appeared to be equal approximately to 200%. The value similar to the FWHM for the first short-wavelength excess appeared to be equal to 100$\mathring{A}$ ± 20$\mathring{A}$. The center of symmetry of the second excess is obviously less than the short-wavelength edge of our observed wavelength region i.e. it is < 4540$\mathring{A}$. Therefore it is difficult to estimate precisely the increase of the flux in the second excess, but most likely it is not less than 200% at its maximum.

2. Discussion

During observations of variations in the spectrum of Mrk 474 the fact was registered that the short-wavelength excess of the flux appeared 77 minutes later than the long-wavelength excess. In addition, the period of motion of the center of symmetry of the flux excess near the Hβ emission line is also equal to 77 minutes. Thus one can conclude that these two events are interrelated. The reason for the origin of one long-wavelength and two short-wavelength excesses and a flux excess near the Hβ emission line may be a reverberation of the hydrogen clouds from a sudden flare in the UV continuum (see e.g., Blandford & McKee 1982; Peterson 1997). This assumption may explain also the shift of the center of symmetry of the long-wavelength flux excess towards shorter wavelength. With propagation of a flare from the central source of continuum, the flare process would gradually involve the more distant clouds of hydrogen, the velocities of which are decreasing with distance from the central source. Specifically, while the intensity of radiation on rapidly moving hydrogen clouds located close to the source of continuum decreases, the intensity of radiation on the more remote, and therefore, slower hydrogen clouds increases. The effects mentioned above will result in a redistribution of the spectral flux and a shift of the center of symmetry of the radiation excess towards a wavelength corresponding to zero velocity.

If our assumption is correct, a model can be considered, where three large hydrogen clouds revolve along elliptic orbits around the source of continuum. For the observer two of them are behind the source of continuum and are moving towards the observer with radial velocities $-7480 \pm 600 \, km \, s^{-1}$ and $-19814 \pm ? \, km \, s^{-1}$ respectively, and one cloud is in front of the central source and moves away from the observer with a radial velocity $12250 \pm 600 \, km \, s^{-1}$. These assigned values are related only to that part of the hydrogen which produces radiation near the maximum of the flux excess. There is a dispersion of velocities inside each hydrogen cloud. The range of velocities may reach several thousand $km \, s^{-1}$ and the more distant fragments would have the smaller velocities. The presence of the slower parts of the clouds is necessary for explaining the shift of the center of symmetry of the radiation excess towards the wavelength of the relative zero velocity of hydrogen.

References

Blandford, R. D., & McKee, C. F. 1982, ApJ, 255, 419
Denissyuk, E. K. 1974, Astron. Zh., No 809, 1
Peterson, B.M. 1997, An introduction to Active Galactic Nuclei (Cambridge: Cambridge Univ. Press), 82

Part 6
Future Projects

Ed Khachikian, Caryl Gronwall, Joe Mazzarella

The Conference Banquet: Areg Mickaelian, Wolfgang Voges

Using the NASA/IPAC Extragalactic Database (NED) and Federated Virtual Observatory Archives for Multiwavelength Studies of AGNs

Joseph M. Mazzarella and the NED Team

California Institute of Technology, Jet Propulsion Laboratory, MS 100-22, Pasadena, CA 91125

Abstract.
We live in an exciting era that offers increasing opportunities for people all over the world to make discoveries about the Universe using interconnected archives on the Internet as a primary research tool. We review how NED (*http://ned.ipac.caltech.edu*) can be used in concert with globally distributed online archives to perform multi-wavelength, cross-correlated studies of AGNs and other galaxy types. The present status and planned evolution of NED capabilities are discussed.

1. Introduction

The NASA/IPAC Extragalactic Database (NED) is an online research facility designed to support scientists, educators, space missions and observatories in the planning, execution and publication of research on extragalactic objects. The foundation of NED is a growing database of galaxies, quasars and all types of extragalactic objects that can be searched by positions, redshifts, object types, references, authors, and multi-wavelength cross-identifications produced from thousands of catalogs and journal articles. The primary goal of NED is to maintain an up-to-date panchromatic fusion of basic data for all known (cataloged and published) extragalactic objects, including pointers to the astrophysical literature and to relevant distributed archive resources. Scientists working in observational extragalactic astronomy use NED in their research at nearly every step, from proposal planning, through data collection, analysis, and publication.

2. Cross-Identifications and Data Integration

2.1. Multi-wavelength Cross-Identifcations and Associations

Cross-identification refer to the process of establishing which observation in one survey catalog (e.g., FIRST) corresponds to the same astrophysical source in surveys at other wavelengths (e.g., SDSS and 2MASS). The process is much more difficult than it may first appear, because observations taken with different telescopes and at various wavelengths often differ in substantial ways including different positional uncertainties, systematic errors in astrometry (some catalogs), different survey resolutions (e.g., IRAS and APM compared to ROSAT), and problems such as matching double-lobe radio sources with their parent galaxies. Objects also populate a hierarchical Universe: AGNs, supernovae, and HII regions reside in their host galaxies; galaxies are members of pairs and groups; pairs and groups are typically members of clusters; and galaxy clusters reside in superclusters separated by vast voids. Therefore, complex relationships between objects are needed (one-to-one, one-to-many, many-to-one, many-to-many), in addition to *statistical associations* for cases in which firm cross-identifications cannot be established. NED activities revolve around a systematic process of constructing and revising cross-identifications and statistical associations between millions of entries in multi-wavelength catalogs and publications.

2.2. Database Contents

The database content of NED is updated periodically on the home page (Figure 1). To date NED contains: 5.9 million cross-identifications in thousands of multi-wavelength surveys and journal articles; 4.7 million unique extragalactic objects; 4.5 million photometric measurements covering gamma-rays through radio wavelengths; 2.0 million detailed position measurements with uncertainties; 1.7 million bibliographic references to 50,000 articles; 291,000 redshifts; 748,000 science-grade FITS images and remote links; 54,000 detailed notes from catalogs and journal publications; and 28,000 abstracts of journal articles and Ph.D. theses.

The essential data for sources in NED include positions, redshifts, morphological types, nuclear spectral types, panchromatic photometry, and images. When available, uncertainties in the measurements are also stored and provided in the query reports. Photometric data are stored in original units and converted to common frequency (Hz) units and flux density units ($W\ m^{-2}\ Hz^{-1}$) for construction of Spectral Energy Distributions (see Figure 5); the data are also tagged with their aperture sizes or status as a "total flux" measurement.

3. Query Services

3.1. Web Interface

There is insufficient space here for a comprehensive history and complete technical review of NED. A discussion of the motivation, initial design, and early history of NED was given by Helou et al. (1991). NED is accessible on the World Wide Web at *http://ned.ipac.caltech.edu*. Figure 1 shows the NED Web interface main menu (home page). Following is a review of the primary services.

NED can be searched for extragalactic objects using menus designed for searches 'By Name', 'Near Name', 'Near Position', 'IAU Format', or 'By Reference Code' (a 19 digit code developed jointly by NED, ADS, and CDS[1] to uniquely identify publications in astronomy). Figure 2 illustrates an example of the Essential Data presented after a query of the galaxy NGC 4151.

Figure 3 shows hyperlinks to 'External Archives and Services', easily found when the user scrolls down below the Essential Data that come directly from

---

[1] *http://adswww.harvard.edu, http://cdsweb.u-strasbg.fr*

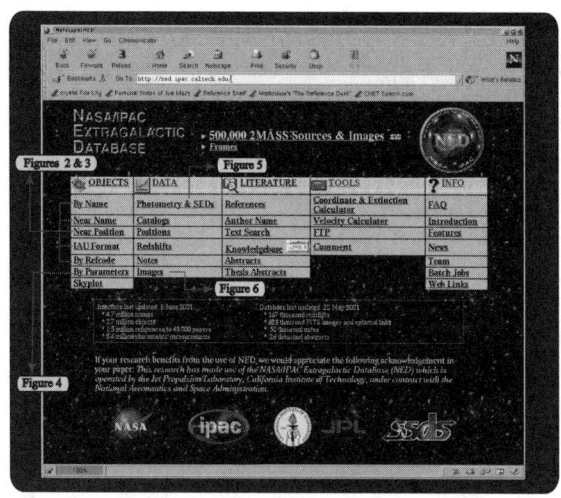

Figure 1. The NED Web interface main menu available at *http://ned.ipac.caltech.edu*. The annotations refer to example query results displayed in Figures 2-6.

the NED database. Links here allow the user to retrieve images and query original catalog data or observation log entries. The first set of hyperlinks are to data related directly to an object name; the second set of links are to services that can be queried at the object's coordinates. A summary of the available resources includes: original catalog record entries in VizieR at CDS; the NVSS, and FIRST catalog and image servers and the Observation Log of the VLA telescope from the *National Radio Astronomy Observatories (NRAO)*; infrared mission archives at *IRSA/IPAC* (2MASS, IRAS, etc.); visual and UV mission archives at *MAST/STScI* (HST, IUE, etc.); and high energy mission archives at *HEASARC/GSFC* (ASCA, CGRO, Einstein, etc.) New archive services are being added as they become available. This tool makes it very easy for researchers to locate existing observations in one or more of the major surveys or observatories, providing a major step toward federating distributed archives.

The 'By Parameters' (recently renamed 'Advanced All-Sky') menu allows the user to query NED using joint constraints on redshift range, sky area, object types, or survey/catalog name prefixes. With this feature one can dynamically regenerate a classic sample that contains not simply the original measurements (available elsewhere, such as VizieR), but rather the most precise and currently available source positions and redshifts, with links to up-to-date references, multi-wavelength photometry, images, etc. For example, today anyone can use NED to generate an up-to-date compilation analogous to the '*Catalog of Markarian Galaxies*' (Mazzarella & Balzano 1986), or generate a current data set for the entire *Third Cambridge (3C)* radio galaxy sample, with the click of a mouse. Since entries in NED are continuously updated through a synthesis

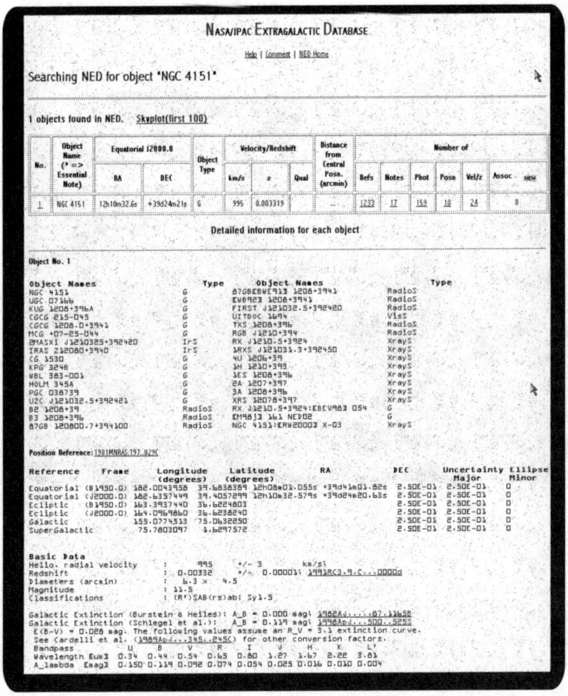

Figure 2. Essential Data returned by NED from a query of NGC 4151: includes coordinates and redshift (with uncertainties), multi-wavelength survey cross-identifications and object types, size, magnitude, classifications, Galactic extinction along the line of site, as well as links to query references, notes, photometry, positions, velocities, and images. This information is followed by links to **External Archives** (Figure 3).

of the literature and large surveys, errors in original catalogs are often found and documented. Therefore, *for many studies it is more efficient and effective to cross-correlate new observations against the data fusion in NED rather than against catalogs in their original published forms.* A visualization of one example of this powerful feature is shown in Figure 4. Other queries that can be performed using this tool, including the recent introduction of photometric constraints, are demonstrated in a new 'Tutorial Examples' feature on the interface menu.

The *DATA* column allows the user to enter an object name (e.g., 'NGC 4151', 'SN 1993G', '2MASXi J1132350+582422', 'SDSS J1044-0125', 'Antennae') and query NED for 'Photometry & SEDs', 'Catalogs', 'Positions', 'Redshifts', 'Notes', or 'Images'. Figure 5 illustrates an example of multi-wavelength photometric data and Spectral Energy Distribution (SED) plots.

Figure 6 shows an example of the rich variety of multi-wavelength FITS images available for immediate download and visualization using the Aladin Java

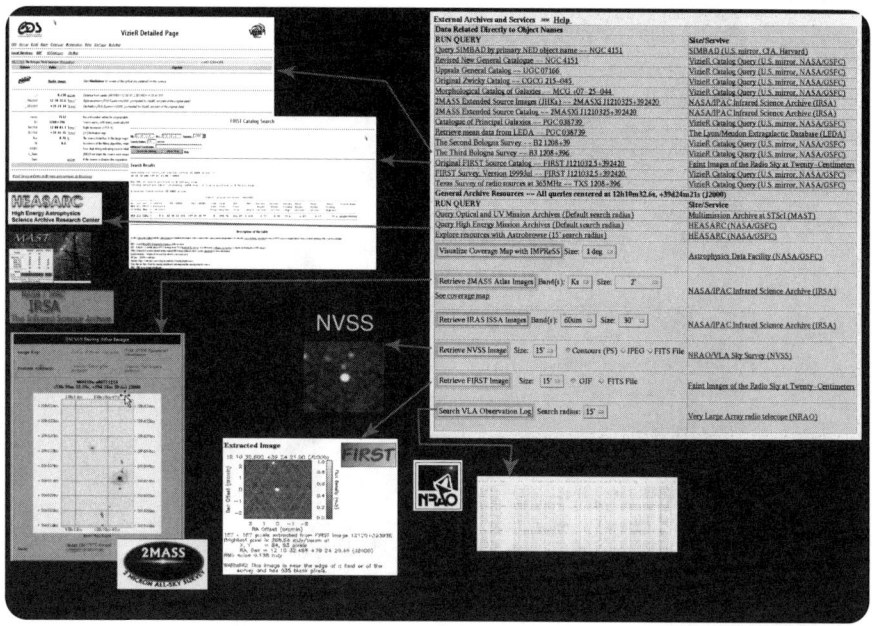

Figure 3. Queries to globally distributed archive data are available with a single mouse click. The first set of links are anchored to data related directly to an object name; the second set of links query services at the coordinates of the NED object. This example is a continuation of the report from a query of NGC 4151 (Figure 2).

applet. Aladin's interoperability with NED and other distributed data services provides a visual summary of the multi-wavelength sky. Aladin was developed at the CDS and configured with NED through a cooperative agreement.

In the *LITERATURE* column the user can: (1) enter an object name and access the 'References' related to that object; (2) enter an 'Author Name' and retrieve a list of references; (3) search journal article 'Abstracts'; (4) search 'Thesis Abstracts'; (5) use the 'Text Search' tool to perform keyword searches on the journal and thesis abstracts in NED or the full text content of LEVEL5; and (6) access the LEVEL5 'Knowledgebase'. LEVEL5 provides hyperlinked review articles and documents of current and lasting interest to cosmologists and extragalactic astronomers. Contents include a glossary of terms, essays, research papers, and reviews. Cited extragalactic objects are cross-linked to NED Basic Data queries, and all available citations are hyperlinked to NASA's Abstract Data Service (ADS), to on-line NED abstracts, or to preprints on astro-ph.

The *TOOLS* column of the main NED menu contains a 'Coordinate & Extinction Calculator' that performs coordinate conversions and precession and displays line-of-sight Galactic extinction estimates using two techniques (Schlegel et al. 1998; Burstein & Heiles 1982). The 'Velocity Calculator' computes conversions between redshifts for extragalactic objects in different reference frames: heliocentric, Local Group, Galactic Standard of Rest, and 3K Microwave Back-

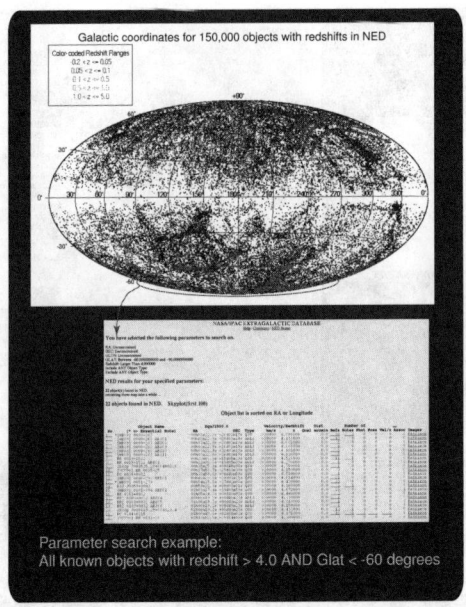

Figure 4. Objects that satisfy the criteria $[Redshift > 4.0\ AND\ Glat < -60°]$ returned after submitting a query using the 'By Parameters' menu (see Figure 1).

ground. There is also a link to NED's public FTP site, primarily used by users to pick up their output from the NED batch mode.

The *INFO* column of the main menu contains links to a Frequently Asked Questions ('FAQ') document, an 'Introduction', summary of 'Features', information about the NED development 'Team', forms and documentation for submitting NED 'Batch jobs', and finally a page of useful 'Web Links' relevant to extragalactic astronomy. Information about new content and capabilities is available at the top of the main interface.

3.2. Batch Mode

NED can process requests for large amounts of data as a 'Batch Job.' Using this mode simply involves submitting to NED via email a "batch form" containing a list of objects or positions, or other constraint parameters (e.g., redshift, object type, or survey/catalog name prefix).[2] After the request has been processed, NED sends the user a notice by return email indicating where the resulting data files may be copied via FTP.

---

[2]The forms are available from the 'Batch Jobs' link on the main NED menu, and they should be emailed to *nedbatch@ipac.caltech.edu* for processing.

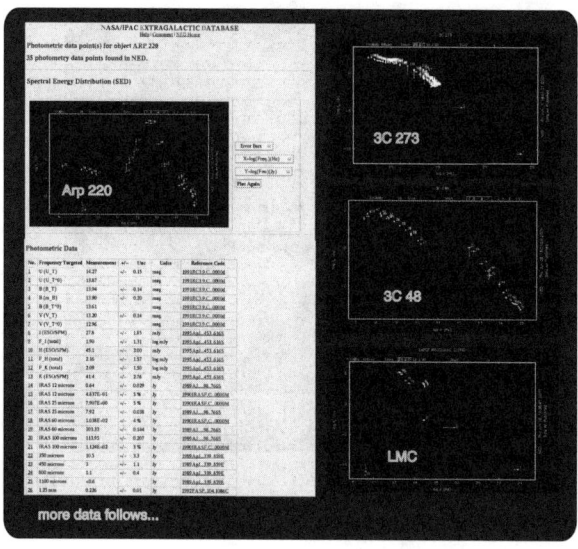

Figure 5. Multi-wavelength photometry and spectral energy distributions (SEDs) covering gamma-ray through radio wavelengths. The data are available in original units as published, and also in common units ($Hz, Wm^{-2}Hz^{-1}$) used to construct SEDs. The data include uncertainties and references. The SED plots are dynamic, with configurable axis units (e.g., wavelength or frequency for the abscissa and f_ν, νf_ν or f_λ for the ordinate) and optional error bars.

3.3. Client/Server Mode Connectivity

For many years NED has provided a 'server mode' with custom client (C) software [3] that has been used by computer programmers all over the world to build applications that issue queries and retrieve data from the NED database in a format that can be integrated into their services. Astrophysics archive centers and observatories use NED's server mode extensively to resolve extragalactic object names into celestial coordinates, and to retrieve lists of objects by specifying an input position and search radius. A number of sky visualization tools also make use of NED's client/server capabilities.

4. NED in the Era of a Global Virtual Observatory

The vision of a 'virtual observatory' (VO) involves interconnected, globally distributed archives from observatories and large-scale sky surveys which are fed-

---

[3] The NED client C code is available at *ftp://ned.ipac.caltech.edu/pub/ned/client.3/*.

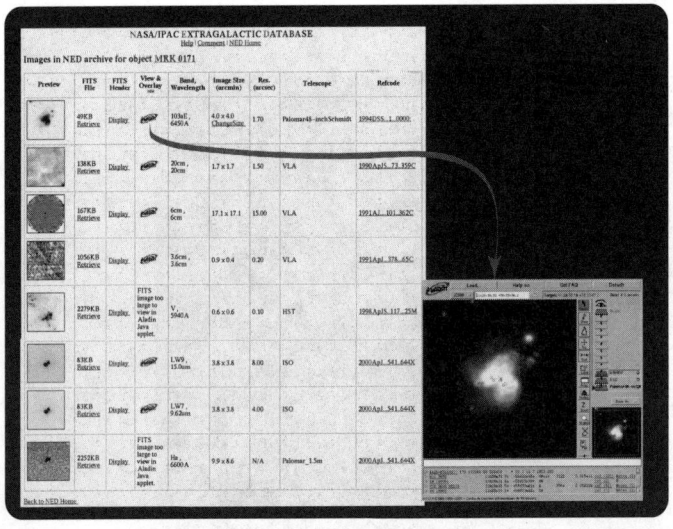

Figure 6. Multi-wavelength galaxy images, including previews (GIF) and science-grade data (FITS). Image overlays and graphical inter-activity between sky coordinates (from information in the FITS image header) and object markers (from NED and other databases) are available using the Aladin Java applet (CDS).

erated using common database query standards and data interchange protocols, combined with user interfaces and data mining tools that integrate and analyze the fused, multi-wavelength data sets to facilitate making new discoveries about the Universe (regardless of the location of the data or the investigators). A popular level description of the concept is given in an article by Cowen (2001). Helou et al. (1991) pointed out that there is a dual challenge presented by the explosive growth in astronomical data: "dealing with the sheer volume, but also inter-connecting intelligently the huge variety of information available." The NED team shares a common vision regarding what a VO can enable for all fields of astrophysics, and we are actively involved in collaborations designed to lay the foundation of the VO and extend its functionality[4]. There is much work to be done on all fronts by a broad community of astronomers teamed up with computer scientists and programmers. It is useful to summarize the role of NED in the emerging global VO in the context of current capabilities and future enhancements that will inter-operate with related projects.

---

[4]The U.S. National Virtual Observatory (NVO) development Web site is *http://www.us-vo.org*.

4.1. Current Capabilities

NED will continue to establish and improve high fidelity relationships between multi-wavelength data with anchors to the literature for millions of extragalactic objects, using a combination of computer software that utilizes positional uncertainties and astronomical source properties, followed by human inspection to resolve important, complex cases that cannot be fully automated. In addition, the NED team is participating in the collaborative development, testing, and deployment of the next generation of catalog cross-ID software tool-kits. We will also continue to enhance NED's object-based portal into distributed data sets (Figure 3). This work will involve keeping up with evolving technology for interoperability between archive query services, primarily XML and the associated family of protocols for implementing modern Web Services[5] as they are adopted by the community. The galaxy spectral energy distributions will also continue to grow to include data from large-scale surveys (2MASS, SDSS, etc.), space missions (SIRTF, GALAX, Chandra, etc.) and of course literature articles.

4.2. Future Enhancements

In addition to staying current with the literature and extragalactic source observations in new large-scale sky surveys, the NED team is committed to providing new functionality to extend the usefulness of NED as a research tool for astronomers. The newer NED capabilities with direct relevance to the VO concept were reviewed above. Over the next few years NED plans to provide the following enhanced capabilities as resources permit: (1) development of a spectral archive for extragalactic sources, including data reports and queries that involve nuclear spectral classifications (Seyfert 1 & 2, LINER, HII/starburst, etc.) and spectroscopic line measurements; (2) enhancements to the 'By Parameters' (Advanced All-Sky) tool to support queries based on multi-wavelength flux and color criteria; (3) upgrades to the 'Batch Mode' to support larger result sets with output content and formats that can be configured by users for easy input into data analysis applications; (4) production of an XML server mode to support people who want to write software to analyze NED data or integrate query results into new VO services and tools.

5. Summary

NED provides data and relationships between multi-wavelength observations of millions of extragalactic objects. The goal is a comprehensive panchromatic census of objects in the extragalactic sky. Project activities revolve around an ongoing fusion of data from sky survey catalogs and the literature, focusing on established and candidate extragalactic sources, and maintaining cross-identifications, statistical associations, and anchors to online references and pointers. NED serves as an interface into its own database and now also as a portal for the extragalactic research community into an emerging federation of astrophysics data centers and service providers with queries indexed by object names and

---

[5]Namespaces, DTD, Schema, XSL, SOAP, WSDL, etc. See *http://www.w3.org*.

coordinates synthesized by NED. As a key participant in the global Virtual Observatory (VO), NED will continue evolving the core functions that it provides today, while supporting new initiatives through use of XML standards to establish higher degrees of interoperability with other archives, collaboration in the development and application of new tools for bulk dynamic catalog cross-identifications, serving large multidimensional data streams to interface with data mining and visualization applications, and in general enabling new opportunities for discovery by leveraging information technologies.

Acknowledgments. The NED Team consists of Kay Baker, Judy Bennett, Harold Corwin, Cren Frayer, George Helou, Anne Kelly, Cheryl Lague, Joseph Mazzarella, Barry Madore, Olga Pevunova, Marion Schmitz, Brian Skiff, and Dianna Schettini. The Apache Web servers are maintained by Rick Ebert, and Informix database administration is performed by Nian-Ming Chiu. We also thank the IPAC Systems Group for their fine computer and network infrastructure support. This work was carried out by the Jet Propulsion Laboratory, California Institute of Technology, under contract with the National Aeronautics and Space Administration (NASA). NED is funded by the Mission Operations and Data Analysis program of NASA's Office of Space Science.

References

Helou, G. et al. 1991, in Databases and Online Astronomy, eds. M. A. Albrecht & D. Egret (Kluwer), 89

Cowen, R. 2001, Science News, 159, 124

Burstein, D. & Heiles, C. 1982, AJ, 87, 1165

Schlegel, D. J., Finkbeiner, D. P., & Davis, M. 1998, ApJ, 500, 525

Mazzarella, J. M., & Balzano, V. A. 1986, ApJS, 62, 751

New Statistical Methods for Analysis of Large Surveys: Distributions and Correlations

Vahé Petrosian

Center for Space Science and Astrophysics, Varian 302c, Stanford University, Stanford, CA 94305-4060

Abstract. The aim of this paper is to describe new statistical methods for determination of the correlations among and distributions of physical parameters from a multivariate data with general and arbitrary truncations and selection biases. These methods, developed in collaboration with B. Efron of Department of Statistics at Stanford, can be used for analysis of combined data from many surveys with different and varied observational selection criteria. For clarity I will use the luminosity function of AGNs and its evolution to demonstrate the methods. I will first describe the general features of data truncation and present a brief review of past methods of analysis. Then I will describe the new methods and results from simulations testing their accuracy. Finally I will present the results from application of the methods to a sample of quasars.

1. INTRODUCTION

One of the important ways of testing the models of AGN, or any other astrophysical source, is through the investigations of the distributions, ranges, and more importantly the correlations among, the relevant physical characteristics, such as luminosity, spectrum, redshifts or distances. A reliable determination of these features requires large samples. As evident from papers presented in this proceedings the samples are becoming larger and larger. Combining the samples, however, is a very challenging task, because different samples are obtained by different instruments and techniques, so that they suffer from different and varied selection biases and data truncations. Overcoming these biases requires care.

The primary goal of this paper is to describe some of the relatively new methods we have developed at Stanford over the past decade (Efron & Petrosian 1992, 1999). These methods are very general and are applicable to any data with well defined but arbitrary truncations. We have applied these to various astrophysical data such as solar flares (Lee, Petrosian, & McTiernan 1993, 1995), gamma-ray burst (see e.g. Lloyd, Petrosian, & Mallozzi 2000) and quasars (Meloney & Petrosian 1999). Instead of using abstract mathematical symbols, the method will be demonstrated using the luminosity function of AGNs and its cosmological evolution, *i.e.* its variation with redshift z; $\Psi(L,z)$. Without loss of generality, we can write the luminosity function as

$$\Psi(L,z) = \rho(z)\psi(L/g(z),\alpha_i)/g(z), \qquad (1)$$

where $\rho(z)$ describes the co-moving density evolution and $g(z)$ (with $g(0) = 1$) describes the luminosity evolution of the population with $L_o = L/g(z)$ as the luminosity adjusted to its present epoch value; $\psi(L_o, \alpha_i)$ gives the local luminosity function. Here I explicitly include the shape parameters α_i, which could also vary with redshift. A surprising result has been the absence of evidence for a strong shape evolution. In this paper I ignore such effects and concentrate on the determination of the the density and luminosity evolution functions $\rho(z)$ and $g(z)$.

In the next section we describe various kind of truncations and in §3 give a brief summary of some of the past methods used for this kind of analysis. The new methods and their accuracy are described in §4 and a sample of results from application to AGNs are summarized in §5.

2. Types of Truncations

The left panel of Figure 1 shows a set of arbitrary data points labeled as luminosity L and redshift z and several generic truncations. The distribution may be truncated parallel to the axis as shown by the dotted lines. This only limits the observed ranges of the variables but does not introduce any bias. This kind of data will be referred to as untruncated data. However, this is rarely the case for astronomical data, and in general, one is dealing with cases where the truncation is not parallel to the axis. The simplest and most common case is when the data suffers a one-sided truncation from below as shown by the solid curve. This is the case for magnitude or flux limited samples; $L \geq 4\pi d_L^2(\Omega_i, z) f_{min}$, where f_{min} is the limiting flux and $d_L(\Omega_i, z)$ is the luminosity distance at z for an assumed cosmological model represented by Ω_i. In some cases the data may be truncated from above as shown by the dashed curve. The statistical methods do not distinguish between truncation from above or below. However, the data analysis is affected when there are truncations both from **above and below**. This is the case for some AGN samples. The situation becomes even more complex when the truncation boundaries are not monotonic, or when one tries to combine samples with different truncations, say different upper and lower flux limits. Such variation may be present even within a given catalog where data taken at different times and directions in the sky may have different limits. The most general truncation is when each data point, say $[L_i, z_i]$, has its individual upper and lower limits, $L_i^- < L_i < L_i^+$ and $z_i^- < z_i < z_i^+$, as shown by the large cross for one point on Figure 1. The methods we have developed can treat this most general truncation.

3. A Brief Historical Review

The process of determination of the distribution of physical characteristics from truncated data has a long history starting with first and rudimentary observations of stars in the disk of our galaxy. Here I will touch upon some of the relevant highlights. A more detailed discussion can be found in the references cited below and in a review article (Petrosian 1992). I will limit the discussion to works aimed at determination of the luminosity function and spatial distribution of sources; $\Psi(L, r)$ from a magnitude limited data (one sided truncation)

New Statistical Methods

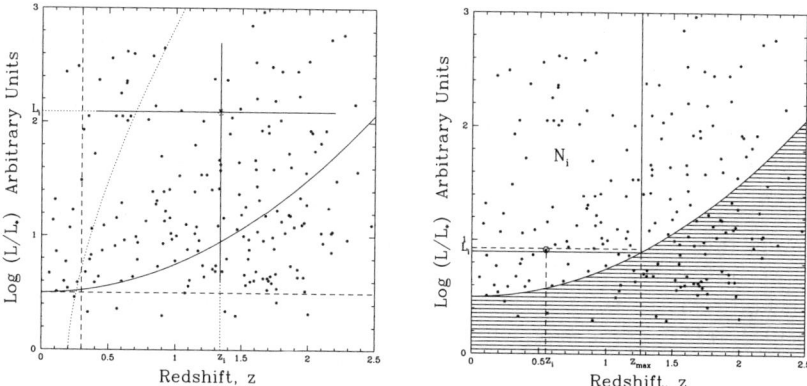

Figure 1. **Left Panel:** Demonstration of various types of data truncations: Parallel to axis (dotted lines), from below (the solid curve), from above (the dashed curve), and a general truncation when each data point has its specific observable range (shown by the cross for only one of the points). **Right Panel:** Description of the constructs used for evaluation of the univariate distribution from uncorrelated data truncated from below. The large box contains N_i sources in the *comparable or associated set* of the source i located in the thin rectangle defined by the dashed horizontal line.

as shown by the right panel in Figure 1. Almost all the methods I will describe ignore possible correlations between the variables and assume that these are independent. In the case of the luminosity function this means that $g(z) = 1$, so that the bivariate distribution is separable; $\Psi(L,r) = \psi(L)\rho(r)$. Unfortunately this often unjustified simplification is prevalent even today. This unnecessary assumption often can lead to erroneous conclusions. As stressed below the first task must be testing the bivariate distribution for correlations between the variables.

In general, most of the methods can be divided into two categories, *parametric* or *non parametric*. In the former one assumes a parametric form for the two functions $\psi(L)$ and $\rho(r)$. In the non parametric methods one often ends up with a description for the cumulative functions

$$\phi(L) = \int_L^\infty \psi(L')dL', \quad \sigma(r) = \int_0^r (dV/dr)\rho(r')dr', \tag{2}$$

where $V(r)$ is the volume of space (included in the observation) from the origin to distance r. In these relation r could be any measure of distance including redshift, look back time etc.

3.1. Parametric Methods

The first methods were developed for investigation of the luminosity function and spatial distribution of stars perpendicular to the galactic disk. These and subsequent applications to other sources have come to be known as correction for **Malmquist Bias** after Malmquist (1922); Eddington (1915, 1940) also describes this method. A more recent description can be found in Trumpler & Weaver (1953). These early works dealing with the distribution of **stars** assume a Gaussian distribution of absolute magnitudes (*i.e.* a log-normal luminosity function) and a Gaussian spatial distribution perpendicular to the disk. Because of the truncation there is absence of low luminosity stars at large distance so that the raw observed distributions of L and r are biased. The essence of the method was to correct for this bias. It turns out that the method used was sound but the final results were incorrect because of the erroneous assumptions about the forms of the distributions. We know today that the luminosity function of stars is best represented by a broken power law and that the fall of the stars perpendicular to the disk is exponential and not Gaussian. This demonstrates the shortcoming of this and other parametric methods. Similar parametric methods have been used for extragalactic sources (**galaxies and quasars**) first by Neymann & Scott (1959) and in numerous works ever since.

3.2. Non-parametric Methods

One of the most commonly non parametric methods used is the so-called V/V_{max} method first described by Kafka (1966) soon after the discovery of the quasars but used most successfully by Schmidt (1968). Independence is again assumed and the presence or absence of density evolution is tested by a single moment, namely the average value of V/V_{max}, where V_{max} is the volume up to the maximum redshift (or distance) that a source of luminosity L can be visible given the limiting flux f_{min}; $L = 4\pi d_L^2(\Omega_i, z_{max})$ (see Figure 1). In the absence of evolution one expects a value of 0.5 for this average. A more general method was described later by Avni & Bahcall (1980). Of course, one need not be limited only to one moment of the distribution. More information can be obtained by examining the distribution of $(V/V_{max})_i$ of the whole data set.

Schmidt (1968) also described a method for determination of $\phi(L)$, which was later dubbed as the "Schmidt Estimator" by Felten (1976). It is straightforward to show that in absence of evolution, *i.e.* for a uniform spatial distribution, the contribution to this cumulative luminosity function of each source is proportional to $V_{max,i}^{-1}$. It is also easy to show that if there is evolution this contribution is proportional to $\sigma_{max,i}^{-1}$ defined in equation (2); σ was denoted as V' by Schmidt.

Most other methods employ binning, which simplifies the problem conceptually but has several shortcomings, the primary being loss of data points in the incomplete bins at the truncation boundaries. Examples of these are anlysis by Nicole & Segal (1978, 1983), Turner (1979) and Choloniewski (1986, 1987). As shown by Petrosian (1986), it turns out that all these procedures, in the limit of one source per bin reduce to Lynden-Bell's (1979) C^- method. For a detailed comparison of these methods see Petrosian (1992).

3.3. The General Method

The right panel of Figure 1 depicts two boxes. Let us assume that number of data points in the narrow box is $n(L)dL$ and the number in the big box, excluding the narrow region is $N(L)$. It is easy to show that if the variables are independent, *i.e.* the luminosity function is separable in L and z, then

$$\frac{n(L)dL}{N(L)} = \frac{\sigma(z_{max})\psi(L)dL}{\sigma(z_{max})\phi(L)} = -d\ln\phi(L). \qquad (3)$$

In the limiting case of one object per (narrow) bin, $n(L) = \delta(L - L_i)$ and $N(L) = N_i + \Theta(L - L_i)$, where Θ is the Heviside step function.

Integration of the above equation then shows that the cumulative luminosity function $\phi(L)$ increases by $\delta\ln\phi(L) = \ln(1 + N_i^{-1})$ going across the source at L_i. Thus, the cumulative luminosity function can be build in increments starting, say from the highest observed luminosity L_1, as

$$\phi(L) = \phi(L_1) \prod_{L_i > L} \ln(1 + N_i^{-1}). \qquad (4)$$

The set of N_i sources in the big box are referred to as the *comparable* or the *associated* set of the source with luminosity L_i. It is clear that because of the complete mathematical symmetry between the two variables we can get an identical expression for the cumulative distribution $\sigma(z)$. The (different) associated sets are defined in a similar manner.

In the next section I will describe the extension of this method to more complex truncations. I will consider only bivariate distributions. The generalizations to multivariate distributions is straightforward.

4. The New Methods

The complete description of the distributions is a two step process. The first step is to determine whether the two variables are *correlated* or they are *independent*, and if correlated, then find a way to account for this correlation. The latter step can be done only parametrically. For the luminosity function it entails the determination of the form of the luminosity evolution function $g(z)$ such that a redefined luminosity $L_0 = L/g(z)$ is uncorrelated with, or is independent of, z. The second step is to determine the univariate distributions of the independent variables z and L_0.

4.1. Untruncated Data

If z and L are independent then the rank R_i of z_i (or L_i) in an untruncated sample (i.e. a sample truncated parallel to the axes; $z > z_{min}$ and $L > L_{min}$) will be distributed uniformly between 1 and N with an expected mean $E = \frac{1}{2}(N+1)$ and variance $V = \frac{1}{12}(N^2 - 1)$. We may then normalize R_i to have a mean of 0 and a variance of 1 by defining the statistic $T_i = (R_i - E)/V$. The hypothesis of independence is then rejected or accepted using a statistics based on the distribution of the T_i. The quantity

$$\tau = \frac{\sum_i (R_i - E)}{\sqrt{\sum_i V}} \qquad (5)$$

is one choice of such a test statistic with a mean of 0 and a variance of 1. The hypothesis of independence is rejected if $|\tau_{data}|$ is too large (e.g. $|\tau_{data}| \geq 1$ for rejection of independence at the 1 σ level). This τ is equivalent to Kendell's τ statistic (see, *e.g.* Press et al. 1990)

If it turns out that $|\tau_{data}| \leq 1$ and that the variables are independent, then the determination of the univariate distributions in each variable is obtained ignoring the value of the other. For example, the cumulative luminosity function will be described by the histogram $\phi(L_{i+1} > L > L_i) = i$. However, if the variables are correlated, one must then carry out a transformation to remove the correlation by some parametric function, say $L_0 = L/g(z)$. It is important to note that the transformed data will now appear truncated in the $L_0 - z$ plane with $L_{0,min} = L_{min}/g(z)$, which is no longer parallel to the z axis. The use of the method described by equation (3) is then required to obtain the univariate distributions.

4.2. Data with One-sided Truncation

A straightforward application of the above method to a truncated data will clearly give a false correlation signal. Efron & Petrosian (1992), and independently Tsai (1990), describe how this method can be applied to data with one-sided truncation. The above procedure is modified as follows. For each object define a new *comparable* or *associated* set

$$J_i = \{j : L_j > L_i,\ L_j^- < L_i\}, \tag{6}$$

where $L_j^- = 4\pi d_L^2(\Omega_i, z_j) f_{min}$. It is easy to see that this is the same set defined above (Figure 1) as the big box containing N_i sources. (Note that the set defined in Efron & Petrosian (1992) includes the object i in question.) This is the largest subset of luminosity and volume limited data that can be constructed for each point (L_i, z_i). If z and L are independent then we expect the rank R_i of z_i (or L_i) in this limited set, not in the whole sample to be uniformly distributed between 1 and N_i. The rest of the procedure follows the steps described above.

Similarly the determination of the univariate distributions will require the use of the method described in the previous section for the transformed and uncorrelated variables. Note again that the truncation boundary also gets transformed in case there is a correlation between L and z.

4.3. Complex Truncations

A generalization of the above method to doubly (or multiply) truncated data was developed by Efron & Petrosian (1999), which is valid for the most general truncation $L_i^- < L_i < L_i^+$ and $z_i^- < z_i < z_i^+$. The method is equivalent to the previous method, with the associated set defined as

$$J_i = \{j : L_j > L_i,\ L_i \in (L_j^-, L_j^+)\}. \tag{7}$$

In this case, however, the distribution of the rankings (or of τ) is unknown. If the data are uncorrelated then τ must still have a mean of zero. But a bootstrap method using simulations based on the $\psi(L)$ obtained from the data, as described below, is required for the purpose of the estimation of the variance.

For doubly (or more complexly) truncated data the comparable set is not completely observed, thus a simple analytic method such as the one described in equation (4) is not possible. However, it turns out that a simple iterative procedure can lead to a maximum likelihood estimate of the distributions. Efron and Petrosian (1999) give a thorough description of this method; for a more brief and transparent description see Maloney & Petrosian (1999).

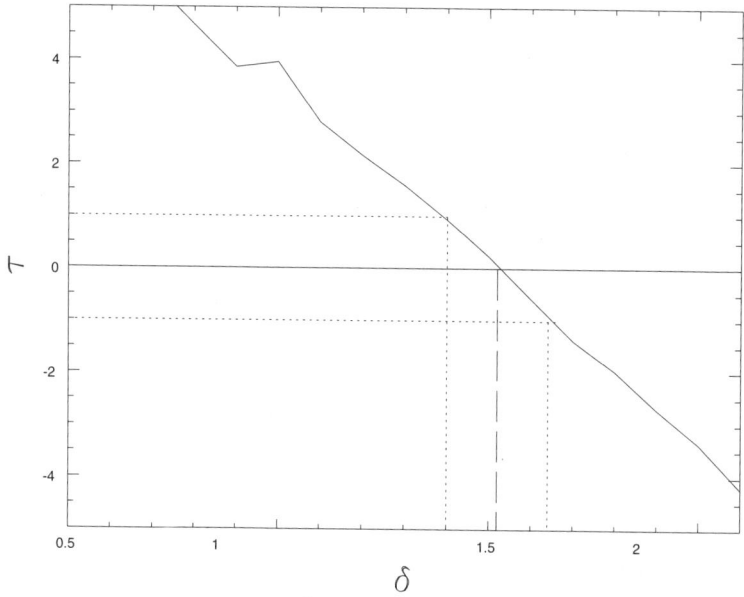

Figure 2. Variation of the τ statistics with exponent δ of the assumed correlation $\bar{y} \propto x^{-\delta}$. For the parent sample $\delta = 1.5$.

4.4. Tests of the Correlation Algorithm

Lloyd et al. (2000) describe two simulations which test how well the above procedures reproduce known distributions. In one simulation the method was applied to a single sided truncation of an uncorrelated bivariate parent distribution. The rank test applied to the untruncated simulated points gave a value of $\tau = 0.9$ But when applied to the truncated data without considering the effects of the truncation resulted in $\tau > 5.0$, indicating the presence of a strong (of course false) correlation that is introduced by the truncation. However, when the test was carried out correctly by accounting for the effects of the truncation (as described above) it was found that $\tau = 0.6$, recovering the fact that the variables were independent. In a second set the simulated parent sample had a strong anti correlation, with variables y and x correlated with average value of $\bar{y} \propto x^{-1.5}$. Application of the correlation to the untruncated randomly selected sample, as expected, gave a negative value for the statistic; $\tau < -5$. The data was truncated and the method applied blindly. This resulted in the value of $\tau = +2.5$, or a (false) positive correlation. However, when applied properly it gave a value of $\tau = -4.0$ indicating the presence of the anti correlation at 4 sigma level. Assuming a correlation of $\bar{y} \propto x^{-\delta}$, the data was transformed accordingly and the method applied to the new data. Figure 2 shows the variation of τ with the exponent δ, from which we deduce a mean value and one sigma range of $\delta = 1.51 \pm 0.11$. This is a strong support for the accuracy of the method.

Further simulations are required to test the methods for determination of the distribution and correlations for more complex truncations.

5. Some Results from AGN Data

Our application of these methods to a combine set of several optically selected samples of quasars are described in Maloney & Petrosian (1999). Here is a brief summary.

- We found a strong correlation between luminosity and redshift, indicating the presence of a rapid luminosity evolution.
- The parametric model of luminosity evolution $(1+z)^{k'}$ provides a better description of the data than the model $e^{kt(z)}$, where $t(z)$ is the look back time. Neither parameterization perfectly removes the correlation in all areas of the $L-z$ plane. In order to better model this evolution future analyses of quasar evolution could consider parametric forms, with more than one free parameter. Some of the more complex forms of evolution suggested in the literature, *e.g.* the used by La Franca & Cristiani (1996) are equivalent with a luminosity evolution form that contains two independent free parameters.

Given the form of the luminosity evolution we make the simple transformation of all luminosities to their hypothetical present epoch values, $L_0 = L/g(z)$, so that L_0 and z are uncorrelated. This allows us to use our methods to determine the univariate distributions of z (the density evolution) and L_0 (the local luminosity function).

- We find that the co-moving density of quasars also evolves, but its extent depends on the cosmological model. For example, for the Einstein−de Sitter model $\rho \sim (1+z)^{2.5}$ for low redshifts and rapidly declines as $\rho \sim (1+z)^{-5}$ for $z > 2$. This is much slower evolution than is obtained when one (incorrectly) assumes a pure density evolution model; $g(z) = 1$ (Schmidt 1968; Miyaji, Hasionger & Schmidt 1998).
- The cumulative local luminosity function $\phi(L_o)$ has the double power law form found previously (Caditz & Petrosian 1990, La Franca & Cristiani 1996), with a break luminosity of $L_* = 6 \times 10^{29}$ erg / (sec Hz), in the Einstein−de Sitter model. The power-law indices at the low and high luminosity ends are -1.5 and -2.3 in rough agreement with previous estimates. There is however, some evidence for evolution of the shape of the luminosity function. More data is required for a quantitative description of this evolution.

References

Avni, Y., & Bahcall, J. N. 1980, ApJ, 235, 694

Caditz, D., & Petrosian, V. 1990, Ap.J., 357, 326

Choloniewski, J. 1986, M.N.R.A.S., 192, 61

Choloniewski, J. 1987, M.N.R.A.S., 226, 273

Eddington, A. 1915, M.N.R.A.S., 73, 359

Eddington, A. 1940, M.N.R.A.S., 100, 354

Efron, B., & Petrosian, V. 1992, Ap.J., 399, 345

Efron, B., & Petrosian, V. 1999, JASA, 94, 447

Felten, J.E. 1976, Ap.J., 207, 700

Kafka, P. 1967, Nature, 213, 346

La Franca, F., & Cristiani, S. 1996, invited talk in *Wide Field Spectroscopy* (20-24 May 1996, Athens), Eds. M. Kontizas et al. astro-ph/9610017

Lee, T. T., Petrosian, V., & McTiernan, J. M. 1993, 412, 401

Lee, T. T., Petrosian, V., & McTiernan, J. M. 1995, 448, 915

Lloyd, M. N., Petrosian, V., & Mallozzi, R. S. 2000, 534, 227

Lynden-Bell, D. 1971, M.N.R.A.S., 155, 95

Malmquist, K. G. 1920, Medd. Lund. Obs., Ser. 2, No. 22

Maloney, A. & Petrosian, V. 1999, Ap.J., 518, 32

Miyaji, T. Husinger, G. & Schmidt, M. 1998, A&A, 369, 49

Neyman, J. & Scott, E. L. 1959, *Handbuch Der Physik*, Vol 53, ed. S. Flugge, Springer-Verlag, Berlin, p. 416

Nicole, J. F. & Segal, I. E. 1978, Ann. Phys. 113, 1

Nicole, J. F. & Segal, I. E. 1983, Astr. Ap., 118, 180

Petrosian, V. 1986, *Structure and Evolution of Active Galactic Nuclei*, eds. Giuricin et al., D. Reidel Publ. Co., pp. 353-381

Petrosian, V. 1992, in *Statistical Challenges in Modern Astronomy*, Eds. E. D. Feigelson & G. J. Babu, (New York: Springer-Verlag), p. 173

Press, W. H., Teukolsky, S. A., Vetterling W. T., & Flannery, B. P., 1992. *Numerical Recipes in FORTRAN*, (Cambridge: Cambridge University Press)

Schmidt, M. 1968, Ap.J., 151, 393

Trumpler, R. J. & Weaver, H. F. 1953 *Statistical Astronomy*, Dover, N.Y.

Tsai, W. 1990, Biometrika, 77, 169.

Digitization of the FBS: Its Future Use and Expected Results

Areg M. Mickaelian

Byurakan Astrophysical Observatory (BAO), Byurakan 378433, Armenia, and Isaac Newton Institute of Chile, Armenian Branch.
E-mail: aregmick@bao.sci.am

Abstract. The First Byurakan Survey (FBS) is the largest spectral survey in the Northern sky. Its plates contain low-dispersion spectra for some 20,000,000 objects. The FBS spectra allow selection of objects by their color, broad emission or absorption lines, or SED; to discover, classify and investigate them. The FBS was conducted originally to search for UVX galaxies (1500 Markarian galaxies were discovered). Selection of blue stellar objects, red stars, and identification of IRAS sources have also been done by means of the FBS spectra. All these projects have been carried out by visual inspection of the plates. Digitization will give new possibilities to search for many new objects: new bright QSOs (m<18^m), new Markarian (UVX) galaxies, BCDGs, optical counterparts of IR, radio, X-ray and other sources, late-type stars, planetary nebulae, emission-line stars, and white dwarfs; and to study star clusters and clusters of galaxies. The digitized FBS will be available via the Internet and on CDs. Software and an appropriate interface for working with the data will be provided.

1. Introduction

The Digitized Sky Survey (DSS) is widely used by all astronomers. It gives good astrometric, photometric and morphological data on all the objects with $\delta > -30°$ to 21^m. However, it cannot give more information on the nature of the objects. Such information is provided by spectroscopic data, but unfortunately, there isn't any all-sky spectroscopic survey. The Hamburg Survey was digitized recently, giving low-dispersion spectra for objects to 18^m at high galactic latitudes ($|b| > 20°$) (Engels et al. 2001). The best data for an area of 10,000 deg^2 of the Northern Sky will be provided by the Sloan Digital Sky Survey (SDSS). However, at present there is the possibility to provide such information to 17.5^m-18^m. The First Byurakan Survey (FBS) is a large spectral survey, covering 17,000 deg^2 of all the Northern Sky and part of the Southern Sky at high galactic latitudes (Markarian et al. 1989). It provides unique observational material for discovery, classification and investigation of various types of objects.

FBS low-dispersion spectra are rather useful for selection of objects by different criteria (such as their color, broad emission or absorption lines, spectral energy distribution), quick study of separate fields for the content of different types of objects, etc. The number and classes of new objects discovered in the FBS emphasize the necessity for open, free access to this information for the astronomical community.

As many astronomers are engaged in searching for new objects and identification of different sources, the FBS low-dispersion spectra may be useful for such work for better understanding of their nature, before making detailed investigations with higher dispersions and resolutions. There is a need for access to this database for all the astronomical community. Digitized information may be widely distributed on CDs, as well as by the Internet. For this reason it is desirable first to create the digitized copies of the FBS plates with a high resolution.

This project will allow use of the First Byurakan Survey together with the DSS for selection, classification and investigation of many new interesting objects, as well as analysis of sample composition in definite areas of interest combining all the data in a given region. The FBS will provide a unique database of homogeneous data on some 20,000,000 objects at high galactic latitudes.

2. Description of the FBS and its Low-Dispersion Spectra

The First Byurakan Survey was carried out by B.E. Markarian, V.A. Lipovetsky and J.A. Stepanian in 1965-1980 with the Byurakan Observatory 102/132/213 cm ($40''/52''/84''$) Schmidt telescope using the 1.5° prism (Markarian et al. 1989). 2050 Kodak IIAF, IIaF, IIF, and 103aF photographic plates in 1133 fields (4°×4° each, the size being 16cm×16cm) were taken. The FBS covers 17,000 deg^2 of all the Northern sky and part of the Southern sky ($\delta >-15°$) at high galactic latitudes ($|b| >15°$). In some regions, it even goes down to $\delta=-19°$ and $|b|=10°$. The limiting magnitude on different plates varies in the range of 16.5-19.5 in V, however for the majority it is 17.5^m-18^m. The scale is $96.8''$/mm and the dispersion is 1800 Å/mm near Hγ and 2500 Å/mm near Hβ (mean spectral resolution being about 50 Å). Low-dispersion spectra cover the range 3400-6900 Å, and there is a sensitivity gap near 5300 Å, dividing the spectra into red and blue parts. It is possible to compare the red and blue parts of the spectrum (easily separating red and blue objects), follow the spectral energy distribution, notice some emission and absorption lines (such as broad Balmer lines, molecular bands, He, [OIII] N_1+N_2 and [OII] lines; broad emission lines of QSOs and Seyferts, etc.), thus gaining some understanding about the nature of the objects (Fig. 1). The FBS is made up of zones (strips), each covering 4° in declination and all right ascensions except the Galactic plane regions. In all there are 28 zones, which are named by their central declination (e.g., zone +27° covers +25°< δ <+29°, zone +63° has +61°< δ <+65°, etc). The zones and the neighboring plates in right ascension overlap about 0.1° (as the exact size of a plate is 4.1°×4.1°) thus making the whole area complete. Each FBS plate contains low-dispersion spectra of some 15,000-20,000 objects, and there are some 20,000,000 objects in the whole survey.

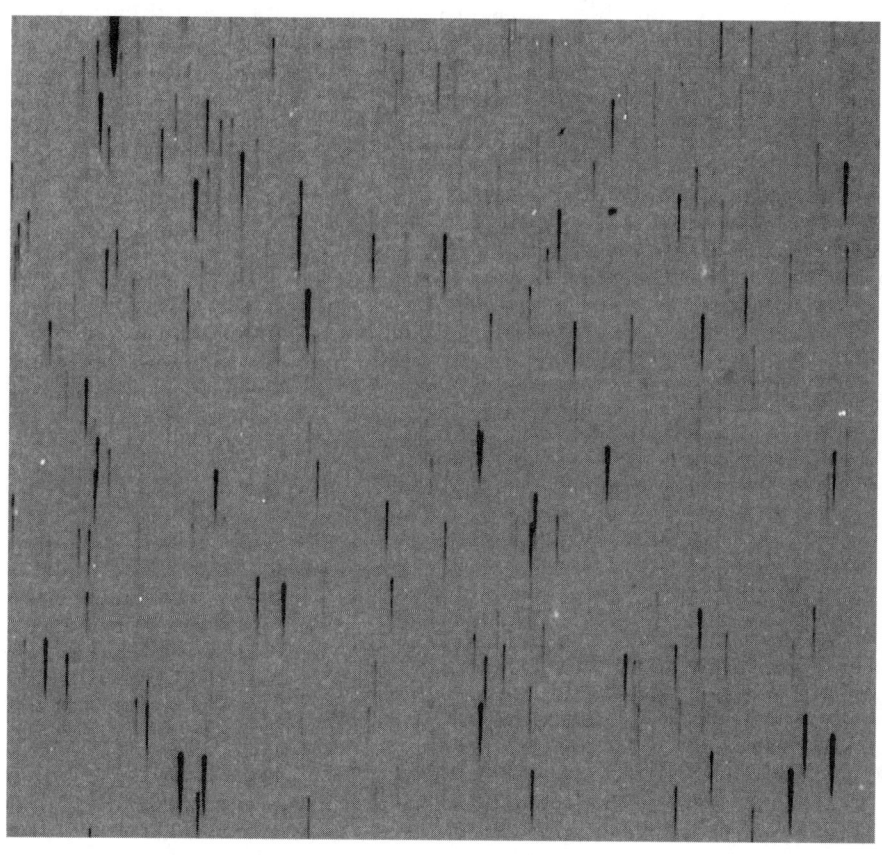

Figure 1. A piece of an FBS plate with its low-dispersion spectra.

3. Results Obtained on the Basis of the FBS Plates

3.1. Markarian Galaxies

The FBS was conducted originally to search for galaxies with UV-excess. The discovery of 1515 UV-excess (UVX) galaxies by Markarian and colleagues (later called Markarian galaxies) was the first and the most important work based on the FBS plates (Markarian 1967; Mazzarella & Balzano 1986; Markarian et al. 1989). Spectral observations of these objects were carried out first by Weedman & Khachikian (1968), and by Arakelian, Dibay & Esipov (1970), and then by hundreds of other authors. The study of Markarian galaxies led to the discovery of many new Seyferts and the first spectral classification of this type of object (Weedman & Khachikian 1971). Among them, there are more than 200 Seyferts, dozens of QSOs, 17 LINERs, a few hundred starburst and isolated HII galaxies, 3 BL Lac objects, radio, IR, X- and gamma-ray sources, interacting and merging objects, and galaxies with double and multiple nuclei. During the last 30 years, more than 3500 scientific papers have been published on the study of these objects. There are many Markarian galaxies, which are the subject of study for understanding the AGN phenomenon, starburst activity and evolution of galaxies, high-luminosity IR radiation, AGN variability, double and multiple structure of the nucleus, composite spectrum AGNs, galaxy interactions and merging, connections between different types of active galaxies, early stages of evolution of galaxies, and other important topics of modern extragalactic astronomy. Some Markarian galaxies are crucial for solving different important problems: Mrk 231 is the most luminous infrared galaxy (ULIG) in the Local Universe (e.g., Weedman 1999); some Markarian galaxies are blue compact dwarf galaxies (BCDGs), such as the remarkable BCD galaxy Mrk 116 (I Zw 18), the most metal-deficient BCDG (most of the known objects of this type are Markarian and SBS galaxies (Second Byurakan Survey; Markarian & Stepanian 1983)); Mrk 501 is one of the 5 known TeV sources; Mrk 938 is the first dynamical merger discovered observationally; Mrk 766 is one of the famous NLS1 galaxies; Mrk 273 is a wonderful double-double nuclei galaxy; Mrk 6 shows variations of spectral lines typical of different types of objects and very high column density of H in X-rays; Mrk 421 and Mrk 501 (BL Lac objects) are among the strongest X-ray sources; Mrk 926 is one of the rare Sy1 galaxies having LINER properties (like 3C 390.3); Mrk 266 has a multiple structure nuclear region; Mrk 110 is crucial for understanding differences between NLS1s and BLS1s; Mrk 231 and Mrk 507 are among the 5 known superstrongest FeII emitters (ex. Véron-Cetty & Véron 2000); Mrk 530, Mrk 993 and Mrk 1018 change their spectra from Sy2 to Sy1 thus being important for the unified scheme of AGN. The Markarian survey was the first systematic survey for AGNs.

3.2. FBS Stellar Objects

The second part of the FBS was devoted to the discovery and investigation of blue (UVX) stellar objects. It was carried out by H.V. Abrahamian and A.M. Mickaelian in 1987-1996 (Abrahamian & Mickaelian 1996; Mickaelian 1994, and references therein) in 278 FBS fields, in a 4009 deg^2 area of the FBS ($+33°< \delta <+45°$ and $+61°< \delta <+90°$). The main purpose of this work was discovery of new bright QSOs (as Markarian's survey was aimed only at extended objects), Seyferts, other compact galaxies, planetary nebula nuclei, cataclysmic variables (CV), white dwarfs (WD), subdwarfs, and other peculiar stellar objects. This project is similar to the Palomar-Green survey (Green et al. 1986), however it was done on the basis of the low-dispersion spectra. 1103 objects have been selected, including 716 new blue stellar objects and many new objects were revealed. Subsamples of candidate QSOs, WDs, CVs, continuous spectrum objects (possible BL Lacs and DCs) have been made up and observations have been carried out for spectral identification of these objects. 40 new bright AGNs have been discovered already among them (Mickaelian et al. 2001a; Véron-Cetty & Véron 2001), proving that the FBS can provide data for enriching our knowledge of the Local Universe. The local density of QSOs and the completeness of the Bright Quasar Survey (BQS) (Schmidt & Green 1983) have been re-estimated. Moreover, cross-correlations of NVSS/USNO and ROSAT/USNO objects, and checking their low-dispersion spectra in the FBS with further follow-up spectroscopy identified new QSOs in the FBS plates which had not been selected before. A number of interesting WDs and CVs have been discovered too, such as the new bright SW Sex type cataclysmic variable (Mickaelian et al. 2001b). In all, 11 lists of the 2nd part of the FBS have been published. The full catalog of the FBS blue stellar objects is available at CDS (Abrahamian et al. 1999).

A survey for late-type stars on the FBS plates has been carried out since 1987 by H.V.Abrahamian and K.S.Gigoyan (Gigoyan et al. 2001 and references therein). 813 late M-type and carbon stars have been selected already in an area of 6640 deg^2 in 14 zones of the FBS, including 560 new ones. Surveys for carbon stars at high galactic latitudes are very rare. Discovery and study of such objects is necessary for the study of the kinematics and chemical composition of the Galactic halo. Besides, many objects with extended dust shells, as well as two new extremely rare dwarf carbon stars have been discovered. Twelve lists of FBS late-type stars have been published.

3.3. Identifications of *IRAS* Sources

Work on the selection of galaxies, blue stellar objects (including QSOs and Seyferts) and late-type stars led to a new program of optical identifications of IRAS sources on the basis of the FBS plates (Mickaelian 1995). Since 1995, 1577 previously unidentified IRAS point sources (IRAS 1988, 1989) have been optically identified in the zone $+61° < \delta < +90°$ with a surface area of 1487 deg^2. In this area, FBS plates have better limiting magnitudes (on average: 18.1 in V). 1178 of the identified sources are galaxies. The identification program produced two samples of objects: BIS (Byurakan-IRAS Stars), and BIG (Byurakan-IRAS Galaxies). Seven lists of galaxies (Mickaelian 2001, and references therein) and 5 lists of stars (Mickaelian & Gigoyan 2001, and references therein) have been published already. The BIS and BIG objects are present in many databases, including CDS catalogs, SIMBAD and NED (extragalactic objects). These samples are the subjects of multi-faceted investigations. Observations revealed many interesting objects, including Seyferts, LINERs and composite spectrum AGNs (e.g. Balayan et al. 2001). The IRAS galaxy sample contains AGNs, high-luminosity IR galaxies (LIGs, ULIGs, HLIGs, Sanders & Mirabel 1996), groups of galaxies (including compact ones), interacting and merging galaxies, etc. Some 30 IRAS sources showing galaxy colors have no optical counterparts even in the DSS fields, being new candidates for obscured IRAS galaxies (probable ULIGs) and important for revealing the real IR galaxy population. Study of the new sample of IRAS galaxies, especially the high-luminosity IR galaxies, may lead to understanding of the interrelationship between AGN and starburst activity induced by galaxy interactions and merging. The subsample of galactic objects is also interesting, as it contains new planetary nebulae, AGB stars, late M and carbon stars. Some of them have IR excesses, suggesting extended dust shells; in addition, a number of stars show evidence of variability.

Table 1 presents a summary of projects undertaken on the basis of the FBS observational material with brief data: name of the project, years, main authors, area of investigation, number of objects involved, and references.

Table 1. Projects Undertaken on the Basis of the FBS

Project	Years	Authors	Area (deg^2)	Objects	Ref.
UVX galaxies (Markarian galaxies)	1965-1980	Markarian, Lipovetsky, Stepanian	17000	1515	(1)
UVX stars (2nd part of the FBS)	1987-1996	Abrahamian, Mickaelian	4009	1103	(2)
Late-type stars	1987-2001	Abrahamian, Gigoyan	6640	813	(3)
IRAS galaxies (BIG)	1995-2001	Mickaelian	1487	1178	(4)
IRAS stars (BIS)	1995-2001	Mickaelian, Gigoyan	1487	370	(5)

References: (1) Markarian et al. 1989; (2) Abrahamian & Mickaelian 1996; (3) Gigoyan et al. 2001; (4) Mickaelian 2001; (5) Mickaelian & Gigoyan 2001.

4. What Information is Still in the Plates?: New Possibilities

Some five thousand interesting objects have already been found in the First Byurakan Survey plates. Nevertheless, its observational material is still useful for new studies. If digitized, FBS low-dispersion spectra can offer the possibility of searching for many new objects, such as:

- New bright QSOs (m<18^m). As the visual selection has missed many objects, and creation of numerical criteria for selection of candidate QSOs will be much more efficient, one can expect some 200 new bright QSOs (m<16.1^m, for a comparison with the BQS) from the FBS, and some 1000 fainter objects, and it will be the most complete sample for study of their local density;

- New Markarian (UVX) galaxies. The same situation applies for UVX galaxies, as visual selection has missed objects near the limits of the plates. It is necessary to have 1-D records for all Markarian galaxies to be able to continue a homogeneous selection of such objects;

- New blue stellar objects. The 2nd part of the FBS will be continued for the whole area and a total number of some 4500 objects is expected. The techniques of selection and preliminary classification have been worked out during 1987-1996 and the digitization will make possible quicker and better quality completion;

- New BCDGs. 22 Markarian galaxies and 135 SBS galaxies are candidate BCDGs. These two surveys have provided the vast majority of the sample of these objects. A search for compact UVX galaxies in the FBS will provide more BCDGs, making possible the understanding of the early stages of galaxy evolution;

- Optical counterparts of IR, radio, X-ray and other sources (optical identifications). The program of identification of IRAS sources proved that this kind of work can be very efficient with the help of low-dispersion spectra. The Hamburg/ROSAT identification program (Bade et al. 1998) comes to prove again the importance of such work on the basis of low-dispersion spectra. The digitization will allow continuation of the program of IRAS identifications to other regions of the FBS, as well as identifications for radio and X-ray sources, too;

- Late-type stars (M, S, carbon stars). The faintest late-type stars look like dots on the low-dispersion plates. It is rather low confidence to select such objects, as they may appear to be defects as well. For automatic selection it will be easier to separate real spectra from artificial ones;

- Planetary nebulae. There still may be faint undetected planetary nebulae. As their number approaches the limit, it seems interesting to find out all PNs in the Galaxy, particularly at high galactic latitudes;

- All emission-line stars (CV, etc). Their emission lines are sometimes of low contrast with respect to the continuum. That is why one needs to deal with 1-D spectral cuts and computer analysis for better selection;

- Non-UVX white dwarfs. During the completion of the Second part of the FBS, the authors have noticed that there are a lot of white dwarfs without UVX. Many such objects have been marked but not selected. However, they may be picked out easily by their broad absorption lines and complement the sample of high-latitude WDs for further studies of stellar evolution and evolution of the Galactic halo;

- Studies of star clusters and clusters of galaxies. It is rather convenient to study clusters using homogeneous observational material, especially having the low-dispersion spectra of all the objects. Estimates of the distribution of brightness, colors and types of objects may be made. Rough estimate of redshifts (membership in the clusters) for galaxies may be made, too.

The FBS will serve as a database for the astronomical community via the Internet, and many new tasks may be put forward by other astronomers, too.

5. Technical Description of the Program of Digitization

Taking into account the size of the photographic grains (25-30 μm), it is reasonable to scan the plates with a resolution of some 10 μm (about 2500 dpi). Therefore, it is planned to scan the FBS plates with a scanner with a resolution of 2400 dpi (for quick completion of the project). The files containing the scanned plates at first will be saved on HDD. For this, large volumes of HDD are needed. Corresponding software for giving coordinate information, the possibility of comparing with the corresponding DSS fields, making 1-D cuts for each spectrum, etc. will be created. Some 170 pixels of data will be obtained for each FBS spectrum (the length of each FBS spectrum is 1700 μm).

It will be useful to offer users a numerical classification scheme. Classification schemes based on criteria worked out during the selection of blue stellar objects and identification of IRAS sources will be developed. The classification principles are based on the relation of magnitudes and widths of the spectra (for separation of stellar and diffuse objects), spectral energy distribution (color), and presence or absence of broad spectral lines. Classification will be linked to general classification schemes using standard objects in the fields. We plan also to make some rough photometric calibration using photometric standards to make possible a quick estimate of magnitudes from the FBS. It is estimated that up to 0.3^m accuracy may be reached. However, we don't pursue photometric precision. It should be done using the DSS database (rough photometry by APS) (Pennington et al. 1993).

Each spectrum will be presented as a small table (of 170 lines), and each (FBS) plate contains some 20,000 such spectra. Taking into account the difficulties concerning the use of full image data to identify and measure objects while dealing with an FBS plate, it makes sense to create a catalog of objects with all available data (positional, photometric, and spectral), and recommend it to users, too, for quicker access to the data.

In all (if we digitize all the plates) 40,000,000 spectra from the FBS plates, or (if we digitize only the best plates in each field) 20,000,000 FBS spectra are to be recorded. Data from each plate will be about 450Mb, and after archiving - some 45Mb. Appropriate media are necessary to store the large volumes of data from the digitized images. CD-ROMs are the most convenient and cheap units for this purpose. Therefore the data from the full plate scans, the 1-D spectra, the catalog of objects, and corresponding software will be copied onto CDs. Data for 14 plates (45Mb each) will be written on each CD (650Mb). In all, 81 CDs will be needed for the FBS database (in the case of digitization of all the plates - 147). The whole FBS database will take some 50Gb.

A web page for the FBS will be created and all the information will be available to astronomers. It will also be possible to extract separately 1-D spectra of objects by searching for them in the FBS catalog (like APS for the DSS), without taking larger volumes of information for the corresponding fields. Users will be able to compare the fields and data with corresponding ones from other surveys, too, including DSS, the Hamburg Survey, etc. All this will make it very efficient to work with unknown objects, especially for the people working in extragalactic astronomy.

By our estimates, the FBS database will be available at the end of 2002 or at the beginning of 2003.

References

Abrahamian, H.V., & Mickaelian, A.M. 1996, Ap, 39, 531
Abrahamian, H.V., Mickaelian, A.M., Lipovetsky, V.A., & Stepanian, J.A. 1999, Catalog No.II/223 at CDS, Strasbourg, at http://vizier.u-strasbg.fr/cgi-bin/VizieR?-source=II/223/fbs2
Arakelian, M.A., Dibay, E.A., & Esipov, V.F. 1970, Afz, 6, 39
Bade, N., Engels, D., Voges, W. et al. 1998, A&AS, 127, 145
Balayan, S.K., Hakopian, S.A., Mickaelian, A.M., & Burenkov, A.N. 2001, AstL, 27, 635
DSS web page at STScI: http://stdatu.stsci.edu/dss/
Engels, D. et al. 2001, Proceedings of the Conference: The New Era of Wide-Field Astronomy, ASP Conf. Series, Vol. 232, Eds. R.G. Clowes, A.J. Adamson, & G.E. Bromage (in press)
Gigoyan, K.S., Abrahamian, H.V., Azzopardi, M., & Russeil, D. 2001, Ap, 44, 88
Green, R.F., Schmidt, M., & Liebert, J. 1986, ApJS, 61, 305
IRAS Catalogs and Atlases 1988, 2. The Point Source Catalog. Declination $90° > \delta > 30°$, Joint IRAS Science Working Group, NASA, Washington, DC: US GPO
IRAS Faint Source Catalog 1989, $|b| > 10°$, Version 2, On Optical Disc, "Selected Astronomical Catalogs", Supplied by NASA, Vol. 1
Markarian, B.E. 1967, Ap, 3, 55
Markarian, B.E., & Stepanian, J.A. 1983, Ap, 19, 639
Markarian, B.E., Lipovetski, V.A., Stepanian, J.A., Erastova, L.K., & Shapovalova, A.I. 1989, Commun. Special Astrophys. Obs., 62, 5
Mickaelian, A.M. 1994, Discovery and Investigation of the Blue Stellar Objects of the First Byurakan Survey, Ph.D. Thesis, Byurakan, 284 p.
Mickaelian, A.M. 1995, Ap, 38, 349
Mickaelian, A.M. 2001, Ap, 44, 185
Mickaelian, A.M., & Gigoyan, K.S. 2001, Ap, 44, 222
Mickaelian, A.M, Gonçalves, A.C., Véron-Cetty, M.P., & Véron, P. 2001a, Ap, 44, 14
Mickaelian, A.M., Balayan, S.K., Ilovaisky, S.A., Chevalier, C., Véron-Cetty, M.P., & Véron, P. 2001b, astro-ph/0108377, OHP Preprint No. 155

Pennington, R.L., Humphreys, R.M., Odewahn, S.C., Zumach, W., & Thurmes, P.M. 1993, PASP, 105, 521
Sanders, D.B., & Mirabel, I.F. 1996, ARA&A, 34, 749
Schmidt, M., & Green, R.F. 1983, ApJ, 269, 352
SDSS web page at Johns Hopkins Univ.: http://tarkus.pha.jhu.edu/
Véron-Cetty, M.-P., & Véron, P. 2000, A&ARv, 10, 81
Véron-Cetty, M.-P., & Véron, P. 2001, A&A, 374, 92; and the electronic version of the Catalog of QSOs (10th edition) at http://cdsweb.u-strasbg.fr/cgi-bin/qcat?J/A+A/374/92 or http://www.obs-hp.fr
Weedman, D.W. 1999, Proceedings of the IAU Symp. 194: Active Galactic Nuclei and Related Phenomena, Eds. Terzian Y., Weedman D., Khachikian E., Astron. Soc. Pacific, 191
Weedman, D.W. & Khachikian, E.Ye., 1968, Afz, 4, 587
Weedman, D.W. & Khachikian, E.Ye., 1971, Ap, 7, 389

Part 7
Summary

Conference Dinner: Areg Mickaelian, Ed Khachikian

Conference Banquet:
Phil Outram, Lily Hovhanessyan, Caryl Gronwall, James Manners,
Omar Almaini, Vicki Sarajedini, Kate Isaak

AGN SURVEYS
ASP Conference Series, Vol. 284, 2002
R.F. Green, E.Ye. Khachikian, D.B. Sanders

Conference Summary

D. B. Sanders[1]

Institute for Astronomy, University of Hawaii, 2680 Woodlawn Drive, Honolulu, HI 96822

Abstract. This conference on *AGN Surveys* has proved to be a significant milestone in our understanding of the redshift distribution of optically selected QSOs, and in our initial understanding of the cosmic distribution of AGN from the first far-infrared and X-ray deep fields. It has also set the stage for continuing debates concerning the multiwavelength properties of AGN, the cosmological distribution of "obscured" AGN, and the "orientation versus evolution" debate on the nature of the sources discovered at different wavelengths. Much of this debate could have been anticipated from previous studies of the complete samples of optically-selected AGN provided to us by the pioneering work carried out by the staff of the Byurakan Astronomical Observatory over the past 40 years.

1. Introductory Remarks

On behalf of the Scientific Organizing Committee I would like to sincerely thank the Local Organizing Committee and the entire staff of the Byurakan Astrophysical Observatory (BAO) for their hard work in preparing for this IAU Colloquium and for their gracious hospitality in hosting this week's conference on the observatory grounds. To have been able to hold this meeting on *AGN Surveys* at the BAO is indeed a fitting tribute to the memory of Beniamin Markarian.

I would also like to thank all of the speakers and poster presenters for their collective effort in making this meeting a great success. Each of the meeting sessions has been well attended and the large number of oral contributions and posters presentations have provoked lively discussion. In the four full days of talks we have managed to thoroughly cover the full wavelength spectrum of AGN Surveys – past, present, and future – and to provide what I believe is the first truly comprehensive picture of the AGN phenomenon out to redshifts $z \sim 6$, and covering a wide range of total luminosity ($M_{\rm B} \sim -7$ to -30). Although the breadth and depth of modern all-sky surveys now routinely carried out from space and on the ground has provided us with tens of thousands of cataloged AGN, it is clear from the talks at this meeting that the "Markarian Lists" generated from the First and Second Byurakan Surveys remain as one of the most important sources of objects for *detailed* studies of the AGN phenomenon.

---

[1] also, Max-Planck-Institut für extraterrestrische Physik, D-85748 Garching, Germany

For many of the participants at this meeting this has been a first visit to the BAO as well as to Armenia in general. The opportunity to explore the Observatory grounds and to talk with many of the staff who were directly involved with the early FBS has reminded many of us both of the energy and dedication required to carry out such a pioneering survey, and also of what can still be accomplished with modest-sized telescopes when they are equipped with modern wide-field detectors.

On a more personal note, I suspect that my introduction to astronomy in Armenia and to the work carried out at the BAO parallels that of many of the non-Armenian astronomers at this meeting. Although I was aware of Ambartsumian through his famous text – "Stellar Evolution and Astrophysics" (1947) – my first real connection with the Byurakan Astrophysical Observatory came both from reading the proceedings of IAU Symposium 29, which was held at the BAO in May, 1966, and from reading the Conference Proceedings of the 1968 Tucson meeting on AGN, where Ed Khachikian presented the now famous first lists of objects from the BAO survey for "Galaxies with Ultraviolet Continuum", which he termed "Markarian galaxies" (e.g. Khachikian 1968).

During the mid 1980's I spent considerable time studying the complete lists of Markarian galaxies, which were published in *Astrofizika* (25 papers) over a sixteen year span (Markarian 1967; ...; Markarian, Lipovetskii & Stepanian 1982). I did this because these objects were turning up in significant numbers as cross-IDs with luminous infrared galaxies (LIGs), then being compiled from the *Infrared Astronomical Satellite* (*IRAS*) all-sky survey. I quickly learned the benefit of having access to a wide-field, systematic, well-calibrated, extragalactic survey, where the majority of the objects had published redshifts. I also found it somewhat surprising that objects discovered in a survey for UV-excess galaxies would have significant overlap with luminous, far-infrared-selected galaxies.

2. Scientific Highlights

The oral presentations given on Days 1, 2 and 4 of this meeting (Day 3 was reserved for the Conference excursion) each covered a specific wavelength range(s) of AGN surveys, while Day 5 presented a more general discussion of AGN phenomena and ended with several talks on Future Projects relating to AGN surveys. I found it most useful to keep to the same chronology in this closing talk.

For each of the four full days of oral presentations I will first give a brief summary of the scientific highlights presented in the invited and contributed talks, followed by a bulleted list of the items that stood out to me as being the most interesting new results. I then list what I feel are some of the major outstanding questions raised by the presentations, followed by my suggested answers (when the weight of the data seemed to provide an answer).

2.1. Day 1 - Optical Surveys for AGN

Ed Khachikian opened the meeting with a brief history of Markarian's scientific career and reminded us of the importance of the First Byurakan Survey (FBS), begun in 1965. Using some of the original objective prism spectra, he then reminded us of the spectral classification schemes developed for describing the different spectral types discovered among these UV-excess galaxies. He also used specific examples to point out the variety of morphologies found in the central regions of these objects, including double nucleus galaxies, galaxies with radio jets, and "star-like" nuclei.

The "final results" (after 27 years of work !) from the Second Byurakan Survey (SBS) that was begun in 1974 in parallel with the FBS, were summarized by *Jivan Stepanian*. The SBS was designed to provide a large well-defined sample of AGNs and QSOs down to $B \sim 19.5$, which is $\sim 2-3\times$ fainter than the FBS. Follow-up spectroscopy of SBS galaxies using 4-6 m telescopes provided evidence for a continuity of properties among the various spectral types – from classical BLSy1s through NLSy1s, Sy2s, to LINERs and starbursts. Additional results from multiwavelength follow-up studies of SBS fields, including comparisons with *ROSAT, IRAS, NVSS,* and *FIRST* were also presented.

More modern objective prism surveys of AGN continue to succeed in discovering substantial numbers of new AGN. *Caryl Gronwell* summarized the results from the KPNO International Spectroscopic Survey (KISS), which is less biased against redder Seyfert 2 and LINER galaxies, and provides a well-defined sample of nearby ($z < 0.1$) AGN for multiwavelength studies. Results include the finding that Sy2s and LINERs are significantly redder in $(B - V)$ colors than Sy1s, plus the rather surprising finding that the majority of AGN in the KISS sample follow the same radio-IR correlation as found for starburst galaxies.

Louis Ho brought us up to date on the latest in AGN statistics from studies of complete samples of nearby galaxies using the Palomar 200-inch telescope, and reminded us of the importance of studying the faint end of the AGN luminosity function. AGNs are most common in early-type systems: $\sim$60% in E-Sbc galaxies compared to $\sim$15-20% in late types. At least 40% of all galaxies with $B_T < 12.5$ mag exhibit AGN-like spectra, i.e. Seyferts, LINERs, or transition objects (LINER/HII). LINERs make up nearly 2/3 of all AGN. They are further divided into two types depending of the detection of a weak broad-line component, with those of type-1 almost certainly being low-luminosity AGN while those of type-2 likely being a mixture of starburst and AGN.

Lutz Wisotzki summarized the highlights from the large objective prism Hamburg All-Sky Bright QSO surveys (the HQS and HES in the north and south, respectively). Over the past decade, the Hamburg surveys have greatly improved our knowledge of both the faint and bright end of the QSO luminosity function (LF). I found the comparisons with other surveys, showing general agreement in the surface density of QSOs, to be reassuring, including the 1.5× higher surface density compared to the Palomar-Green survey, which finally rules out previous claims of extreme incompleteness of the PG sample. In a second invited talk (also delivered by Wisotzki) the evolution of the QSO luminosity function was discussed using a merged sample of six optical QSO surveys. The statistics now appear sufficient to discriminate between pure-luminosity evolution (PLE) and pure-density evolution (PDE). At low redshift ($z < 1$), PLE seems clearly favored, while at high redshift ($z = 1 - 3$), some form of luminosity-dependent density evolution seems to be required.

Additional properties of optically selected QSOs were presented in several contributed talks. *Phil Outram* discussed QSO power spectra from the 2DF Survey and found that the shape and amplitude of the QSO power spectrum are similar to galaxies at the present day, and are a factor of ~ 10 lower than that measured for Abell clusters. *Ken Mitchell* reviewed the properties of the 19 QSOs in the US Bright Quasar Sample and discussed the importance of this sample for helping define the extreme bright end of the QSO LF. *Helmut Meusinger* and *Vicki Sarajedini* showed how optical variability and proper motion studies could be used to efficiently discover new, primarily lower luminosity AGN, while *Dario Trevese* presented an analysis of the variability of QSOs in the optical band.

Highlights from Day 1 included:

- The "classical Seyfert" fraction of optically selected galaxies is $\sim$10-15%.
- For optically selected galaxies the LINER fraction is $\sim$35% (i.e. $\gtrsim 2/3$ of all AGN). This result is largely independent of galaxy luminosity.
- The LF for optically selected AGN appears to be very well determined over the range $M_B = -8$ to -21 in the "local volume" and for $M_B = -14$ to -28 out to $z \sim 1.5$. The LF exhibits strong positive evolution over the range $z = 0-1$ [i.e. $\propto (1+z)^{3.3}$ assuming PLE], is nearly invariant over the range $z = 1-3$, and shows a single power-law decline over the range $z = 3-6$.

Major questions raised (plus consensus answers) included:

Q1.1 <u>Do all major optical AGN surveys give consistent LFs ?</u>

Yes, within the statistical uncertainties. Claims from a few years ago of large differences, e.g. between the HES and the PG surveys seem to have largely disappeared.

Q1.2 <u>Are the majority of LINERs classified as LLAGN true AGN ?</u>

Yes, given the results presented by Luis Ho.

Q1.3 <u>PLE vs. PDE vs. Z ?</u>

Although the majority of published papers discuss the QSO LF in terms of PLE at all redshift, the results presented by Wisotzki seem to favor PDE at $z \lesssim 1$. At higher redshift the statistics are insufficient to distinguish between PLE and PDE, plus the shortened timescales would seem to argue for a mixture of both (i.e. PLE + PDE).

2.2. Day 2 - Infrared/Submillimeter Surveys for AGN

AGN *discovered* at infrared and Submillimeter wavelengths (as opposed to the infrared/submillimeter properties of optically selected AGN), is a relatively new area of AGN research. Although *IRAS* provided the first large samples of infrared-selected AGN over a decade ago, this subfield has received several new "shots in the arm", most recently from the all-sky near-infrared survey 2MASS, and over the past half-decade from the small deep-fields and targeted observations made by both *The Infrared Space observatory (ISO)* and the Submillimeter Common User Bolometer Array (SCUBA) on the James Clerk Maxwell Telescope (JCMT).

It is probably fair to say that, in total, the results presented on Day 2 were the least familiar to the majority of those in the audience, and raised some of the longest discussions during the conference. Aside from the intrigue associated with the moniker "obscured AGN", these objects seem to be providing important new information related to the origin and evolution of the AGN population as a whole. My own personal hope is that these obscured AGN may provide a key link that will allow us to unify what has often been referred to as the "AGN zoo".

Sylvain Veilleux began the day by providing a refresher course on AGN emission line diagnostics (primarily optical/near-infrared). He reminded us that the field has evolved considerably from the early Seyfert 1/2 classification schemes (and several decimal classifications in-between !), which were based primarily on line-width, through the line-ratio classification schemes of Baldwin-Phillips-Terlevich, to the more recent line-ratio diagnostic diagrams of Veilleux-Osterbrock that attempt to decrease the effects of reddening by using line-pairs with small $\Delta\lambda$. New line ratios in the near-infrared have also been added, and the importance of high-ionization lines (e.g. CIV, FeVII, NeV, ...) was stressed. But problems still remain; aperture effects, shocks, metallicity, morphological bias were some of those mentioned, along with the most obvious - heavy obscuration.

Roc Cutri then provided dramatic new evidence from the Two Micron All Sky Survey (2MASS) showing the existence of a substantial new population of "red AGN". He pointed out that nearly all AGN discovered in optical/UV surveys have near-infrared colors, $(J - K) < 2$. In a follow-up redshift survey of 413 2MASS extragalactic sources with $(J - K) > 2$, **72%** were found to be either Seyfert 1 or 2 ! He stated that the new 2MASS results suggest that 50–80% of all AGNs have been missed in existing optical/UV surveys. The implication for *SIRTF* is that *SIRTF* should discover ~25,000 "red AGN" deg^{-2}. The 2MASS "red AGN" contain the most polarized objects (11/89 had $P > 3\%$), are systematically underluminous in X-rays (nearly 2 orders of magnitude less than the mean for PGQSOs), and have inferred line-of-sight gas columns $N(H)\sim 10^{22}$ cm^{-2}. "Red AGN" appear to have ~5 mag of extinction on top of what is present in the "average" PGQSO. It was also pointed out, perhaps not surprisingly given the above description, that $< 1\%$ of the 2MASS "red AGN" have been discovered in surveys at other wavelengths. And finally, a word of caution that the 2MASS "red AGN" search was not sensitive to heavily obscured objects, implying that the 2MASS "red AGN" may only represent a fraction of the true size of the obscured AGN "iceberg".

Zeljko Ivezić then gave a summary of the exciting new results that have emerged from the early analyses of data from the Sloan Digital Sky Survey (SDSS) by matching SDSS sources with sources from 2MASS and the Faint Images of the radio Sky at Twenty-cm (FIRST) survey. Preliminary results include the finding that the optical colors of radio-loud quasars (RLQs) are ~ 0.05mag redder than the optical colors of radio-quiet quasars (RQQs), and that the fraction of quasars *with stellar colors* missed by the SDSS spectroscopic survey is probably not larger than 10%. There was some debate concerning the ability of SDSS to both detect and identify the "red AGN" population seen by 2MASS, and this issue was not clearly resolved during the conference.

Ski Antonucci gave a somewhat provocative presentation focusing on what he termed "questionable interpretations" of AGN surveys. In particular, he focused on searches for polarized broad Hα lines in Sy 2s as a descriminant for various beaming models and models with circumnuclear tori. He pointed out (correctly in my opinion) that many previous statements claiming that Sy 1s and Sy 2s are intrinsically different, are flawed due to failure to select the samples by an isotropic property. Concerning the nature of the dominant energy source powering ultraluminous infrared galaxies (ULIGs), he noted the disparate results obtained from optical, near-infrared, mid-infrared and X-rays on the *same objects*, and pointed out that optical-infrared line emission is a diagnostic only to the depth penetrated by light of these wavelengths. In heavily obscured AGN this is likely often to be only the thin outer onion skin of a nuclear source which is buried behind column densities of order $\sim 10^{23} - 10^{24}\,\mathrm{cm}^{-2}$ (i.e. $A_V = 100 - 1000$). He also offered the opinion that "there is no accurate conversion from any particular starburst spectral feature to the bolometric luminosity of the associated starburst population".

Gene Smith brought us up to date on plans for the *SIRTF* Wide-area Infrared Extragalactic Legacy Survey (SWIRE) that will survey $\sim 67\,\mathrm{deg}^2$ in all 7 SIRTF imaging bands to much fainter flux levels than obtained by *IRAS* and *ISO*. SWIRE expects to detect $\sim$25,000 "classical AGN" and perhaps several times as many dust-enshrouded AGN, and will allow us to trace cosmic evolution in the AGN population(s) out to $z \sim 3$.

Hervé Aussel summarized the results of *ISO* deep mid-infrared surveys at 15μm. Using data primarily from the Hubble Deep Field (HDF), he claimed that by making reasonable assumptions for the infrared/submillimeter spectral energy distributions (SEDs), he could show that the faint 15μm sources could account for the bulk of the infrared/submillimeter cosmic infrared background (CIB), which actually peaks between 100μm and 250μm. He went on to further state that the contribution of "type-2 AGN" to the CIB is likely to be $< 15\%$.

Yoshi Taniguchi presented results from both deep mid-infrared and far-infrared surveys with *ISO*, and pointed out the relatively large number of far-infrared sources being discovered at intermediate redshifts ($z \sim 0.3 - 0.8$), indicating strong evolution in the far-infrared source population at these redshifts.

Martin Haas presented a new mid-infrared/submillimeter diagnostic – the PAH 7.7μm-line-to-850μm-continuum flux ratio – that could be used to test for high mid-infrared extinction. He then used this ratio to show evidence for a hidden QSO in the archetypal ULIG Arp 220.

Additional properties of infrared/submillimeter-selected QSOs were presented in several contributed papers. *Israel Matute* presented evidence for very strong evolution, $L_{15}(z) \propto (1+z)^3$, in the faint 15$\mu$m source population from analysis of an ISO deep survey in the ELAIS S1 region. *Henrique Schmitt* (in the spirit of Ski Antonucci's directive !) used a 60μm selection criteria to show evidence in support of the "Unified Model" in which Sy 2s contain a torus seen more edge-on than in Sy 1s. *Maria Marcha* presented preliminary results from the CLASS blazar survey of "Type 0" objects (weak or no emission lines) showing that, under the assumption that passive elliptical galaxies (PEGs) contain hidden BL Lac nuclei, that the flattening of the radio luminosity function (RLF) at low luminosities disappears. *Omar Almaini* then presented new evidence for a very strong cross-correlation signal between Chandra faint X-ray sources and SCUBA 8 mJy 850μm sources, suggesting that both populations may be tracing the same large-scale structure at high redshift.

Highlights from Day 2 included:

- The number of new, near-IR-selected "red AGN" (2MASS vs. SDSS)
- AGN Diagnostic Lines - *"Veilleux's Ten Commandments"*
- The true number of new MIR/FIR/Submm-selected AGN

The major questions raised (+ consensus answers) included:

Q2.1 <u>What is the new contribution from $(J - K)$-selected AGN ?</u>
The results from ground-based follow-up spectroscopy of 2MASS extragalactic objects with $(J - K) > 2$ seems to clearly indicate that the great majority of these objects are Seyferts (many of type 1). Given that these objects have a comparable space density to optically-selected QSOs, this would indeed seem to indicate that the QSO population may be at least a factor of 2 larger than previously estimated from optical surveys.

Q2.2 <u>What is the AGN fraction from MIR/FIR/Submm surveys ?</u>
Optical and near-IR spectroscopy of luminous infrared galaxies (i.e. $L_{ir} > 10^{11} L_\odot$) discovered in far-infrared/submillimeter surveys finds that only ~ 10-$15\,\%$ are Seyferts, consistent with optically selected luminous galaxies. Luminous galaxies discovered in the mid-infrared tend to have a higher percentage of Seyferts (~ 20–$25\,\%$); plus the most luminous galaxies (i.e. $L_{ir} > 10^{12.3} L_\odot$) at all infrared/submillimeter wavelengths tend to have the highest fraction of Seyferts (i.e. $\gtrsim 50\,\%$).

Q2.3 <u>What are infrared-selected LINERs (AGN vs *B) ?</u>
The evidence seems to favor a mixture of both *B and AGN powering these sources.

Q2.4 <u>Is there evidence for an IR/Optical AGN-fraction change vs. z ?</u>
None at present. Given the limited sensitivity of current FIR/Submm surveys it is not possible to separate trends with luminosity from trends with redshift.

2.3. Day 4 - X-Ray and Radio Surveys for AGN

Kazushi Iwasawa reviewed the X-Ray properties of "obscured AGN" using recent results from *ASCA, BeppoSax* and *Chandra*, and reminded us that studies of the X-ray background suggest that obscured AGN should significantly outnumber their unobscured counterparts. The surprising new findings include the fact that a significant number of X-Ray selected AGN have been discovered in objects where there is no optical signature of an AGN, or in the case of the most heavily obscured objects, no optical emission lines at all. The possibility that a significant number of "Compton thick" AGN (i.e. sources with obscuring columns, $N(H) > 10^{25}\,cm^{-2}$) may exist can not be ruled out from current data.

Andrea Comastri then discussed the AGN content of hard X-Ray surveys, focussing on sources discovered by *BeppoSax* and *XMM-Newton*. He stressed that their optical properties are quite varied, and that there is apparently no simple relation between optical and X-Ray absorption in individual objects.

Wolfgang Voges presented initial results from a comparison of the ROSAT All-Sky Survey (RASS) and SDSS concerning X-ray-variable AGN, where the goal is to obtain optical IDs for ∼20,000 X-ray sources in the RASS. Results from the first follow-up observations at Apache Point covering ∼200 deg^2 have detected more than 2000 new AGN of which many are also detected in X-rays.

Contributed talks by *Masayuki Akiyama* and *James Manners* presented results on ASCA and ROSAT sources, respectively. The ASCA results suggest a lack of absorbed luminous AGNs (i.e. a deficiency of type-2 QSOs). The ROSAT results suggest that QSOs are more variable at $z > 2$, suggesting that high-redshift QSOs may be accreting at a higher efficiency than local AGN.

Leonid Gurvits reviewed recent results from global and Space VLBI studies of quasars and other types of AGN, by showing the wealth of new data on the detailed sub-milliarcsecond radio structure in hundreds of sources. He presented convincing evidence that with large enough samples it may soon be possible to disentangle intrinsic evolutionary phenomena of parsec-scale radio structures, and thus to use VLBI radio data to possibly "age date" AGNs.

A contributed talk by *Carlos DeBreuck* presented preliminary results for a sample of 669 Ultra Steep Spectrum (USS) radio sources, a sample designed to find significant numbers of $z > 3$ radio galaxies. Initial ground-based follow-up spectroscopy finds a sample mean redshift of $z \sim 2.5$, suggesting that the USS selection technique indeed works. The sample also included the most distant radio galaxy known, at $z = 5.19$. *Pedro Augusto* then discussed a technique for selecting radio sources with the objective of studying the size and properties of the narrow-line emission region (NLR) in AGN, showing its promise for probing the evolution of the NLR in AGN as a function of redshift.

Highlights from day 4 included:

- The number of new heavily obscured X-ray selected AGN
- VLBI surveys and evolution of parsec-scale structures in the cores of AGN
- The efficiency of USS selection methods for selecting high-z RGs
- Evidence for BLAZAR unification (FRSQs vs. BL Lacs)

Major questions raised (+ consensus answers) included:

> **Q4.1** <u>The relative fraction of Compton-thick vs. Compton-thin AGN ?</u>
> Although surveys of optically selected Seyferts show evidence for unexpectedly large mean absorbing columns, i.e. $N(\mathrm{H}) \sim 10^{22-23}\,\mathrm{cm}^{-2}$, there is as yet no clear evidence to suggest a large population of Compton-thick objects, i.e. $N(\mathrm{H}) > 10^{25}\,\mathrm{cm}^{-2}$.
> **Q4.2** <u>Can radio VLBI help unify/trace the evolution of AGN ?</u>
> Perhaps, given large enough samples.
> **Q4.3** <u>Is it possible to select complete samples of high-z RGs and RLQs ?</u>
> USS selection techniques seem to provide such samples.

2.4. Day 5 - AGN Phenomena

John Hutchings reviewed current knowledge of the properties of the host galaxies of high-redshift QSOs, and pointed out the difficulties of PSF-removal in both space as well as ground-based AO imaging. Surveying statements in the literature, he gave several caveats to "host galaxy folklore", which included the oft repeated statements such as "all QSO hosts are ellipticals", "radio-loud QSOs have the most massive black holes", and "many hosts do not look irregular or peculiar (i.e. interactions are rare)". All such statements seem to be suspect.

Brigitte Rocca-Volmerange presented new results from 3D spectroscopy observations designed to study the ionisation processes in extended distant radio sources, with the intent of constraining the role of AGNs in the star formation history of the Universe.

Richard Green then summarized the demographics of supermassive black holes from studies of ellipticals with the *Hubble Space Telescope (HST)*. The "Magorrian relation", which shows that M_{BH} is strongly correlated with the velocity dispersion of the whole bulge, suggests that black hole formation may be an intrinsic aspect of bulge formation.

Wolfgang Duschl presented theoretical arguments to show that gas-rich galaxy-galaxy mergers can provide not only the "fuel" for quasar activity, but can also build a super-massive black hole on the timescale of the merger (i.e. in less than 10^9 yrs). The key theme in his scenario is that viscosity in self-gravitating accretion disks will naturally lead to a radial inflow of material to feed/build the central engine.

A contributed talk by *Igor Karachentsev* presented the results of an imaging study of a distance limited sample of ~ 300 galaxies within 7 Mpc designed to search for AGN in nearby dwarf galaxies. *Michael Hawkins* then presented results from a 23-year monitoring program of 610 confirmed QSOs, which favor an accretion disk model for low luminosity AGN, but which suggest that the variations of more luminous AGN may be dominated by microlensing.

Highlights from Day 5 included:

- Although QSO hosts appear to have a substantial bulge component, there is also evidence in many objects for a disk component as well as evidence for tidal features such as rings, loops and tails.

- The " Magorrian relation" strongly suggests that black hole growth and bulge growth are both linked through a common overall process (e.g. galaxy mergers)

- Theoretical evidence suggesting that self-gravitating gas disks provide a natural fuel source for feeding the growth of MBH

Day 5 raised a number of important questions that must eventually be answered if we are to finally understand the origin and evolution of AGN:

> **Q5.1** Do all QSOs reside in E-like hosts ?
> Published evidence is contradictory, but some QSOs clearly contain a disk component.
> **Q5.2** What is the difference in RLQ vs. RQQ hosts ?
> Current evidence is contradictory.
> **Q5.3** Is there strong evidence in support of coeval Bulge/MBH formation?
> Evidence at present is circumstantial.
> **Q5.4** Does the "luminous" QSO phase last $\sim$ few $\times 10^8$ years ?
> There is not enough evidence to date upon which to base accurate ages.

2.5. Future Projects

Joe Mazzarella presented the "new reality" that is now offered to the international astronomical community in the form of electronic services such as the National Extragalactic Database (NED). He showed how these powerful new services will, in the near future, likely evolve into a larger National Virtual Observatory (NVO).

Vahé Petrosian described new statistical methods that can be used for determining correlations among and distributions of physical parameters from multivariate data sets with general and arbitrary truncations and selection biases.

Serguei Dodonov proposed a photometric method for identifying stars, galaxies and QSOs in multi-band color surveys, using a library of color templates for comparison with observed objects.

Finally, *Areg Mickaelian* discussed plans to digitize the full FBS, which is the largest spectral survey in the Northern sky, containing low-dispersion spectra for $\sim$20,000,000 objects. Current plans are to make the digitized FBS available via the Internet and on CDs along with the appropriate interface for accessing the data. Digitization is clearly needed if the full discovery potential of the FBS is to be realised.

3. Closing Remarks

The history of the BAO is closely tied to the stature and foresight of its first Director, Victor Ambartsumian. Its importance as an Observatory is in large part the result of the great effort carried out by its staff in producing the systematic wide-scale surveys of UV-excess extragalactic objects and in carrying out the important follow-up spectroscopic studies of the "Markarian" galaxies. Its importance to world astronomy was helped early on by the international exposure provided by Khachikian and his collaborators who carried out spectroscopic follow-up observations of the UV-excess galaxies using large telescopes, thereby demonstrating the importance of AGN in powering many of these objects.

This meeting has come at a propitious time when the availability of new large all-sky surveys now provides a more comprehensive understanding of the AGN phenomenon and of the cosmic evolution of AGN. It also comes at a time when the BAO appears to be at an important crossroads. In the future the BAO will need to choose carefully which projects it will pursue with its current suite of telescopes, and how to best enter into collaborative projects with other observatories.

The BAO must also find the means to insure access to the large data sets that are now freely available electronically through data centers such as NED and SIMBAD, and eventually through the NVO. The cost of the required terabytes of disk space and GHz-workstations continues to decrease, and a way must be found to provide such computer power to the BAO. Perhaps more problematic will be attaining a high speed link to the outside world, but resources *must* be found so that the staff of the BAO are not isolated from the international astronomical community. Once key projects are defined, the commitment of BAO staff and telescope resources can once again make its mark on world astronomy.

Acknowledgments. On behalf of all the conference participants, the Scientific Organizing Committee extends its deep appreciation to our hosts at the Byurakan Astrophysical Observatory and to the members of the Local Organizing Committee who worked hard to ensure the success of this international colloquium. The author also gratefully acknowledges support from a Senior Award from the Alexander van Humboldt-Foundation and the hospitality of the Max-Planck-Institut fur extraterrestriche Physik. Additional financial support was provided by the Research Council of the University of Hawaii, and by NASA JPL contract 961566.

References

Ambartsumian, V. A. 1947, Stellar Evolution and Astrophysics (Yerevan: USSR)
Ambartsumian, V. A. 1968, in IAU Symp. 29, held 4-12 May, 1966, in Byurakan, Armenia, Non-stable Phenomena in Galaxies, (Yerevan: Academy of Sciences of Armenia), 11
IAU Symposium 121, held 3-7 June 1986, in Byurakan, Armenia, Observational Evidence of Activity in Galaxies, eds. E. Khachikian, K. Fricke & J. Melnick (Dordrecht: Reidel), 1987

IAU Symposium 194, held 17-21 Aug. 1998, in Yerevan, Armenia, Active Galactic Nuclei and Related Phenomena, eds. Y. Terzian, E. Khachikian & D. Weedman (San Francisco: ASP), 1999

Khachikian, E. Ye. 1968, AJ, 73, 891

Khachikian, E. Ye. & Weedman, D. W. 1971, S&T, 41, 217

Markarian, B. E. 1967, Afz, 3, 24 (Mrk 1-70)

Markarian, B. E., Lipovetskii, V. A. & Stepanian, J. A. 1982, Afz, 17, 321

Weedman, D. W. & Khachikian, E. Ye. 1968, Afz, 4, 587

Author Index

Afanasiev, V.L. 367
Akiyama, M. **245**
Almaini, O. 251
Amirkhanian, A.S. **87**
Andernach, H. 306
Antón, S. **289**
Antonelli, A. 235
Antonucci, R. **147**
Antonucci, R.R.J. 173
Appenzeller, I. 259
Artyukh, V.S. 291
Asatrian, N.S. **357, 359**
Augusto, P. **281**
Aussel, H. **179**

Balayan, S.K. 97, 217, 220
Baldi, A. 235
Becker, R.H. 137
Blanton, M. 137
Bochkarev, N.G. 369
Bogdantsov, A.V. 297, 299
Boschetti, C.S. **89**
Bower, G. 335
Browne, I. 289
Brunzendorf, J. 69, 99, 215
Brusa, M. 235

Caccianiga, A. 189, 257
Carilli, C. 275
Carrasco, L. 293
Chavushyan, V.H. 23, 259, **293**, 308
Chen, Y.J. 304
Chernenkov, V.N. 312
Ciliegi, P. 235
Ciroi, S. 89
Clarke, C.J. 173
Comastri, A. **235**
Cowie, L.L. 213
Cutri, R.M. **127**

Della Ceca, R. 257
Denissyuk, E.K. **361**
De Breuck, C. **275**
Dodonov, S.N. 367
Dressler, A. 335
Duschl, W.J. **343**

Edge, A.C. 281
Egikian, A.G. 87
Erastova, L.K. **92**

Faber, S. 335
Fan, X. 137
Filippenko, A. 335
Finlator, K. 137
Fiore, F. 235
Francis, P.J. 127
Funes, J. 89

Gaisina, V.N. 361
Gebhardt, K. 335
Gioia, I.M. 257
Giommi, P. 235
Gonzalez-Serrano, J.I. 281
Green, R.F. **335**
Grillmair, C. 335
Gronwall, C. **43**
Gruppioni, C. 167
Gunn, J.E. 137
Gurvits, L.I. **265**
Gyulzadian, M.V. **94**

Haas, M. **205**
Hagen-Thorn, A.V. 363
Hagen-Thorn, V.A. **363**
Hakopian, S.A. **97**, 217, 220
Hall, P. 137
Hawkins, M.R.S. **351**
Ho, L.C. **13**, 335
Hovhannisyan, M.A. 291
Hutchings, J.B. **317**, 335

Irwin, M. 69
Iskudarian, S.G. **365**
Ivezić, Ž **137**
Iwasawa, K. **225**

Jamrozy, M. **295**
Jiang, D.R. 304
Jorstad, S.G. 363
Joseph, C. 335
Joseph, R.D. 213
Ju'arez, Y. 259

Kaiser, M.E. 335

Kakazu, Y. 213
Karachentsev, I.D. **325**
Karachentseva, V.E. 325
Kawara, K. 213
Khachikian, E.Ye. **3**, 357, 359
Khrapov, S.S. 367
Kim, R.S.J. 137
Kinney, A.L. 173
Knapp, G.R. 137
Kopylov, A.I. 310, 312
Kormendy, J. 335
Kovalev, Y.Y. 297, **299**
Kovalev, Yu.A.. **297**, 299
Krautter, J. 259

Lari, C. 167
Larionov, V.M. 363
Lauer, T. 335
Lawrence, A. 251
La Franca, F. 167, 235
Loveday, J. 137
Lupton, R.H. 137

Mújica, R. **259**, 293, 308
Maccacaro, T. 257
Machalski, J. 295
Magorriam, J. 335
Mahtesyan, A.P. 291
Maiolino, R. 235
Manners, J. **251**
Marchã, M.J.M. **189**, 257
Matsuhara, H. 213
Matsumoto, T. 213
Matt, G. 235
Matute, I. **167**
Mazzarella, J.M. **379**
Menou, K. 137
Meusinger, H. **69**, **99**, **215**
Mickaelian, A.M. **101, 217, 220, 399**
Mignoli, M. 235
Mitchell, K.J. **53**
Moiseev, A.V. **367**
Molendi, S. 235
Movsesyan, V.H. 291
Murayama, T. 213
Mustsevoi, V.V. 367

Narayanan, V. 137
Nelson, B.O. 127

Nelson, C. 335
Nizhelsky, N.A. 299
Nizhelsky, N.A.. 297
Notni, P. 357, 359
Novikova, O.A. **369**

Ohanian, G.A. **301**
Ohta, K. 245
Okuda, H. 213
Omizzolo, A. 89

Parijskij, Yu.N. 310, 312
Pedlar, A. 289
Perez-Fournon, I. 281
Perola, G.C. 235
Petrosian, V. **389**
Pinkney, J. 335
Pozzi, F. 167
Pringle, J.E. 173

Röttgering, H. 275
Rafanelli, P. 89
Reimers, D. **33**
Richards, G.R. 137
Richstone, D. 335
Richter, G.M. 89
Rifatto, A. 89
Rockosi, C.M. 137

Sadoyan, A.A. **371**
Salzer, J.J. 43
Sanders, D.B. 213, **411**
Sarajedini, V.L. 43, **75**
Sato, Y. 213
Schlegel, D. 137
Schmitt, H.R. **173**
Schneider, D.P. 137
Scholz, R.-D. 69
Serrano, A. 259
Severgnini, P. 235
Shen, Z.-Q. **304**
Siopis, C. 335
Smith, H.E. **157**
Smith, P.S. 127
Soboleva, N.S. 310, 312
Sofue, Y. 213
Stecklum, B. 215
Stepanian, J.A. **23**, 293
Stoll, D. 87
Strateva, I. 137
Strauss, M.A. 137

Strittmatter, P.A. 343

Taniguchi, Y. **195**, 213
Temirova, A.V. 310, 312
Temnov, G.O. 363
Thean, A. 289
Tiersch, H. 87
Torosyan, O.Kh. **103, 106**
Trèvese, D. **81**
Tremaine, S. 335
Trushkin, S.A. 308

Ueda, Y. 245
Ulvestad, J.S. 173
Usher, P.D. 53

Vagnetti, F. 81
Valdés, J.R. 293, 308
Valiullin, R.R. 361, **375**
Vanden Berk, D. 137
van Breugel, W. 275
Veilleux, S. **111**, 213
Verkhodanov, O.V. 293, **306, 308, 310, 312**
Verkhodanova, N.V. 306, 312
Vignali, C. 235
Voges, W. 137

Wakamatsu, K. 213
Wan, T.-S. 304
Weistrop, D. 335
Wisotzki, L. 33, **59**
Wolter, A. 257
Woodgate, B. 335

Xue, S. **261**

Yanny, B. 137
Yun, M.S. 213

Zhang, H. 261
Zhekanis, G.V. 299
Zhelenkova, O.P. 310, 312
Zhou, X. 261
Zickgraf, F.J. 259

A LIST OF THE VOLUMES

Published
by

THE ASTRONOMICAL SOCIETY OF THE PACIFIC
(ASP)

An international, nonprofit, scientific and educational organization founded in 1889

All book orders or inquiries concerning

THE ASTRONOMICAL SOCIETY OF THE PACIFIC CONFERENCE SERIES (ASP - CS)

and

INTERNATIONAL ASTRONOMICAL UNION VOLUMES (IAU)

should be directed to the:

The Astronomical Society of the Pacific Conference Series
390 Ashton Avenue
San Francisco   CA   94112-1722   USA

Phone:	800-335-2624	(Within USA)
Phone:	415-337-2126	
Fax:	415-337-5205	

E-mail: service@astrosociety.org
Web Site: http://www.astrosociety.org

Complete lists of proceedings of past IAU Meetings are maintained at the
IAU Web site at the URL: http://www.iau.org/publicat.html

Volumes 32 - 189 in the IAU Symposia Series may be ordered from:

Kluwer Academic Publishers
P. O. Box 117
NL 3300 AA Dordrecht
The Netherlands

Kluwer@wKap.com

ASP CONFERENCE SERIES VOLUMES
Published by the Astronomical Society of the Pacific

PUBLISHED: 1988 (* asterisk means OUT OF STOCK)

Vol. CS-1   PROGRESS AND OPPORTUNITIES IN SOUTHERN HEMISPHERE OPTICAL ASTRONOMY: CTIO 25TH Anniversary Symposium
eds. V. M. Blanco and M. M. Phillips
ISBN 0-937707-18-X

Vol. CS-2   PROCEEDINGS OF A WORKSHOP ON OPTICAL SURVEYS FOR QUASARS
eds. Patrick S. Osmer, Alain C. Porter, Richard F. Green, and Craig B. Foltz
ISBN 0-937707-19-8

Vol. CS-3   FIBER OPTICS IN ASTRONOMY
ed. Samuel C. Barden
ISBN 0-937707-20-1

Vol. CS-4   THE EXTRAGALACTIC DISTANCE SCALE:
Proceedings of the ASP 100th Anniversary Symposium
eds. Sidney van den Bergh and Christopher J. Pritchet
ISBN 0-937707-21-X

Vol. CS-5   THE MINNESOTA LECTURES ON CLUSTERS OF GALAXIES AND LARGE-SCALE STRUCTURE
ed. John M. Dickey
ISBN 0-937707-22-8

PUBLISHED: 1989

Vol. CS-6 *   SYNTHESIS IMAGING IN RADIO ASTRONOMY: A Collection of Lectures from the Third NRAO Synthesis Imaging Summer School
eds. Richard A. Perley, Frederic R. Schwab, and Alan H. Bridle
ISBN 0-937707-23-6

PUBLISHED: 1990

Vol. CS-7   PROPERTIES OF HOT LUMINOUS STARS: Boulder-Munich Workshop
ed. Catharine D. Garmany
ISBN 0-937707-24-4

Vol. CS-8 *   CCDs IN ASTRONOMY
ed. George H. Jacoby
ISBN 0-937707-25-2

Vol. CS-9   COOL STARS, STELLAR SYSTEMS, AND THE SUN: Sixth Cambridge Workshop
ed. George Wallerstein
ISBN 0-937707-27-9

Vol. CS-10 *   EVOLUTION OF THE UNIVERSE OF GALAXIES:
Edwin Hubble Centennial Symposium
ed. Richard G. Kron
ISBN 0-937707-28-7

Vol. CS-11   CONFRONTATION BETWEEN STELLAR PULSATION AND EVOLUTION
eds. Carla Cacciari and Gisella Clementini
ISBN 0-937707-30-9

Vol. CS-12   THE EVOLUTION OF THE INTERSTELLAR MEDIUM
ed. Leo Blitz
ISBN 0-937707-31-7

PUBLISHED: 1991

Vol. CS-13   THE FORMATION AND EVOLUTION OF STAR CLUSTERS
ed. Kenneth Janes
ISBN 0-937707-32-5

ASP CONFERENCE SERIES VOLUMES
Published by the Astronomical Society of the Pacific

PUBLISHED: 1991 (* asterisk means OUT OF STOCK)

Vol. CS-14   ASTROPHYSICS WITH INFRARED ARRAYS
ed. Richard Elston
ISBN 0-937707-33-3

Vol. CS-15   LARGE-SCALE STRUCTURES AND PECULIAR MOTIONS IN THE UNIVERSE
eds. David W. Latham and L. A. Nicolaci da Costa
ISBN 0-937707-34-1

Vol. CS-16   Proceedings of the 3rd Haystack Observatory Conference on ATOMS, IONS, AND MOLECULES: NEW RESULTS IN SPECTRAL LINE ASTROPHYSICS
eds. Aubrey D. Haschick and Paul T. P. Ho
ISBN 0-937707-35-X

Vol. CS-17   LIGHT POLLUTION, RADIO INTERFERENCE, AND SPACE DEBRIS
ed. David L. Crawford
ISBN 0-937707-36-8

Vol. CS-18   THE INTERPRETATION OF MODERN SYNTHESIS OBSERVATIONS OF SPIRAL GALAXIES
eds. Nebojsa Duric and Patrick C. Crane
ISBN 0-937707-37-6

Vol. CS-19   RADIO INTERFEROMETRY: THEORY, TECHNIQUES, AND APPLICATIONS, IAU Colloquium 131
eds. T. J. Cornwell and R. A. Perley
ISBN 0-937707-38-4

Vol. CS-20   FRONTIERS OF STELLAR EVOLUTION:
50th Anniversary McDonald Observatory (1939-1989)
ed. David L. Lambert
ISBN 0-937707-39-2

Vol. CS-21   THE SPACE DISTRIBUTION OF QUASARS
ed. David Crampton
ISBN 0-937707-40-6

PUBLISHED: 1992

Vol. CS-22   NONISOTROPIC AND VARIABLE OUTFLOWS FROM STARS
eds. Laurent Drissen, Claus Leitherer, and Antonella Nota
ISBN 0-937707-41-4

Vol CS-23 *   ASTRONOMICAL CCD OBSERVING AND REDUCTION TECHNIQUES
ed. Steve B. Howell
ISBN 0-937707-42-4

Vol. CS-24   COSMOLOGY AND LARGE-SCALE STRUCTURE IN THE UNIVERSE
ed. Reinaldo R. de Carvalho
ISBN 0-937707-43-0

Vol. CS-25   ASTRONOMICAL DATA ANALYSIS, SOFTWARE AND SYSTEMS I - (ADASS I)
eds. Diana M. Worrall, Chris Biemesderfer, and Jeannette Barnes
ISBN 0-937707-44-9

Vol. CS-26   COOL STARS, STELLAR SYSTEMS, AND THE SUN:
Seventh Cambridge Workshop
eds. Mark S. Giampapa and Jay A. Bookbinder
ISBN 0-937707-45-7

Vol. CS-27   THE SOLAR CYCLE: Proceedings of the
National Solar Observatory/Sacramento Peak 12th Summer Workshop
ed. Karen L. Harvey
ISBN 0-937707-46-5

ASP CONFERENCE SERIES VOLUMES
Published by the Astronomical Society of the Pacific

PUBLISHED: 1992 (asterisk means OUT OF STOCK)

Vol. CS-28   AUTOMATED TELESCOPES FOR PHOTOMETRY AND IMAGING
eds. Saul J. Adelman, Robert J. Dukes, Jr., and Carol J. Adelman
ISBN 0-937707-47-3

Vol. CS-29   Viña del Mar Workshop on CATACLYSMIC VARIABLE STARS
ed. Nikolaus Vogt
ISBN 0-937707-48-1

Vol. CS-30   VARIABLE STARS AND GALAXIES
ed. Brian Warner
ISBN 0-937707-49-X

Vol. CS-31   RELATIONSHIPS BETWEEN ACTIVE GALACTIC NUCLEI
AND STARBURST GALAXIES
ed. Alexei V. Filippenko
ISBN 0-937707-50-3

Vol. CS-32   COMPLEMENTARY APPROACHES TO DOUBLE
AND MULTIPLE STAR RESEARCH, IAU Colloquium 135
eds. Harold A. McAlister and William I. Hartkopf
ISBN 0-937707-51-1

Vol. CS-33   RESEARCH AMATEUR ASTRONOMY
ed. Stephen J. Edberg
ISBN 0-937707-52-X

Vol. CS-34   ROBOTIC TELESCOPES IN THE 1990's
ed. Alexei V. Filippenko
ISBN 0-937707-53-8

PUBLISHED: 1993

Vol. CS-35 *   MASSIVE STARS: THEIR LIVES IN THE INTERSTELLAR MEDIUM
eds. Joseph P. Cassinelli and Edward B. Churchwell
ISBN 0-937707-54-6

Vol. CS-36   PLANETS AROUND PULSARS
ed. J. A. Phillips, S. E. Thorsett, and S. R. Kulkarni
ISBN 0-937707-55-4

Vol. CS-37   FIBER OPTICS IN ASTRONOMY II
ed. Peter M. Gray
ISBN 0-937707-56-2

Vol. CS-38   NEW FRONTIERS IN BINARY STAR RESEARCH: Pacific Rim Colloquium
eds. K. C. Leung and I.-S. Nha
ISBN 0-937707-57-0

Vol. CS-39   THE MINNESOTA LECTURES ON THE STRUCTURE
AND DYNAMICS OF THE MILKY WAY
ed. Roberta M. Humphreys
ISBN 0-937707-58-9

Vol. CS-40   INSIDE THE STARS, IAU Colloquium 137
eds. Werner W. Weiss and Annie Baglin
ISBN 0-937707-59-7

Vol. CS-41   ASTRONOMICAL INFRARED SPECTROSCOPY:
FUTURE OBSERVATIONAL DIRECTIONS
ed. Sun Kwok
ISBN 0-937707-60-0

ASP CONFERENCE SERIES VOLUMES
Published by the Astronomical Society of the Pacific

PUBLISHED: 1993 (* asterisk means OUT OF STOCK)

Vol. CS-42	GONG 1992: SEISMIC INVESTIGATION OF THE SUN AND STARS ed. Timothy M. Brown ISBN 0-937707-61-9
Vol. CS-43	SKY SURVEYS: PROTOSTARS TO PROTOGALAXIES ed. B. T. Soifer ISBN 0-937707-62-7
Vol. CS-44	PECULIAR VERSUS NORMAL PHENOMENA IN A-TYPE AND RELATED STARS, IAU Colloquium 138 eds. M. M. Dworetsky, F. Castelli, and R. Faraggiana ISBN 0-937707-63-5
Vol. CS-45	LUMINOUS HIGH-LATITUDE STARS ed. Dimitar D. Sasselov ISBN 0-937707-64-3
Vol. CS-46	THE MAGNETIC AND VELOCITY FIELDS OF SOLAR ACTIVE REGIONS, IAU Colloquium 141 eds. Harold Zirin, Guoxiang Ai, and Haimin Wang ISBN 0-937707-65-1
Vol. CS-47	THIRD DECENNIAL US-USSR CONFERENCE ON SETI -- Santa Cruz, California, USA ed. G. Seth Shostak ISBN 0-937707-66-X
Vol. CS-48	THE GLOBULAR CLUSTER-GALAXY CONNECTION eds. Graeme H. Smith and Jean P. Brodie ISBN 0-937707-67-8
Vol. CS-49	GALAXY EVOLUTION: THE MILKY WAY PERSPECTIVE ed. Steven R. Majewski ISBN 0-937707-68-6
Vol. CS-50	STRUCTURE AND DYNAMICS OF GLOBULAR CLUSTERS eds. S. G. Djorgovski and G. Meylan ISBN 0-937707-69-4
Vol. CS-51	OBSERVATIONAL COSMOLOGY eds. Guido Chincarini, Angela Iovino, Tommaso Maccacaro, and Dario Maccagni ISBN 0-937707-70-8
Vol. CS-52	ASTRONOMICAL DATA ANALYSIS SOFTWARE AND SYSTEMS II - (ADASS II) eds. R. J. Hanisch, R. J. V. Brissenden, and Jeannette Barnes ISBN 0-937707-71-6
Vol. CS-53	BLUE STRAGGLERS ed. Rex A. Saffer ISBN 0-937707-72-4

PUBLISHED: 1994

Vol. CS-54	THE FIRST STROMLO SYMPOSIUM: THE PHYSICS OF ACTIVE GALAXIES eds. Geoffrey V. Bicknell, Michael A. Dopita, and Peter J. Quinn ISBN 0-937707-73-2
Vol. CS-55	OPTICAL ASTRONOMY FROM THE EARTH AND MOON eds. Diane M. Pyper and Ronald J. Angione ISBN 0-937707-74-0
Vol. CS-56 *	INTERACTING BINARY STARS ed. Allen W. Shafter ISBN 0-937707-75-9

ASP CONFERENCE SERIES VOLUMES
Published by the Astronomical Society of the Pacific

PUBLISHED: 1994 (* asterisk means OUT OF STOCK)

Vol. CS-57　　STELLAR AND CIRCUMSTELLAR ASTROPHYSICS
eds. George Wallerstein and Alberto Noriega-Crespo
ISBN 0-937707-76-7

Vol. CS-58 *　　THE FIRST SYMPOSIUM ON THE INFRARED CIRRUS
AND DIFFUSE INTERSTELLAR CLOUDS
eds. Roc M. Cutri and William B. Latter
ISBN 0-937707-77-5

Vol. CS-59　　ASTRONOMY WITH MILLIMETER AND SUBMILLIMETER WAVE
INTERFEROMETRY,
IAU Colloquium 140
eds. M. Ishiguro and Wm. J. Welch
ISBN 0-937707-78-3

Vol. CS-60　　THE MK PROCESS AT 50 YEARS: A POWERFUL TOOL FOR ASTROPHYSICAL
INSIGHT, A Workshop of the Vatican Observatory --Tucson, Arizona, USA
eds. C. J. Corbally, R. O. Gray, and R. F. Garrison
ISBN 0-937707-79-1

Vol. CS-61　　ASTRONOMICAL DATA ANALYSIS SOFTWARE AND SYSTEMS III - (ADASS III)
eds. Dennis R. Crabtree, R. J. Hanisch, and Jeannette Barnes
ISBN 0-937707-80-5

Vol. CS-62　　THE NATURE AND EVOLUTIONARY STATUS OF HERBIG Ae/Be STARS
eds. Pik Sin Thé, Mario R. Pérez, and Ed P. J. van den Heuvel
ISBN 0-9837707-81-3

Vol. CS-63　　SEVENTY-FIVE YEARS OF HIRAYAMA ASTEROID FAMILIES:
THE ROLE OF COLLISIONS IN THE SOLAR SYSTEM HISTORY
eds. Yoshihide Kozai, Richard P. Binzel, and Tomohiro Hirayama
ISBN 0-937707-82-1

Vol. CS-64 *　　COOL STARS, STELLAR SYSTEMS, AND THE SUN:
Eighth Cambridge Workshop
ed. Jean-Pierre Caillault
ISBN 0-937707-83-X

Vol. CS-65 *　　CLOUDS, CORES, AND LOW MASS STARS:
The Fourth Haystack Observatory Conference
eds. Dan P. Clemens and Richard Barvainis
ISBN 0-937707-84-8

Vol. CS-66 *　　PHYSICS OF THE GASEOUS AND STELLAR DISKS OF THE GALAXY
ed. Ivan R. King
ISBN 0-937707-85-6

Vol. CS-67　　UNVEILING LARGE-SCALE STRUCTURES BEHIND THE MILKY WAY
eds. C. Balkowski and R. C. Kraan-Korteweg
ISBN 0-937707-86-4

Vol. CS-68 *　　SOLAR ACTIVE REGION EVOLUTION:
COMPARING MODELS WITH OBSERVATIONS
eds. K. S. Balasubramaniam and George W. Simon
ISBN 0-937707-87-2

Vol. CS-69　　REVERBERATION MAPPING OF THE BROAD-LINE REGION
IN ACTIVE GALACTIC NUCLEI
eds. P. M. Gondhalekar, K. Horne, and B. M. Peterson
ISBN 0-937707-88-0

Vol. CS-70 *　　GROUPS OF GALAXIES
eds. Otto-G. Richter and Kirk Borne
ISBN 0-937707-89-9

ASP CONFERENCE SERIES VOLUMES
Published by the Astronomical Society of the Pacific

PUBLISHED: 1995 (* asterisk means OUT OF STOCK)

Vol. CS-71	TRIDIMENSIONAL OPTICAL SPECTROSCOPIC METHODS IN ASTROPHYSICS, IAU Colloquium 149 eds. Georges Comte and Michel Marcelin ISBN 0-937707-90-2
Vol. CS-72	MILLISECOND PULSARS: A DECADE OF SURPRISE eds. A. S Fruchter, M. Tavani, and D. C. Backer ISBN 0-937707-91-0
Vol. CS-73	AIRBORNE ASTRONOMY SYMPOSIUM ON THE GALACTIC ECOSYSTEM: FROM GAS TO STARS TO DUST eds. Michael R. Haas, Jacqueline A. Davidson, and Edwin F. Erickson ISBN 0-937707-92-9
Vol. CS-74	PROGRESS IN THE SEARCH FOR EXTRATERRESTRIAL LIFE: 1993 Bioastronomy Symposium ed. G. Seth Shostak ISBN 0-937707-93-7
Vol. CS-75	MULTI-FEED SYSTEMS FOR RADIO TELESCOPES eds. Darrel T. Emerson and John M. Payne ISBN 0-937707-94-5
Vol. CS-76	GONG '94: HELIO- AND ASTERO-SEISMOLOGY FROM THE EARTH AND SPACE eds. Roger K. Ulrich, Edward J. Rhodes, Jr., and Werner Däppen ISBN 0-937707-95-3
Vol. CS-77	ASTRONOMICAL DATA ANALYSIS SOFTWARE AND SYSTEMS IV - (ADASS IV) eds. R. A. Shaw, H. E. Payne, and J. J. E. Hayes ISBN 0-937707-96-1
Vol. CS-78	ASTROPHYSICAL APPLICATIONS OF POWERFUL NEW DATABASES: Joint Discussion No. 16 of the 22nd General Assembly of the IAU eds. S. J. Adelman and W. L. Wiese ISBN 0-937707-97-X
Vol. CS-79 *	ROBOTIC TELESCOPES: CURRENT CAPABILITIES, PRESENT DEVELOPMENTS, AND FUTURE PROSPECTS FOR AUTOMATED ASTRONOMY eds. Gregory W. Henry and Joel A. Eaton ISBN 0-937707-98-8
Vol. CS-80 *	THE PHYSICS OF THE INTERSTELLAR MEDIUM AND INTERGALACTIC MEDIUM eds. A. Ferrara, C. F. McKee, C. Heiles, and P. R. Shapiro ISBN 0-937707-99-6
Vol. CS-81	LABORATORY AND ASTRONOMICAL HIGH RESOLUTION SPECTRA eds. A. J. Sauval, R. Blomme, and N. Grevesse ISBN 1-886733-01-5
Vol. CS-82 *	VERY LONG BASELINE INTERFEROMETRY AND THE VLBA eds. J. A. Zensus, P. J. Diamond, and P. J. Napier ISBN 1-886733-02-3
Vol. CS-83 *	ASTROPHYSICAL APPLICATIONS OF STELLAR PULSATION, IAU Colloquium 155 eds. R. S. Stobie and P. A. Whitelock ISBN 1-886733-03-1
ATLAS	INFRARED ATLAS OF THE ARCTURUS SPECTRUM, 0.9 - 5.3 µm eds. Kenneth Hinkle, Lloyd Wallace, and William Livingston ISBN: 1-886733-04-X

ASP CONFERENCE SERIES VOLUMES
Published by the Astronomical Society of the Pacific

PUBLISHED: 1995 (* asterisk means OUT OF STOCK)

Vol. CS-84   THE FUTURE UTILIZATION OF SCHMIDT TELESCOPES, IAU Colloquium 148
eds. Jessica Chapman, Russell Cannon, Sandra Harrison, and Bambang Hidayat
ISBN 1-886733-05-8

Vol. CS-85 * CAPE WORKSHOP ON MAGNETIC CATACLYSMIC VARIABLES
eds. D. A. H. Buckley and B. Warner
ISBN 1-886733-06-6

Vol. CS-86   FRESH VIEWS OF ELLIPTICAL GALAXIES
eds. Alberto Buzzoni, Alvio Renzini, and Alfonso Serrano
ISBN 1-886733-07-4

PUBLISHED: 1996

Vol. CS-87   NEW OBSERVING MODES FOR THE NEXT CENTURY
eds. Todd Boroson, John Davies, and Ian Robson
ISBN 1-886733-08-2

Vol. CS-88 * CLUSTERS, LENSING, AND THE FUTURE OF THE UNIVERSE
eds. Virginia Trimble and Andreas Reisenegger
ISBN 1-886733-09-0

Vol. CS-89   ASTRONOMY EDUCATION: CURRENT DEVELOPMENTS,
FUTURE COORDINATION
ed. John R. Percy
ISBN 1-886733-10-4

Vol. CS-90   THE ORIGINS, EVOLUTION, AND DESTINIES OF BINARY STARS
IN CLUSTERS
eds. E. F. Milone and J. -C. Mermilliod
ISBN 1-886733-11-2

Vol. CS-91   BARRED GALAXIES, IAU Colloquium 157
eds. R. Buta, D. A. Crocker, and B. G. Elmegreen
ISBN 1-886733-12-0

Vol. CS-92 * FORMATION OF THE GALACTIC HALO .... INSIDE AND OUT
eds. Heather L. Morrison and Ata Sarajedini
ISBN 1-886733-13-9

Vol. CS-93   RADIO EMISSION FROM THE STARS AND THE SUN
eds. A. R. Taylor and J. M. Paredes
ISBN 1-886733-14-7

Vol. CS-94   MAPPING, MEASURING, AND MODELING THE UNIVERSE
eds. Peter Coles, Vicent J. Martinez, and Maria-Jesus Pons-Borderia
ISBN 1-886733-15-5

Vol. CS-95   SOLAR DRIVERS OF INTERPLANETARY AND TERRESTRIAL DISTURBANCES:
Proceedings of 16th International Workshop National Solar
Observatory/Sacramento Peak
eds. K. S. Balasubramaniam, Stephen L. Keil, and Raymond N. Smartt
ISBN 1-886733-16-3

Vol. CS-96   HYDROGEN-DEFICIENT STARS
eds. C. S. Jeffery and U. Heber
ISBN 1-886733-17-1

Vol. CS-97   POLARIMETRY OF THE INTERSTELLAR MEDIUM
eds. W. G. Roberge and D. C. B. Whittet
ISBN 1-886733-18-X

ASP CONFERENCE SERIES VOLUMES
Published by the Astronomical Society of the Pacific

PUBLISHED: 1996 (* asterisk means OUT OF STOCK)

Vol. CS-98   FROM STARS TO GALAXIES: THE IMPACT OF STELLAR PHYSICS ON GALAXY EVOLUTION
eds. Claus Leitherer, Uta Fritze-von Alvensleben, and John Huchra
ISBN 1-886733-19-8

Vol. CS-99   COSMIC ABUNDANCES:
Proceedings of the 6th Annual October Astrophysics Conference
eds. Stephen S. Holt and George Sonneborn
ISBN 1-886733-20-1

Vol. CS-100   ENERGY TRANSPORT IN RADIO GALAXIES AND QUASARS
eds. P. E. Hardee, A. H. Bridle, and J. A. Zensus
ISBN 1-886733-21-X

Vol. CS-101   ASTRONOMICAL DATA ANALYSIS SOFTWARE AND SYSTEMS V – (ADASS V)
eds. George H. Jacoby and Jeannette Barnes
ISBN 1080-7926

Vol. CS-102   THE GALACTIC CENTER, 4th ESO/CTIO Workshop
ed. Roland Gredel
ISBN 1-886733-22-8

Vol. CS-103   THE PHYSICS OF LINERS IN VIEW OF RECENT OBSERVATIONS
eds. M. Eracleous, A. Koratkar, C. Leitherer, and L. Ho
ISBN 1-886733-23-6

Vol. CS-104   PHYSICS, CHEMISTRY, AND DYNAMICS OF INTERPLANETARY DUST,
IAU Colloquium 150
eds. Bo Å. S. Gustafson and Martha S. Hanner
ISBN 1-886733-24-4

Vol. CS-105   PULSARS: PROBLEMS AND PROGRESS, IAU Colloquium 160
ed. S. Johnston, M. A. Walker, and M. Bailes
ISBN 1-886733-25-2

Vol. CS-106   THE MINNESOTA LECTURES ON EXTRAGALACTIC NEUTRAL HYDROGEN
ed. Evan D. Skillman
ISBN 1-886733-26-0

Vol. CS-107   COMPLETING THE INVENTORY OF THE SOLAR SYSTEM:
A Symposium held in conjunction with the 106th Annual Meeting of the ASP
eds. Terrence W. Rettig and Joseph M. Hahn
ISBN 1-886733-27-9

Vol. CS-108   M.A.S.S. -- MODEL ATMOSPHERES AND SPECTRUM SYNTHESIS:
5th Vienna - Workshop
eds. Saul J. Adelman, Friedrich Kupka, and Werner W. Weiss
ISBN 1-886733-28-7

Vol. CS-109   COOL STARS, STELLAR SYSTEMS, AND THE SUN: Ninth Cambridge Workshop
eds. Roberto Pallavicini and Andrea K. Dupree
ISBN 1-886733-29-5

Vol. CS-110   BLAZAR CONTINUUM VARIABILITY
eds. H. R. Miller, J. R. Webb, and J. C. Noble
ISBN 1-886733-30-9

Vol. CS-111   MAGNETIC RECONNECTION IN THE SOLAR ATMOSPHERE:
Proceedings of a Yohkoh Conference
eds. R. D. Bentley and J. T. Mariska
ISBN 1-886733-31-7

ASP CONFERENCE SERIES VOLUMES
Published by the Astronomical Society of the Pacific

PUBLISHED: 1996 (* asterisk means OUT OF STOCK)

Vol. CS-112   THE HISTORY OF THE MILKY WAY AND ITS SATELLITE SYSTEM
eds. Andreas Burkert, Dieter H. Hartmann, and Steven R. Majewski
ISBN 1-886733-32-5

PUBLISHED: 1997

Vol. CS-113   EMISSION LINES IN ACTIVE GALAXIES: NEW METHODS AND TECHNIQUES, IAU Colloquium 159
eds. B. M. Peterson, F.-Z. Cheng, and A. S. Wilson
ISBN 1-886733-33-3

Vol. CS-114   YOUNG GALAXIES AND QSO ABSORPTION-LINE SYSTEMS
eds. Sueli M. Viegas, Ruth Gruenwald, and Reinaldo R. de Carvalho
ISBN 1-886733-34-1

Vol. CS-115   GALACTIC CLUSTER COOLING FLOWS
ed. Noam Soker
ISBN 1-886733-35-X

Vol. CS-116   THE SECOND STROMLO SYMPOSIUM:
THE NATURE OF ELLIPTICAL GALAXIES
eds. M. Arnaboldi, G. S. Da Costa, and P. Saha
ISBN 1-886733-36-8

Vol. CS-117   DARK AND VISIBLE MATTER IN GALAXIES
eds. Massimo Persic and Paolo Salucci
ISBN-1-886733-37-6

Vol. CS-118   FIRST ADVANCES IN SOLAR PHYSICS EUROCONFERENCE:
ADVANCES IN THE PHYSICS OF SUNSPOTS
eds. B. Schmieder. J. C. del Toro Iniesta, and M. Vázquez
ISBN 1-886733-38-4

Vol. CS-119   PLANETS BEYOND THE SOLAR SYSTEM
AND THE NEXT GENERATION OF SPACE MISSIONS
ed. David R. Soderblom
ISBN 1-886733-39-2

Vol. CS-120   LUMINOUS BLUE VARIABLES: MASSIVE STARS IN TRANSITION
eds. Antonella Nota and Henny J. G. L. M. Lamers
ISBN 1-886733-40-6

Vol. CS-121   ACCRETION PHENOMENA AND RELATED OUTFLOWS, IAU Colloquium 163
eds. D. T. Wickramasinghe, G. V. Bicknell, and L. Ferrario
ISBN 1-886733-41-4

Vol. CS-122   FROM STARDUST TO PLANETESIMALS:
Symposium held as part of the 108th Annual Meeting of the ASP
eds. Yvonne J. Pendleton and A. G. G. M. Tielens
ISBN 1-886733-42-2

Vol. CS-123   THE 12th 'KINGSTON MEETING': COMPUTATIONAL ASTROPHYSICS
eds. David A. Clarke and Michael J. West
ISBN 1-886733-43-0

Vol. CS-124   DIFFUSE INFRARED RADIATION AND THE IRTS
eds. Haruyuki Okuda, Toshio Matsumoto, and Thomas Roellig
ISBN 1-886733-44-9

Vol. CS-125   ASTRONOMICAL DATA ANALYSIS SOFTWARE AND SYSTEMS VI
eds. Gareth Hunt and H. E. Payne
ISBN 1-886733-45-7

ASP CONFERENCE SERIES VOLUMES
Published by the Astronomical Society of the Pacific

PUBLISHED: 1997 (* asterisk means OUT OF STOCK)

Vol. CS-126 FROM QUANTUM FLUCTUATIONS TO COSMOLOGICAL STRUCTURES
eds. David Valls-Gabaud, Martin A. Hendry, Paolo Molaro, and Khalil Chamcham
ISBN 1-886733-46-5

Vol. CS-127 PROPER MOTIONS AND GALACTIC ASTRONOMY
ed. Roberta M. Humphreys
ISBN 1-886733-47-3

Vol. CS-128 MASS EJECTION FROM AGN (Active Galactic Nuclei)
eds. N. Arav, I. Shlosman, and R. J. Weymann
ISBN 1-886733-48-1

Vol. CS-129 THE GEORGE GAMOW SYMPOSIUM
eds. E. Harper, W. C. Parke, and G. D. Anderson
ISBN 1-886733-49-X

Vol. CS-130 THE THIRD PACIFIC RIM CONFERENCE ON
RECENT DEVELOPMENT ON BINARY STAR RESEARCH
eds. Kam-Ching Leung
ISBN 1-886733-50-3

PUBLISHED: 1998

Vol. CS-131 BOULDER-MUNICH II: PROPERTIES OF HOT, LUMINOUS STARS
ed. Ian D. Howarth
ISBN 1-886733-51-1

Vol. CS-132 STAR FORMATION WITH THE INFRARED SPACE OBSERVATORY (ISO)
eds. João L. Yun and René Liseau
ISBN 1-886733-52-X

Vol. CS-133 SCIENCE WITH THE NGST (Next Generation Space Telescope)
eds. Eric P. Smith and Anuradha Koratkar
ISBN 1-886733-53-8

Vol. CS-134 BROWN DWARFS AND EXTRASOLAR PLANETS
eds. Rafael Rebolo, Eduardo L. Martin, and Maria Rosa Zapatero Osorio
ISBN 1-886733-54-6

Vol. CS-135 A HALF CENTURY OF STELLAR PULSATION INTERPRETATIONS:
A TRIBUTE TO ARTHUR N. COX
eds. P. A. Bradley and J. A. Guzik
ISBN 1-886733-55-4

Vol. CS-136 GALACTIC HALOS: A UC SANTA CRUZ WORKSHOP
ed. Dennis Zaritsky
ISBN 1-886733-56-2

Vol. CS-137 WILD STARS IN THE OLD WEST: PROCEEDINGS OF THE 13th NORTH
AMERICAN WORKSHOP ON CATACLYSMIC VARIABLES
AND RELATED OBJECTS
eds. S. Howell, E. Kuulkers, and C. Woodward
ISBN 1-886733-57-0

Vol. CS-138 1997 PACIFIC RIM CONFERENCE ON STELLAR ASTROPHYSICS
eds. Kwing Lam Chan, K. S. Cheng, and H. P. Singh
ISBN 1-886733-58-9

Vol. CS-139 PRESERVING THE ASTRONOMICAL WINDOWS:
Proceedings of Joint Discussion No. 5 of the 23rd General Assembly of the IAU
eds. Syuzo Isobe and Tomohiro Hirayama
ISBN 1-886733-59-7

ASP CONFERENCE SERIES VOLUMES
Published by the Astronomical Society of the Pacific

PUBLISHED: 1998 (* asterisk means OUT OF STOCK)

Vol. CS-140 SYNOPTIC SOLAR PHYSICS --18th NSO/Sacramento Peak Summer Workshop
eds. K. S. Balasubramaniam, J. W. Harvey, and D. M. Rabin
ISBN 1-886733-60-0

Vol. CS-141 ASTROPHYSICS FROM ANTARCTICA:
A Symposium held as a part of the 109th Annual Meeting of the ASP
eds. Giles Novak and Randall H. Landsberg
ISBN 1-886733-61-9

Vol. CS-142 THE STELLAR INITIAL MASS FUNCTION: 38th Herstmonceux Conference
eds. Gerry Gilmore and Debbie Howell
ISBN 1-886733-62-7

Vol. CS-143 * THE SCIENTIFIC IMPACT OF THE GODDARD HIGH RESOLUTION
SPECTROGRAPH (GHRS)
eds. John C. Brandt, Thomas B. Ake III, and Carolyn Collins Petersen
ISBN 1-886733-63-5

Vol. CS-144 RADIO EMISSION FROM GALACTIC AND EXTRAGALACTIC COMPACT
SOURCES, IAU Colloquium 164
eds. J. Anton Zensus, G. B. Taylor, and J. M. Wrobel
ISBN 1-886733-64-3

Vol. CS-145 ASTRONOMICAL DATA ANALYSIS SOFTWARE AND SYSTEMS VII – (ADASS VII)
eds. Rudolf Albrecht, Richard N. Hook, and Howard A. Bushouse
ISBN 1-886733-65-1

Vol. CS-146 THE YOUNG UNIVERSE GALAXY FORMATION
AND EVOLUTION AT INTERMEDIATE AND HIGH REDSHIFT
eds. S. D'Odorico, A. Fontana, and E. Giallongo
ISBN 1-886733-66-X

Vol. CS-147 ABUNDANCE PROFILES: DIAGNOSTIC TOOLS FOR GALAXY HISTORY
eds. Daniel Friedli, Mike Edmunds, Carmelle Robert, and Laurent Drissen
ISBN 1-886733-67-8

Vol. CS-148 ORIGINS
eds. Charles E. Woodward, J. Michael Shull, and Harley A. Thronson, Jr.
ISBN 1-886733-68-6

Vol. CS-149 SOLAR SYSTEM FORMATION AND EVOLUTION
eds. D. Lazzaro, R. Vieira Martins, S. Ferraz-Mello, J. Fernández, and C. Beaugé
ISBN 1-886733-69-4

Vol. CS-150 NEW PERSPECTIVES ON SOLAR PROMINENCES, IAU Colloquium 167
eds. David Webb, David Rust, and Brigitte Schmieder
ISBN 1-886733-70-8

Vol. CS-151 COSMIC MICROWAVE BACKGROUND
AND LARGE SCALE STRUCTURES OF THE UNIVERSE
eds. Yong-Ik Byun and Kin-Wang Ng
ISBN 1-886733-71-6

Vol. CS-152 FIBER OPTICS IN ASTRONOMY III
eds. S. Arribas, E. Mediavilla, and F. Watson
ISBN 1-886733-72-4

Vol. CS-153 LIBRARY AND INFORMATION SERVICES IN ASTRONOMY III -- (LISA III)
eds. Uta Grothkopf, Heinz Andernach, Sarah Stevens-Rayburn,
and Monique Gomez
ISBN 1-886733-73-2

ASP CONFERENCE SERIES VOLUMES
Published by the Astronomical Society of the Pacific

PUBLISHED: 1998 (* asterisk means OUT OF STOCK)

Vol. CS-154   COOL STARS, STELLAR SYSTEMS AND THE SUN: Tenth Cambridge Workshop
eds. Robert A. Donahue and Jay A. Bookbinder
ISBN 1-886733-74-0

Vol. CS-155   SECOND ADVANCES IN SOLAR PHYSICS EUROCONFERENCE:
THREE-DIMENSIONAL STRUCTURE OF SOLAR ACTIVE REGIONS
eds. Costas E. Alissandrakis and Brigitte Schmieder
ISBN 1-886733-75-9

PUBLISHED: 1999

Vol. CS-156   HIGHLY REDSHIFTED RADIO LINES
eds. C. L. Carilli, S. J. E. Radford, K. M. Menten, and G. I. Langston
ISBN 1-886733-76-7

Vol. CS-157   ANNAPOLIS WORKSHOP ON MAGNETIC CATACLYSMIC VARIABLES
eds. Coel Hellier and Koji Mukai
ISBN 1-886733-77-5

Vol. CS-158   SOLAR AND STELLAR ACTIVITY: SIMILARITIES AND DIFFERENCES
eds. C. J. Butler and J. G. Doyle
ISBN 1-886733-78-3

Vol. CS-159   BL LAC PHENOMENON
eds. Leo O. Takalo and Aimo Sillanpää
ISBN 1-886733-79-1

Vol. CS-160   ASTROPHYSICAL DISCS: An EC Summer School
eds. J. A. Sellwood and Jeremy Goodman
ISBN 1-886733-80-5

Vol. CS-161   HIGH ENERGY PROCESSES IN ACCRETING BLACK HOLES
eds. Juri Poutanen and Roland Svensson
ISBN 1-886733-81-3

Vol. CS-162   QUASARS AND COSMOLOGY
eds. Gary Ferland and Jack Baldwin
ISBN 1-886733-83-X

Vol. CS-163   STAR FORMATION IN EARLY-TYPE GALAXIES
eds. Jordi Cepa and Patricia Carral
ISBN 1-886733-84-8

Vol. CS-164   ULTRAVIOLET–OPTICAL SPACE ASTRONOMY BEYOND HST
eds. Jon A. Morse, J. Michael Shull, and Anne L. Kinney
ISBN 1-886733-85-6

Vol. CS-165   THE THIRD STROMLO SYMPOSIUM: THE GALACTIC HALO
eds. Brad K. Gibson, Tim S. Axelrod, and Mary E. Putman
ISBN 1-886733-86-4

Vol. CS-166   STROMLO WORKSHOP ON HIGH-VELOCITY CLOUDS
eds. Brad K. Gibson and Mary E. Putman
ISBN 1-886733-87-2

Vol. CS-167   HARMONIZING COSMIC DISTANCE SCALES IN A POST-HIPPARCOS ERA
eds. Daniel Egret and André Heck
ISBN 1-886733-88-0

Vol. CS-168   NEW PERSPECTIVES ON THE INTERSTELLAR MEDIUM
eds. A. R. Taylor, T. L. Landecker, and G. Joncas
ISBN 1-886733-89-9

ASP CONFERENCE SERIES VOLUMES
Published by the Astronomical Society of the Pacific

PUBLISHED: 1999 (* asterisk means OUT OF STOCK)

Vol. CS-169	11th EUROPEAN WORKSHOP ON WHITE DWARFS eds. J.-E. Solheim and E. G. Meištas ISBN 1-886733-91-0
Vol. CS-170	THE LOW SURFACE BRIGHTNESS UNIVERSE, IAU Colloquium 171 eds. J. I. Davies, C. Impey, and S. Phillipps ISBN 1-886733-92-9
Vol. CS-171	LiBeB, COSMIC RAYS, AND RELATED X- AND GAMMA-RAYS eds. Reuven Ramaty, Elisabeth Vangioni-Flam, Michel Cassé, and Keith Olive ISBN 1-886733-93-7
Vol. CS-172	ASTRONOMICAL DATA ANALYSIS SOFTWARE AND SYSTEMS VIII eds. David M. Mehringer, Raymond L. Plante, and Douglas A. Roberts ISBN 1-886733-94-5
Vol. CS-173	THEORY AND TESTS OF CONVECTION IN STELLAR STRUCTURE: First Granada Workshop ed. Álvaro Giménez, Edward F. Guinan, and Benjamín Montesinos ISBN 1-886733-95-3
Vol. CS-174	CATCHING THE PERFECT WAVE: ADAPTIVE OPTICS AND INTERFEROMETRY IN THE 21st CENTURY, A Symposium held as a part of the 110th Annual Meeting of the ASP eds. Sergio R. Restaino, William Junor, and Nebojsa Duric ISBN 1-886733-96-1
Vol. CS-175	STRUCTURE AND KINEMATICS OF QUASAR BROAD LINE REGIONS eds. C. M. Gaskell, W. N. Brandt, M. Dietrich, D. Dultzin-Hacyan, and M. Eracleous ISBN 1-886733-97-X
Vol. CS-176	OBSERVATIONAL COSMOLOGY: THE DEVELOPMENT OF GALAXY SYSTEMS eds. Giuliano Giuricin, Marino Mezzetti, and Paolo Salucci ISBN 1-58381-000-5
Vol. CS-177	ASTROPHYSICS WITH INFRARED SURVEYS: A Prelude to SIRTF eds. Michael D. Bicay, Chas A. Beichman, Roc M. Cutri, and Barry F. Madore ISBN 1-58381-001-3
Vol. CS-178	STELLAR DYNAMOS: NONLINEARITY AND CHAOTIC FLOWS eds. Manuel Núñez and Antonio Ferriz-Mas ISBN 1-58381-002-1
Vol. CS-179	ETA CARINAE AT THE MILLENNIUM eds. Jon A. Morse, Roberta M. Humphreys, and Augusto Damineli ISBN 1-58381-003-X
Vol. CS-180	SYNTHESIS IMAGING IN RADIO ASTRONOMY II eds. G. B. Taylor, C. L. Carilli, and R. A. Perley ISBN 1-58381-005-6
Vol. CS-181	MICROWAVE FOREGROUNDS eds. Angelica de Oliveira-Costa and Max Tegmark ISBN 1-58381-006-4
Vol. CS-182	GALAXY DYNAMICS: A Rutgers Symposium eds. David Merritt, J. A. Sellwood, and Monica Valluri ISBN 1-58381-007-2
Vol. CS-183	HIGH RESOLUTION SOLAR PHYSICS: THEORY, OBSERVATIONS, AND TECHNIQUES eds. T. R. Rimmele, K. S. Balasubramaniam, and R. R. Radick ISBN 1-58381-009-9

ASP CONFERENCE SERIES VOLUMES
Published by the Astronomical Society of the Pacific

PUBLISHED: 1999 (* asterisk means OUT OF STOCK)

Vol. CS-184	THIRD ADVANCES IN SOLAR PHYSICS EUROCONFERENCE: MAGNETIC FIELDS AND OSCILLATIONS eds. B. Schmieder, A. Hofmann, and J. Staude ISBN 1-58381-010-2
Vol. CS-185	PRECISE STELLAR RADIAL VELOCITIES, IAU Colloquium 170 eds. J. B. Hearnshaw and C. D. Scarfe ISBN 1-58381-011-0
Vol. CS-186	THE CENTRAL PARSECS OF THE GALAXY eds. Heino Falcke, Angela Cotera, Wolfgang J. Duschl, Fulvio Melia, and Marcia J. Rieke ISBN 1-58381-012-9
Vol. CS-187	THE EVOLUTION OF GALAXIES ON COSMOLOGICAL TIMESCALES eds. J. E. Beckman and T. J. Mahoney ISBN 1-58381-013-7
Vol. CS-188	OPTICAL AND INFRARED SPECTROSCOPY OF CIRCUMSTELLAR MATTER eds. Eike W. Guenther, Bringfried Stecklum, and Sylvio Klose ISBN 1-58381-014-5
Vol. CS-189	CCD PRECISION PHOTOMETRY WORKSHOP eds. Eric R. Craine, Roy A. Tucker, and Jeannette Barnes ISBN 1-58381-015-3
Vol. CS-190	GAMMA-RAY BURSTS: THE FIRST THREE MINUTES eds. Juri Poutanen and Roland Svensson ISBN 1-58381-016-1
Vol. CS-191	PHOTOMETRIC REDSHIFTS AND HIGH REDSHIFT GALAXIES eds. Ray J. Weymann, Lisa J. Storrie-Lombardi, Marcin Sawicki, and Robert J. Brunner ISBN 1-58381-017-X
Vol. CS-192	SPECTROPHOTOMETRIC DATING OF STARS AND GALAXIES ed. I. Hubeny, S. R. Heap, and R. H. Cornett ISBN 1-58381-018-8
Vol. CS-193	THE HY-REDSHIFT UNIVERSE: GALAXY FORMATION AND EVOLUTION AT HIGH REDSHIFT eds. Andrew J. Bunker and Wil J. M. van Breugel ISBN 1-58381-019-6
Vol. CS-194	WORKING ON THE FRINGE: OPTICAL AND IR INTERFEROMETRY FROM GROUND AND SPACE eds. Stephen Unwin and Robert Stachnik ISBN 1-58381-020-X

PUBLISHED: 2000

Vol. CS-195	IMAGING THE UNIVERSE IN THREE DIMENSIONS: Astrophysics with Advanced Multi-Wavelength Imaging Devices eds. W. van Breugel and J. Bland-Hawthorn ISBN 1-58381-022-6
Vol. CS-196	THERMAL EMISSION SPECTROSCOPY AND ANALYSIS OF DUST, DISKS, AND REGOLITHS eds. Michael L. Sitko, Ann L. Sprague, and David K. Lynch ISBN: 1-58381-023-4
Vol. CS-197	XVth IAP MEETING DYNAMICS OF GALAXIES: FROM THE EARLY UNIVERSE TO THE PRESENT eds. F. Combes, G. A. Mamon, and V. Charmandaris ISBN: 1-58381-24-2

ASP CONFERENCE SERIES VOLUMES
Published by the Astronomical Society of the Pacific

PUBLISHED: 2000 (* asterisk means OUT OF STOCK)

Vol. CS-198 EUROCONFERENCE ON "STELLAR CLUSTERS AND ASSOCIATIONS: CONVECTION, ROTATION, AND DYNAMOS"
eds. R. Pallavicini, G. Micela, and S. Sciortino
ISBN: 1-58381-25-0

Vol. CS-199 ASYMMETRICAL PLANETARY NEBULAE II: FROM ORIGINS TO MICROSTRUCTURES
eds. J. H. Kastner, N. Soker, and S. Rappaport
ISBN: 1-58381-026-9

Vol. CS-200 CLUSTERING AT HIGH REDSHIFT
eds. A. Mazure, O. Le Fèvre, and V. Le Brun
ISBN: 1-58381-027-7

Vol. CS-201 COSMIC FLOWS 1999: TOWARDS AN UNDERSTANDING OF LARGE-SCALE STRUCTURES
eds. Stéphane Courteau, Michael A. Strauss, and Jeffrey A. Willick
ISBN: 1-58381-028-5

Vol. CS-202 * PULSAR ASTRONOMY – 2000 AND BEYOND, IAU Colloquium 177
eds. M. Kramer, N. Wex, and R. Wielebinski
ISBN: 1-58381-029-3

Vol. CS-203 THE IMPACT OF LARGE-SCALE SURVEYS ON PULSATING STAR RESEARCH, IAU Colloquium 176
eds. L. Szabados and D. W. Kurtz
ISBN: 1-58381-030-7

Vol. CS-204 THERMAL AND IONIZATION ASPECTS OF FLOWS FROM HOT STARS: OBSERVATIONS AND THEORY
eds. Henny J. G. L. M. Lamers and Arved Sapar
ISBN: 1-58381-031-5

Vol. CS-205 THE LAST TOTAL SOLAR ECLIPSE OF THE MILLENNIUM IN TURKEY
eds. W. C. Livingston and A. Özgüç
ISBN: 1-58381-032-3

Vol. CS-206 HIGH ENERGY SOLAR PHYSICS – *ANTICIPATING HESSI*
eds. Reuven Ramaty and Natalie Mandzhavidze
ISBN: 1-58381-033-1

Vol. CS-207 NGST SCIENCE AND TECHNOLOGY EXPOSITION
eds. Eric P. Smith and Knox S. Long
ISBN: 1-58381-036-6

ATLAS VISIBLE AND NEAR INFRARED ATLAS OF THE ARCTURUS SPECTRUM 3727-9300 Å
eds. Kenneth Hinkle, Lloyd Wallace, Jeff Valenti, and Dianne Harmer
ISBN: 1-58381-037-4

Vol. CS-208 POLAR MOTION: HISTORICAL AND SCIENTIFIC PROBLEMS, IAU Colloquium 178
eds. Steven Dick, Dennis McCarthy, and Brian Luzum
ISBN: 1-58381-039-0

Vol. CS-209 SMALL GALAXY GROUPS, IAU Colloquium 174
eds. Mauri J. Valtonen and Chris Flynn
ISBN: 1-58381-040-4

Vol. CS-210 DELTA SCUTI AND RELATED STARS: Reference Handbook and Proceedings of the 6th Vienna Workshop in Astrophysics
eds. Michel Breger and Michael Houston Montgomery
ISBN: 1-58381-043-9

ASP CONFERENCE SERIES VOLUMES
Published by the Astronomical Society of the Pacific

PUBLISHED: 2000 (* asterisk means OUT OF STOCK)

Vol. CS-211　MASSIVE STELLAR CLUSTERS
　　　　　　eds. Ariane Lançon and Christian M. Boily
　　　　　　ISBN: 1-58381-042-0

Vol. CS-212　FROM GIANT PLANETS TO COOL STARS
　　　　　　eds. Caitlin A. Griffith and Mark S. Marley
　　　　　　ISBN: 1-58381-041-2

Vol. CS-213　BIOASTRONOMY `99: A NEW ERA IN BIOASTRONOMY
　　　　　　eds. Guillermo A. Lemarchand and Karen J. Meech
　　　　　　ISBN: 1-58381-044-7

Vol. CS-214　THE Be PHENOMENON IN EARLY-TYPE STARS, IAU Colloquium 175
　　　　　　eds. Myron A. Smith, Huib F. Henrichs and Juan Fabregat
　　　　　　ISBN: 1-58381-045-5

Vol. CS-215　COSMIC EVOLUTION AND GALAXY FORMATION:
　　　　　　STRUCTURE, INTERACTIONS AND FEEDBACK
　　　　　　The 3rd Guillermo Haro Astrophysics Conference
　　　　　　eds. José Franco, Elena Terlevich, Omar López-Cruz, and Itziar Aretxaga
　　　　　　ISBN: 1-58381-046-3

Vol. CS-216　ASTRONOMICAL DATA ANALYSIS SOFTWARE AND SYSTEMS IX
　　　　　　eds. Nadine Manset, Christian Veillet, and Dennis Crabtree
　　　　　　ISBN: 1-58381-047-1　　　　ISSN: 1080-7926

Vol. CS-217　IMAGING AT RADIO THROUGH SUBMILLIMETER WAVELENGTHS
　　　　　　eds. Jeffrey G. Mangum and Simon J. E. Radford
　　　　　　ISBN: 1-58381-049-8

Vol. CS-218　MAPPING THE HIDDEN UNIVERSE: THE UNIVERSE BEHIND THE MILKY WAY
　　　　　　THE UNIVERSE IN HI
　　　　　　eds. Renée C. Kraan-Korteweg, Patricia A. Henning, and Heinz Andernach
　　　　　　ISBN: 1-58381-050-1

Vol. CS-219　DISKS, PLANETESIMALS, AND PLANETS
　　　　　　eds. F. Garzón, C. Eiroa, D. de Winter, and T. J. Mahoney
　　　　　　ISBN: 1-58381-051-X

Vol. CS-220　AMATEUR - PROFESSIONAL PARTNERSHIPS IN ASTRONOMY:
　　　　　　The 111th Annual Meeting of the ASP
　　　　　　eds. John R. Percy and Joseph B. Wilson
　　　　　　ISBN: 1-58381-052-8

Vol. CS-221　STARS, GAS AND DUST IN GALAXIES: EXPLORING THE LINKS
　　　　　　eds. Danielle Alloin, Knut Olsen, and Gaspar Galaz
　　　　　　ISBN: 1-58381-053-6

PUBLISHED: 2001

Vol. CS-222　THE PHYSICS OF GALAXY FORMATION
　　　　　　eds. M. Umemura and H. Susa
　　　　　　ISBN: 1-58381-054-4

Vol. CS-223　COOL STARS, STELLAR SYSTEMS AND THE SUN:
　　　　　　Eleventh Cambridge Workshop
　　　　　　eds. Ramón J. García López, Rafael Rebolo, and María Zapatero Osorio
　　　　　　ISBN: 1-58381-056-0

Vol. CS-224　PROBING THE PHYSICS OF ACTIVE GALACTIC NUCLEI
　　　　　　BY MULTIWAVELENGTH MONITORING
　　　　　　eds. Bradley M. Peterson, Ronald S. Polidan, and Richard W. Pogge
　　　　　　ISBN: 1-58381-055-2

ASP CONFERENCE SERIES VOLUMES
Published by the Astronomical Society of the Pacific

PUBLISHED: 2001 (* asterisk means OUT OF STOCK)

Vol. CS-225   VIRTUAL OBSERVATORIES OF THE FUTURE
eds. Robert J. Brunner, S. George Djorgovski, and Alex S. Szalay
ISBN: 1-58381-057-9

Vol. CS-226   12th EUROPEAN CONFERENCE ON WHITE DWARFS
eds. J. L. Provencal, H. L. Shipman, J. MacDonald, and S. Goodchild
ISBN: 1-58381-058-7

Vol. CS-227   BLAZAR DEMOGRAPHICS AND PHYSICS
eds. Paolo Padovani and C. Megan Urry
ISBN: 1-58381-059-5

Vol. CS-228   DYNAMICS OF STAR CLUSTERS AND THE MILKY WAY
eds. S. Deiters, B. Fuchs, A. Just, R. Spurzem, and R. Wielen
ISBN: 1-58381-060-9

Vol. CS-229   EVOLUTION OF BINARY AND MULTIPLE STAR SYSTEMS
A Meeting in Celebration of Peter Eggleton's 60th Birthday
eds. Ph. Podsiadlowski, S. Rappaport, A. R. King, F. D'Antona, and L. Burderi
IBSN: 1-58381-061-7

Vol. CS-230   GALAXY DISKS AND DISK GALAXIES
eds. Jose G. Funes, S. J. and Enrico Maria Corsini
ISBN: 1-58381-063-3

Vol. CS-231   TETONS 4: GALACTIC STRUCTURE, STARS, AND
THE INTERSTELLAR MEDIUM
eds. Charles E. Woodward, Michael D. Bicay, and J. Michael Shull
ISBN: 1-58381-064-1

Vol. CS-232   THE NEW ERA OF WIDE FIELD ASTRONOMY
eds. Roger Clowes, Andrew Adamson, and Gordon Bromage
ISBN: 1-58381-065-X

Vol. CS-233   P CYGNI 2000: 400 YEARS OF PROGRESS
eds. Mart de Groot and Christiaan Sterken
ISBN: 1-58381-070-6

Vol. CS-234   X-RAY ASTRONOMY 2000
eds. R. Giacconi, S. Serio, and L. Stella
ISBN: 1-58381-071-4

Vol. CS-235   SCIENCE WITH THE ATACAMA LARGE MILLIMETER ARRAY (ALMA)
ed. Alwyn Wootten
ISBN: 1-58381-072-2

Vol. CS-236   ADVANCED SOLAR POLARIMETRY: THEORY, OBSERVATION, AND
INSTRUMENTATION, The 20th Sacramento Peak Summer Workshop
ed. M. Sigwarth
ISBN: 1-58381-073-0

Vol. CS-237   GRAVITATIONAL LENSING: RECENT PROGRESS AND FUTURE GOALS
eds. Tereasa G. Brainerd and Christopher S. Kochanek
ISBN: 1-58381-074-9

Vol. CS-238   ASTRONOMICAL DATA ANALYSIS SOFTWARE AND SYSTEMS X
eds. F. R. Harnden, Jr., Francis A. Primini, and Harry E. Payne
ISBN: 1-58381-075-7

Vol. CS-239   MICROLENSING 2000: A NEW ERA OF MICROLENSING ASTROPHYSICS
ed. John Menzies and Penny D. Sackett
ISBN: 1-58381-076-5

ASP CONFERENCE SERIES VOLUMES
Published by the Astronomical Society of the Pacific

PUBLISHED: 2001 (* asterisk means OUT OF STOCK)

Vol. CS-240 GAS AND GALAXY EVOLUTION,
A Conference in Honor of the 20th Anniversary of the VLA
eds. J. E. Hibbard, M. P. Rupen, and J. H. van Gorkom
ISBN: 1-58381-077-3

Vol. CS-241 CS-241 THE 7TH TAIPEI ASTROPHYSICS WORKSHOP ON
COSMIC RAYS IN THE UNIVERSE
ed. Chung-Ming Ko
ISBN: 1-58381-079-X

Vol. CS-242 ETA CARINAE AND OTHER MYSTERIOUS STARS:
THE HIDDEN OPPORTUNITIES OF EMISSION SPECTROSCOPY
eds. Theodore R. Gull, Sveneric Johannson, and Kris Davidson
ISBN: 1-58381-080-3

Vol. CS-243 FROM DARKNESS TO LIGHT:
ORIGIN AND EVOLUTION OF YOUNG STELLAR CLUSTERS
eds. Thierry Montmerle and Philippe André
ISBN: 1-58381-081-1

Vol. CS-244 YOUNG STARS NEAR EARTH: PROGRESS AND PROSPECTS
eds. Ray Jayawardhana and Thomas P. Greene
ISBN: 1-58381-082-X

Vol. CS-245 ASTROPHYSICAL AGES AND TIME SCALES
eds. Ted von Hippel, Chris Simpson, and Nadine Manset
ISBN: 1-58381-083-8

Vol. CS-246 SMALL TELESCOPE ASTRONOMY ON GLOBAL SCALES, IAU Colloquium 183
eds. Wen-Ping Chen, Claudia Lemme, and Bohdan Paczyński
ISBN: 1-58381-084-6

Vol. CS-247 SPECTROSCOPIC CHALLENGES OF PHOTOIONIZED PLASMAS
eds. Gary Ferland and Daniel Wolf Savin
ISBN: 1-58381-085-4

Vol. CS-248 MAGNETIC FIELDS ACROSS THE HERTZSPRUNG-RUSSELL DIAGRAM
eds. G. Mathys, S. K. Solanki, and D. T. Wickramasinghe
ISBN: 1-58381-088-9

Vol. CS-249 THE CENTRAL KILOPARSEC OF STARBURSTS AND AGN:
THE LA PALMA CONNECTION
eds. J. H. Knapen, J. E. Beckman, I. Shlosman, and T. J. Mahoney
ISBN: 1-58381-089-7

Vol. CS-250 PARTICLES AND FIELDS IN RADIO GALAXIES CONFERENCE
eds. Robert A. Laing and Katherine M. Blundell
ISBN: 1-58381-090-0

Vol. CS-251 NEW CENTURY OF X-RAY ASTRONOMY
eds. H. Inoue and H. Kunieda
ISBN: 1-58381-091-9

Vol. CS-252 HISTORICAL DEVELOPMENT OF MODERN COSMOLOGY
eds. Vicent J. Martínez, Virginia Trimble, and María Jesús Pons-Bordería
ISBN: 1-58381-092-7

PUBLISHED: 2002

Vol. CS-253 CHEMICAL ENRICHMENT OF INTRACLUSTER AND INTERGALACTIC MEDIUM
eds. Roberto Fusco-Femiano and Francesca Matteucci
ISBN: 1-58381-093-5

ASP CONFERENCE SERIES VOLUMES
Published by the Astronomical Society of the Pacific

PUBLISHED: 2002 (* asterisk means OUT OF STOCK)

Vol. CS-254   EXTRAGALACTIC GAS AT LOW REDSHIFT
eds. John S. Mulchaey and John T. Stocke
ISBN: 1-58381-094-3

Vol. CS-255   MASS OUTFLOW IN ACTIVE GALACTIC NUCLEI: NEW PERSPECTIVES
eds. D. M. Crenshaw, S. B. Kraemer, and I. M. George
ISBN: 1-58381-095-1

Vol. CS-256   OBSERVATIONAL ASPECTS OF PULSATING B AND A STARS
eds. Christiaan Sterken and Donald W. Kurtz
ISBN: 1-58381-096-X

Vol. CS-257   AMiBA 2001: HIGH-Z CLUSTERS, MISSING BARYONS, AND CMB POLARIZATION
eds. Lin-Wen Chen, Chung-Pei Ma, Kin-Wang Ng, and Ue-Li Pen
ISBN: 1-58381-097-8

Vol. CS-258   ISSUES IN UNIFICATION OF ACTIVE GALACTIC NUCLEI
eds. Roberto Maiolino, Alessandro Marconi, and Neil Nagar
ISBN: 1-58381-098-6

Vol. CS-259   RADIAL AND NONRADIAL PULSATIONS AS PROBES OF STELLAR PHYSICS, IAU Colloquium 185
eds. Conny Aerts, Timothy R. Bedding, and Jørgen Christensen-Dalsgaard
ISBN: 1-58381-099-4

Vol. CS-260   INTERACTING WINDS FROM MASSIVE STARS
eds. Anthony F. J. Moffat and Nicole St-Louis
ISBN: 1-58381-100-1

Vol. CS-261   THE PHYSICS OF CATACLYSMIC VARIABLES AND RELATED OBJECTS
eds. B. T. Gänsicke, K. Beuermann, and K. Reinsch
ISBN: 1-58381-101-X

Vol. CS-262   THE HIGH ENERGY UNIVERSE AT SHARP FOCUS: CHANDRA SCIENCE, held in conjunction with the 113th Annual Meeting of the ASP
eds. Eric M. Schlegel and Saeqa Dil Vrtilek
ISBN: 1-58381-102-8

Vol. CS-263   STELLAR COLLISIONS, MERGERS AND THEIR CONSEQUENCES
ed. Michael M. Shara
ISBN: 1-58381-103-6

Vol. CS-264   CONTINUING THE CHALLENGE OF EUV ASTRONOMY: CURRENT ANALYSIS AND PROSPECTS FOR THE FUTURE
eds. Steve B. Howell, Jean Dupuis, Daniel Golombek, Frederick M. Walter, and Jennifer Cullison
ISBN: 1-58381-104-4

Vol. CS-265   ω CENTAURI, A UNIQUE WINDOW INTO ASTROPHYSICS
eds. Floor van Leeuwen, Joanne D. Hughes, and Giampaolo Piotto
ISBN: 1-58381-105-2

Vol. CS-266   ASTRONOMICAL SITE EVALUATION IN THE VISIBLE AND RADIO RANGE, IAU Technical Workshop
eds. J. Vernin, Z. Benkhaldoun, and C. Muñoz-Tuñón
ISBN: 1-58381-106-0

Vol. CS-267   HOT STAR WORKSHOP III: THE EARLIEST STAGES OF MASSIVE STAR BIRTH
ed. Paul A. Crowther
ISBN: 1-58381-107-9

Vol. CS-268   TRACING COSMIC EVOLUTION WITH GALAXY CLUSTERS
eds. Stefano Borgani, Marino Mezzetti, and Riccardo Valdarnini
ISBN: 1-58381-108-7

ASP CONFERENCE SERIES VOLUMES
Published by the Astronomical Society of the Pacific

PUBLISHED: 2002 (* asterisk means OUT OF STOCK)

Vol. CS-269 THE EVOLVING SUN AND ITS INFLUENCE ON PLANETARY ENVIRONMENTS
eds. Benjamín Montesinos, Álvaro Giménez, and Edward F. Guinan
ISBN: 1-58381-109-5

Vol. CS-270 ASTRONOMICAL INSTRUMENTATION AND THE BIRTH AND GROWTH OF ASTROPHYSICS: A Symposium held in honor of Robert G. Tull
eds. Frank N. Bash and Christopher Sneden
ISBN: 1-58381-110-9

Vol. CS-271 NEUTRON STARS IN SUPERNOVA REMNANTS
eds. Patrick O. Slane and Bryan M. Gaensler
ISBN: 1-58381-111-7

Vol. CS-272 THE FUTURE OF SOLAR SYSTEM EXPLORATION, 2003-2013
Community Contributions to the NRC Solar System Exploration Decadal Survey
ed. Mark V. Sykes
ISBN: 1-58381-113-3

Vol. CS-273 THE DYNAMICS, STRUCTURE AND HISTORY OF GALAXIES
eds. G. S. Da Costa and H. Jerjen
ISBN: 1-58381-114-1

Vol. CS-274 OBSERVED HR DIAGRAMS AND STELLAR EVOLUTION
eds. Thibault Lejeune and João Fernandes
ISBN: 1-58381-116-8

Vol. CS-275 DISKS OF GALAXIES: KINEMATICS, DYNAMICS AND PERTURBATIONS
eds. E. Athanassoula, A. Bosma, and R. Mujica
ISBN: 1-58381-117-6

Vol. CS-276 SEEING THROUGH THE DUST:
THE DETECTION OF HI AND THE EXPLORATION OF THE ISM IN GALAXIES
eds. A. R. Taylor, T. L. Landecker, and A. G. Willis
ISBN: 1-58381-118-4

Vol. CS 277 STELLAR CORONAE IN THE CHANDRA AND XMM-NEWTON ERA
eds. Fabio Favata and Jeremy J. Drake
ISBN: 1-58381-119-2

Vol. CS 278 NAIC–NRAO SCHOOL ON SINGLE-DISH ASTRONOMY:
TECHNIQUES AND APPLICATIONS
eds. Snezana Stanimirovic, Daniel Altschuler, Paul Goldsmith, and Chris Salter
ISBN: 1-58381-120-6

Vol. CS 279 EXOTIC STARS AS CHALLENGES TO EVOLUTION, IAU Colloquium 187
eds. Christopher A. Tout and Walter Van Hamme
ISBN: 1-58381-122-2

Vol. CS 280 NEXT GENERATION WIDE-FIELD MULTI-OBJECT SPECTROSCOPY
eds. Michael J. I. Brown and Arjun Dey
ISBN: 1-58381-123-0

Vol. CS 281 ASTRONOMICAL DATA ANALYSIS SOFTWARE AND SYSTEM XI
eds. David A. Bohlender, Daniel Durand, and Thomas H. Handley
ISBN: 1-58381-124-9   ISSN: 1080-7926

Vol. CS 282 GALAXIES: THE THIRD DIMENSION
eds. Margarita Rosado, Luc Binette, and Lorena Arias
ISBN: 1-58381-125-7

Vol. CS 283 A NEW ERA IN COSMOLOGY
eds. Nigel Metcalfe and Tom Shanks
ISBN: 1-58381-126-5

ASP CONFERENCE SERIES VOLUMES
Published by the Astronomical Society of the Pacific

PUBLISHED: 2002 (* asterisk means OUT OF STOCK)

Vol. CS 284   AGN SURVEYS
eds. R. F. Green, E. Ye. Khachikian, and D. B. Sanders
ISBN: 1-58381-127-3

A LISTING OF IAU VOLUMES MAY BE FOUND ON THE NEXT PAGE

INTERNATIONAL ASTRONOMICAL UNION (IAU) VOLUMES
Published by the Astronomical Society of the Pacific

PUBLISHED: 1999 (* asterisk means OUT OF STOCK)

Vol. No. 190   NEW VIEWS OF THE MAGELLANIC CLOUDS
eds. You-Hua Chu, Nicholas B. Suntzeff, James E. Hesser, and David A. Bohlender
ISBN: 1-58381-021-8

Vol. No. 191   ASYMPTOTIC GIANT BRANCH STARS
eds. T. Le Bertre, A. Lèbre, and C. Waelkens
ISBN: 1-886733-90-2

Vol. No. 192   THE STELLAR CONTENT OF LOCAL GROUP GALAXIES
eds. Patricia Whitelock and Russell Cannon
ISBN: 1-886733-82-1

Vol. No. 193   WOLF-RAYET PHENOMENA IN MASSIVE STARS AND STARBURST GALAXIES
eds. Karel A. van der Hucht, Gloria Koenigsberger, and Philippe R. J. Eenens
ISBN: 1-58381-004-8

Vol. No. 194   ACTIVE GALACTIC NUCLEI AND RELATED PHENOMENA
eds. Yervant Terzian, Daniel Weedman, and Edward Khachikian
ISBN: 1-58381-008-0

PUBLISHED: 2000

Vol. XXIVA   TRANSACTIONS OF THE INTERNATIONAL ASTRONOMICAL UNION REPORTS ON ASTRONOMY 1996-1999
ed. Johannes Andersen
ISBN: 1-58381-035-8

Vol. No. 195   HIGHLY ENERGETIC PHYSICAL PROCESSES AND MECHANISMS FOR EMISSION FROM ASTROPHYSICAL PLASMAS
eds. P. C. H. Martens, S. Tsuruta, and M. A. Weber
ISBN: 1-58381-038-2

Vol. No. 197   ASTROCHEMISTRY: FROM MOLECULAR CLOUDS TO PLANETARY SYSTEMS
eds. Y. C. Minh and E. F. van Dishoeck
ISBN: 1-58381-034-X

Vol. No. 198   THE LIGHT ELEMENTS AND THEIR EVOLUTION
eds. L. da Silva, M. Spite, and J. R. de Medeiros
ISBN: 1-58381-048-X

PUBLISHED: 2001

IAU SPS   ASTRONOMY FOR DEVELOPING COUNTRIES
Special Session of the XXIV General Assembly of the IAU
ed. Alan H. Batten
ISBN: 1-58381-067-6

Vol. No. 196   PRESERVING THE ASTRONOMICAL SKY
eds. R. J. Cohen and W. T. Sullivan, III
ISBN: 1-58381-078-1

Vol. No. 200   THE FORMATION OF BINARY STARS
eds. Hans Zinnecker and Robert D. Mathieu
ISBN: 1-58381-068-4

Vol. No. 203   RECENT INSIGHTS INTO THE PHYSICS OF THE SUN AND HELIOSPHERE: HIGHLIGHTS FROM SOHO AND OTHER SPACE MISSIONS
eds. Pål Brekke, Bernhard Fleck, and Joseph B. Gurman
ISBN: 1-58381-069-2

Vol. No. 204   THE EXTRAGALACTIC INFRARED BACKGROUND AND ITS COSMOLOGICAL IMPLICATIONS
eds. Martin Harwit and Michael G. Hauser
ISBN: 1-58381-062-5

INTERNATIONAL ASTRONOMICAL UNION (IAU) VOLUMES
Published by the Astronomical Society of the Pacific

PUBLISHED: 2001 (* asterisk means OUT OF STOCK)

Vol. No. 205   GALAXIES AND THEIR CONSTITUENTS
AT THE HIGHEST ANGULAR RESOLUTIONS
eds. Richard T. Schilizzi, Stuart N. Vogel, Francesco Paresce, and Martin S. Elvis
ISBN: 1-58381-066-8

Vol. XXIVB   TRANSACTIONS OF THE INTERNATIONAL ASTRONOMICAL UNION
REPORTS ON ASTRONOMY
ed. Hans Rickman
ISBN: 1-58381-087-0

PUBLISHED: 2002

Vol. No. 12   HIGHLIGHTS OF ASTRONOMY
ed. Hans Rickman
ISBN: 1-58381-086-2

Vol. No. 199   THE UNIVERSE AT LOW RADIO FREQUENCIES
eds. A. Pramesh Rao, G. Swarup, and Gopal-Krishna
ISBN: 58381-121-4

Vol. No. 206   COSMIC MASERS: FROM PROTOSTARS TO BLACKHOLES
eds. Victor Migenes and Mark J. Reid
ISBN: 1-58381-112-5

Vol. No. 207   EXTRAGALACTIC STAR CLUSTERS
eds. Doug Geisler, Eva K. Grebel, and Dante Minniti
ISBN: 1-58381-115-X

Ordering information is available at the beginning of the listing

OHIO UNIVERSITY LIBRARY

Please return this book as soon as you have finished with it. In order to avoid a fine it must be returned by the latest date stamped below. All books are subject to recall after two weeks or immediately if needed for reserve.

JUN 1 6 2006

CF